SYMPOSIA OF THE
SOCIETY FOR EXPERIMENTAL BIOLOGY

NUMBER XLVI

PROCEEDINGS OF A MEETING
HELD AT THE UNIVERSITY OF BIRMINGHAM,
ENGLAND
8-12 SEPTEMBER 1991

SYMPOSIA OF THE
SOCIETY FOR EXPERIMENTAL BIOLOGY

SYMPOSIA OF THE
SOCIETY FOR EXPERIMENTAL BIOLOGY

NUMBER XLVI

MOLECULAR BIOLOGY OF MUSCLE

EDITED BY

ALICIA EL HAJ

SOCIETY FOR EXPERIMENTAL BIOLOGY SYMPOSIA

SEB Symposia form a long-standing series of volumes first published in 1947. The series is annual and each publication is a collection of authoritative articles on an aspect of modern experimental Biology. The contributors are all invited and speak on specific topics within the chosen field. Meetings are held annually, in September, over a period of two or three days.

   The aims of the Symposium series are to stimulate discussion and communication between scientists of all nationalities, to foster the development of, and research on, modern aspects of plant, animal, and cell biology.

Published for the Society for Experimental Biology
by The Company of Biologists Limited
Department of Zoology, University of Cambridge
Downing Street, Cambridge CB2 3EJ

Typeset, Printed and Published by The Company of Biologists Limited,
Department of Zoology, University of Cambridge
Downing Street, Cambridge CB2 3EJ

**A CIP Catalogue record for this
book is available from The British
Library**

Cover photograph

Adult rat cardiomyocytes cultured for 10 days, have undergone extensive reorganization of their cytoarchitecture and contain cytoskeletal as well as myofibrillar regions. One cell reexpresses α-smooth muscle actin (red; Eppenberger et al., 1990, *Dev. Biol.* **130**, 1-15), while both cells contain myofibrils stained by an antibody against heart C-protein (green). The cells were subjected to confocal microscopy and the resulting data stack was processed by the software IMARIS (Messerli et al., 1992; submitted to Cytobiology) which is marketed by BioRad, using an amSFP (achromatic multi-channel simulated fluorescence process) routine for the reconstruction of the 3-dimensional representation. The diameter of the myofibrillar region in the cell expressing α-smooth muscle actin measures about 30 μm. The voxel dimensions are 0.41 μm $\times$ 0.41 μm $\times$ 0.50 μm. Further information can be found in the article by Perriard et al. in this volume.

CONTENTS

Contents

PREFACE

This symposium was generated out of a new section of the Society for Experimental Biology, the animal molecular biology group. The group sets out to investigate molecular mechanisms underlying many of the physiological processes of interest to the society in comparative animal biology. The field of muscle biology presents an area which traditionally has been covered by the society in the areas of development, physiology, mechanics and comparative structure and function. Molecular biology of muscle is a rapidly expanding field and has demonstrated many of the key processes by which genes are regulated during differentiation and growth of a tissue with comparative animal models employed.

The symposium set out to cover four major themes: molecular biology of muscle cell differentiation - with the major advances in the DNA proteins which are involved in determining muscle committment; comparative muscle genes - a rapidly expanding area with many animal models used to elucidate basic mechanisms in muscle biology at the molecular level; molecular biology of muscle development - covering aspects of heart and skeletal muscle embryonic development and adult plasticity; and transcriptional control of muscle genes - papers concerned with unraveling the complex regulatory sequences and their actions in controlling muscle gene expression.

The symposium was a success in gathering experts in the field to a small and informal meeting, an original aim of the S.E.B. symposiums, which provided great opportunity for discussion. The animal molecular biology group is now active and organising once yearly meetings on other topics as part of the annual S.E.B. meeting in April.

Many thanks are due to the Society for Experimental Biology, The Wellcome Trust, British Heart Foundation, The Royal Society, Muscular Dystrophy Group and The Royal Microscopical Society for providing support for the meeting and to the Steering committee, Professor Geoffrey Goldspink, Dr Paul Barton, Dr Yvonne Edwards, Dr Paul Loughna and Professor John Bryant for their large contribution in setting up the meeting.

Alicia J. El Haj
Convener, Animal Molecular Biology Group,
Society for Experimental Biology
1992

Printed in Great Britain © Society for Experimental Biology 1992

ECLIPSE OF THE IDENTIFIABLE MYOBLAST

FRANK E. STOCKDALE

Stanford University School of Medicine, Stanford, California 95304-5306

Summary

During development of the limb musculature there is an eclipse of the myoblasts from the time first formed until the limb musculature begins to develop. The formation of the muscles of the limb is from cells of somite origin that migrate into the forming limb bud. While it is possible to identify myogenic cells in the somite prior to when migration occurs, and to determine the time frame during which migration occurs, there is a hiatus of about 34 hours during which it is difficult to identify or culture myogenic cells from the limb. In part this is because it is difficult to distinguish myogenic cells that migrate from those that do not. There are no markers that distinguish the myogenic cells of the somite that form the axial muscles and therefore stay in place, and from the myogenic cells that migrate to form the peripheral muscles of the body. Thus it unclear if the cell commitment to different myogenic fates occurs during somite formation, or during their sojourn to the limb bud.

Introduction

While there have been cell culture studies on myoblasts isolated from very early embryos (Dienstman, Biehl, Holtzer and Holtzer, 1974; Miller, Crow and Stockdale, 1985) and studies on the first appearance of transcripts of the MyoD family of determination genes in the early embryo (Charles de la Brousse and Emerson, 1990; Sassoon, Lyons, Wright, Lin, Lassar, Weintraub and Buckingham, 1989), there remain many questions about the cells set aside for myogenesis in the limb. In part this is because it has been difficult to isolate myogenic cells from limbs of the embryo at times during development when they are thought to first arrive in the limb. Myogenic cells of the avian embryo enter the limb bud at stage 15 (about 2 days of development), but they can not be isolated in cell culture as myoblasts from the limb bud until stage 21 (3.5 days) (Bonner and Hauschka, 1974; Rutz, Haney and Hauschka, 1982; Seed and Hauschka, 1984; White, Bonner, Nelson and Hauschka, 1975). Schramm and Solursh (1990) have recently shown that the dorsal and ventral pre-muscle masses, the first sign of myogenesis in the limb bud, appear between stage 21 and 23. Though the transplantation work of Hauschka and colleagues (Seed and Hauschka, 1984) indicated that myogenic cells are present in the limb bud at stage 15, identification of myoblasts in the limb at this time by any one of several criteria generally have been unsuccessful. For example it is difficult demonstrate cells in the limb bud that

Key words: myoblasts, myogenesis, somites.

express MyoD until stage 24 (de la Brousse et al. state that *in situ* hybridization with a probe to *qmf*1 is not positive in the quail hind limb bud until stage 24) (Charles de la Brousse and Emerson, 1990; Sassoon, et al., 1989); myosin is not demonstrated until about 4 days of chick development (Stage 26 of quail hind limb development is the first time troponin T transcripts are found) (Charles de la Brousse and Emerson, 1990; Holtzer, Marshall and Finck, 1957; Stockdale and Holtzer, 1961); and muscle fiber colony formation from isolated cells can not be demonstrated until stage 21 (Bonner and Hauschka, 1974; Seed and Hauschka, 1984). Thus, there is a hiatus of about 34 hours in limb development between when cells have arrived in the limb from the somites and when they can be detected. There is a period of eclipse in the sojourn of the myoblast from the somite to the dorsal and ventral pre-muscle masses of the limb bud. What transpires in the interval from a cell first becoming committed to myogenesis and being isolated as a myoblast in a peripheral anatomic site of the embryo?

The early limb bud myogenic cell population

The early myoblasts in the limb bud are different than myoblasts at later times during development. The limb bud appears as a lateral bulge on the somatopleura at about stage 13 (48-52 hrs) (Hamburger and Hamilton, 1951) and becomes a more clearly elongated structure by about stage 15 (50-55 hrs) when myogenic cells arrive. The dorsal and ventral pre-muscle masses are can be identified by stage 23 (Schramm and Solursh, 1990) and is readily identifiable by immunostaining of fibers at day 4 of development (Page, Miller, Hager and Stockdale, 1992). Dissociation of the limb bud at this time (Stage 21 to 24) and plating of the cells at low density results in colonies containing small myotubes (Miller and Stockdale, 1986a; Miller and Stockdale, 1986b). These small myotubes have an average number of nuclei between two and three, and are irregular in shape. First described by Hauschka and colleagues (White and Hauschka, 1971), they contain different isoforms of myosin heavy chain (Miller, Crow and Stockdale, 1985) which serve to distinguish the embryonic myoblasts from which they form. These myoblasts also differ from those at other times in limb development because they will not form fibers in the presence of TPA (Cossu, Pacifici, Adamo, Bouche and Molinaro, 1982); they require special types of media if they are to form fibers (White, et al., 1975; White and Hauschka, 1971); and they respond differently to growth factors (Seed and Hauschka, 1988). Myoblasts with these features - those that form fiber of special morphology, degree of nucleation, sensitivity to TPA, and sensitivity to growth factors - are designated, as early or embryonic myoblasts. Embryonic myoblasts constitute a relatively small number of cells of the limb bud as judged from the number of fibers formed in mass cultures of limb bud at ED 4 (stage 24). They are also a transient population because they are difficult to isolate from the limb beyond day 7-8 of chick development (Miller and Stockdale, 1986b). What is the origin of this transient myoblast population found within the dorsal and ventral pre-muscle masses of the early limb bud?

The origins of the myoblasts that form the muscles of the limb

Myoblasts that form the muscles of the limb or for that matter, the muscles of the abdominal wall or the intercostal spaces, originate in the dermatome of the early somite (Christ, Jacob and Jacob, 1974; Chevallier, Kieny and Mauger, 1977). It is envisioned that myogenic precursor cells migrate into several anatomic regions - the neck, head, forelimb, chest, abdomen, and hindlimb. Furthermore, the dermatome of one somite does not appear to be different than that of any other somite at the time migration is to begin. Somites from the forelimb region can provide myogenic cells to the hindlimb musculature of the bird when they are transplanted into regions adjacent to the hindlimb bud (Chevallier, et al., 1977). The peripheral structures dictates the final form of the muscles that develop. But cells from a particular somite are not randomly distributed in the muscles of the limb - particular somite contribute muscle cells to specific muscles (Beresford, 1983). The first cells that express the muscle determination genes have been described in the medial portion of the somite and later in the lateral regions of the somite, the dermatome (Charles de la Brousse and Emerson, 1990), though grafts of paraxial mesenchyme transplanted prior to somite segmentation do form muscle fibers (HH stage 5) (Wachtler, Christ and Jacob, 1982).

Some transplantation experiments of early stages of chick somite development suggest that the cells from different regions within a somite are not necessarily restricted in their fate. For example, if somites are dissociated into individual cells from stage 17-19 embryos, mixed and centrifuged into a 'somite' lacking compartments before being explanted and recombined with notochord, a cartilage mass always forms in the center of such clustered 'somites' (Stockdale, Holtzer and Lash, 1961). While migration of somite cells can not be ruled out, these studies suggest that cells in all locations within a somite can form cartilage. The position of a cell relative to other cells does appear to influence and ultimately stabilize commitment to specific fates. The importance of compartmentalization of the muscle forming cells is less well studied. Recent analysis by Selleck and Stern (1991) indicates the lateral and medial parts of the somite have different origins. It will be important to determine the correspondence of the dual origin of the somite with the the two major cell types that are formed from somite cells - muscle and cartilage.

The origins of the axial versus the limb musculature

It is conceptually difficult to understand how cells become separated into those that form the truncal musculature as opposed to the muscles of the limb. The first evidence of striated myogenesis in the bird is the heart, with skeletal muscle first forming in the somites. Within each somite a thin plate of highly aligned bundles of fibers, the myotome, are formed and these fibers span the entire length of each somite. This highly ordered arrangement is particularly apparent with fluorescent staining of the myotome with antibodies to myosin (Stockdale and Holtzer, 1961).

The fibers of the limb bud, on the other hand, form in two separate regions at the base of the limb bud - the dorsal and ventral pre-muscle masses. There is a zone of clear demarcation devoid of muscle fibers that separates the somite muscle fibers from those of the limb pre-muscle masses (Page, et al., 1992). These first limb bud fibers, undoubtedly, are forerunners of the pelvic and limb musculature and differ from those of the somite, because they are not all of the same length nor are they highly aligned. The first fibers of the limb bud run in the longitudinal axis of the limb, but fibers are not bundled into parallel arrays as in the myotome of the somite. Thus the initial phases of myogenesis in the somite - those fibers that form the axial musculature - differ in appearance from those of the musculature of the limb. While this may merely indicate local developmental influences, it may reflect that from the outset the myogenic cells that form the axial musculature differ from those that will migrate and form the muscles of the periphery.

When in development does diversity among myoblasts appear

The first myogenic cells of the limb bud are a diverse population because when cloned, they form colonies of fibers that synthesize different types of MHC (Miller and Stockdale, 1986a; Miller and Stockdale, 1986b). Cloning studies of limb buds suggest that at least by day 4 of development there has been diversification of embryonic myoblasts. Developmental diversity of the myoblasts that form the limb muscles could emerge in at least three locations. Diversity could emerge prior to migration of myogenic cells from the somite, during the traverse of myogenic cells from the somite to the forming limb bud, or after myogenic cells arrive in the limb bud. There must be features of the myogenic cells or their precursors that distinguish them from other cells in the somite, otherwise, why would some myogenic cells migrate to the forming limb bud while others remain in the somite to form the axial muscles? Some 'difference' must exist in those cells that remain and those that migrate. That 'difference' may merely be the consequence that they become mobile when confronting specific extracellular materials, other cell types, or guidance systems that only exist in specific regions of the somite. On the other hand specific myogenic precursor cell types may exist to form the axial versus the muscles of the limb. The lack of specific markers of myoblasts has hampered studies of this possibility.

Perhaps there are pathways that lead to the the distant pre-muscle mass that provide selective cues to myogenic populations that are of specific types. For example, it is likely that the ventral pre-muscle mass has more myoblasts of the fast/slow type than does the dorsal pre-muscle mass because more fast/slow fibers form in ventral than the dorsal pre-muscle mass (Page, et al., 1992). Such a distribution of myoblasts may suggest selective migration of a particular myogenic cell type to a particular pre-muscle mass. However, this distribution may only indicate that cells of unspecified type arrive in the limb, and that their characteristics or ultimate fate, is determined by local processes. While somite cells neutral to their ultimate myoblast type may begin their sojourn to the forming limb bud, there may be diversity in the features of the pathway (matrix factors, growth factors or cell surfaces) which lead to diversification of myogenic cells intrinsic to the ventral and dorsal portion of the limb bud. There have been limited experimental approaches to defining pathways taken by myogenic cells so it is not known if such a pathway(s) exists. Formation of the ventral and dorsal pre-muscle masses may merely be a consequence of the central cartilaginous condensation of the limb bud splitting the migrating precursor population into two compartments.

Amplification of myoblasts

Once myogenic cells arrive in the limb bud there must be a mechanism for selective replication of some, but not all of the myogenic cell types that arrive from the somite. The cellularity of the limb increases rapidly during embryonic and fetal life (Herrmann, 1957). Between day 2 and 6 in the chick embryo, the increase of myogenic cell number must be confined to embryonic myoblasts that form primary or embryonic fibers of the limb, because this is the only type of myoblast that can be isolated through day 6. There must be selective proliferation of embryonic myoblasts among other resident myogenic cells because Seed and Hauschka (1984) have shown that late myoblasts, or fetal myoblasts, as well as embryonic myoblasts are resident in the muscle forming regions of the limb at day 2 of development. Presumably, between day 2 and 4 of development, when there is an eclipse of the myoblast, there must be proliferative forces in the limb that focus on that portion of the myogenic population that is of the embryonic type. Embryonic myoblasts do not continue to proliferate because they are virtually absent beyond day 7 or 8 of limb development, when there is a marked increase in myoblasts of the fetal type. Fetal myoblasts therefore are in a cryptic state during embryonic develop. Those mechanism responsible for proliferation of the fetal myoblasts must emerge during the latter half of the first week of development.

Events that establish the first fibers of the limb bud appear to be driven by intrinsic properties of myogenic cells and perhaps local factors that control movement, morphogenesis, and myoblast proliferation. After day 8 of chick development, innervation of fibers, and hormone effects (those indigenously produced) may become more important for myogenesis (Stockdale, 1990a; Stockdale, 1990b; Stockdale, 1990c). It is perhaps the addition of the innervation and hormones that permits the emergence of the fetal myoblast and the subsequent adult or satellite cell populations.

Conclusions

There remain many unresolved questions about the early events in skeletal muscle myogenesis. The importance of the myogenic determination genes can not be minimized. There are, however, important questions that relate to the location, distribution and proliferation of myogenic cells that need to be resolved. In limb development, there is an eclipse of identifiable myoblasts. This period is an important period of myogenesis that encompasses the time from when the myogenic cells destined to form the muscles of the limb first are set aside until they can be isolated from the limb bud. What transpires within myogenic cells in the interval from when a cell first becomes committed to myogenesis and when a myoblast can be isolated in a peripheral anatomic site of the embryo? When and where does diversification of myoblasts occur in the embryo? Does the expression of the helix-loop-helix family of myogenic determination genes provide answers for these questions? Are the first myogenic cells 'generically' committed to myogenesis with 'further' refinements added later to produce stable or semi-stable subsets of myoblasts? How early in development do those skeletal muscle cell precursors of the limb become set aside from those that form the muscles of the trunk? What 'pathway' is followed by the myogenic cells as they migrate from the somite to the limb? And is it a single 'pathway', or are there 'pathways' that lead specifically to where the dorsal and the ventral pre-muscle masses will form? After reaching the limb how does the fetal myoblast population remain quiescent for days and then begin to proliferate as the organism enters fetal life? The answers to these questions will shed light on the eclipse of the myoblast.

References

Beresford, B. (1983). Brachial muscles in the chick embryo: the fate of individual somites. *J Embryol Exp Moroph.* **77**, 99-116.

Bonner, P. H. and Hauschka, S. D. (1974). Clonal analysis of vertebrate myogenesis. I. Early developmental events in the chick limb. *Dev Biol.* **37**, 317-328.

Charles de la Brousse, F. C. and Emerson, C. P., Jr (1990). Localized expression of a myogenic regulatory gene, *qmf1*, in the somite dermatome of avian embryos. *Genes Dev.* **4**, 567-581.

Chevallier, A., Kieny, M. and Mauger, A. (1977). Limb-somite relationships: origin of the limb musculature. *J Embryol Exp Morphol.* **41**, 245-258.

Christ, B., Jacob, H. J. and Jacob, M. (1974). Experimentelle Untersuchungen zur Entwicklung der Brustwand beim Huhnerembryo. *Experientia* **30**, 1449-1451.

Cossu, G., Pacifici, M., Adamo, S., Bouche, M. and Molinaro, M. (1982). TPA-induced inhibtion of the expression of differentiative traits in cultured myotubes: Dependence on protein synthesis. *Differentiation.* **21**, 62-65.

Dienstman, S. R., Biehl, J., Holtzer, S. and Holtzer, H. (1974). Myogenic and chondrogenic lineages in developing limb buds grown in vitro. *Dev Biol.* **39**, 83-95.

Hamburger, V. and Hamilton, H. (1951). A series of normal stages in the development of the chick embryo. *J. Morphol.* **88**, 49-92.

Herrmann, H., White, B. and Cooper, M. (1957). The accumulation of tissue components in the leg muscle of the developing chick. *J Cell Comp Phys.* **49**, 227.

Holtzer, H., Marshall, J. M. and Finck, H. (1957). An analysis of myogenesis by the use of fluorescent antimyosin. *J Biophysic Biochem Cytol.* **3**, 705-729.

Miller, J. B., Crow, M. T. and Stockdale, F. E. (1985). Slow and fast myosin heavy chain content defines three types of myotubes in early muscle cell cultures. *J Cell Biol.* **101**, 1643-1650.

Miller, J. B. and Stockdale, F. E. (1986a). Developmental origins of skeletal muscle fibers: Clonal analysis of myogenic cell lineages based on fast and slow myosin heavy chain expression. *Proc Natl Acad Sci USA.* **83**, 3860-3864.

Miller, J. B. and Stockdale, F. E. (1986b). Developmental regulation of the multiple myogenic cell lineages of the avian embryo. *J. Cell Biol.* **103**, 2197-2208.

Page, S., Miller, J. B., Hager, E. J. and Stockdale, F. E. (1992). Developmentally regulated expression of multiple isoforms of slow myosin heavy chain: evidence for fiber diversity among the first fibers to form in avian muscle. *Devl Biol.* in press.

Rutz, R., Haney, C. and Hauschka, S. (1982). Spatial analysis of limb bud myogenesis: A proximodistal gradient of muscle colony-forming cells in chick embryo leg buds. *Dev Biol.* **90**, 399-411.

Sassoon, D., Lyons, G., Wright, W. E., Lin, V., Lassar, A., Weintraub, H. and Buckingham, M. (1989). Expression of two myogenic regulatory factors myogenin and MyoD1 during mouse embryogenesis. *Nature.* **341**, 303-307.

Schramm, C. and Solursh, M. (1990). The formation of premuscle masses during chick wing bud development. *Anat Emb* **182**, 235-247.

Seed, J. and Hauschka, S. D. (1984). Temporal separation of the migration of distinct myogenic precursor populations into the developing chick wing bud. *Dev Biol.* **106**, 389-393.

Seed, J. and Hauschka, S. D. (1988). Clonal analysis of vertebrate myogenesis. VIII. Fibroblast growth factor (FGF)-dependent and FGF-independent muscle colony types during chick wing development. *Dev Biol.* **128**, 40-49.

Selleck, M. J. and Stern, C. D. (1991). Fate mapping and cell lineage of Hensen's node in the chick embryo. *Develop* **112**, 615-626.

Stockdale, F. E. (1990a). Muscle fiber development and origins of fiber diversity. In 'The Dynamic State of Muscle Fibers' (D. Pette, Ed.), pp. 121-125. Walter de Gruyter and Co.

Stockdale, F. E. (1990b). Myoblast diversity and the formation of the early limb musculature. *Ann New York Acad Sci.* **599**, 111-118.

Stockdale, F. E. (1990c). The myogenic lineage: evidence for multiple cellular precursors during avian limb development. *Proc Exp Biol Med.* **194**, 71-75.

Stockdale, F. E. and Holtzer, H. (1961). DNA synthesis and myogenesis. *Exp Cell Res.* **24**, 508-520.

Stockdale, F. E., Holtzer, H. and Lash, J. (1961). An experimental analysis of the development of the spinal column. Response of dissociated somite cells. *Acta Embryol Morphol Exp.* **4**, 40-46.

Wachtler, F., Christ, B. and Jacob, H. J. (1982). Grafting experiments on determination and migratory behaviour of presomitic, somitic an somatopleural cells in avian embryos. *Anat Embryol.* **164**, 369-378.

White, N. K., Bonner, P. H., Nelson, D. R. and Hauschka, S. D. (1975). Clonal analysis of vertebrate myogenesis. IV. Medium-dependent classification of colony-forming cells. *Dev Biol.* **44**, 346-361.

White, N. K. and Hauschka, S. D. (1971). Muscle development in vitro: A new conditioned medium effect on colony differentiation. *Exp Cell Res.* **67**, 479-482.

Printed in Great Britain © Society for Experimental Biology 1992   9

REGULATING THE MYOGENIC REGULATORS

HELEN M. BLAU

Department of Pharmacology, Stanford University School of Medicine, Stanford, CA 94305

Introduction

As vertebrates develop, totipotent cells in the early embryo give rise to cell types specialized for function in tissues. Although the lineage, or progression, from totipotent cell to differentiated cell is not well defined, at some point a cell is destined to specialize along a given pathway, such as myogenesis. This process entails a series of stages during which the fate of a cell is influenced by extrinsic and intrinsic signals. Soon after blastula formation, cells are generated for each of three distinct embryonic layers: endoderm, ectoderm, and mesoderm. Eventually, an array of differentiated cell types emerges. Much remains to be elucidated regarding the molecular regulatory mechanisms underlying the generation of these differentiated states.

Waddington's epigenetic landscape suggested that differentiated cells are destined for a groove, or valley, and that changing that valley is not easily achieved (Waddington, 1940). Gurdon's experiments (Gurdon, Laskey and Reeves, 1975) provided perhaps the first indication that the destiny of a differentiated cell could be altered; when nuclei from amphibian intestine were introduced into enucleated eggs, feeding tadpoles developed. This dramatic change in nuclear function was questioned at first, because the relatively low frequency of occurrence could have been characteristic of a subpopulation of cells, possibly residual stem cells. However, later experiments by Gurdon et al. (1975) and DiBerardino and Hoffner (1983) confirmed the hypothesis. These experiments clearly showed that, when transplanted, the nuclei of well-differentiated keratinocytes and even non- cycling erythrocytes could display multipotentiality. Indeed, feeding tadpoles were obtained at a high frequency of 75% when the nuclei were initially injected into oocytes and conditioned by oocyte cytoplasm prior to transplantation into enucleated eggs (DiBerardino, Orr and McKinnell, 1986; for review see DiBerardino, 1988). That genetic material was not lost or permanently inactivated during vertebrate differentiation was apparent from these experiments. Moreover, they suggested that the specialized state of a cell is achieved by regulating the activity of its genes. However, the possibility remained that this degree of reactivation of dormant genes depended upon nuclei being exposed to a sequence of cues that accompanies the progression from undifferentiated zygote to specialized tissue.

Somatic cell hybrid experiments suggest that the differentiated state of a cell could be altered without recourse to the regulatory hierarchy characteristic of development. Moreover, cell fusion to form non-dividing heterokaryons showed

Key words: myogenic regulation, heterokaryons, gene activation, helix-loop-helix proteins, *MyoD*.

that the differentiated state of a cell could be changed in the absence of changes in chromatin requiring DNA replication or cell division. This plasticity of nuclear function is likely to result from a dynamic interaction of the combination of proteins the fused cells contain. Although, in muscle, the MyoD family of helix-loop-helix transcriptional regulators probably plays a major role, heterokaryons provide evidence that other critical regulators are responsible for inducing or repressing their function. Evidence for such regulators and approaches to their isolation are discussed below.

Evidence from somatic cell hybrids

Cell fusion experiments allow two differentiated cell types to be combined so that the influence of one on the function of the other can be studied. Such somatic cell hybrids yielded early evidence that mammalian gene expression could be altered by diffusible trans-acting regulators. When fused with another cell type, the expression of differentiated functions frequently ceased (Mevel-Ninio and Weiss, 1981; for review see Weiss, 1982). This repression was reversed upon chromosome loss and, in some cases, was ascribed to specific genetic loci that act in trans (Killary and Fournier, 1984; Petit, Levilliers, Ott and Weiss, 1986). In a few cases, trans-activation of genes was also reported (Davidson, 1972; Peterson and Weiss, 1972; Rankin and Darlington, 1979). Studies with hybrids also provided early evidence that the malignant state is, in some cases, recessive to the differentiated state (Harris and Klein, 1969; Peehl and Stanbridge, 1982; for review see Harris, 1988), a prediction confirmed at the molecular level for retinoblastoma (Friend et al., 1986). Taken together, these experiments suggested that the balance of positive and negative regulators might dictate the pattern of genes expressed in the differentiated cell.

Evidence from heterokaryons

To analyze changes in gene activity in the context of the whole cell, a heterokaryon system is advantageous. The key feature of this system is that it is stable for the duration of the experiment. Since after fusion, there is no nuclear fusion or cell division, all genetic material remains intact within its own nucleus. This type of short-term nondividing fusion product makes possible an assessment of the influence of two or more sets of cytoplasmic and nuclear components on gene expression with minimal disruption. Moreover, since growth and genetic selection are not required to obtain the fusion product of interest, changes in gene expression can be monitored immediately following the fusion event and at well-defined time intervals thereafter.

The use of non-dividing stable interspecific heterokaryons in the study of regulation of gene expression, was pioneered by Harris and Ringertz. Following fusion with human cells, differentiated chick erythrocyte nuclei swelled, resumed RNA synthesis, and contained human nuclear proteins (Harris and Watkins, 1965;

Harris, Sidebottom, Grace and Bramwell, 1969; Ringertz, Carlsson, Ege and Bolund, 1971; for review see Ringertz and Savage, 1976). Either coexpression or extinction of genes contributed by both cell types were frequent outcomes of heterokaryon experiments produced with cells of different species and differentiated states, including muscle (Carlsson, Luger, Ringertz and Savage, 1974; Lawrence and Coleman, 1983; Wright and Aronoff, 1983; Konieczny, Lawrence and Coleman, 1983). However, in these early experiments gene activation was not detected.

The first evidence that previously silent genes could be activated in heterokaryons was provided by Blau, Chiu and Webster (1983). Following fusion with mouse muscle cells, muscle gene expression was induced in human amniotic fibroblasts. A combination of three features was probably responsible for the success of these experiments: (1) Culture media -- the heterokaryons were maintained in media that was mitogen-poor and low in serum, thus promoting differentiation, not proliferation. (2) The choice of cell type -- primary diploid cells, rather than aneuploid transformed cells were employed. This was critical because conditions that stimulate proliferation are antagonistic to muscle differentiation. Moreover, this medium prevented heterokaryons from dividing for up to two weeks, permitting an analysis of gene expression over much longer periods of time than was previously possible. (3) The differentiated state -- the muscle cells used were multinucleated, cells in which differentiation was well underway. These findings were soon corroborated for muscle cells by Wright (1984a,b) and extended to other cell types such as erythroid cells by Baron and Maniatis (1986), pancreatic cells by Wu et al. (1990) and hepatic cells by Spear and Tilghman (1990).

Gene activation was detected in heterokaryons by assaying ten different human muscle gene products using species specific assays (Blau, Pavlath, Hardeman, Chiu, Silberstein, Webster, Miller and Webster, 1985). These encompassed a range of muscle functions including structural proteins of the contractile apparatus, membrane components, and enzymes (Blau et al., 1985). The relative amounts produced and the time course of appearance of these gene products was typical of human myogenesis (Hardeman, Chiu, Minty and Blau, 1986). In addition to muscle gene activation, repression of the production of proteins and transcripts typical of the non-muscle cell type was observed and the distribution of organelles within the cell rapidly changed to be like muscle (Miller, Pavlath, Blakely and Blau, 1988). Within hours of cell fusion, the Golgi apparatus and the microtubule organizing center moved from a polar to a circumnuclear location and the centrioles were dispersed. This complexity of changes suggested that fusion with muscle cells in heterokaryons induced a fundamental alteration, or phenotypic change, in the differentiated function of non-muscle cells.

Differences among cell types implicate additional regulators

In the analysis of differentiation, a particular advantage of heterokaryons is that changes in gene expression can be induced and studied in cells that would normally never express those genes. Thus, this type of analysis has advantages over the majority of studies of differentiation in tissue culture which typically utilize precursor cells such as myoblasts, adipoblasts, or erythroleukemia cells which are already destined for the differentiated state they ultimately express. Studies using such precursors are likely to miss key regulatory steps, because the cells have already undergone a substantial number of the changes in protein composition and DNA conformation required by their particular cell type. By contrast, heterokaryons allow an analysis of a greater complexity of regulatory steps. These steps are necessary to induce a novel differentiated state in cells previously committed to a different fate.

The effect of cell origin and its influence on the ability to express previously silent muscle genes was analyzed with heterokaryons produced with representatives of the three embryonic lineages, fibroblasts (mesoderm), keratinocytes (ectoderm) and hepatocytes (endoderm) (Blau et al., 1985; Miller et al., 1988; Pavlath, Chiu and Blau, 1989). Marked differences were detected among cell types in the time course and efficiency of muscle gene expression in heterokaryons. For heterokaryons containing fibroblasts, keratinocytes and hepatocytes, the average frequency of expression of human muscle N-CAM on the cell surface was 95, 60, and 25% respectively. In addition, the onset of expression of N-CAM differed by at least two days. To understand the regulatory mechanisms that control the differentiated state, knowledge of the molecular events that constitute the lag period prior to muscle gene expression is critical. The differences among cell types are likely to reflect a requirement for different numbers or concentrations of regulators necessary to achieve muscle gene activation.

Regulatory mechanisms for activating silent genes

Muscle gene activation was generally observed in heterokaryons with primary diploid cells, but not in heterokaryons with transformed aneuploid cell types. In one case, the Hela cell, we showed that the lack of response could be reversed (Chiu and Blau, 1985). If treated with 5-azacytidine (5AC) prior to fusion, muscle gene expression was induced in Hela cells in heterokaryons. Thus, muscle gene activation in HeLa cells required changes induced by 5AC, presumably in the level of DNA methylation of structural genes or the genes encoding their regulators. Thus, unlike normal cell types, muscle genes are silenced by a mechanism that is not as easily reversed in this transformed cell type. It remains to be determined whether this mechanism is generally employed to repress tissue specific genes in transformed cell types.

Activation of previously silent genes occurs without DNA replication

We designed experiments to determine whether the lag period prior to muscle gene activation reflected a requirement for DNA replication. Possibly, negative regulation was alleviated by changes in chromatin structure during DNA synthesis. To address this possibility, the fibroblasts, keratinocytes and hepatocytes were each exposed to a DNA synthesis inhibitor, cytosine arabinoside, prior to and continuously after cell fusion (Chiu and Blau, 1985; Blau et al., 1985; Miller et al., 1988) and assayed for the novel activation of several muscle genes. No significant differences in the expression of the muscle gene encoding human N-CAM were observed in the presence or absence of DNA replication. These results indicate that conformational changes in chromatin either do not accompany gene activation in heterokaryons or do not require extensive DNA replication.

A network of positive and negative regulators mediates muscle differentiation

In hybrids, gene dosage, or the relative genetic contribution of the two fused cell types, is critical to the suppression of malignancy (for review see Harris, 1988). Similarly, in heterokaryons the relative contribution of muscle and non-muscle nuclei is critical to novel gene activation (Blau et al., 1985). Both the existence of negative and positive regulators is implicated in these cell fusion experiments. Their concentration and role differs among cell types. For example, irrespective of nuclear ratio, 95% of heterokaryons containing fibroblasts from three different sources (fetal lung, fetal skin, and adult skin) ultimately expressed human muscle genes (Pavlath and Blau, 1986; Pavlath et al., 1989). These results suggest that if fibroblasts produce inhibitors of muscle differentiation, like the well-documented extinguishers of liver gene activation (Killary and Fournier, 1984; Petit et al., 1986), they are readily overridden by the positive regulators contributed by differentiated muscle cells, due to differences in their nature or concentration. By contrast, in heterokaryons produced with hepatocytes the ultimate probability of activating muscle genes was dependent on the proportion of muscle and hepatocyte nuclei at all times (Miller et al., 1988). When hepatocyte nuclei outnumbered muscle cell nuclei, muscle genes were repressed presumably due to the relative levels of negative regulators that overwhelmed the positive regulators contributed by the muscle cell type.

Effect of cell history on response to helix-loop-helix family of myogenic regulators

To examine the molecular basis for the muscle-specific gene expression observed in multinucleated heterokaryons formed from the fusion of differentiated muscle cells to either hepatocytes or fibroblasts, we tested the role of the MyoD family of regulators. MyoD and its relatives are regulators of muscle-specific gene expression that have a helix-loop-helix motif (Davis, Weintraub and Lassar, 1987). We tested whether these regulators alone could induce the phenotypic conversion observed in heterokaryons. MyoD or myogenin were stably or transiently introduced into fibroblasts or hepatocytes by microinjection, transfection or retroviral infection with complementary DNA in expression vectors. Fibroblasts expressed muscle-specific genes, whereas hepatocytes did not. However, following fusion of hepatocytes stably expressing MyoD to fibroblasts activation in the heterokaryon of muscle-specific genes of both cell types was detected. These results imply that other regulators, present in fibroblasts but not in hepatocytes, are necessary for the activation of muscle-specific genes. They also show that the differentiated state of a cell is dictated by its history and a dynamic interaction among the proteins that it contains.

Negative control of the helix-loop-helix family of myogenic regulators in the NFB mutant

A nondifferentiating mouse muscle cell line, NFB, that represses the activity of the helix-loop-helix (HLH) family of myogenic regulators, yet expresses sarcomeric actins was analyzed. In NFB cells, the MyoD gene is silent, but can be activated upon transfection of a long terminal region-controlled chicken MyoD cDNA, resulting in myogenesis. When NFB cells were fused with H9c2 rat muscle cells in heterokaryons, the level of rat MyoD transcripts declines. Thus, the stoichiometry of MyoD and the putative repressor controls myogenesis. Although NFB cells express myogenin and Myf-5 transcripts (other members of the MyoD family), the activity of these regulators is also repressed: myogenesis is not induced in 10T1/2 fibroblasts and is repressed in L6 muscle cells upon fusion with NFB cells. These findings led us to conclude that: (1) the myogenic HLH regulators are not required for sarcomeric actin gene activation, and (2) that myogenesis is subject to dominant-negative control.

Conclusions and questions raised

The major conclusion from the heterokaryon studies described above is that the differentiated state is not fixed but can readily be altered. However, the differentiated state *in vivo* appears relatively stable. Thus, an understanding of the regulators of cell differentiation by mechanisms that allow for the type of plasticity observed in heterokaryons yet stably maintain a differentiated state is of particular interest. How is the change in differentiated state achieved? Heterokaryon experiments do not support molecular models which suggest that DNA replication is required for the activation of previously silent genes (Holtzer et al., 1983; Brown, 1984; Weintraub, 1985). Changes in chromatin structure such as the creation of DNase hypersensitive sites, removal of histones or changes in DNA methylation could still occur. However, if they occur, these changes are mediated by mechanisms that are independent of DNA replication. The elucidation of these mechanisms is now possible.

How, then, is a heritable differentiated state achieved? Heterokaryon studies suggest that the differentiated state is largely governed by the dynamic interaction of the combination of proteins a cell contains. This is particularly clear from the striking effect on gene activation observed when the relative contribution of components, or dosage, of the two fused cell types is altered. How does this pertain to differentiation *in vivo*? The protein composition of a cell is part of its heritage, the product of a history of responses to cues in the course of development. The cell transmits these proteins to progeny through division. The response of cells to a single regulator such as MyoD depends on the cell context, or set of proteins, it encounters. As a result, MyoD induces muscle gene expresssion in some cell types, but not in others. In cases where it fails to do so, such as liver cells, additional regulators, both positive and negative, are implicated. A threshold concentration of regulators may be critical to the expression and maintenance of the differentiated state. For some proteins this may be achieved by activating the expression of their own promoters, like c-jun (Angel, Hattori, Smeal and Karin, 1988). Positive autoregulation and feedback loops are levels of control that operate during the commitment of phage lambda to lysogeny (for review, see Ptashne, 1986) and are likely to be characteristic of early acting regulators in mammalian cell differentiation.

Protein-protein interactions provide another means of controlling regulator concentration. Recent findings that transcriptional regulators with 'leucine zipper' or 'helix-loop-helix' motifs bind DNA sequences as heterodimers (Landschulz, Johnson and McKnight, 1989; Murre, McCaw and Baltimore, 1989), suggest that a range of protein-protein combinations is possible. Indeed, the activity of the MyoD family of regulators, helix-loop-helix proteins, is determined, in part, by the concentration and nature (inhibitory or facilitating) of their protein partners. In heterokaryons, different partners lead to novel heterodimers that act either as positive or negative regulators. These heterodimers could act on the same or on different promoters from those recognized by the protein complexes originally present in each of the parental cells.

The future: isolation of novel regulators of muscle differentiation by a genetic approach

Regulators in addition to the MyoD family of regulators are clearly implicated in heterokaryon experiments. How can they be identified and isolated? One approach to the identification of regulators that precede MyoD is to purify the proteins that bind to the DNA sequences responsible for its tissue-specific expression as determined by deletion or mutagenesis. Alternatively, a 'genetic approach' could be used to isolate regulators that activate or repress myogenesis which entails transfection of DNA from a donor cell into recipient cells that are then assayed for the heritable expression of muscle gene products (for review see Blau, 1988). As described elsewhere (Hardeman et al., 1988), heterokaryons provide a crucial step in identifying the appropriate recipient clone for use in this approach. The genetic approach is particularly attractive because it permits identification of genes that act both directly and indirectly to induce tissue-specific gene expression, for example by modifying or interacting cooperatively with transcription factors. Moreover, the approach should lend itself to the identification of intermediate points in the signal transduction pathway. This strategy has been fruitfully exploited in elucidating hierarchies of regulatory genes in Drosophila, C. elegans, and yeast, as well as mammalian oncogenes and tumor suppressor genes (for review, see Blau, 1988). Its further application should lead to the identification of additional genes with a regulatory function and increased knowledge of the pathways and interactions at a molecular level that are responsible for the differentiated state.

References

Angel, P., Hattori, K., Smeal, T. and Karin, M. (1988). The *jun* proto-oncogene is positively autoregulated by its product. *Cell* **55**, 875-885.

Baron, M. H. and Maniatis, T. (1986). Rapid reprogramming of globin gene expression in transient heterokaryons. *Cell* **46**, 591-602.

Blau, H. M. (1988). Hierarchies of regulatory genes may specify mammalian development. *Cell* **53**, 673-674.

Blau, H. M., Chiu, C.-P. and Webster, C. (1983). Cytoplasmic activation of human nuclear genes in stable heterokaryons. *Cell* **32**, 1171-1180.

Blau, H. M., Pavlath, G. K., Hardeman, E. C., Chiu, C.-P., Silberstein, L., Webster, S. G., Miller, S. C. and Webster, C. (1985). Plasticity of the differentiated state. *Science* **230**, 758-766.

Brown, D. D. (1984). The role of stable complexes that repress and activate eucaryotic genes. *Cell* **37**, 359-365.

Carlsson, S.-A., Luger, O., Ringertz, N. R. and Savage, R. E. (1974). Phenotypic expression in chick erythrocyte × rat myoblast hybrids and in chick myoblast × rat myoblast hybrids. *Exp. Cell Res.* **84**, 47-55.

Chiu, C.-P. and Blau, H. M. (1984). Reprogramming cell differentiation in the absence of DNA synthesis. *Cell* **37**, 879-887.

Chiu, C.-P. and Blau, H. M. (1985). 5-Azacytidine permits gene activation in a previously non-inducible cell type. *Cell* **40**, 417-424.

Davidson, R. L. (1972). Regulation of melanin synthesis in mammalian cells: effect of gene dosage on the expression of differentiation. *Proc. Natl. Acad. Sci. USA* **69**, 951-955.

Davis, R. L., Weintraub, H. and Lassar, A. B. (1987). Expression of a single transfected cDNA converts fibroblasts to myoblasts. *Cell* **51**, 987-1000.

DiBerardino, M. A. (1988). Genomic multipotentiality of differentiated somatic cells. *Cell Diff. and Devel.* **25**, 129-136.

DiBerardino, M. A. and Hoffner, N. J. (1983). Gene reactivation in Erythrocytes: Nuclear transplantation in oocytes and eggs of Rana. *Science* **219**, 862-864.

DiBerardino, M. A., Orr, N. H. and McKinnell, R. G. (1986). Feeding tadpoles cloned from Rana erythrocyte nuclei. *Proc. Natl. Acad. Sci. USA* **83**, 8231-8234.

Eguchi, G. (1988). Cellular and molecular background of wolffian lens regeneration. *Cell Diff. and Devel.* **25**, 147-158.

Friend, S. H., Bernards, R., Rogelj, S., Weinberg, R. A., Rapaport, J. M., Albert, D. M. and Dryja, T. P. (1986). A human DNA segment with properties of the gene that predisposes to retino blastoma and osteosarcoma. *Nature* **323**, 643-646.

Gurdon, J. B. (1962). The developmental capacity of nuclei taken from intestinal epithelium cells of feeding tadpoles. *J. Embryol. Exp. Morphol.* **10**, 622-640.

Gurdon, J. B., Laskey, R. A. and Reeves, O. R. (1975). The developmental capacity of nuclei transplanted from keratinized skin cells of adult frogs. *J. Embryol. Exp. Morph.* **34**, 93-112.

Hardeman, E., Chiu, C.-P., Minty, A. and Blau, H. M. (1986). The pattern of actin expression in human fibroblast × mouse muscle heterokaryons suggests that human muscle regulatory factors are produced. *Cell* **47**, 123-130.

Harris, H. (1988). The analysis of malignancy by cell fusion: The position in 1988. *Cancer Res.* **48**, 3302-3306.

Harris, H. and Klein, G. (1969). Malignancy of somatic cell hybrids. *Nature* **224**, 1314-1316.

Harris, H., Sidebottom, E., Grace, D. M. and Bramwell, M. E. (1969). The expression of genetic information: A study with hybrid animal cells. *J. Cell Sci.* **4**, 499-525.

Harris, H. and Watkins, J. F. (1965). Hybrid cells derived from mouse and man: Artificial heterokaryons of mammalian cells from different species. *Nature* **205**, 640-646.

Holtzer, H., Biehl, J., Payette, R., Sasse, J. and Holtzer, S. (1983). Cell diversification: Differing roles of cell lineages and cell-cell interactions, in *Limb Development and Regeneration.* Alan R. Liss, New York, pp. 272-280.

Killary, A. M. and Fournier, R. E. K. (1984). A genetic analysis of extinction: Trans-dominant loci regulate expression of liver-specific traits in hepatoma hybrid cells. *Cell* **38**, 523-534.

Konieczny, S. F., Lawrence, J. B. and Coleman, J. R. (1983). Analysis of muscle protein expression in polyethylene glycol-induced chicken: rat myoblast heterokaryons. *J. Cell Biol.* **97**, 1348-1355.

Landschulz, W. H., Johnson, P. F. and McKnight, S. L. (1989). The DNA binding domain of the rat liver nuclear protein C/EBP is bipartite. *Science* **243**, 1681-1688.

Lawrence, J. B. and Coleman, J. R. (1984). Extinction of muscle-specific proterites in somatic cell heterokaryons. *Dev. Biol.* **101**, 463-476.

Mevel-Ninio, M. and Weiss, M. C. (1981). Immunofluorescence analysis of the time-course of extinction, reexpression, and activation of albumin production in rat hepatoma-mouse fibroblast heterokaryons and hybrids. *J. Cell Biol.* **90**, 339-350.

Miller, S. C., Pavlath, G. K., Blakely, B. T. and Blau, H. M. (1988). Muscle cell components dictate hepatocyte gene expression and the distribution of the Golgi apparatus in heterokaryons. *Genes and Development* **2**, 330-340.

Murre, C., McCaw, P. S. and Baltimore, D. (1989). Interactions between heterologous helix-loop-helix proteins generate complexes that bind specifically to a common DNA sequence. *Cell* **56**, 777-783.

Pavlath, G. K. and Blau, H. M. (1986). Expression of muscle genes in heterokaryons depends on gene dosage. *J. Cell Biol.* **102**, 124-130.

Pavlath, G. K., Chiu, C.-P. and Blau, H. M. (1989). *In vivo* aging of human fibroblasts does not alter nuclear plasticity in heterokaryons. *Som. Cell and Molec. Genetics* **15**, 191-202.

Peehl, D. M. and Stanbridge, E. J. (1982). The role of differentiation in the suppression of tumorigenicity in human cell hybrids. *Int. J. Cancer* **30**, 113-120.

Peterson, C. A., Gordon, H., Hall, Z. W., Paterson, B. M. and Blau, H. M. (1990). Negative control of helix-loop-helix family of myogenic regulators in NFB mutant. *Cell* **62**, 493-502.

Peterson, J. A. and Weiss, M. C. (1972). Expression of differentiated functions in hepatoma cell hybrids: Induction of mouse albumin production in rat hepatoma - mouse fibroblast hybrids. *Proc. Natl. Acad. Sci. USA* **69**, 571-575.

Petit, C., Levilliers, J., Ott, M.-O. and Weiss, M. C. (1986). Tissue-specific expression of the rat albumin gene: Genetic control of its extinction in microcell hybrids. *Proc. Natl. Acad. Sci. USA* **83**, 2561-2565.

Ptashne, M. (1986). *A Genetic Switch: Gene Control and Phage λ.* Cell Press and Blackwell Scientific Publications, Cambridge, MA.

Rankin, J. K. and Darlington, G. J. (1979). Expression of human hepatic genes in mouse hepatoma-human amniocyte hybrids. *Somatic Cell Genet.* **5**, 1-10.

Ringertz, N. R., Carlsson, S.-A., Ege, T. and Bolund, L. (1971). Detection of human and chick nuclear antigens in nuclei of chick erythrocytes during reactivation in heterokaryons with HeLa cells. *Proc. Natl. Acad. Sci. USA* **68**, 3228-3232.

Ringertz, N. R. and Savage, R. E. (1976). *Cell Hybrids.* Academic Press, New York.

Schafer, B. W., Blakely, B. T., Darlington, G. J. and Blau, H. M. (1990). Effect of cell history on response to helix-loop-helix family of myogenic regulators. *Nature* **344**, 454-458.

Spear, B. T. and Tilghman, S. M. (1990). The role of the α-fetoprotein regulatory elements in transcriptional activation in transient heterokaryons. *Mol. Cell Biol.* **10**, 5047-5054.

Waddington, C. H. (1940). Organisers and Genes. *Cambridge University Press, London, England.* pp. 91-93.

Weintraub, H. (1985). Assembly and propagation of repressed and derepressed chromosomal states. *Cell* **42**, 705-711.

Weiss, M. C. (1982). in *Somatic Cell Genetics* (Caskey, T.C. and Robbins, D.C., eds), Plenum Press, NY, pp. 169-179.

Wright, W. E. (1984). Induction of muscle genes in neural cells. *J. Cell Biol.* **98**, 427-435.

Wright, W. E. (1984). Expression of differentiated functions in heterokaryons between skeletal myocytes, adrenal cells, fibroblasts and glial cells. *Exp. Cell Res.* **151**, 55-69.

Wright, W. E. and Aronoff, J. (1983). The suppression of myogenic functions in heterokaryons formed by fusing chick myocytes to diploid rat fibroblasts. *Cell Differ.* **12**, 299-306.

Wu, K. J., Samuelson, L. C., Howard, G., Meisler, M. H. and Darlington, G. H. (1990). Transactivation of pancreas-specific gene sequence in somatic cell hybrids. *Mol. Cell Biol.* **11**, 4423-4430.

THE CONSEQUENCES OF A CONSTITUTIVE EXPRESSION OF MyoD1 IN ES CELLS AND MOUSE EMBRYOS

MOSHE SHANI, ALEXANDER FAERMAN, CHARLES P. EMERSON*, SONIA PEARSON-WHITE*, ITZHAK DEKEL and YONAT MAGAL

The Institute of Animal Science, The Volcani Center, Bet Dagan 50250, Israel and *Department of Biology, University of Virginia, Charlottesville, VA 22901, USA

Abstract

A variety of differentiated cell types can be converted to skeletal muscle following transfection with the myogenic regulatory gene MyoD1. To determine whether multipotent embryonic stem (ES) cells respond similarly, cultures of two ES cell lines were electroporated with a MyoD1 cDNA driven by the β-actin promoter. All transfected clones tested, carrying single copy of the exogenous gene, expressed high levels of MyoD1 mRNA. Surprisingly, although maintained in mitogen-rich medium, this ectopic expression was associated with a trans-activation of the endogenous myogenin and myosin light chain 2 genes but not the endogenous MyoD1, MRF4, myf5, skeletal muscle actin or myosin heavy chain genes. Preferential myogenesis and the appearance of contracting skeletal muscle fibers was observed only when the transfected cells were allowed to differentiate, via embryoid bodies, in low mitogen-containing medium. Myogenesis was associated with the activation of MRF4 and myf5 genes and in a significant increase in the level of myogenin mRNA. Not all cells were converted to skeletal muscle, indicating that only a subset of stem cells can respond to MyoD1. Moreover, the continued expression of MyoD1 was not required for myogenesis. Interestingly, no preferential myogenesis was observed when the transfected ES cells were allowed to differentiate *in vivo* to teratocarcinomas. These results show that ES cells can respond to MyoD1, but environmental factors control the expression of its myogenic differentiation function. Second, MyoD1 function in ES cells, even under environmental conditions that favour differentiation, is not dominant (incomplete penetrance). Third, that the exogenous MyoD1 trans-activates the endogenous myogenin and MLC2 genes in ES cells.

No live transgenic mice could be produced following microinjection of the β-actin/MyoD1 gene into the pronuclei of fertilized eggs. Transgenic embryos died before mid gestation. The majority of tested embryos between 7.5 and 9.5 days, although retarded compared to control litermates, differentiated into tissues representative of all three germ layers. The expression of the introduced gene was detected in all ectodermal and mesodermal tissues but was absent in all endodermal cells. These results demonstrate again that MyoD1 is not a dominant regulatory factor.

Key words: ES cells, MyoD1, myogenesis, transgenic embryos, embryonic lethalities.

Introduction

At least 5 genes (MyoD1, myogenin, myf5, myd, and herculin-MRF4) involved in the determination of the myogenic program have already been identified (Konieczny and Emerson, 1984; Davis, Weintraub and Lassar, 1987; Pinney, Pearson-White, Konieczny, Latham and Emerson, 1988; Wright, Sassoon and Lin, 1989; Edmonson and Olson, 1989; Braun, Buschausen-Denker, Bober, Tannich and Arnold, 1989; Lin, Dechesner, Eldridge and Paterson, 1989; Rhodes and Konieczny, 1989; Miner and Wold, 1990). The three common features of four of these regulatory proteins are that their expression is restricted to skeletal muscle (Davis et al., 1987; Wright et al., 1989; Edmondson and Olson, 1989; Rhodes and Konieczny, 1989; Miner and Wold, 1990), that they share an helix-loop-helix structure (Murre, Schonleber McCaw and Baltimore, 1989a) with extensive similarity to a myc domain (Davis et al., 1987; Murre, Schonleber McCaw, Vasessin, Caudy, Cabrera, Bushkin, Hauschka, Weintraub and Baltimore, 1989b), and that they can convert non-muscle cells to skeletal muscle cells (for recent reviews see Olson, 1990; Emerson, 1990; Weintraub, Davis, Tapscott, Thayer, Krause, Benezra, Blackwell, Turner, Rupp, Hollenberg, Zhuang and Lassar, 1991).

MyoD1, the most extensively studied regulator, is a nuclear phosphoprotein requiring the myc homology region and the adjacent basic domain to convert fibroblastic to myogenic cells (Tapscott, Davis, Thayer, Cheng, Weintraub and Lassar, 1988). Through this region it binds two of the muscle specific creatine kinase enhancer regions as well as the promoter region of the myosin light chain 1/3, the avian acetylcholine receptor α-subunit and actin genes (Murre et al., 1989b; Piette, Bessereaux, Hutchet and Changeoux, 1990; Rosenthal, Berglund, Wenthworth, Donoughue, Winter, Bober, Braun and Arnold, 1990; Sartorelli, Webster and Kedes, 1990).

Transfection experiments have demonstrated that MyoD1 can convert a variety of cell types to skeletal muscle. This includes primary cells or cell lines derived from the three germ layers: adipocytes and fibroblasts (mesoderm), hepatocytes (endoderm) and neuroblastoma and melanocytes (ectoderm) (Choi, Costa, Mermelstein, Chaga, Holtzer and Holtzer, 1990). This and other data led Weintraub, Tapscott, Davis, Thayer, Adam, Lassar and Miller (1989) to suggest that MyoD1 is a master regulatory gene for skeletal myogenesis, and that no other factors are needed to activate the downstream program for terminal differentiation. A different view about the role of MyoD1 has been suggested by Schafer, Blakely, Darlington and Blau (1990), claiming that the differentiated state of a cell is dictated by its history and a dynamic interaction among the proteins that it contains. Thus, despite the high level of MyoD1 expression in hepatocytes, the cells did not express muscle specific genes, unless they were first fused to fibroblasts, implying that other regulators necessary for the phenotypic conversion to skeletal muscle are present in fibroblasts but are absent in hepatocytes.

A major problem with the transfection of somatic cells is that they test only a narrow portion of the spectrum of cell types and developmental states that might be accessible to myogenic conversion by these factors. Therefore, we have studied the consequences of the constitutive expression of MyoD1 when introduced into the multipotent early embryonic stem (ES) cells (for reviews see Robertson, 1987; Bradley, 1990) or mouse fertilized eggs. If MyoD1 is the master regulator for skeletal myogenesis, its expression in undifferentiated ES cells and early mouse embryos would be sufficient to convert them to skeletal muscle. However, if the myogenic regulatory function of MyoD1 is environmentally controlled and it is only one link in the cascade leading to terminal differentiation of muscle cells, then its ectopic expression at the stem cell stage would not alter their multipotentiality. The results indicate that high level of MyoD1 expression in ES cells can convert a subpopulation of differentiated cells to skeletal muscle but only under specific environmental conditions, that MyoD1 expression in undifferentiated ES cells grown in mitogen-rich medium is associated with the transactivation of myogenin and the muscle specific MLC2 but not the endogenous MyoD1, myf5, MRF4, the skeletal muscle actin or myosin heavy chain genes, that the continued expression of MyoD1 is not required for the maintenance of the myogenic program, and that there is no preferential myogenesis upon differentiation *in vivo* to teratocarcinomas. Transgenic embryos expressing the exogenous β-actin/MyoD1 gene die around mid-gestation. Although the majority of these embryos are retarded compared to normal littermates, all tissues representative of the three germ layers are formed normally. These results therefore demonstrate that in ES cells and early mouse embryos MyoD1 is not dominant in its ability to convert all embryonic cells to skeletal muscle.

Methods

Cell culture

The established ES cell lines D3 (Doetschman, Eistetter, Katz, Schmidt and Kemler, 1985) and B2B2 (Robertson, Kaufman, Bradley and Evans, 1983) were grown in 40-60% of Buffalo Rat Liver (BRL) conditioned medium with 15% fetal calf serum and 10^{-4} M β-mercaptoethanol (mitogen-rich medium or growth medium) on gelatin coated plates, as described (Smith and Hooper, 1987). In vitro differentiation was induced essentially as described (Shinar, Yoffe, Shani and Yaffe, 1989). Briefly, the cells were first plated in the absence of conditioned medium or gelatin, for 3 days. The aggregates formed were then detached from the plates and grown in suspension for 10 days to form embryoid bodies (EB), with daily changes of medium. Finally, the aggregates were allowed to settle onto gelatin coated plates. The EB attached to the plates, and differentiated cell types migrated out of the aggregates. Medium was changed every 2 days and the cultures were inspected for the appearance of differentiated cell types. To induce preferential myogenesis, the EB were grown in the presence of 10% horse serum and $0.1 \mu g/ml$ insulin (2HI-muscle-specific differentiation medium). To induce teratocarcinomas, 5×10^6 ES cells were injected subcutaneously into athymic mice.

Tumors appeared at the site of injection within 3-5 weeks. The tumors were excised and subjected to histological observation and RNA analysis.

DNA electroporation

Prior to electroporation, the plasmid DNA sequences were removed. The fragment containing the MyoD1 and the neomycin resistance genes (10 μg) was electroporated at 600 V as described (Shinar, Yoffe, Shani and Yaffe, 1989). Clones resistant to Geneticin G-418 (300 μg/ml) were picked individually and expanded into mass cultures.

Production of transgenic embryos

Microinjection of the plasmid-free β-actin/MyoD1 fragment into mouse fertilized eggs was performed as described previously (Hogan, Costantini and Lacy, 1986; Shani, 1985; Shani, 1986).

Analysis of RNA

Total cellular RNA was prepared by the urea-LiCl method (Auffray and Rougeon, 1980). Northern blots were prepared as described in Sambrook, Fritsch and Maniatis (1989). The blots were hybridized with DNA fragments labelled by random priming. The MyoD1 probe was the full length cDNA (Pearson-White and Emerson, unpublished), the MLC2 and the skeletal muscle actin probes were obtained from plasmid p103 and p106, respectively (Katcoff, Nudel, Zevin-Sonkin, Carmon, Shani, Lehrach, Frischauf and Yaffe, 1980), the MRF4, myf5 and myogenin specific probes were derived from the full length cDNA clones kindly provided by Dr. S. Konieczny. Insitu hybridization of ^{35}S-labelled riboprobes to parafin sections of embryos was performed essentially as described by Sassoon, Lyon, Wright, Lin, Lassar, Weintraub and Buckingham (1989).

Results

Transfection of ES cells with a β-actin/MyoD1 gene construct

Cultures of the multipotent embryonic stem cell lines ES-D3 (Doetschman, Eistetter, Katz, Schmidt and Kemler, 1985) and B2B2 (Robertson, Kaufman, Bradley and Evans, 1983) were transfected by electroporation with a β-actin/MyoD1 containing DNA fragment. This fragment included the MyoD1 cDNA driven by the β-actin promoter and the SV40 polyadenylation signal, and also the PMC1Neo DNA, carrying the bacterial NeoR gene driven by a promoter that is efficiently expressed in ES cells (Thomas and Chapecchi, 1987) (Fig. 1). The choice of the β-actin promoter was to ensure that the exogenous MyoD1 would be expressed constitutively, in the stem cells and their differentiated derivatives. Fig. 1 shows Southern blot analysis of DNA isolated from several MyoD1-transfected clones digested with HindIII and hybridized to the β-actin/MyoD1 DNA. The 1.2 Kb hybridized fragment corresponds to the introduced MyoD1 cDNA fragment

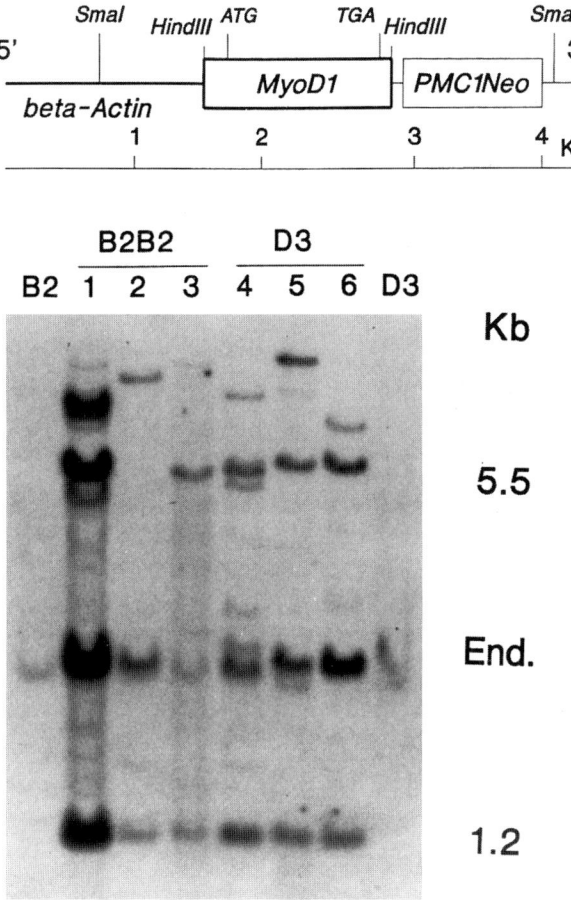

Fig. 1. The structure of the β-actin/MyoD1 construct and a southern blot analysis of transfected ES cells. The SV40 polyadenylation addition site is included in the MyoD1 box. About 2-10 μg of DNA of the indicated clones was digested with HindIII, and the fragments were blot-hybridized to ^{32}P-labelled plasmid p340 containing the β-actin/MyoD1/PMC1neo fragment.

(Fig. 1). From the intensity of the hybridized fragments, relative to the endogenous MyoD1 gene, it was concluded that a single copy of the exogenous gene had integrated in each clone.

All MyoD1-transfected ES clones tested expressed the exogenous MyoD1 gene. The mRNA of the exogenous gene could easily be distinguished from that of the endogenous MyoD1 transcripts since its 3' untranslated region was significantly shorter than that of the endogenous gene (Fig. 2). In most cases, the level of expression in the different clones was higher than that found in the established

Activation of Myogenin and MLC2 in ES Cells

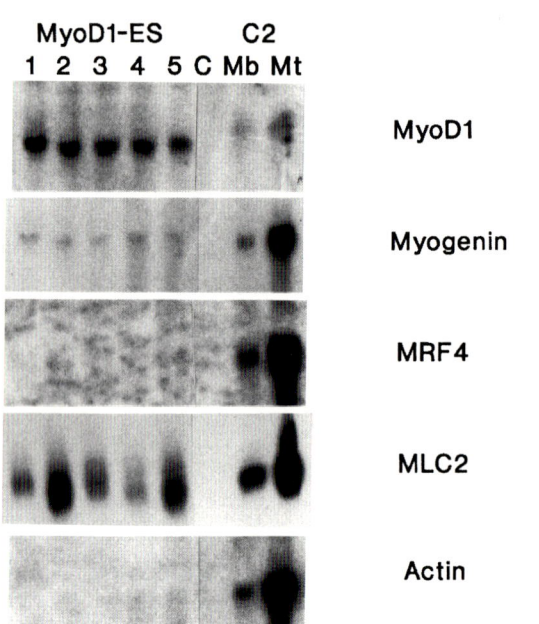

Fig. 2. Expression of MyoD1, myogenin and MLC2 genes in transfected ES cells. Ten ug of total RNA from the indicated transfected ES clones and control (C) untransfected ES cells (grown in mitogen-rich medium) and 0.5 μg (for the detection of MLC2 and α-actin) or 10 μg (for the detection of MyoD1, myogenin myf5 and MRF4) of total RNA from cultures of proliferating myoblasts (Mb) or differentiated myotubes (Mt) of the myogenic cell line C2 cells were electrophoresed in a formaldehyde/MOPS gel, transferred into Nytran nylon membrane and hybridized to the indicated ^{32}P labelled probes. Note the size difference between the endogenous and introduced MyoD1 mRNAs.

myogenic cell line C2. Surprisingly, the forced expression of the exogenous MyoD1 gene did not activate the endogenous gene.

Expression of MyoD1 at the stem cell stage did not alter the morphology or the developmental potential (see below) of these cells. As long as the cells were maintained in mitogen-rich medium (see Materials and Methods), they retained their characteristic embryonic morphology. When undifferentiated ES cells, both transfected and untransfected, were cultured in differentiation medium (2HI), they stopped proliferating, increased in size and after about 48 hours were detached from the plates. During the 48 hours we have never seen the formation of muscle fibers.

Activation of the endogenous myogenin and MLC2 genes in the MyoD1 transfected ES cells

The constitutive expression of MyoD1 in the transfected ES cells, grown in mitogen-rich medium, was associated with the activation of myogenin and the skeletal muscle myosin light chain 2 genes in all MyoD1-transfected clones that were tested, but not that of the endogenous MyoD1, MRF4, myf5 and skeletal muscle actin genes (Fig. 2). In addition, the cells remained unstained with specific monoclonal antibodies for myosin heavy chain (data not shown). The level of the MLC2 transcripts in the transfected MyoD1-ES cells was approximately 10 to 100 times lower than in differentiated C2 cells, whereas the level of myogenin mRNA was significantly lower than that found in proliferating myoblasts of the C2 cells (Fig. 2). These results indicate that MyoD1 can act as a trans-activator, even in the presence of growth factors.

In vitro differentiation of ES-MyoD1 cells

MyoD1-transfected ES clones were induced to differentiate via embryoid body (EB) formation (Martin, Wiley and Damjanov, 1977). When differentiation was induced in Eagle's medium supplemented with 15% foetal calf serum (GM) no skeletal muscle fibers were observed, although, beating heart muscle (another striated muscle) is one of the first differentiated tissues to appear. However, when differentiation was induced in the presence of muscle-specific differentiation-inducing medium (10HI-10% horse serum and 0.1 μg/ml insulin), the MyoD1-transfected cells differentiated preferentially into skeletal muscle fibers, that in most cases contracted spontaneously, whereas mock-transfected cells did not. About 20-50% of all differentiated cell types that migrated out of the EB differentiated into muscle fibers. Fig. 3 shows the morphology of skeletal muscle fibers and some of the differentiated cell types that appeared in such cultures. These fibers were also stained specifically with the antimyosin monoclonal antibodies MF-20 (data not shown). Similar skeletal myogenesis was observed with 6 independent MyoD1-transfected clones, derived from the ES cell line D3 and 5 transfected clones derived from cell line B2B2. Future experiments with lineage-specific markers will be needed to identify the embryonic origin of the non-myogenic cell types.

Every MyoD1-transfected clone tested differentiated to contracting muscle fibers. However only 7 out of the 9 clones tested retained MyoD1 expression in the differentiated cultures. Fig. 4A shows the level of expression in the stem cells and the differentiated derivatives of several clones. While clones 3 and 7 expressed the transfected MyoD1 mRNA both in differentiation conditions and in undifferentiated ES cells, differentiated clones 1 and 2 had a strong skeletal muscle phenotype but had lost expression of its exogenous MyoD1 gene. These results suggest that once committed, the cells no longer need the continued expression of MyoD1 for the maintenance of the differentiated state. Consistent with the results with the undifferentiated ES cells, there was no evidence for the activation of the

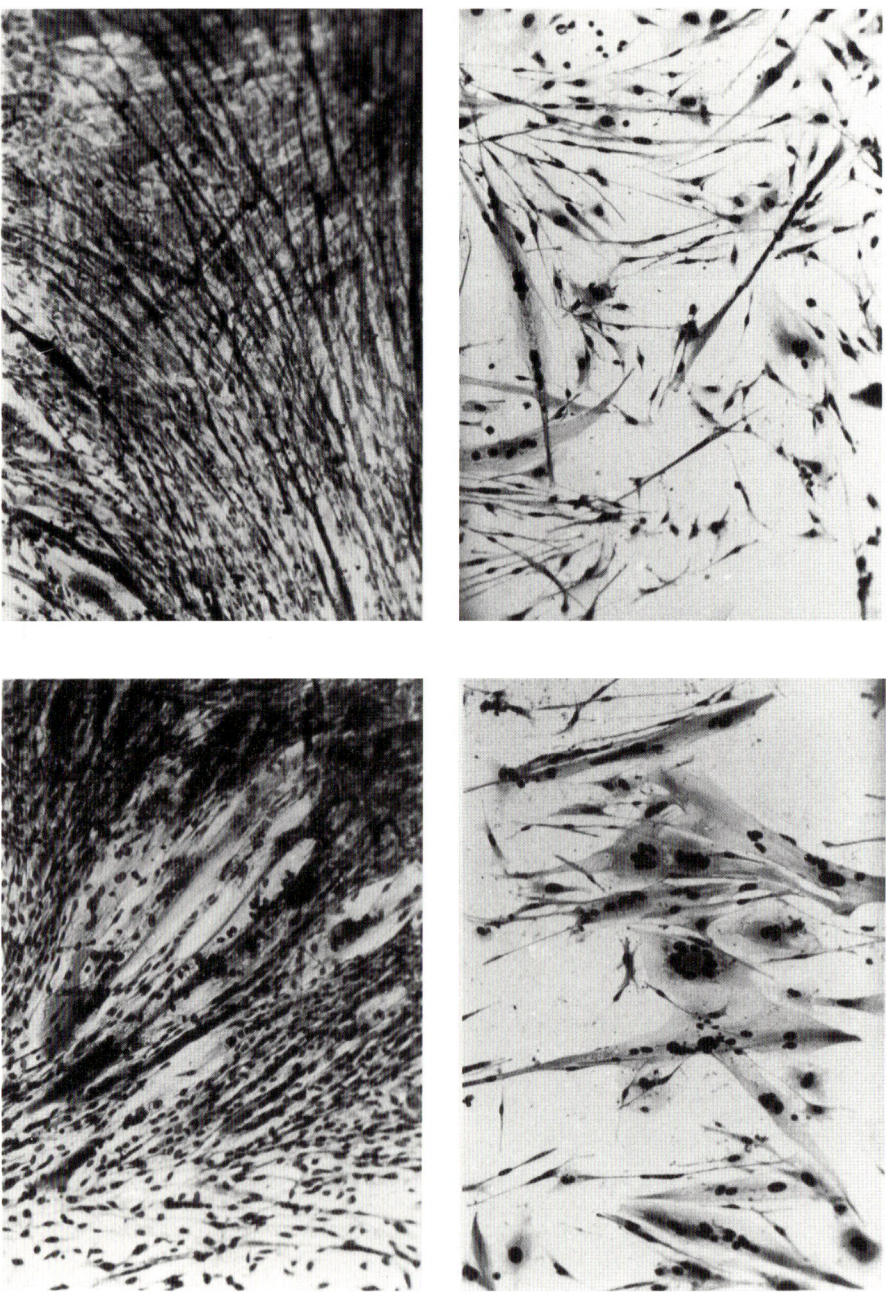

Fig. 3. The morphology of skeletal muscle fibers that appear in cultures of transfected ES cells differentiated in muscle-specific differentiation medium (see Materials and Methods). Since cultures of differentiated ES cells are multilayered, pictures were taken from different regions of a single differentiated embryoid body.

endogenous MyoD1 gene in the differentiated cultures. However, skeletal myogenesis was associated with the activation of the MRF4 and myf5 genes and in an increase in the level of expression of myogenin (Fig. 4B). In contrast, when MyoD1-ES clone 3 was induced to differentiate in growth medium (GM), no muscle fibers were formed, no transcripts of MRF4 and myf5 could be detected, and there was no increase in the level of myogenin mRNA (Fig. 4B).

In vivo differentiation of ES-MyoD1 cells

To induce differentiation in-vivo we have injected the transfected cells subcutaneously into syngeneic (nude) mice. This resulted in the development of teratocarcinoma tumors at the site of injection, consisting of differentiated tissues representing the 3 germ layers. These results, therefore, indicate that despite the constitutive expression of MyoD1 in the stem cells and during the critical stages of their determination and commitment to the different embryonic lineages, the tumors displayed the typical array of differentiated tissues of teratocarcinomas. In situ hybridization analysis of tumors produced with two transfected ES cells and control mock transfected ES cells revealed that the only tissue expressing MyoD1 is skeletal muscle (data not shown). Therefore, in view of previous results with the expression of genes introduced into ES cells (Shinar et al., 1990) it is reasonable to assume that this reflects transcription of the endogenous gene and that the exogenous gene was inactivated.

Ectopic expression of MyoD1 during early mouse development

The onset of MyoD1 expression in the myotomes of somites is at day 10.5 (Sassoon et al., 1989). To better understand the function of MyoD1 as a muscle regulator we have induced MyoD1 expression constitutively from the one-cell stage, by microinjecting the same β-actin/MyoD1 construct into mouse fertilized eggs. Of more than 80 mice born from this set of microinjection experiments none contained the injected DNA, indicating that ectopic expression of MyoD1 during mouse development is deleterious to early embryos.

To determine the stage of development at which transgenic conceptuses were dying and the possible cause of death, we examined embryos from 7.0 to 9.0 days after microinjection histologically and by in situ hybridization.

Fig. 5 shows a transgenic embryos isolated 7.0 days after microinjection (equivalent to 7.5 days of gestation). This embryo is slightly retarded compared to its normal littermates. Developmentally, it resembles day 6.5-7.0 normal embryo, since the proamniotic cavity is not yet separated into the amniotic and ectoplacental cavities. This embryo is also bent along the border between the embryonic and extraembryonic regions. Hybridization to MyoD1 riboprobe shows a strong signal in giant cells, the ectoplacental cone and the ectodermal and mesodermal layers of the embryonic region. Interestingly, the endodermal layer is negative.

Fig. 6 shows a sagital section of a transgenic embryo isolated 8 days after

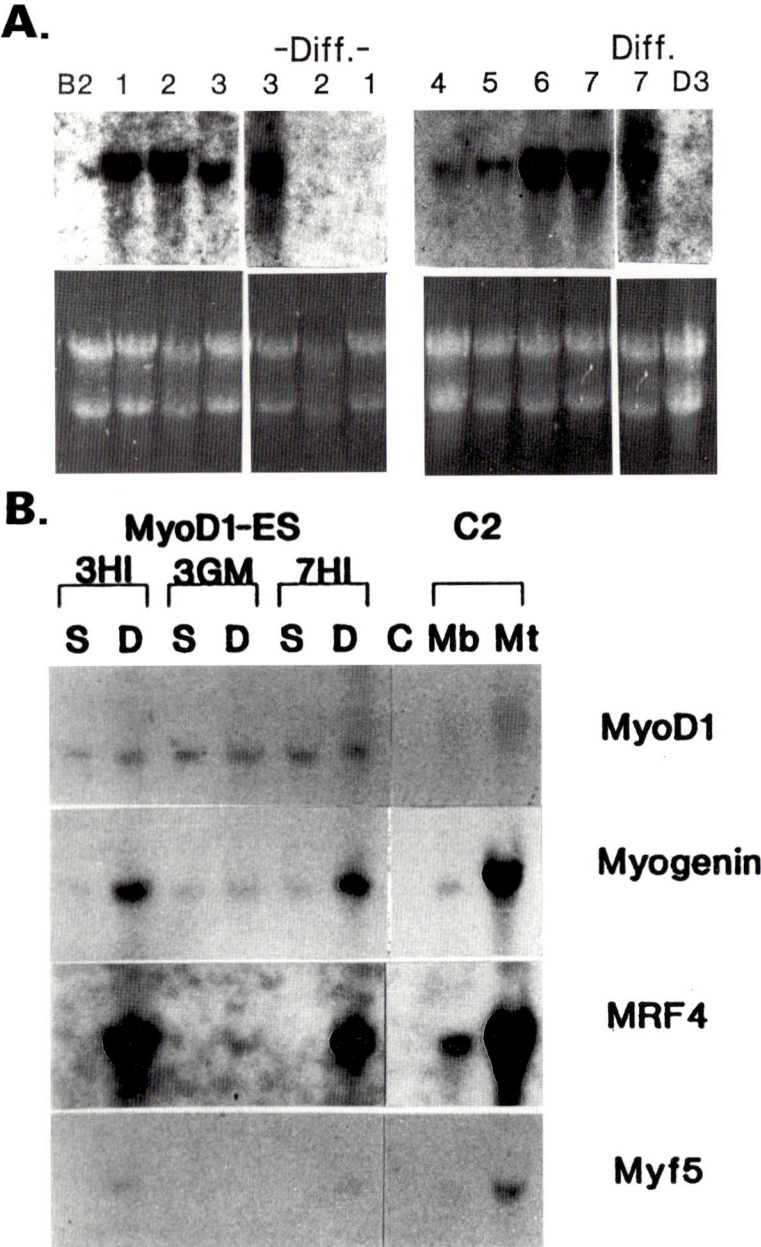

microinjection (equivalent to day 8.5 of gestation). Morpholgically this embryo resembles day 7-7.5 non-transgenic embryos, since there is no sign of heart or somite formation. Expression of the introduced gene is confined to the embryonic

Fig. 4. Expression of MyoD1, myogenin, MRF4 and myf5 genes in transfected ES cells and their differentiated derivatives. (A) 20 μg of total RNA from the indicated transfected ES clones and their differentiated derivatives were electrophoresed in a formaldehyde/MOPS gel, transferred into Zeta nylon membrane (BioRad) and hybridized to ^{32}P labelled MyoD1 riboprobe. B2 and D3 are the untransfected ES cell lines. Bottom, ethidium bromide staining of the gel is shown to confirm that intactness of the RNA samples. (B) 10 μg of total RNA from the indicated transfected ES clones and their differentiated derivatives, induced to differentiate either in 2HI or in GM, hybridized to random-primed MyoD1, myogenin, MRF4 and myf5 probes. S, undifferentiated stem cells; D, differentiated cultures. Note that in the presence of GM, MRF4 and myf5 genes are not activated and the level of myogenin mRNA does not increase.

ectodermal and mesodermal layers and there is no detectable signal in the endodermal layer.

Fig. 7 shows an example of a day 9.5 transgenic embryo. This embryo is also slightly retarded, since the turning of the embryo is not completed and the neural tube is closed only at the midbody region. Thus the developmental stage of this embryo is somewhere between day 8.5 to 9.0 days of mouse development. The major sites of MyoD1 expression are; neural crest cells at the dorsal part of the neural tube; the somites; the dorsal part of the heart, mainly the heart wall and the mesodermal cells lining the pericardial cavity. No signal can be detected in endodermal cells lining the hindgut and foregut diverticulums.

Discussion

The present results demonstrate that under specific culture conditions even multipotent ES cells can respond to the myogenic regulatory gene MyoD1. When undifferentiated MyoD1-expressing ES cells were maintained in mitogen-rich medium, the cells retained their typical embryonic morphology and developmental potential, and did not convert to skeletal muscle cells. Even when induced to differentiate, via embryoid body formation, in 15% foetal calf serum, no skeletal muscle fibers appeared. However, in the presence of the muscle-specific differentiation medium, the majority of the embryoid bodies gave rise to contracting muscle fibers. Thus, the ability to convert ES cells to skeletal muscle depends on specific environmental conditions. This is in agreement with recent work in Xenopus (Hopwood and Gordon, 1990). In this study transient ectopic expression of XMyoD1 in embryonic cells did not induce presumptive ectoderm to differentiate into definitive muscle, as assayed histologically. However, it lead to the activation of the cardiac and skeletal muscle actin genes as well as the endogenous Xmyod gene (Hopwood and Gordon, 1990). Similarly, expression of MyoD1 couldn't overcome the repression of myogenesis by fibroblast growth factor (Vaidya, Rhodes, Taparowsky and Konieczny, 1989), and the forced expression of MyoD1 in hepatocytes did not result in the activation of muscle specific genes unless they were fused to fibroblasts (Schafer et al., 1990).

Not all MyoD1-ES cells induced to differentiate in low-mitogen containing

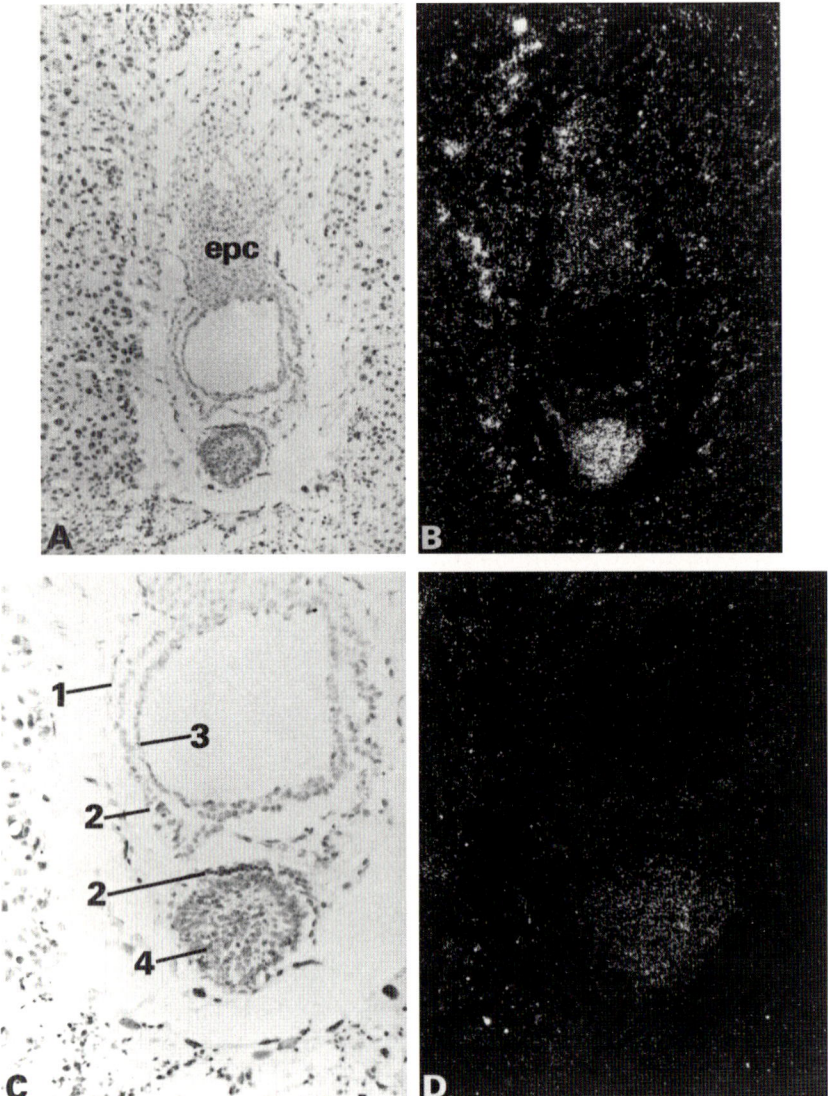

Fig. 5. A section of 7.5 day transgenic embryo. (A) Bright-field micrograph; (B) Dark-field micrograph of (A) hybridized with the MyoD1 probe. (C) and (D) are brightfield and dark field, respectively, of a higher magnification of the embryonic region. epc-ectoplacental cone; 1-distal endoderm; 2-proximal endoderm; 3-extraembryonic ectoderm; 4-embryonic ectoderm/mesoderm.

medium were converted to muscle cells. Many other cell types that normally appear in such cultures continue to appear even with MyoD1 expressed, implying that only a subset of stem cells are susceptible to MyoD1 control. It would, therefore, be of importance to identify the responding stem cell, since it would

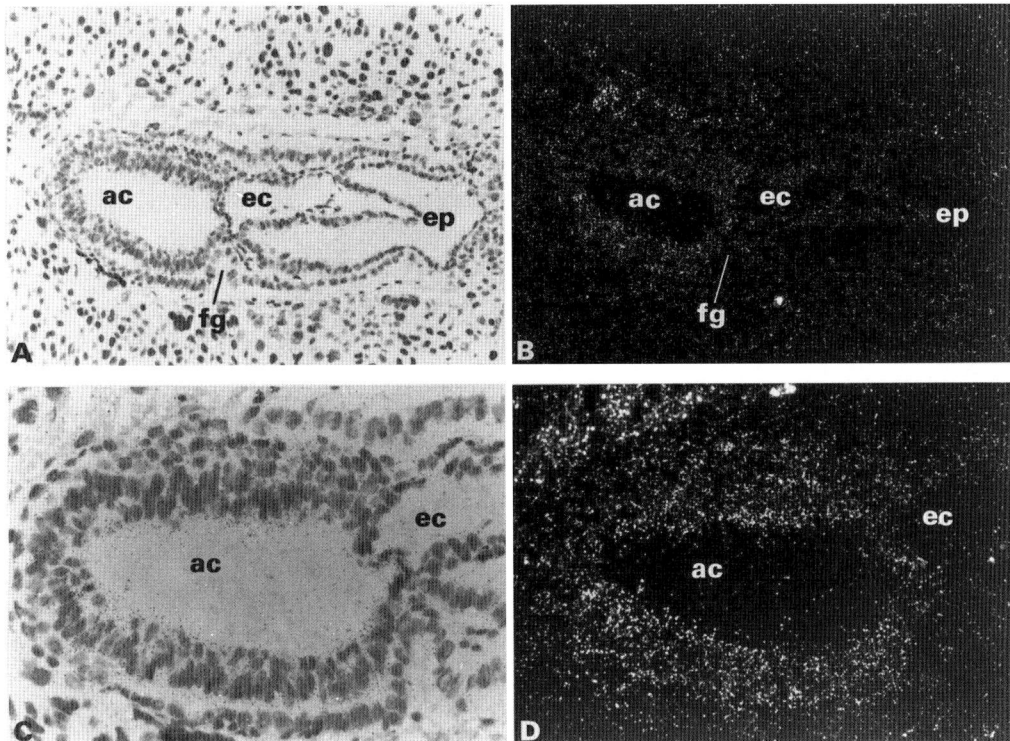

Fig. 6. A section of 8.5 day transgenic embryo. (A) and (B), brightfield and darkfield. (C) and (D), higher magnification of brightfield and darkfield of the embryonic region. ac-amniotic cavity; ec-exocoelom; ep-ectoplacental cavity; fg-foregut.

indicate whether responsiveness to MyoD1 is restricted to a specific embryonic layer and to specific cell types within that layer.

Surprisingly, when the transfected ES cells were induced to differentiate into teratocarcinomas in vivo, no such preferential myogenesis was observed. The tumors contained the typical differentiated tissues derived from the three germ layers. These results suggest that there is a fundamental difference between the effect of MyoD1 in vitro and in vivo. While in vitro differentiation could be manipulated by the immediate environment, no such manipulations are possible in vivo. It could be that myogenic conversion in vivo requires even higher levels of MyoD1 expression. Alternatively, MyoD1 could have a subtle effect on these tumors that was overlooked by the histological examination of tumor sections.

The forced expression of MyoD1 in the transfected ES cells resulted in the activation of the endogenous myogenin and the skeletal muscle MLC2 genes, even when the cells were maintained in mitogen-rich medium, but failed to activate MRF4, myf5, myosin heavy chain and skeletal muscle actin genes. It is of interest that of the 3 muscle specific structural protein genes that were examined in this study only the MLC2 gene responded to the myogenic regulatory gene. Previous

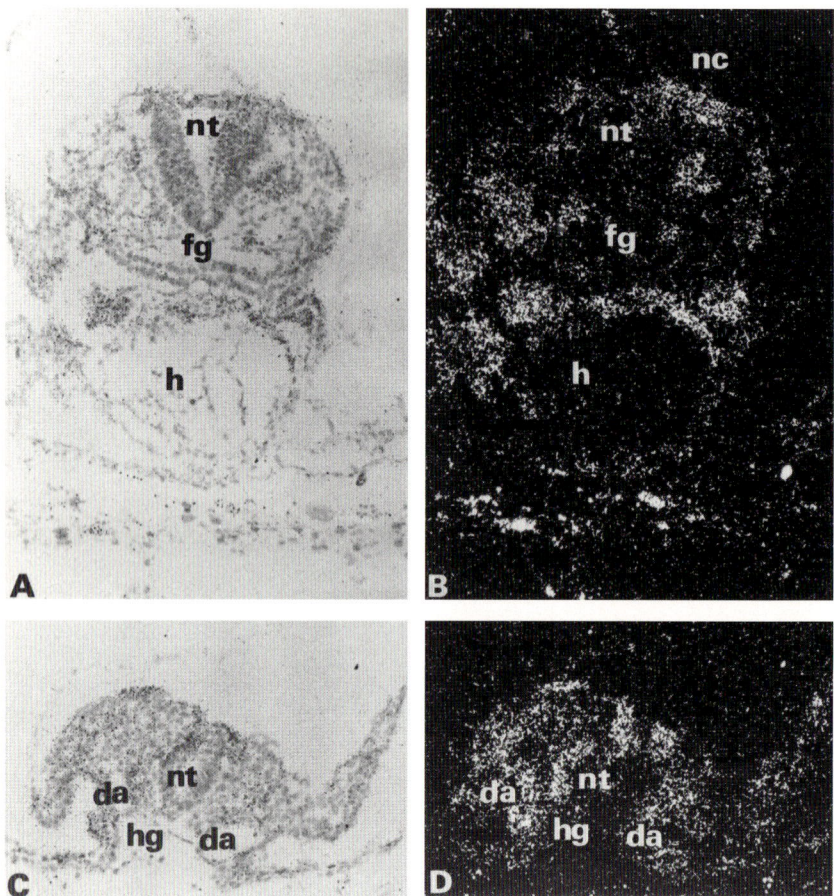

Fig. 7. A section of 9.5 day transgenic embryo. (A) and (B) mid-body region. (C) and (D) posterior part of the embryo. nt-neural tube; fg-foregut; h-heart; da-dorsal auorta; hg-hindgut.

results have indicated that MLC2 and skeletal muscle actin genes are DNaseI sensitive in skeletal muscle and in differentiated myogenic cells in culture. However, unlike the muscle actin gene, the MLC2 gene is moderately sensitive to DNAaseI in the brain (Carmon, Czosnek, Nudel, Shani and Yaffe, 1982) and liver (unpublished results), indicating that the chromatin organization of this gene in non-muscle tissues differs than other muscle-specific genes. This could be associated with the enhanced susceptibility of this gene to MyoD1. It still remains to be determined what type of MyoD1 heterodimers can activate specifically one but not other muscle specific genes, despite the unfavourable culture conditions. Alternatively, each muscle regulator could activate different target muscle protein genes despite their in vitro binding specificity for related sequence motifs in these genes.

One of the characteristics of the muscle regulatory genes is their ability to autoregulate their own and activate one another's expression. It is, therefore, of interest that the endogenous MyoD1 wasn't activated in the transfected ES cells and/or their differentiated derivatives. It has been reported that exogenous MyoD1 can activate the endogenous gene in some, but not all, transfected cell types (Weintraub et al., 1989; Thayer, Tapscott, Davis, Wright, Lassar and Weintraub, 1989). Cell types in which the endogenous MyoD1 wasn't activated included NIH3T3, Swiss 3T3 C2, adipoblast F442A, the melanoma B16 and the rat neuroblastoma B50 C5. A common feature of these cell types is the low level of myogenic conversion using either 5 Azacytidine treatment or forced expression of an exogenous MyoD1 gene. Thus, it appears that other factors, not present in these cell types, play a role in the activation of the endogenous gene.

None of the control, untransfected ES cells, differentiated into skeletal muscle fibers under the conditions used. Therefore, the appearance of contracting muscle cells in the transfected cultures is most likely due to the expression of the exogenous MyoD1. However, once the myogenic program has been established, the continued expression of MyoD1 is no longer necessary, as indicated by the fact that about 30% (4/11) of the transfected clones tested did not retain MyoD1 expression, yet they differentiated into skeletal muscle fibers. This may imply that MyoD1 is a determination rather than a differentiation protein. This issue has been challenged recently (Montarras, Pinset, Chelly, Kahan and Gross, 1989). In experiments using permissive vs. inducible C2 cells it was demonstrated that in the permissive cells MyoD1 is expressed constitutively and the cells differentiate spontaneously. In contrast, the inducible cells require insulin to undergo terminal differentiation and MyoD1 is activated following this trigger. It may well be that MyoD1 plays different roles at different stages of development, depending upon the interaction with different partner proteins (Olson, 1990; Weintraub et al., 1991). Future experiments will be directed towards analyzing the interplay between the different muscle regulatory genes in the transfected ES cell system.

The fact that no live transgenics could be produced following the microinjection of the β-actin/MyoD1 construct suggests that ectopic expression of MyoD1 during early stages of development is lethal. The in situ hybridization analyses of transgenic embryos revealed that in most cases normal looking embryos were formed, but they were significantly retarded compared to their normal littermates. Thus, it could be that the lethality observed is not directly related to the myogenic conversion ability of this factor, but rather to its role in the control of cell division (Crescenzi, Fleming, Lassar, Weintraub and Aaronson, 1990).

The majority of transgenic embryos express the exogenous gene in most tissues derived from the ectoderm and mesoderm, with no detectable morphological consequences. These results again indicate that in the absence of the appropriate partner regulator(s) and in the wrong microenvironment MyoD1 is not a dominant myogenic factor. Noteworthy, is the observation that none of the endodermal cell types expressed the introduced gene. Since the β-actin promoter is active in most cell types tested, a possible interpretation of these finding is that embryos expressing MyoD1 in endodermal cells are selected against and do not survive to this stage.

The majority of transgenic embryos looked normal in terms of gross morphology. However, we have found several abnormal transgenic embryos in which embryogenesis was blocked prior to neurulation. Of particular interest was a 9.5 day transgenic embryo that consisted of two embryonic layers in the dorsal region and the three embryonic layers in the ventral part. Some cells in the mesodermal layer resembled blood cell precursors while a large number of cells had the typical spindle shaped moprphology and they hybridized to the skeletal muscle MLC2 probe (not shown). Thus, ectopic expression of MyoD1 can convert a large part of the mesoderm to skeletal muscle, demonstrating that only a subset of stem cells can respond to this regulator. Further in situ hybridization analyses of the abnormal and normal transgenic embryos with other members of this family should provide more information on the cross-regulation of these genes during mouse development.

This study was supported, in part, by the Israel Academy of Sciences and Humanities, the United States-Israel Binational Agricultural Research and Development Fund and by the Israeli Minstry of Science and Development.

References

Auffray, C. and Rougeon, F. (1980). Purification of immunoglobulin heavy chain mRNA from total myeloma tumor RNA. *Eur. J. Biochem.* **107**, 303-313.

Braun, T., Buschausen-Denker, G., Bober, E., Tannich, E. and Arnold, H. H. (1989). A novel human muscle factor related to but distinct from MyoD1 induces myogenic conversion in 10T1/2 fibroblasts. *EMBO J.* **8**, 701-709.

Brennan, T. J. and Olson, E. N. (1990). Myogenin resides in the nucleus and acquires high affinity for a conserved enhancer element on heterodimerization. *Genes and Devel.* **4**, 582-595.

Carmon, Y., Czosnek, H., Nudel, U., Shani, M. and Yaffe, D. (1982). DNAaseI sensitivity of gene expression during myogenesis. *Nucl. Acids Res.* **10**, 3085-3098.

Choi, J., Costa, M. L., Mermelstein, C. S., Chagas, C., Holtzer, S. and Holtzer, H. (1990). MyoD converts primary dermal fibroblasts, chondroblasts, smooth muscle and retinal pigmented epithelial cells into striated mononucleated myoblasts and multinucleated myotubes. *Proc. Natl. Acad. Sci. USA* **87**, 7988-7992.

Crescenzi, M., Fleming, T. P., Lassar, A. B., Weintraub, H. and Aaronson, S. A. (1990). MyoD induces growth arrest independent of differentiation in normal and transformed cells. *Proc. Natl. Acad. Sci. USA* **87**, 8442-8446.

Davis, R. L., Weintraub, H. and Lassar, A. B. (1987). Expression of a single transfected cDNA converts fibroblasts to myoblasts. *Cell* **51**, 987-1000.

Doetschman, T., Eistetter, H., Katz, M., Schmidt, W. and Kemler, R. (1985). The in vitro development of blastocyst-derived embryonic stem cell line. *J. Embryol. Exp. Morph.* **87**, 27-45.

Edmonson, D. G. and Olson, E. N. (1989). A gene with homology to the myc similarity region of MyoD1 is expressed during myogenesis and is sufficient to activate the muscle differentiation program. *Genes and Devel.* **3**, 628-640.

Emerson, C. P. (1990). Myogenesis and developmentla control genes. Current Opinion in Cell Biology **2**, 1065-1075.

Hogan, B., Costantini, F. and Lacy, E. (1986). *Manipulating the mouse embryo: A laboratory manual.* Cold Spring Harbor Laboratory, New York.

Hopwood, N. D. and Gordon, J. B. (1990). Activation of muscle genes without myogenesis by ectopic expression of MyoD in frog embryo cells. *Nature* **347**, 197-200.

Katcoff, D., Nudel, U., Zevin-Sonkin, D., Carmon, Y., Shani, M., Lehrach, H., Frischauf, A. M. and Yaffe, D. (1980). Construction of recombinant plasmids containing rat muscle actin and myosin light chain DNA sequences. *Proc. Natl. Acad. Sci. USA* **77**, 960-964.

Konieczny, S. F. and Emerson, C. P. (1984). 5-azacytidine induction of stable mesodermal stem cell lineages from 10T1/2 cells: evidence for regulatory genes controlling determination. *Cell* **38**, 791-800.

Lin, Z. Y., Dechesner, C. A., Eldridge, J. and Paterson, B. M. (1989). An avian muscle factor related to MyoD1 activates muscle-specific promoters in nonmuscle cells of different germ-layer origin and in BrdU-treated myoblasts. *Genes and Devel.* **3**, 986-996.

Martin, E., Wiley, L. M. and Damjanov, I. (1977). The development of cystic embryoid bodies in vitro from clonal teratocarcinoma stem cells. *Devel. Biol.* **61**, 230-244.

Miner, K. H. and Wold, B. (1990). Herculin, a fourth member of the MyoD family of myogenic regulatory genes. *Proc. Natl. Acad. Sci. USA*, **87**, 1089-1093.

Montarras, D., Pinset, C., Chelly, J., Kahan, A. and Gros, F. (1989). Expression of MyoD1 coincides with terminal differentiation in determined but inducible muscle cells. *EMBO J.* **8**, 2203-2207.

Murre, C., Schonleber McCaw, P. and Baltimore, D. (1989a). A new DNA binding and dimerization motif in immunoglobulin enhancer binding, daughterless, MyoD and myc proteins. *Cell* **56**, 777-783.

Murre, C., Schonleber McCaw, P., Vasessin, H., Caudy, M., Jan, L. Y., Cabrera, C. V., Bushkin, J. N., Hauschka, S. D., Lassar, A. B., Weintraub, H. and Baltimore, D. (1989b). Interactions between heterologous helix-loop-helix proteins generate complexes that bind specifically to a common DNA sequence. *Cell* **58**, 537-544.

Olson, E. N. (1990). MyoD family: a paradigm for development. *Genes and Devel.* **4**, 1454-1461.

Piette, J., Bessereaux, J. L., Hutchet, M. and Changeoux, J. P. (1990). Two adjacent MyoD1-binding sites regulate expression of the acetylcholine receptor α-subunit gene. *Nature* **345**, 353-355.

Pinney, D. F., Pearson-White, S. M., Konieczny, S. F., Latham, K. E. and Emerson, C. P. (1988). Myogenic lineage determination and differentiation: evidence for a regulatory gene pathway. *Cell* **53**, 781-793.

Rhodes, S. J. and Konieczny, S. F. (1989). Identification of MRF4: a new member of the muscle regulatory gene family. *Genes and Devel.* **3**, 2050-2061.

Robertson, E. J. (1987). Embryo-derived stem cell lines. In Robertson, E.J. (ed.), *Teratocarcinomas and embryonic stem cells. A practical approach.* IRL Press, pp. 71-112.

Robertson, E. J., Kaufman, M. H., Bradley, A. and Evans, M. J. (1983). In Silver, L.M. and Martin, G.R. and Strickland, S. (eds.) *Teratocarcinoma stem cells,* Cold Spring Harbor, vol. 10, pp. 647-664.

Rosenthal, N., Berglund, E. B., Wentworth, B. M., Donoughue, M., Winter, B., Bober, E., Braun, T. and Arnold, H. H. (1990). A Highly conserved enhancer downstream of human MLC1/3 locus is a target for multiple myogenic determination factors. *Nucl. Acids Res.* **18**, 6239-6246.

Sambrook, J., Fritsch, E. F. and Maniatis, T. (1989). Molecular cloning. A laboratory manual.

Sartorelli, V., Webster, K. A. and Kedes, L. (1990). Muscle specific expression of the cardiac actin gene requires MyoD1, CArG-box biding factor, and SP1. *Genes Dev.* **4**, 1811-1822.

Sassoon, D., Lyon, G., Wright, W. E., Lin, V., Lassar, A., Weintraub, H. and Buckingham, M. (1989). Expression of two myogenic regulatory factors myogenin and MyoD1 during mouse embryogenesis. *Nature* **341**, 303-307.

Schafer, B. A., Blakely, B. T., Darlington, G. J. and Blau, H. (1990). Effect of cell history on response to helix-loop-helix family of myogenic regulators. *Nature* **344**, 454-458.

Shani, M. (1985). Tissue specific expression of rat myosin light chain 2 gene in transgenic mice. *Nature* **314**, 283-286.

Shani, M. (1986). Tissue-specific and developmentally regulated expression of a chimeric actin/globin gene in transgenic mice. *Mol. Cell Biol.* **6**, 2624-2631.

Shinar, D., Yoffe, O., Shani, M. and Yaffe, D. (1989). Regulated expression of muscle- specific genes introduced into mouse embryonal stem cells: inverse correlation with DNA methylation. *Differentiation* **41**, 116-126.

Smith, A. G. and Hooper, M. L. (1987). Buffalo rat liver cells produce a diffusible activity which inhibits the differentiation of murine embryonic stem cells. *Devel. Biol.* **121**, 1-9.

Tapscott, S. J., Davis, R. L., Thayer, M. J., Cheng, P. F., Weintraub, H. and Lassar, A. B. (1988). MyoD1: A nuclear phosphoprotein requiring a myc homology region to convert fibroblasts to myoblasts. *Science* **242**, 405-411.

Thayer, M. J., Tapscott, B. J., Davis, R. L., Wright, W. E., Lassar, A. B. and Weintraub, H. (1989). Positive autoregulation of the myogenic determination gene MyoD1. *Cell* **58**, 241-248.

Thomas, K. J. and Capecchi, M. R. (1987). Site-directed mutagenesis by gene targeting in mouse embryo-derive stem cells. *Cell* **51**, 503-512.

Vaidya, T. B., Rhodes, S. J., Taparowsky, E. J. and Konieczny, S. F. (1989). Fibroblast growth factor and transforming growth factor β repress transcription of the myogenic regulatory gene MyoD1. *Mol. Cell Biol.* **9**, 3576-3579.

Weintraub, H., Davis, R., Tapscott, S., Thayer, M., Krause, M., Benezra, R., Blackwell, T. K., Turner, D., Rupp, R., Hollenberg, S., Zhuang, Y. and Lassar, A. (1991). MyoD1: a nodal point during the specification of the muscle cell lineage. *Science* **251**, 761-766.

Weintraub, H., Tapscott, S. J., Davis, R. L., Thayer, M. J., Adam, M. A., Lassar, A. B. and Miller, A. D. (1989). Activation of muscle-specific genes in pigment, nerve, fat, liver and fibroblast cell lines by forced expression of MyoD. *Proc. Natl. Acad. Sci. USA* **86**, 5434-5438.

Wright, W. E., Sassoon, D. and Lin, V. K. (1989). Myogenin, a factor regulating myogenesis, has a domain homologous to MyoD. *Cell* **56**, 607-617.

Printed in Great Britain © *Society for Experimental Biology 1992*   37

REGULATION OF MYOGENIN EXPRESSION IN NORMAL AND TRANSFORMED MYOGENIC CELL LINES

HANS H. ARNOLD, THOMAS BRAUN, EVA BOBER, ASTRID BUCHBERGER, BARBARA WINTER and ANTERO SALMINEN

Department of Toxicology, University of Hamburg, Medical School, Grindelallee 117, 2000 Hamburg 13

Summary

The control of myogenin (Myf-4), one of the muscle-specific regulatory proteins, is particularly interesting since its expression appears obligatory in myoblasts at the onset of differentiation. We isolated the human Myf-4 (myogenin) gene and determined promoter elements which direct cell type-specific expression and are subject to transactivation by the muscle transcription factors Myf-5 and MyoD1 in fibroblasts. Extrinsic signals such as serum components and purified growth factors or potential intracellular signals such as cAMP down-regulate transcription of the myogenin gene. Constitutive expression of the catalytic subunit of PKA completely suppresses transactivation of the myogenin promoter by Myf-5 or MyoD1 suggesting that cAMP may act via phosphorylation by PKA.

In contrast to normal myogenic cell lines in which differentiation and myogenin expression can be induced by the removal of serum components, retinoic acid (RA) is required for differentiation in the rat rhabdomyosarcoma cell line BA-Han-1C. This model system was utilized to investigate factors which influence the balance between the transformed state and differentiation. Administration of retinoic acid to BA-Han-1C cells leads to the accumulation of myogenin mRNA approximately 48 h after the addition of RA. This late induction requires ongoing protein- and DNA-synthesis suggesting that trans- and cis-acting factors may be involved in the control.

The critical involvement of myogenin in the process of terminal muscle differentiation was also demonstrated in the rat L6 muscle cell line which has been blocked for differentiation by the transforming protein E1a of Ad5 adenovirus. In cells which stably express E1a, myogenin expression is completely suppressed while Myf-5 continues to be synthesized normally. However, E1a inhibits the transactivator function of Myf-5, as demonstrated on GAL4-Myf5 chimeric proteins. A possible interpretation of this result is that Myf-5 or factors activated by Myf-5 are required for the expression of myogenin and myogenin itself is necessary for the terminal differentiation of myoblasts.

Introduction

Myogenesis involves a family of muscle-specific proteins (MyoD1, myogenin,

Key words: myogenesis, helix-loop-helix proteins, growth factors, transcriptional control.

Myf-5, and MRF4/herculin/Myf-6) which have been identified by a common
sequence motif which is thought to form a basic-helix-loop-helix (bHLH) structure
and by their ability to convert 10T1/2 fibroblasts to determined myoblasts (Davis,
Cheng, Lassar and Weintraub, 1987; Edmondson and Olson, 1989; Wright,
Sassoon and Lin, 1989; Braun, Buschhausen-Denker, Bober, Tannich and
Arnold, 1989a; Braun, Bober, Buschhausen-Denker, Kohtz and Grzeschik,
1989b; Rhodes and Konieczny, 1989; Miner and Wold, 1990; Braun, Bober,
Winter, Rosenthal and Arnold, 1990a). All four proteins indistinguishably
heterodimerize with ubiquitously expressed gene products of the E2A gene (HLH
proteins) via the putative amphipathic helix-loop-helix structure and bind to a
DNA consensus motif, referred to as E-box, via a cluster of basic amino acids
(Murre, McCaw and Baltimore, 1989a; Murre, McCaw, Vaessin, Caudy, Jan,
Cabrera, Buskin, Hauschka, Lassar, Weintraub and Baltimore, 1989b; Brennan
and Olson, 1990; Braun, Gearing, Wright and Arnold, 1991; Davis, Cheng, Lassar
and Weintraub, 1990). They furthermore transactivate muscle-specific reporter
genes containing the E-box element within enhancer or promoter regions in
fibroblasts or other nonmuscle cells (Braun et al. 1990a; Rosenthal, Berglund,
Wentworth, Donoghue, Winter, Bober, Braun and Arnold, 1990; Lin, Dechesne,
Eldrige and Paterson, 1989) with the possible exception of MRF-4 which
apparently only moderately activates transcription in tissue culture cells (Yutzey,
Rhodes and Konieczny, 1990; Chakraborty, Brennan and Olson, 1991). Taken
together, these observations have led to the view that muscle-specific HLH
proteins are lineage determining 'master genes' which probably function as cell
type-specific transcription factors. Despite the similar biochemical properties
exerted in vitro, the distinct spatio-temporal expression pattern of the individual
muscle regulatory proteins during mouse embryogenesis suggests that they may
play specific roles in myogenesis (Sassoon, Lyons, Wright, Lin, Lassar, Weintraub
and Buckingham, 1989; Ott, Bober, Lyons, Arnold and Buckingham, 1991;
Bober, Lyons, Braun, Cossu, Buckingham and Arnold, 1991). In particular, the
expression of Myf-5 in premyotomal cells prior to the activation of myogenin,
MRF4 and MyoD1, in this order, may indicate a hierarchy in which Myf-5 controls
directly or indirectly the expression of myogenin (Ott et al. 1991). This hypothesis
is supported by observations in most myogenic cell lines in which accumulation of
Myf-5 precedes the expression of myogenin transcripts which start to accumulate
only at the onset of differentiation (Edmondson and Olson, 1989; Wright et al.
1989). The strict temporal correlation of myogenin synthesis and the subsequent
acquisition of the muscle phenotype suggest that either myogenin plays an
essential role for the induction of terminal muscle cell differentiation or,
alternatively, constitutes a very early event during this process. To answer these
questions, clearly more information on the functional relationship between the
individual myogenic factors and the regulation of their gene activities is needed.

   In this context, we have begun to determine factors which control the activity of
the myogenin gene in normal muscle cell lines and in transformed myogenic cells
which have lost their ability to differentiate spontaneously. In each instance, we

find that agents which block myogenin synthesis in cells also lead to a block of differentiation. Furthermore, malignant transformation either *in vivo* (rhabdomyosarcoma) or *in vitro* by the expression of the transforming viral protein E1a results in cells which no longer synthesize myogenin and also fail to differentiate.

Results

Serum components, growth factors, and cAMP inhibit the accumulation of myogenin mRNA in C2C12 muscle cells

Myoblasts in culture can be induced to differentiate by the replacement of high concentrations of fetal calf serum (FCS) with low concentrations of horse serum (HS). Similarly, purified growth factors such as basic fibroblast growth factor (bFGF), trans-forming growth factor beta (TGFβ) or epidermal growth factor (EGF) prevent muscle cell differentiation. To investigate the effect of these changes in culture conditions on the expression of myogenic factors and to determine possible signal transduction pathways which may be responsible for the proliferative and antimyogenic effect, we have isolated total RNA from C2C12 cells under various growth conditions and analyzed the steady state levels of myogenin mRNA on Northern blots. Additionally, we examined the effect of cAMP which may be involved in the transduction of signals from membrane bound receptors to the nucleus, as well as bacterial toxins that interfere with the activity of G proteins which also may participate in intracellular signalling. As shown in fig. 1, the amount of myogenin mRNA accumulating in cells drastically changes under the various culturing conditions. Proliferative C2C12 myoblasts in the presence of FCS, TGF-β (not shown) and basic FGF express very low levels or no myogenin mRNA whereas the same cells cultured in the presence of HS contain high levels of myogenin mRNA. Interestingly, db-cAMP, forskolin, and choleratoxin prevent the expression of myogenin mRNA and also inhibit differentiation. In contrast, pertussis toxin which interferes with inhibitory G_i proteins effectively stimulates transcription of myogenin, even in the presence of FCS. We conclude from these observations that purified growth factors and unidentified components present in FCS exert negative control on myogenin gene expression and prevent differentiation. Cyclic AMP is a possible candidate for the intracellular signalling involved in these processes since db-cAMP as well as other agents that increase cellular cAMP levels inhibit myogenin expression. It is not clear, however, whether the inhibition observed in the presence of growth factors and that observed with cAMP are related events in one pathway or whether they constitute two independent mechanisms both exerting negative control over myogenin. It is also interesting to note that myogenin expression is under the positive control of a pertussis-sensitive signalling pathway. This control appears to be independent of cAMP levels since we have not found significant changes of intracellular cAMP concentrations in the presence of pertussis toxin (data not shown).

In order to delineate regions of the myogenin gene which are directing muscle

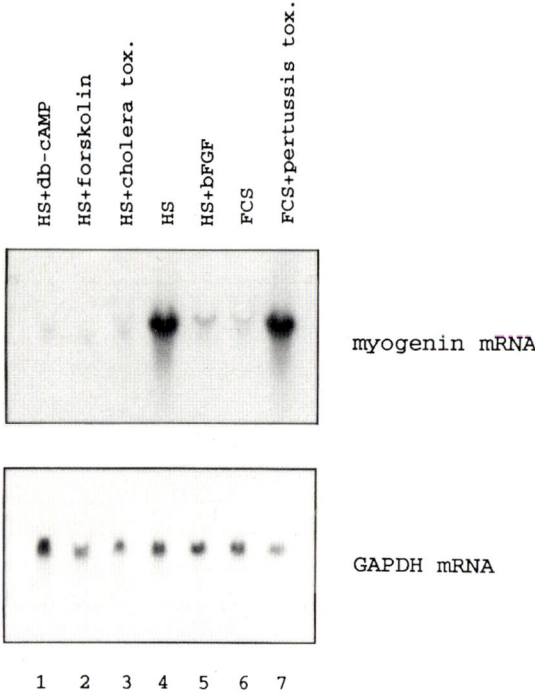

Fig. 1. The accumulation of myogenin mRNA in C2C12 muscle cells is regulated by serum components. Total RNA (20 μg) from C2C12 cells growing under the indicated conditions was isolated and analyzed on Northern blots with a myogenin-specific cDNA fragment used as hybridization probe. A mouse GAPDH cDNA probe served as control for the RNA loading of the blot.

cell-specific expression and may be subject to the modulation by the extrinsic control factors described above, we have isolated the human Myf-4 gene which corresponds to mouse myogenin (Salminen, Braun, Jürs, Winter and Arnold, 1991). Test genes have been constructed with Myf-4 promoter fragments of different length joined to the chloramphenicol acetyltransferase (CAT) gene. As shown in Fig. 2, the various reporter genes containing successively truncated 5′upstream sequences of the Myf-4 gene exhibit strong CAT activity when transiently transfected into primary chick skeletal muscle cells but do not support CAT expression in primary fibroblasts or other nonmuscle cells (data not shown). No major differences in activity have been observed for constructs containing up to 1.1 kb or as little as 200 base pairs of the Myf-4 promoter. This result indicates that cell type-specfic regulation of the myogenin gene is mediated by approximately 200 nucleotides of the proximal promoter region. In agreement with this result is the observation that the described Myf-4/CAT reporter plasmids can be trans-activated in 10T1/2 fibroblasts by the forced expression of MyoD1, Myf-5 and MRF-4 from cotransfected expression vehicles (Salminen et al. 1991). This dependence on muscle-specific transcription factors (MyoD1, Myf-5, MRF-4)

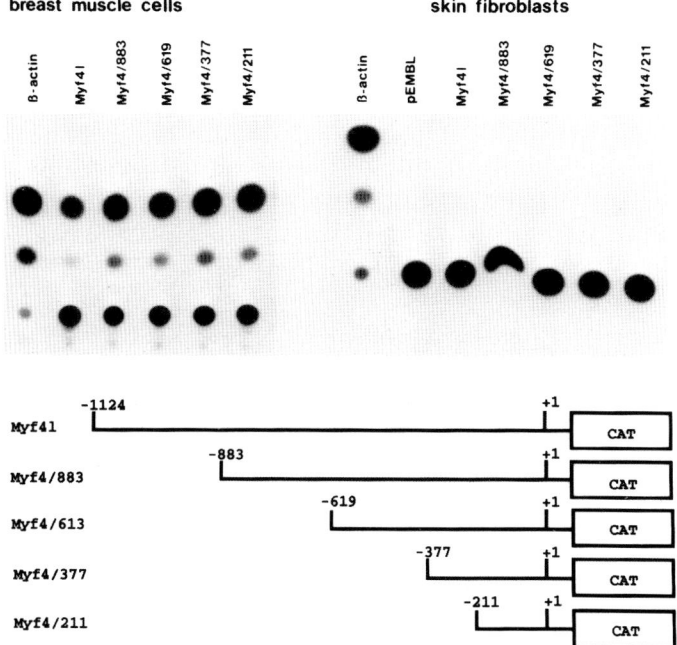

Fig. 2. Cell type-specific expression of the CAT reporter gene directed by 5′ deletion mutatants of the human myogenin (Myf-4) promoter. CAT expression plasmids containing the indicated fragments of the Myf-4 promoter were transiently transfected into fetal chicken (10 day) breast muscle cells and skin fibroblasts. CAT assays were performed 3 days after transfections with cellular extracts that were calibrated according to β-galactosidase activity of a cotransfected RSV-LacZ vector. The β-actin promoter was used as constitutive control plasmid. pEMBL is the CAT vector lacking a promoter.

or gene products which are activated by them and function downstream, may explain the observed tissue-specific expression of myogenin, although one has to keep in mind that myogenin is not expressed in myoblasts even though these cells contain at least one of the myogenic regulatory proteins. Clearly, additional controls must operate on the myogenin gene in myoblasts.

Modulation of Myf-4 gene activity by extrinsic signals is mediated by 5′upstream sequences

The Myf4-CAT reporter gene containing the 5′-upstream sequence from -1124 to $+54$ was transfected into C2C12 myoblasts together with the neomycin resistance gene and clones carrying the stable integrated transgenes were selected. Independently isolated colonies generally exhibited high CAT activity in myotubes cultured in 5% HS and lower or no expression in myoblasts growing in 20% FCS. Fig. 3 shows the influence of various growth conditions on one of the isolated Myf4-CAT clones. When FCS was partially depleted for mitogenic factors by affinity chromatography on heparin-sepharose, CAT activity of the transgene

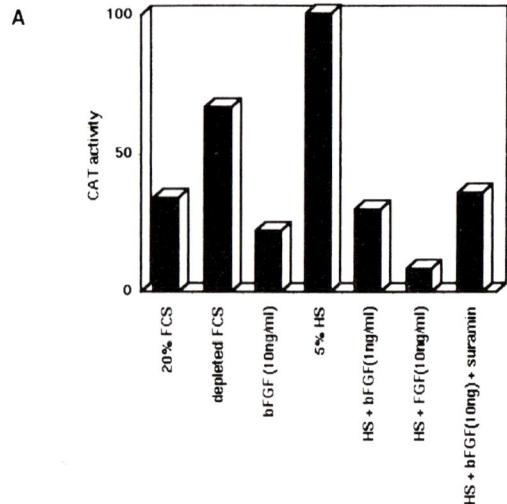

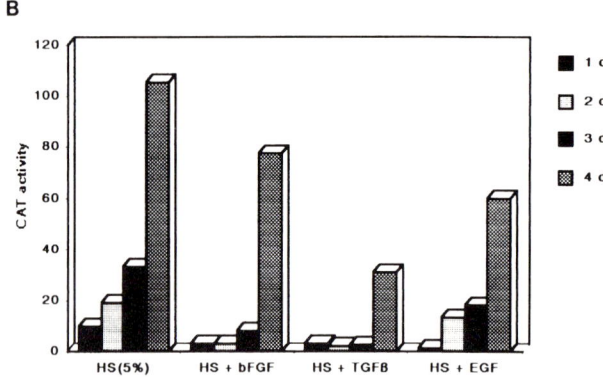

Fig. 3. Activity of the stably integrated Myf4l reporter gene in C2C12 cells in the presence of serum and purified growth factors. A. Cells were cultured under various conditions for 2.5 days and specific CAT activity was determined by standard procedures. Where indicated FCS was depleted by chromatography over heparin-sepharose. Suramin was added at 150 μM. B. Purified growth factors were present up to 4 days in culture; bFGF (15 μg/ml); TGF-β (5 μg/ml); EGF (30 μg/ml).

increased (Fig. 3A). In contrast, addition of bFGF to depleted FCS or to HS inhibited Myf4 promoter driven CAT activity in a concentration dependent fashion (Fig. 3A). In the presence of suramin, a drug that inhibits receptor-ligand interactions, the effect of bFGF was partially obliterated.

In a time-course experiment performed over four days, the purified growth factors bFGF, TGF-β and EGF clearly diminished Myf4-CAT activity, particulary during the early phase of the treatment when cells were still subconfluent (Fig. 3B). The inhibitory effect became weaker and eventually disappeared at higher cell densities. Taken together, these results indicate that a 1.1kb fragment of the Myf4 promoter appears to be sufficient to confer cell type-specific control to a

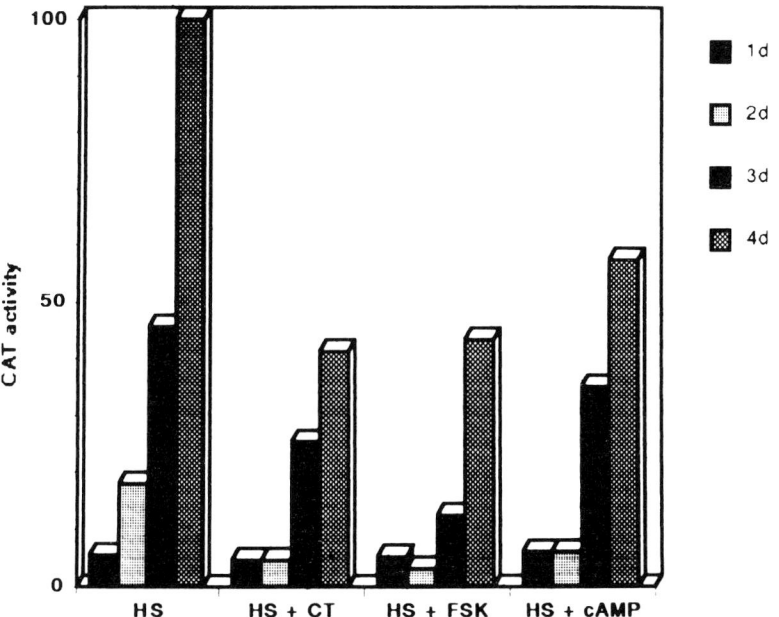

Fig. 4. The myogenin promoter is suppressed by cAMP, forskolin, and choleratoxin. C2C12 cells harboring the Myf4l-CAT reporter gene were cultured in the presence of choleratoxin (CT, 2.5 μg/ml), forskolin (FSK, 34 μM), dibutyryl cAMP (cAMP, 3 mM) and CAT activities were determined daily over a 4 day period.

reporter gene and mediate the down-regulation by purified growth factors. The precise elements involved in this control have not yet been delimited.

To examine the effects of cholera toxin, forskolin and cAMP, the C2C12 cell clone containing the Myf4-CAT transgene has been cultured over 4 days either in HS alone or in the presence of the various drugs. As shown in Fig. 4, all three agents inhibited the transgene in comparison to cells growing in horse serum alone. C2C12 cells carrying RSV-CAT or β-actin CAT genes as nonmuscle control genes did not show any differences in the absence or presence of the drugs (data not shown).

Transactivation of the Myf-4 promoter by Myf-5 can be inhibited by protein kinase A

Regulatory effects of cAMP on the transcriptional activity of target genes may involve protein kinase A which controls the activity of transcription factors by site-specific phosphorylation. Since the Myf4-CAT reporter gene can be activated in 10T1/2 cells by the overexpression of Myf-5, we utilized this notion to examine the effect of a constitutively active protein kinase A on the transactivation of the myogenin promoter. As shown in Fig. 5, coexpression of the catalytic subunit of protein kinase A, repressed the activation of Myf4-CAT by Myf-5. In contrast, the expression of the regulatory subunit of protein kinase A had no inhibitory effect.

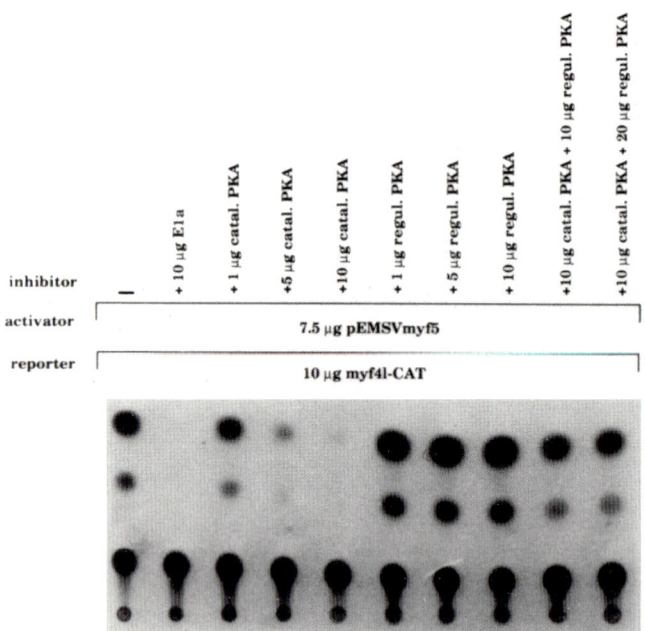

Fig. 5. Protein kinase A (PKA) suppresses transactivation of the myogenin (Myf-4) promoter by Myf-5. The Myf4l-CAT reporter construct was cotransfected with pEMSV-Myf5 in 10T1/2 fibroblasts together with increasing amounts of the expression plasmid for the catalytic subunit of PKA, the regulatory subunit of PKA, or both. E1a in the second lane was used as an unrelated control which also inhibits Myf-5 activity.

Coexpression of both, the regulatory and catalytic subunits abrogates the suppression, as expected in the absence of any protein kinase A activator. The effect of protein kinase A on transactivation by the myogenic transcription factor Myf-5 strongly argues for the important role of cAMP-dependent protein phosphorylation at some level of myogenin gene activation. We have evidence that PKA does not act on the Myf-5 trans-activation domain itself, and probably does not involve Myf-5 as a substrate (unpublished results).

Myogenic induction in the rhabdomyosarcoma line BA-Han-1C by retinoic acid is accompanied by myogenin expression

The clonal rat rhabdomyosarcoma cell line BA-Han-1C, a proliferating mononuclear tumor cell, can be induced to differentiate into post-mitotic myotubes by exposure to retinoic acid (RA). This observation suggests that these cancer cells have not irreversibly lost the functions which control proliferation and differentiation but rather constitute a suitable model to study the regulation of genes involved in these processes. We have analyzed the expression of the myogenic regulatory genes in BA-Han-1C cells growing in the presence and absence of retinoic acid and found that MyoD1 mRNA constitutively accumulates irrespective of RA whereas myogenin mRNA expression strictly depends on RA induction (Fig. 6). The line BA-Han-1B, independently isolated from the same

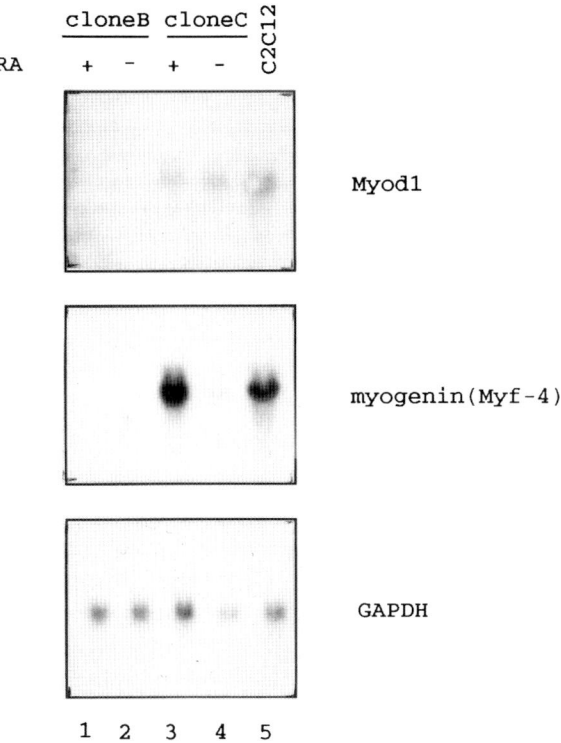

Fig. 6. Induction of myogenin mRNA in the rhabdomyosarcoma cell line Ba-Han-1C (clone C) by retinoic acids (RA). RNA from clones B and C were analyzed on Northern blots using cDNA fragments specific for MyoD1, myogenin and mouse GAPDH as hybridization probes.

tumor, can not be induced to differentiate by RA and does not express any myogenic factor proteins. Thus myogenin expression strictly correlates with the potential for myogenic differentiation even in a tumor cell. In this respect, BA-Han-1C cells behave like normal myoblasts provided that the inducer RA is present. Interestingly, the accumulation of myogenin mRNA starts approximately 48 h after the addition of RA, although the inducer itself needs to be present only for 1 hour or less (data not shown). This relatively long period between RA induction and the actual onset of myogenin synthesis suggested to us that either protein or DNA synthesis or both are required. Indeed, cycloheximide, an inhibitor of protein synthesis, as well as aphidicolin and hydroxyurea, both drugs that block DNA synthesis, prevent the initiation of myogenin transcription in the presence of RA. These results can be interpreted in two ways: First, RA induces the synthesis of protein(s) which are required for myogenin transcription but can only act on the target gene after the cells have undergone at least one round of replication. The necessity of ongoing DNA synthesis could mean that the target gene either has to be demethylated in order to be active or alternatively that a change in chromatin configuration, maybe dependent on an RA induced factor, must occur prior to the activation of the gene. Second, RA induces a protein that leads to the degradation or inactivation of negative control elements, for instance growth factors or the corresponding receptors. Both possibilities are amenable to investigations which are in progress.

Inhibition of myogenesis by the transforming protein E1a from adenovirus is associated with a block in myogenin transcription

As an alternative approach to study the importance of myogenin for the acquisition of the differentiated phenotype in muscle cells, we have applied inhibitors of muscle differentiation and determined their effects on the expression and activity of muscle regulatory proteins. It has been shown previously that the transforming protein E1a from adenovirus inhibits the differentiation of muscle cell lines in culture (Webster, Muscat, and Kedes, 1988). We have established several independent rat L6 myoblast lines which express E1a and fail to differentiate to myotubes under culture conditions which readily support differentiation in the parental L6 cells. Rat L6 myoblasts do not express MyoD1 or MRF-4 but constitutively synthesize Myf-5 and begin to accumulate myogenin after removal of serum, prior to morphological differentiation. In E1a transformed L6 cells we find that Myf-5 mRNA accumulates to normal or slightly enhanced concentrations compared to levels obtained in L6 control cells. In contrast, myogenin mRNA cannot be detected on Northern-blots (Fig. 7). This result suggests that E1a leads to an arrest of L6 cells at the myoblast stage and this arrest is associated with the failure to induce normal myogenin expression after removal of FCS. As we have previously shown, Myf-5 or a gene product induced by Myf-5 functions as a positive control element for the activation of the myogenin promoter. It is therefore tempting to speculate that the block of myogenin expression in L6-E1a cells may be related to the impairment of the Myf-5 activity by E1a. Indeed, we can demonstrate that E1a interferes with the transactivator function of Myf-5. As shown in Fig. 8, transactivation of a GAL4-CAT reporter plasmid by the GAL-Myf5 chimeric protein containing the GAL4 DNA binding domain and the Myf-5 transactivator domain, is severely inhibited by cotransfection of an E1a expressing plasmid. The antimyogenic activity of E1a constitutes a valuable tool to study the relationship of myogenic factors and the relative importance of each factor for the regulation of the other family members. The inhibition of muscle differentiation by E1a and its negative effect on myogenin expression provides additional evidence for the decisive role of myogenin in establishing the terminal differentiation pathway.

Conclusions

Proliferating myoblasts in culture contain the muscle-specific transcription factors MyoD1 or Myf-5 but nevertheless do not express any muscle-specific proteins. The induction of the muscle morphology and the activation of muscle marker genes requires the removal of serum components and conversely can be

A

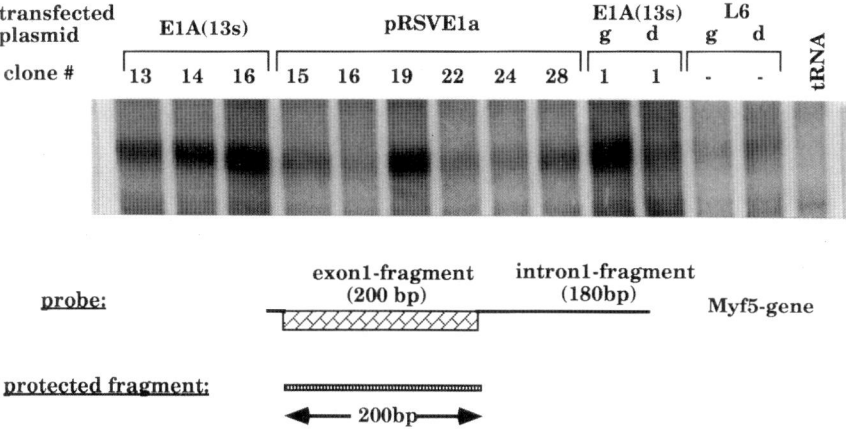

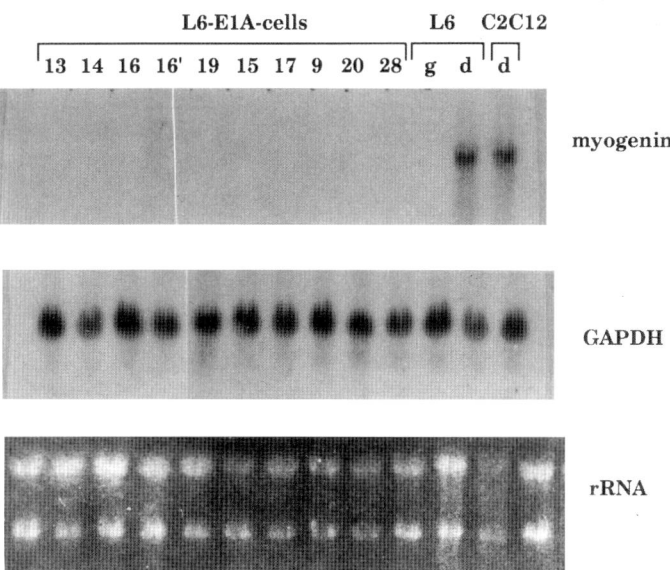

Fig. 7. The adenovirus protein E1a prevents the expression of myogenin in rat L6 muscle cells but does not affect Myf-5 expression. L6 cells were stably transfected with expression plasmids for E1a from its own promoter (E1a(13s)) or from the RSV promoter (pRSV5E1a wt) and RNA was analyzed by RNA protection of a Myf-5 fragment (A) or on Northern blot using a myogenin-specific probe (B). GAPDH was used as constitutively expressed control. g=growing cells; d=differentiated cells.

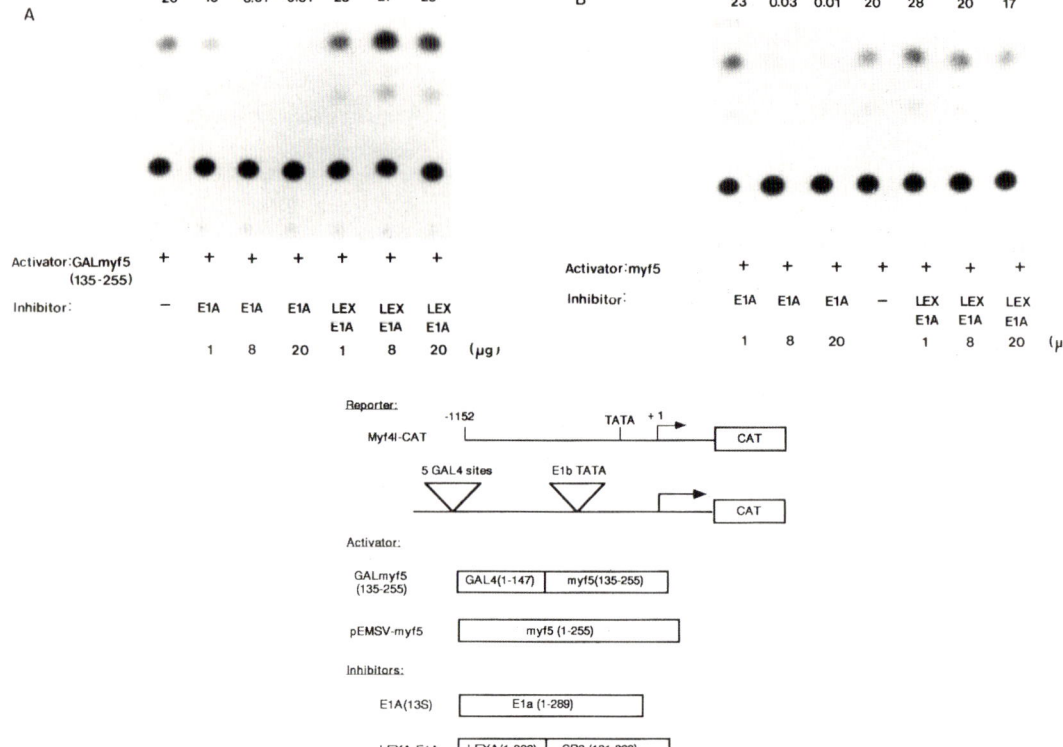

Fig. 8. E1a inhibits the transactivator function of Myf-5 and GAL4-Myf5 hybrid proteins in 10T1/2 fibroblasts. A. A reporter plasmid containg GAL4 binding sites was cotrans-fected with GALmyf5 (135-255) containing the GAL4 DNA binding site and the Myf-5 C-terminal transactivator region. Expression of E1a (13s) (1,8, and 20 μg) results in complete inhibition of trans-activation. LexE1a expressing only the conserved region 3 (CR3) does not inhibit and was used as a control. B. The Myf4l-CAT reporter construct was transactivated by pEMSV-Myf5 and is also inhibited by E1a. The diagram shows the various constructs used in the transfection experiments.

suppressed by purified growth factors such as bFGF, TGFβ, and EGF. It has been suggested that Id, an HLH protein which lacks the basic region and therefore forms complexes with other HLH proteins that fail to bind to DNA confers this negative regulation onto myoblasts (Benezra, Davis, Lockshon, Turner and Weintraub, 1990). In agreement with this suggestion is the observation that Id expression is downregulated in cells deprived of serum. However, there is evidence that purified growth factors do not act via Id (unpublished results, E. Olson personal communication). To study the mechanisms by which growth factors may prevent myoblasts from differentiating even though they contain myogenic regulatory proteins, we have begun to investigate the effects of these antimyogenic signals on the activity of the muscle HLH proteins and their genes. Myogenin appeared to be a particularly interesting candidate as its expression correlates with the onset of muscle differentiation and it is the only muscle regulatory protein which has been found in every myogenic cell line described to date. Here, we report that the expression of myogenin is actually down-regulated by serum and purified growth factors. This negative control is mediated by the gene promoter which is sufficient to confer cell type-specific and regulated expression to a heterologous test gene. cAMP, a potential intracellular signal and choleratoxin which leads to constitutive activation of G proteins also inhibit myogenin expression. The inhibitory effect of cAMP may be translated by PKA since constitutively high levels of this enzyme interfere with the transactivation of the myogenin promoter by Myf-5 in transfection experiments. The substrate for the phosphorylation by PKA, and the mechanism by which this may lead to the inhibition of myogenin expression remain to be determined.

The importance of myogenin for the terminal differentiation of muscle cells in culture has been deduced from observations in a non differentiating rhabdomyosarcoma cell line and in muscle cells which have been blocked for differentiation by the expression of the adenoviral protein E1a.

In the rhabdomyosarcoma cells which constitutively express MyoD1 withdrawal of serum does not lead to differentiation and the expression of myogenin remains suppressed. However, when these cells are cultured in the presence of RA, they activate myogenin synthesis and subsequently acquire the muscle phenotype. The mechanism by which RA induces myogenin expression is not clear but we know that ongoing synthesis of proteins and DNA is required. It seems possible that RA interrupts an autocrine loop which stimulates growth by factors that resemble the antimyogenic signals described above. These tumor cells should allow us to identify regulatory mechanisms which promote unlimited growth and antagonize differentiation. In a different approach, we have utilized the expression of the transforming protein E1a to prevent L6 cells from differentiating which also involves a block in myogenin expression. In contrast to the unknown mechanisms operating in the rhabdomyosarcoma cells, in E1a expressing L6 cells the antimyogenic agent and some of the molecular events underlying the inhibition of myogenesis are known. In these cells, the transactivator function of Myf-5 appears to be the primary target for the interference by E1a. As a working model we would like to suggest that, although Myf-5 alone is not sufficient for muscle differentiation, it is required for the activation of the myogenin gene which either causes the onset of differentiation or is closely associated with it. Abrogation of the myf-5 activity by E1a would result in the prevention of myogenin synthesis and the inhibition of terminal muscle differentiation.

Work described in this article was supported by the Deutsche Forschungsgemeinschaft and Deutsche Muskelschwundhilfe e.V.

References

Benezra, R., Davis, R. L., Lockshon, D., Turner, D. L. and Weintraub, H. (1990). The protein Id: A negative regulator of helix-loop-helix DNA binding proteins. *Cell* **61**, 49-59.
Bober, E., Lyons, G. E., Braun, T., Cossu, G., Buckingham, M. and Arnold, H. H. (1991). The

muscle regulatory gene, Myf-5, has a biphasic pattern of expression during early mouse development. *J. Cell. Biol.* **6**, 1255-1265.

Braun, T., Buschhausen-Denker, G., Bober, E., Tannich, E. and Arnold, H. H. (1989a). A novel human muscle factor related to but distinct from MyoD1 induces myogenic conversion in 10T1/2 fibroblasts. *EMBO J.* **8**, 701-709.

Braun, T., Bober, E., Buschhausen-Denker, G., Kohtz, S., Grzeschik, K. H. and Arnold, H. H. (1989b). Differential expression of myogenic determination genes in muscle cells: possible autoactivation by the myf gene products. *EMBO J.* **8**, 3617-3625.

Braun, T., Bober, E., Winter, B., Rosenthal, N. and Arnold, H. H. (1990a). Myf-6, a new member of the human gene family of myogenic determination factors: Evidence for a gene cluster on chromosome 12. *EMBO J.* **9**, 821-831.

Braun, T., Gearing, K., Wright, W. E. and Arnold, H. H. (1991). Baculovirus-expressed myogenic determination factors require E12 complex formation for binding to the myosin-light-chain enhancer. *Eur. J. Biochem.* **198**, 187-193.

Brennan, T. J. and Olson, E. N. (1990). Myogenin resides in the nucleus and acquires high affinity for a conserved enhancer element on heterodimerization. *Genes Dev.* **4**, 582-595.

Chakraborty, T., Brennan, T. and Olson, E. (1991). Trans-Activation of a muscle-specific enhancer by myogenic helix-loop-helix proteins is separable from DNA binding. *Jour. Biol. Chem.* **266**, 2878-2882.

Davis, R. L., Cheng, P. F., Lassar, A. B. and Weintraub, H. (1990). The MyoD DNA binding domain contains a recognition code for muscle-specific gene activation. *Cell* **60**, 733-746.

Davis, R. L., Weintraub, H. and Lassar, A. B. (1987). Expression of a single transfected cDNA converts fibroblasts to myoblasts. *Cell* **51**, 987-1000.

Edmondson, D. G. and Olson, E. N. (1989). A gene with homology to the myc similarity region of MyoD1 is expressed during myogenesis and is sufficient to activate the muscle differentiation program. *Genes Dev.* **3**, 628-640.

Lin, Z. Y., Dechesne, C. A., Eldrige, J. and Paterson, B. M. (1989). An avian muscle factor related to MyoD1 activates muscle-specific promoters in nonmuscle cells of different germlayer origin and in BrdU-treated myoblasts. *Genes Dev.* **3**, 986-996.

Miner, J. H. and Wold, B. (1990). Herculin, a fourth member of the myoD family of myogenic regulatory genes. *Proc. Natl. Acad. Sci.* **87**, 1089-1093.

Murre, C., McCaw, P. S. and Baltimore, D. (1989a). A new DNA binding and dimerization motif in immunoglobulin enhancer binding, daughterless, MyoD, and myd protein. *Cell* **56**, 777-783.

Murre, C., McCaw, P. S., Vaessin, H., Caudy, M., Jan, L. Y., Jan, Y. N., Cabrera, C. V., Buskin, J. N., Hauschka, S. D., Lassar, A. B., Weintraub, H. and Baltimore, D. (1989b). Interactions between heterologous helix-loop-helix proteins generate complexes that bind specifically to a common DNA sequence. *Cell* **58**, 537-544.

Ott, M.-O., Bober, E., Lyons, G., Arnold, H. H. and Buckingham, M. (1991). Early expression of the myogenic regulatory gene, myf5, in precursor cells of skeletal muscle in the mouse embryo. *Development* **111**, 1097-1107.

Rhodes, S. J. and Konieczny, S. F. (1989). Identification of MRF4: A new member of the muscle regulatory factor gene family. *Genes Dev.* **3**, 2050-2061.

Rosenthal, N., Berglund, E. B., Wentworth, B. M., Donoghue, M., Winter, B., Bober, E., Braun, T. and Arnold, H. H. (1990). A highly conserved enhancer downstream of the human MLC1/3 locus is a target for multiple myogenic determination factors. *Nucleic Acids Res.* **18**, 6239-6245.

Salminen, A., Braun, T., Jürs, S., Winter, B. and Arnold, H. H. (1991). Transcription of the muscle regulatory gene MYF4 is repressed by serum mitogens, fibroblast growth factor, transforming growth factor β, and cyclic AMP. *J. Cell. Biol.* **115**, 905-917.

Sassoon, D., Lyons, G., Wright, W. E., Lin, V., Lassar, A., Weintraub, H. and Buckingham, M. (1989). Expression of two myogenic regulatory factors myogenin and MyoD1 during mouse embryogenesis. *Nature* **341**, 303-307.

Webster, K. A., Muscat, G. E. O. and Kedes, L. (1988). Adenovirus E1a products suppress myogenic differentiation and inhibit transcription from muscle-specific promoters. *Nature* **332**, 553-557.

Wright, W. E., Sassoon, D. A. and Lin, V. K. (1989). Myogenin, a factor regulating myogenesis, has a domain homologous to MyoD1. *Cell* **56**, 607-617.
Yutzey, K. E., Rhodes, S. J. and Konieczny, S. F. (1990). Differential trans-activation associated with the muscle regulatory factors MyoD1, myogenin and MRF4. *Mol. Cell. Biol.* **10**, 3934-3944.

Printed in Great Britain © *Society for Experimental Biology 1992*   53

INHIBITION OF *IN VITRO* MUSCLE DIFFERENTIATION BY THE IMMORTALIZING ONCOGENE Py LT-ag

ROSSELLA MAIONE, GIAN MARIA FIMIA and PAOLO AMATI

Dipartimento di Biopatologia Umana, Sezione di Biologia Cellulare,
Università di Roma La Sapienza, Viale Regina Elena 326, 00161 Roma, Italy

Abstract

The interference of Polyomavirus (Py) early functions with *in vitro* myogenic differentiation is the object of this study. Single cell analysis of C2 myogenic Py infected cells showed a mutual exclusion between Py early functions and muscle gene expression. The morphological and biochemical analysis of clones obtained from C2 cells stably transfected with a plasmid carrying an ORI⁻ Py genome, showed that myogenesis is blocked and cells display the transformed phenotype. By using plasmids separately encoding Middle T or Large T functions, the involvement of individual early viral gene products was determined. Py Middle T alone does not inhibit myotube formation even though cells are morphologically transformed. Myogenic differentiation, on the other hand, is inhibited by Py Large T. This inhibition, which is proportional to the level of Py Large T expression, does not entail to require alteration of cell growth properties and acts by blocking the expression of myogenin and terminal differentiation markers without altering the expression of the regulatory gene MyoD.

Introduction

Muscle differentiation is a multistep process whose stages have been defined and analysed in some detail. Muscle-specific genes are expressed in a developmentally regulated manner. The regulatory regions of several genes coding for muscle-specific functions have been isolated and characterized. More recently, many genes involved in regulating myogenic determination and differentiation have been identified. These regulatory genes, their products, and cell transcription factors interacting with muscle-specific promoters are currently the subject of intensive analysis.

Many agents inhibit myogenic differentiation; among them are growth factors, which seem to act by preventing withdrawal from cell cycle (Linckhart, Clegg, and Hauschka, 1981; Spizz, Roman, Strauss and Olson, 1986), a condition necessary for terminal differentiation in many cell systems. Moreover, the effect of a variety of viral and cellular oncogenes has been extensively studied in many myogenic cell systems. Several observations have been made upon infection of avian myoblasts with retroviruses transducing different oncogenes (reviewed in Alemà and Tatò, 1987). The phenotype of chicken embryo myoblasts infected with Rous sarcoma

Key words: myogenesis, polyoma, Middle T antigen, Large T antigen, MyoD, myogenin.

virus, transducing the v-src oncogene, becomes transformed, and they are no longer able to synthesize muscle-specific products and to fuse into myotubes (Holtzer, Biehl, Yeoh, Meganathan and Keji, 1975; Fiszman and Fuchs, 1975). Similarly, inhibition of myogenesis was observed upon infection of quail myoblasts with retroviruses transducing v-erbB or v-fps, which, like v-src, code for tyrosine kinases, and v-myc, which codes for a nuclear protein (Falcone, Alemà and Tatò, 1985). Other evidence comes from immortalized rodent myoblasts transfected with activated proto-oncogenes (Olson, Spizz and Tainsky, 1987; Payne, Olson, Hsiau, Roberts, Perryman and Schneider, 1987; Schneider, Perryman, Payne, Spizz, Roberts and Olson, 1987).

Oncogene products play an important role in regulating cell growth and differentiation by interfering at different steps of the regulatory cascade primed by growth factors. It has been shown that, as might have been expected, they inhibit muscle differentiation using different mechanisms which involve different steps of the myogenic pathway. For example, in some cases, such as in v-myc transformed quail myoblasts, they act by providing a continuous mitogenic signal (Falcone et al. 1985). The inhibition of myogenesis in these cells can be reversed upon co-cultivation with normal myoblasts, even though v-myc is still being expressed (La Rocca, Grossi, Falcone, Alemá and Tató, 1989). V-src, on the other hand, seems to block differentiation independently of cell proliferation using a mechanism which involves the repression of muscle-specific gene transcription. (Falcone, Tató and Alemá, 1991).

BC3H1 (Payne et al. 1987) and C2 myoblasts (Olson et al. 1987), expressing oncogenic ras, cease to proliferate in the absence of mitogens yet are unable to differentiate. It has been shown that the blocking of myogenesis in ras-transfected C2 cells is due to the inactivation of MyoD expression (Lassar, Thayer, Overell and Weintraub, 1989b).

Evidence of interference with myogenesis was also observed in cells expressing some early products of DNA tumor viruses. Adenovirus E1A products inhibit differentiation without altering cell growth properties and by inhibiting the expression of regulatory genes (Webster, Muscat and Kedes, 1988; Enkeman, Konieczny and Taparowsky, 1990). SV40 Large T antigen behaves quite differently in that it immortalizes primary rat myoblasts without inhibiting morphological and biochemical differentiation. DNA continues to be synthesized in the differentiated structures produced by these myoblasts, a fact not in keeping with all previous observations which seemed to indicate that DNA synthesis and differentiation are mutually exclusive phenomena (Iujvidin, Fuchs, Nudel and Yaffe, 1990).

The early products of DNA tumor viruses appear very interesting in this regard because, in addition to regulating the viral lytic cycle, they are responsible for virus-induced transformation and tumorigenesis. They obtain these effects by interacting with and altering the activities of cell factors, many of which have been shown to be the products of proto-oncogenes and anti-oncogenes.

Polyoma (Py) is a small DNA tumor virus which undergoes a lytic cycle in many mouse cell types and transforms other rodent cells (Tooze, 1981). The virus encodes three early products: Large T (LT), Middle T (MT) and Small T (ST) antigens.

LT is a nuclear phosphoprotein. In addition to its function in the regulation of productive infection (initiation of viral DNA replication and regulation of viral transcription), LT plays an important role in Py transformation. The viral product belongs to the class of immortalizing oncogenes which includes, among others, myc and Adenovirus E1A. These oncogenes can induce the indefinite growth of primary rodent fibroblasts and, in oncogene cooperation assays, complement the action of ras, Adenovirus E1B or Py MT in the transformation of these cells (Rassoulzadegan, Naghashfar, Cowie, Carr, Grisoni, Kamen and Cuzin, 1983; Land, Parada and Weinberg, 1983). The biochemical activities of Py LT which up to now have been positively ascertained are: the binding to double-stranded DNA (Cowie and Kamen, 1984); the transactivation of viral and cellular promoters (Kern, Pellegrini and Basilico, 1986; Kingston, Cowie, Morimoto and Gwinn, 1986); and, as has recently been shown, the ability to bind *in vitro* with the product of the Rb anti-oncogene (Dyson, Bernards, Friend, Gooding, Hassel, Major, Pipas, Vandyke and Harlow, 1990). This latter property is correlated with the ability to immortalize rat embryo fibroblasts (Larose, Dyson, Sullivan, Harlow and Bastin, 1991).

MT is a cytoplasmic protein. It can transform stable cell lines all by itself, and cooperate with immortalizing oncogenes to transform also primary cells (Treisman, Novak, Favaloro and Kamen, 1981; Land et al. 1983). MT forms complexes with several cellular proteins, involved in a number of metabolic pathways like tyrosine kinases $pp60^{c-src}$, $pp60^{c-fyn}$ and $pp60^{c-yes}$ (Courtneidge and Smith, 1983; Kypta, Hemming and Courtneidge, 1988; Kornbluth, Sudol and Hanafusa, 1987), an 85 Kd phosphatidyl inositol kinase (Kaplan, Whitman, Schaffausen, Raptis, Garcea, Pallas, Roberts and Cantley, 1986), and protein phosphatase 2A (Pallas, Shahrik, Martin, Jaspers, Miller, Brautigan and Roberts, 1990).

ST contributes to the mechanism of tumorigenesis *in vivo* (Asselin, Gelinas, Branton and Bastin, 1984) and increases the saturation density of fibroblasts in culture (Noda, Satake, Robins and Ito, 1983). Like MT, ST associates with protein phosphatase 2A (Pallas et al. 1990).

We have previously shown that undifferentiated C2 myoblasts are not susceptible to wild type Py growth and that, during differentiation, they acquire the ability to transcribe and replicate the viral genome (Felsani, Maione, Ricci and Amati, 1985; Maione, Felsani, Pozzi, Caruso and Amati, 1989). In the present paper Py infected cells were analyzed by immunofluorescence single-cell analysis, which revealed a mutual exclusion between Py early functions and muscle gene expression as a result of the inhibition of myogenesis by Py early products. The apparent contradiction between this finding and previous results will be discussed. The interference with differentiation is very evident in stably transfected cells in which the viral genome is not subjected to differentiation-dependent regulation.

Moreover, by transfecting C2 myoblasts with plasmids separately encoding MT or LT we have shown that it is LT which is responsible for this effect, while MT does not seem to affect myoblast differentiation. The expression of terminal differentiation markers and muscle regulatory genes, as well as the growth properties of LT- expressing cells, is reported.

Materials and methods

Cell cultures and Py infection

The mouse myoblast C2 cell line (Yaffe and Saxel, 1977) clone 7 was obtained from M. Buckingham. Cells were maintained as undifferentiated myoblasts in Dulbecco's modified Eagle medium (DMEM) supplemented with 20% foetal calf serum (FCS) in a 10% CO_2 atmosphere. To induce cell differentiation, myoblasts were grown to confluence and then shifted to DMEM supplemented with 2% FCS; extensive morphological and biochemical differentiation were obtained after 24-48 hours.

Agar selection of MT transformed cells was carried out by seeding 10^3 or 10^4 neomycin resistant cells/60 mm diameter dish in DMEM containing 5% FCS and 0.35% Bacto-Agar (Difco) and layering this suspension in quadruplicate onto a base of 0.7% agar.

Wild type Polyoma virus strain A2 (PyA2) (Ruley and Fried, 1983) was propagated at a low multiplicity of infection in mouse 3T6 fibroblast cells and titrated by plaque assay. Viral infection of myoblasts was performed at a multiplicity of 50 Plaque Forming Units (PFU) per cell.

Indirect immunofluorescence staining

Cells grown on glass coverslips were fixed by immersion in methanol-acetone (3/7 vol/vol) for 15 min at $-20°C$ and then air dried. Coverslips were then incubated at 37°C in a humidified atmosphere with the primary antibody, in some cases diluted in phosphate-buffered saline (PBS) containing 1% bovine serum albumin. After three washes with PBS, the coverslips were incubated with the secondary fluorochrome-conjugated antibody diluted in PBS plus 1% bovine serum albumin, washed repeatedly with PBS and mounted with 70% glycerol in PBS. Double immunofluorescence stainings were performed following the same protocol in four steps, incubating the coverslips sequentially with the two primary antibodies and then with the two secondary ones.

The following primary antibodies were used: a) for Py early proteins (in particular Py LT), an antiserum obtained from Brown Norway rats inoculated with syngenic Py-transformed cells, diluted 1:20; b) for embryonic myosin heavy chain (MHCe), the mouse monoclonal antibody MF20 (Bader, Masaki and Fischman, 1982) as undiluted hybridoma supernatant; c) to detect desmin, the anti-desmin monoclonal antibody from Boehringer Mannheim, diluted 1:4; d) the rabbit anti-MyoD polyclonal antibody kindly provided by Dr. A. Lassar (Fred Hutchinson Cancer Center, Seattle); e) the mouse monoclonal antibody anti-myogenin, from

Dr. W. Wright's lab and made available by Dr. G. Cossu (University of Rome, La Sapienza).

As secondary antibodies we used the following Cappel Immunochemical Products diluted 1:100: i) a goat anti-rat IgG fluorescein-conjugated IgG fraction, ii) a goat anti-mouse IgG rhodamine-conjugated IgG fraction and iii) a sheep anti-rabbit IgG rhodamine-conjugated IgG fraction. For double immunofluorescence staining the fluorochrome-conjugated anti-rat IgG and anti-mouse IgG were adsorbed on mouse serum and rat serum, respectively.

Plasmids and probes

As a selectable marker for stably transfected cells we used plasmid pSV2-hygro, which contains the hygromycin B resistance gene under the control of the SV40 promoter-enhancer region; this plasmid was a gift of Dr. S. Pellegrini (Institut Pasteur, Paris).

The following plasmids coding for Py early functions were used to stably transfect C2 cells: a) p48.19 (Kern et al., 1986), containing an entire Py genome with the origin of replication inactivated by a brief nucleotide sequence insertion (Py ORI⁻); b) p91023LT (Cowie, De Villiers and Kamen, 1986) containing a fragment coding for Py LT under the control of the SV40 enhancer, the adenovirus major late promoter and other eukaryotic regulatory signals and c) pMTZip, coding for Py MT under the control of M-MuLV transcriptional unit (Cepko, Roberts and Mulligan, 1984) (from T. Roberts through the courtesy of B. Schaffhausen, Tufts University School of Medicine, Boston)

Plasmid p3300 MCK CAT, used for transient CAT assays, contains 3.3 kb of the muscle creatine kinase (MCK) 5′ flanking sequence fused to the transcription start site of the CAT gene (Jaynes, Chamberlain, Buskin, Jonson and Hauschka, 1986).

Plasmid pCH110, coding for β-galctosidase activity (Hall, Jacob, Ringold and Lee, 1983), was used to normalize CAT transfection efficiencies.

To detect specific transcripts, the following probes were used: i) the fragment of the Py genome extending from the HindIII site (nt 1656) to the PvuII site (nt 1144), specific for Py LT transcript; ii) plasmid pMHC 2.2 (Weydert, Daubas, Lazaridis, BArton, Garner, Leader, Bonhomme, Catalan, Simon, Guenet, Gros and Buckingham, 1985), specific for embryonic myosin heavy chain (MHCe) transcript, kindly provided by Dr. M. Buckingham; iii) the Eco RI fragment from plasmid pEMC11s (Davis, Weintraub and Lassar, 1987), specific for MyoD transcript; iv) the human cDNA myf4 (Braun, Buschhausen-Denker, Bober, Tannich and Arnold, 1989), highly homologous to the mouse myogenin transcript, kindly provided by Dr. H. Arnold and, v) the SmaI fragment from pE:Id(S) (Benezra, Davis, Lockson, Turner and Weintraub, 1990), specific for Id transcript, kindly provided, as well as p3300 MCK CAT, by Dr. H. Eisen (Fred Hutchinson Cancer Center, Seattle).

Cell transfections and CAT assay

Cells were transfected by the calcium phosphate precipitation method (Wigler,

Silverstein, Lee, Pellicer, Cheg and Axel, 1977), with approximately 10^6 cells per 100-mm-diameter dish.

Stable transfectants were selected by co-transfecting each dish with the following amounts of DNA: 2 μg of p48.19 or p91023LT, 0.2 μg of pSV2hygro and 8 μg of mouse genomic DNA as the carrier; 36 h after transfection cells were split 1:7 in selective medium containing 200 μg of hygromycin (Sigma) per ml. The medium was changed every 2-3 days and, after 15 days, the surviving colonies were isolated and separately amplified or pooled in a mixed culture.

CAT assay was performed in duplicate by transfecting nearly confluent cells in growth medium with 5 μg of MCK CAT and 2 μg of pCH110. 16 hours after transfection cells were re-fed with differentiation medium and, 48 hours later, cell extracts were prepared and assayed for CAT activity following the method of Gorman, Mofat and Howard (1982) and for β-galactosidase activity as described in Herbomel, Bourachot and Yaniv (1984).

RNA extraction and Northern analysis

Total RNA was isolated by acid guanidinium thiocyanate-phenol-chloroform extraction (Chomczynski and Sacchi, 1987). Northern blots (electrophoresis through formaldehyde gels and transfer of RNA to nylon membranes) and hybridizations were carried out by the methods of Sambrook, Fritsch and Maniatis (1989).

To provide evidence that the RNA samples were not degraded and to normalize their amounts, the filters were rehybridized with a probe for a constitutively expressed gene, glyceraldehyde-3-phosphate-dehydrogenase (GAPDH) (Piechaczyk, Blanchard, Marty, Dani, Panabieres, El Sabouti, Fort and Jeanteur, 1984), as an internal control.

Results

Mutual exclusion between Py early functions and muscle gene expression

To simultaneously analyse the expression of Py early functions concomitantly with myogenic functions in single cells we used the double indirect immunofluorescence technique.

Subconfluent C2 cells, growing in high serum medium, were infected with Py virus. Immediately after infection cells were shifted to differentiation medium and, 48 hours later, they were fixed and immunostained. To detect Py early proteins we used an antiserum against viral tumor antigens (which detects mainly Large T antigen in the nucleus) and a secondary fluorescein-conjugated antibody. To detect a muscle differentiation marker we used a monoclonal antibody specific for the embryonic myosin heavy chain and a secondary rhodamine-conjugated antibody. It is clearly evident in Fig. 1 that T antigen positive nuclei were seldom localized in myotubes or in mononucleated MHC positive cells. Similar results were obtained using a monoclonal antibody against desmin, a muscle specific cytoskeleton protein (data not shown). In the experiments reported below we

Fig. 1. Simultaneous determination of Py early functions and muscle gene expression. Fluorescence micrograph of Py infected C2 cells stained by double immunofluorescence. Fluorescein identifies LT-positive nuclei. Rhodamine identifies the cytoplasm of myosin-positive myofibers.

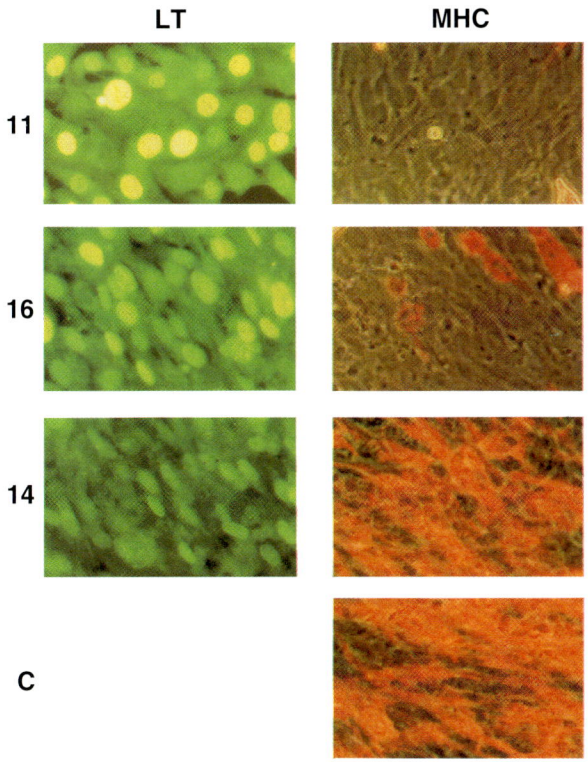

Fig. 4. Expression of Py LT and MHC in p91023LT transfected clones. Immunostaining of Py LT and MHC in three representative clones. Numbers refer to different LT expressing clones, C to control C2H cells. Myosin immunostaining was observed after cells were grown in differentiation medium for 48 h.

Table 1. *Characteristics of Py ORI⁻ transfected clones*

Class	Clone N°	LT expression	Myotube formation	Growth in agar
1	A2, A11, B2, C6, D7, E6, E9, G7, G10, H8, H12	−	++	−
2	B5, B6, C8, F5, G9,	±	+*	+
3	B9, B10, C1, D2, E10, F4, F8, G6,	+	−	+

Myotube formation was observed under a phase contrast microscope and LT expression was determined by immunofluorescence staining. Asterisk indicates the presence of atypical multinucleated cells (see Fig. 2).

tested the hypothesis that this mutual exclusion was determined by an inhibitory effect of viral proteins on myosin and desmin expression.

Py early function interference with myogenic differentiation

To avoid viral replication and cell lysis we used a cloned viral genome defective for replication (Py ORI⁻, plasmid p48.19). The effect of early viral proteins on myogenesis was analyzed in Py Ori⁻ expressing C2 cells.

Selection of Py Ori⁻ stably transfected cells was performed using hygromycin after co-transfection of the p48.19 and pSV2hygro plasmids. Independently isolated hygromycin-resistant clones were amplified separately. The presence of viral T antigens, growth properties and ability to differentiate were analysed in several of them. The characteristics of these clones are summarized in Table 1.

Clones negative for T antigen expression were all able to express muscle differentiation markers and to form multinucleated myotubes (Fig. 2A). Two groups of T antigen positive clones were distinguishable: one (5 clones) that expressed low levels of T-ag, as evaluated by immunofluorescence, and formed several myotubes with an atypical morphology in which cells were multinucleated but their shape was irregular (Fig. 2B). They were not elongated like normal myotubes and their nuclei were not aligned. The other group (8 clones), which displayed higher levels of T-ag expression, showed the morphology of transformed cells: they grew in multilayer and were completely unable to fuse into multinucleated myotubes. Both groups of clones unlike the T-ag negative ones and the untransfected myoblasts, had acquired the property of growing in agar medium. These results confirm the hypothesis that Py early proteins interfere with the correct expression of the myogenic program.

Py Middle T does not alter in vitro myogenic differentiation

Since the three viral early proteins have different biochemical activities and intracellular targets, it was important to determine if one of them was responsible

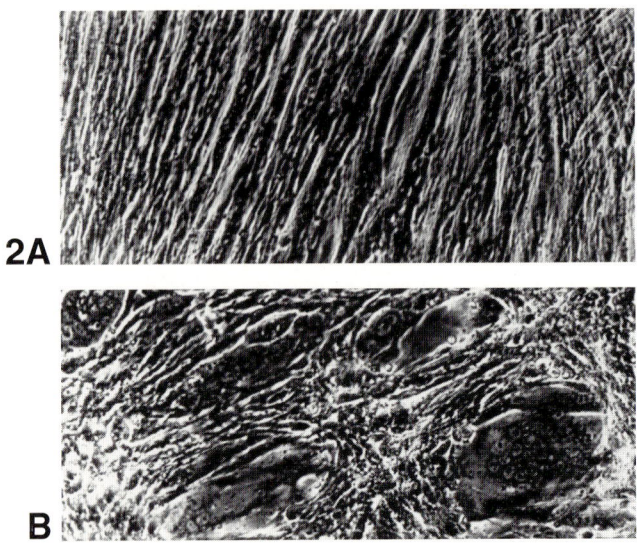

Fig. 2. Phase contrast micrographs showing the altered myotube morphology of Py ORI⁻ transfected clones. A: normal myotubes formed by negative clones (class 1 of Table 1). B: atypical multinucleated cells formed by five positive clones (class 2 of Table 1).

for the inhibition of myogenic differentiation or whether they acted in cooperation. In order to analyse their contributions to the observed phenotype we stably transfected C2 cells with plasmids individually expressing MT and LT proteins. Cells transfected with plasmid pMTzip, selected by neomycin resistance and thereafter by growth in agar medium, were analysed as independent clones. In high serum medium, MT expressing cells show the typical transformed phenotype. These cells, when sparsely seeded, grow in multilayers forming clumps. Two days after the shift to differentiation medium, a number of myosin positive myotubes appear at the edges of the colonies, as shown in Fig. 3 for the representative clone MTzip1.

Py Large T inhibits in vitro myogenic differentiation

The effect on myogenic differentiation of Py Large T was studied by utilizing plasmid p91023 LT. In this plasmid, the sequence coding for the viral protein is under the control of the SV40 enhancer, the Adenovirus major late promoter and other eukaryotic regulatory signals that allow high levels of expression. Transfected cells were selected on the basis of the resistance to hygromycin conferred upon them by the co-transfected pSV2 hygro plasmid. Some of the surviving colonies were amplified as independent clones. The analysis of these clones in comparison with C2 H cells (only expressing hygromycin resistance), performed both by immunofluorescence (Fig. 4) and Northern blot (data not shown), showed that the level of Large T expression was inversely proportional to

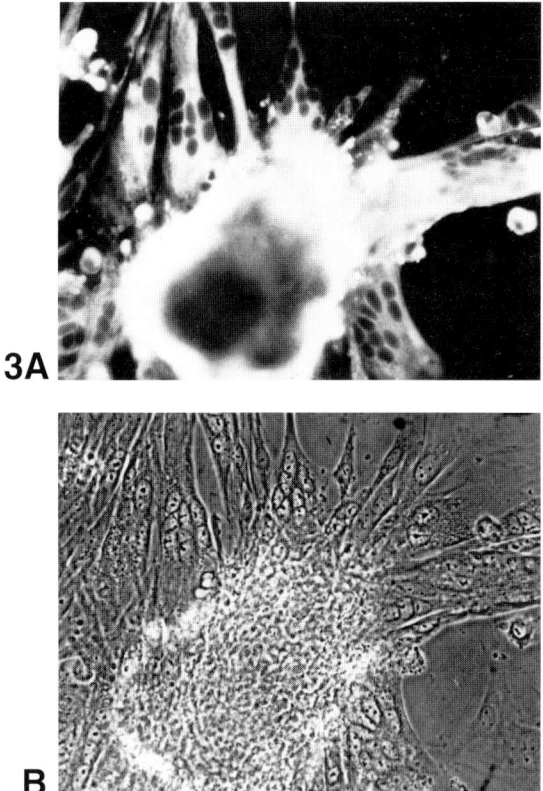

Fig. 3. Differentiation properties of MTzip1 clone. A) myosin immunostaining, and B) phase contrast micrographs of the same field. Cells were fixed and immunostained 48 hours after the shift to differentiation medium.

the degree of differentiation. In particular, clone LT11, which expresses the highest level of the viral protein, was unable to synthesize myosin. Desmin synthesis is also inhibited, though to a lesser extent (data not shown). We found that LT11 cells, if cultured for many generations (about 40) in the absence of hygromycin selection, lost the ability to synthesize LT; if kept in the presence of the antibiotic, they instead stably expressed the viral protein. This property confirmed the direct involvement of Py LT in the blocking of differentiation. We named these cells LT11R to indicate that they represent the reverted LT11 phenotype.

In order to determine whether Py LT inhibits the activity of muscle DNA regulatory elements, we assayed the expression of a reporter gene under the muscle creatine kinase (MCK) promoter-enhancer region, transiently transfected in LT11 cells. As shown in Table 2, the MCK CAT gene, known to be up-regulated during differentiation in normal cells (Jaynes et al., 1986), is much less expressed in LT11 cells than in C2H or LT11R cells.

Table 2. *MCK promoter-enhancer activity in cells expressing LT*

Cells	Percent of chloramphenicol conversion*
C27	38
LT11	1.5
LT11R	40

* Values reported were calculated on the basis of the CPM of acetilated ^{14}C-Chloramphenicol and normalised to B-gal expression of the cotransfected plasmid pCH110. Transient CAT assay was performed as described in Materials and Methods.

Growth properties of T-ag expressing cells

It has been demonstrated that there is a correlation between cell cycle arrest and myogenesis. We investigated if there is a difference in the cell growth properties of cells expressing MT or LT and those of the control C2 cells.

Py Middle T is a transforming protein which, in conjunction with an immortalizing protein, is able to release cells from normal cell cycle control. However, though it confers tumorigenic properties and anchorage independence to the host cells, it is by itself unable to render them independent from mitogenic factors. Myogenic C2 cells expressing MT are still mitogen dependent, even though they are able to grow, at high serum, to cell densities higher than those of the C2H control cells (Fig. 5 A).

To control that the observed myogenic inhibition by Py Large T is not due to an altered cell cycle regulation, as would instead be expected for an immortalizing factor, we measured the rate of cell proliferation and serum requirements also for the LT11 clone. Fig. 5A shows that, in 20% serum, LT11 and C2 cells have similar doubling times and reach similar saturation densities, while MTzip1 cells seem to grow faster and at a higher density. In 2% serum, all the clones cease to proliferate.

The absence of any alteration in the growth properties is also supported by the observation that the LT11 cells, when transferred to differentiation medium, express the same levels of proliferating cell nuclear antigen (PCNA) RNA as the LT11R and control C2 cells (Fig. 5B). This gene function is known to be up-regulated during the cell cycle at the G_1/S transition (Bravo and McDonald-Bravo, 1985).

Unlike Py ORI$^-$ and Py MT transfected cells, none of the Py Large T-expressing clones displayed the property of growing in agar medium. Furthemore, they did not form atypical myotubes, a fact which suggests a role(s) in determining this phenotype for the other Py early proteins, expressed in conjunction with PyLT.

Expression of muscle-specific genes in LT 11 clone

Specific muscle gene products are controlled in their expression by a cascade of regulatory genes, some of which have been characterized. Myo D is a nuclear DNA-binding phosphoprotein. *In vivo* it is expressed only in skeletal muscle, and

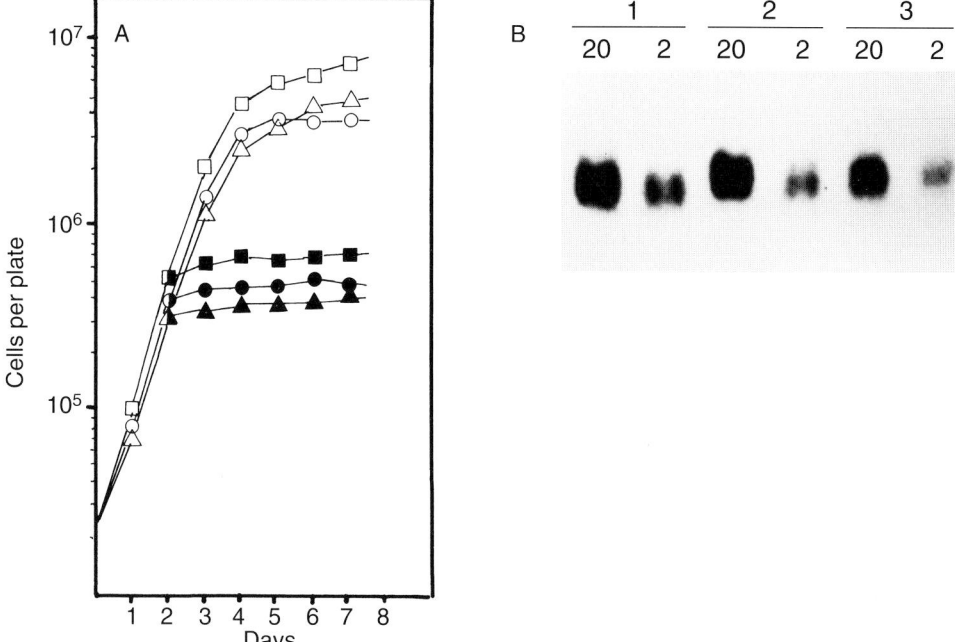

Fig. 5. Cell proliferation of cells expressing MT or LT and expression of PCNA. A) Cells were seeded in DMEM containing 20% FCS and, on day 1, exposed to fresh medium containing either 20% FCS (continuous lines) or 2% FCS (dashed lines). Cell numbers were determined on consecutive days. Open circles: LT11; open triangles: C2H; open squares: MTzip1. Closed symbols represent the counts after shift (at 1 day from seeding plates) to differentiation medium. B) Northern blot of PCNA mRNA from cells grown in medium with 20% or 2% FCS. 1: C2H, 2: LT11 and 3: LT11R cells.

in vitro in a variety of myogenic cell lines both at the myoblast and the myotube stage. The forced expression of Myo D can activate the myogenic program in fibroblast cells (Davis et al. 1987; Weintraub et al. 1989). Myogenin is another nuclear regulatory factor which, like Myo D, can convert fibroblasts to myoblasts. Insofar as myogenin is expressed in C2 cells very soon after the induction of differentiation, in this cells it can be considered an early differentiation marker (Wright, Sassoon and Lin, 1989). ID is a regulatory protein which interacts with MyoD and inhibits its ability to bind to DNA. Furthermore, Id is down regulated upon differentiation in normal cells, probably as a consequence of their withdrawal from the cell cycle (Benezra et al. 1990). Interference of PyLT with one of these gene products may cause inhibition at different stages of myogenesis. Therefore, we were interested in determining the levels of expression of MyoD, myogenin and ID, concomitantly with those of some of the terminal differentiation markers. This analysis was performed by measuring levels of specific mRNA synthesis and by protein detection.

The levels of mRNA synthesis were determined by Northern analysis of RNA extracts of C2H, LT11 and LT11R cells. Results reported in Fig 6 show that LT11 cells, both in growth and in differentiation medium, accumulate the same level of

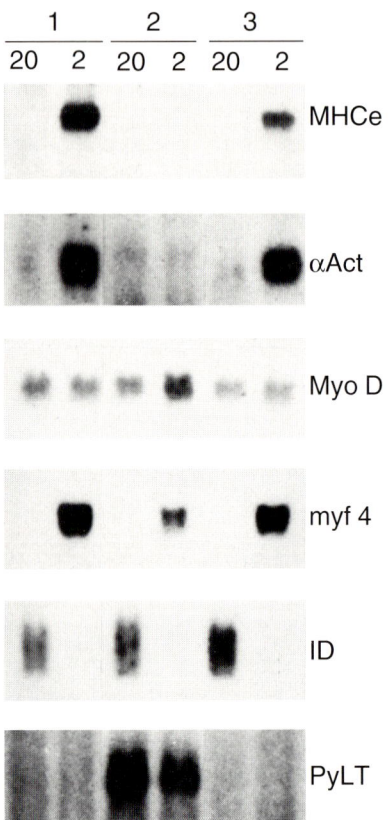

Fig. 6. Northern blot analysis of muscle regulatory genes in LT expressing cells. Northern blot analysis of 20 μg of total RNA extracted from cultures kept in growth medium (20) or differentiation medium (2). Cells were seeded in growth medium (20% FCS) at the density of 2.5×10^5 cells/100 mm diameter dish. Three days later samples were collected for RNA extraction. Four days after seeding the remaining samples were shifted to differentiation medium (2% FCS) and RNA was extracted 48 hours later. 1: C2H; 2: LT11; and 3: LT11R. The six panels represent identical filters hybridized with different probes (indicated on the right of each panel). Filters were re-hybridized with a GAPDH probe to ensure that the same RNA amounts were loaded on the gel (data not shown).

Myo D RNA as C2H and LT11R control cells. Myogenin, on the other hand, is strongly inhibited in LT11 cells grown in differentiation medium. ID RNA is normally down-regulated in all cell lines after the shift to differentiation medium.

Both α-actin and embryonic Myosin Heavy Chain (MHCe) genes are, as expected, strongly inhibited in their transcription in LT11 cells grown in differentiation medium.

To exclude that inhibition of differentiation by PyLT might be due to post-transcriptional inactivation of regulatory factors mRNA, we also tested Myo D and myogenin expression at the protein level. The results of these tests, performed by immunofluorescence staining, are shown in Fig. 7. Again we found that Myo D

| Myo D | Myogenin | MHC |

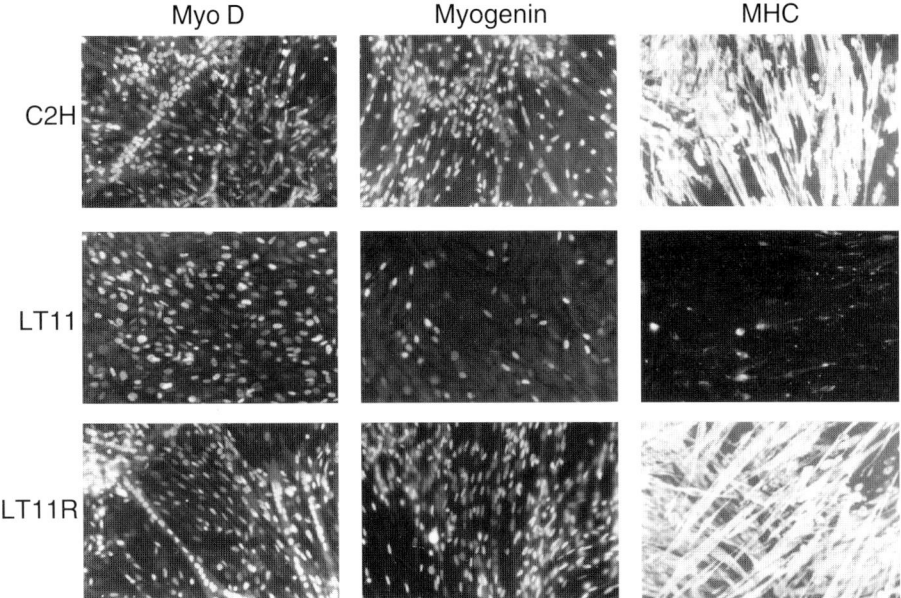

Fig. 7. Immunostaining determination of the expression of muscle regulatory genes. Fluorescence micrographs of cells immunostained with anti-MyoD, anti-myogenin and anti-myosin antibodies of C2H, LT11 and LT11R cells 48 hours after the shift to differentiation medium.

was expressed at the same level both in LT11 and in the control cells. By contrast, myogenin was less frequently expressed in LT11 nuclei than in the nuclei of the other cell types. In comparison with myosin, myogenin is inhibited to a lesser extent.

Discussion

In a previous work (Maione et al. 1989) we showed that the stage of myogenic differentiation affects viral expression. After Py infection the early viral mRNA was co-ordinately transcribed with mRNA for myogenic functions. In the present work, still referring to Py infection, we found cells that express LT but not myosin and cells that express myosin but not LT. A possible explanation of this apparent contradiction is that, despite the fact that Py expression is activated by an event correlated to the induction of differentiation, in a fraction of the cells the accumulation of high viral protein levels precedes that of specific muscle products. In this case the inhibitory effect would prevail. Moreover, we cannot exclude that the infection process also contributes to this phenomenon. In fact, it has been shown (Glenn and Eckhart, 1990) that Py infection of mouse fibroblasts gives rise to a biphasic accumulation of the RNA of several early response genes, the first peak being due to the virus capsid interacting with the virus receptor. Owing to the considerable complexity of this interaction, which results from the reciprocal interference of viral and cellular functions, we chose to analyse the effect of Py early products by stable transfection experiments in which it was possible to maintain constant expression of the viral proteins.

The properties of cell clones expressing individual Py early products suggest that LT can account for Py inhibition of myogenic differentiation. Both Py ORI⁻ and PyLT transfected myoblasts appear unable to differentiate. We found, as expected, a transformed phenotype in the former, as a result of the presence of MT, already known to be a transforming oncogene. We have not further characterized the atypical myotubes formed by the low level Py ORI⁻ expressing clones. Similar aberrant structures have been observed in v-src transformed avian myoblasts (Tatò, Alema, Dlugosz, Boettiger, Holtzer, Cossu and Pacifici, 1983) and could be due to abnormal myofibrils assembly.

The expression of MT does not affect myoblast differentiation. Similar findings have been reported by Cherington, Morgan, Spiegelman and Roberts (1986) who observed that Py LT, but not MT, suppresses preadipocyte differentiation. The inability to inhibit myogenesis could be due to a 'reversion' of the transformed phenotype determined by the removal of serum. This is the hypothesis of Rassoulzadegan, Cowie, Carr, Glaichenhous, Kamen and Cuzin (1982), who reported that MT expressing rat fibroblasts exhibit a fully transformed phenotype only at high serum concentrations or in the presence of LT expression.

The expression of muscle-specific markers in LT myoblasts is blocked both at the protein and the RNA level. The extent to which myogenesis is inhibited depends on viral protein levels. We found some variability of LT expression both between cells of a single clone and between different clones. From a large number of observations we hypothesize a threshold of LT expression required for differentiation to be inhibited. In the clones with reduced LT levels, a variable proportion of cells do not reach this threshold (due to physiological conditions?) and their differentiation is not blocked. Clones homogeneously expressing high LT levels are completely unable to differentiate.

To obtain some insights into the differentiation step affected by LT we assayed the expression of some muscle regulatory genes in differentiation-inhibited cells. This anlysis shows that, among the muscle regulatory genes tested, only myogenin is affected, though to a lesser extent than the terminal differentiation markers. MyoD, which in C2 cells is an earlier element in the regulatory cascade controlling myogenesis, does not appear to be affected at all. This finding suggests that Py LT affects different muscle regulatory events than those affected by many other oncogenes and growth factors. For example, Adenovirus E1A, which like Py LT is classified as an immortalizing oncogene, inhibits both Myo D and myogenin expression (Enkeman et al. 1990). Similarly, activated ras or fos inhibit C2 cell differentiation primarily by blocking Myo D expression (Lassar et al. 1989b). FGF or TGFβ1 treatment of C2 cells also reduces transcription of the regulatory gene, although in this case an additional regulatory step must be involved, since the inhibition of differentiation still persists after the introduction of exogenous MyoD (Vaidya, Rhodes, Taparowsky and Konieczny, 1989). The absence of Myo D repression has been observed in quail myoblasts transformed by the tyrosine kinase pp60[v-src] (Falcone et al. 1990). At the present time we cannot determine whether Myo D is functional or whether it has been inactivated by post-translational mechanisms. It has been shown that this regulatory factor can transactivate the expression of muscle-specific promoters (Yutzey et al. 1990) probably by binding directly to DNA regulatory elements, as has been demonstrated to be the case at least for the MCK regulatory region (Lassar, Buskin, Lockson, Davis, Apone, Hauschka and Weintraub, 1989a). The absence of MCK CAT activity does not demonstrate that Myo D is inactive since the CAT plasmid contains a very large regulatory sequence (3300 bp) from the MCK gene and additional transcription factors may be involved in its expression. We can exclude, in our case, the possibility that Py LT affects the transcription of ID, known to be a negative regulator of Myo D activity.

Py LT is known to regulate several aspects of cell growth. It can confer on primary fibroblasts the ability to grow in long-term cultures and can reduce the mitogen requirements of estabilished rat fibroblasts (Rassouldzadegan et al. 1982; Rassouldzadegan et al. 1983). The immortalization of human primary fibroblasts is accompanied by PCNA induction (Strauss, Hering, Lubbe and Griffin, 1990). On the other hand LT expressing myoblasts show the same serum requirements as control cells, and they down-regulate PCNA transcription when shifted to low serum. Even though the transition to the S phase does not occur, we cannot exclude that LT cells exit from the G_0 phase and arrest in another point of the cell cycle incompatible with myogenesis. This possibility can be tested by analysing the expression of some early growth response genes, like c-myc, known to be up-regulated at the G_0/G_1 transition (Cole, 1986).

As described above, Py LT is a multifunctional protein which suggests several possible mechanisms by which the viral product may alter gene expression. For example, LT is a DNA binding protein. This property is responsible for repression of viral early transcription. In addition, LT can transactivate viral and cellular promoters in the absence of DNA binding. Very exciting are the recent findings about the interaction between Py LT and the product of the Rb anti-oncogene (Dyson et al. 1990; Larose et al. 1991). This gene, known to be altered in a number of tumors and tumor cell lines, seems to be important for the negative regulation of cell growth (Weinberg, 1990). The importance of Rb in cell regulation is supported by the finding that this protein changes its phosphorylation state in the course of the cell cycle and differentiation (Chen, Scully, Shew, Wang and Lee, 1989). Moreover, increased expression of Rb mRNA has been evidenced in a muscle cell line which was induced to differentiate, as well as in other differentiated cell lines (Coppola, Lewis and Cole, 1990). There are many reports about the interaction between the products of DNA tumor virus oncogenes, such as Adenovirus E1A, SV40 LT and Papillomavirus E7 proteins and the Rb product (De Caprio, Ludlow, Figge, Shew, Huang, Lee, Marsilio, Paucha and Livingston, 1988; Whyte, Buchkovich, Horowitz, Friend, Raybuck, Weiberg and Harlow, 1988; Dyson et al. 1989). It has been suggested that this interaction inhibits Rb function and is important for cell transformation. We are currently analysing the effects on myogenesis of several LT mutated in different functional domains, including some mutants impaired in their ability to bind Rb.

We are grateful to Drs. H. Arnold, M. Buckingham, C. Basilico, H. Eisen, A. Lassar, S. Pellegrini, B. Schaffhausen and W. Wright for supplying us with plasmids, probes and antibodies. We thank Dr. O. Sthandier for her collaboration in immunostaining experiments and Mr. L. De Angelis for technical help. This work was supported by PP. FF. Biotecnologia e Biostrumentazione, CNR, Roma, the Associazione Italiana Ricerca sul Cancro (AIRC), Milano, and the Foundation Istituto Pasteur-Cenci Bolognetti and the Telethon 91 research program, Roma. R. Maione holds a AIRC postdoctoral fellowship.

References

Alemà, S. and Tatò, F. (1987). Interaction of retroviral oncogenes with the differentiation programme of myogenic cells. *Adv. Cancer Res.* **49**, 1-28.

Asselin, C., Gelinas, C., Branton, P. E. and Bastin, M. (1984). Polyoma Middle T antigen requires cooperation from another gene to express the malignant phenotype in vivo. *Mol. Cell. Biol.* **4**, 755-760.

Bader, D., Masaki, T. and Fischman, D. A. (1982). Immunochemical analysis of myosin heavy chain during avian myogenesis in vivo and in vitro. *J. Cell. Biol.* **95**, 763-770.

Benezra, R., Davis, R. L., Lockson, D., Turner, D. L. and Weintraub, H. (1990). The protein Id: a negative regulator helix-loop-helix DNA binding protein. *Cell* **61**, 49-59.

Braun, T., Buschhausen-Denker, G., Bober, E., Tannich, E. and Arnold, H. H. (1989). A novel human muscle factor related to but distinct from MyoD1 induces myogenic conversion in 10T1/2 fibroblasts. *EMBO J.* **8**, 701-709.

Bravo, R. and McDonald-bravo, H. (1985). Changes in the nuclear distribution of cyclin (PCNA) but not its synthesis depend on DNA replication. *EMBO J.* **4**, 655-661.

Cepko, C. L., Roberts, B. E. and Mulligan, R. C. (1984). Construction and applications of a highly transmissible murine retrovirus shuttle vector. *Cell* **37**, 1053-1062.

Chen, P., Scully, P., Shew, J., Wang, J. Y. J. and Lee, W. (1989). Phosphorilation of the retinoblastoma gene product is modulated during the cell cycle and celular differentiation. *Cell* **58**, 1193-1198.

Cherington, V., Morgan, B., Spiegelman, B. M. and Roberts, T. M. (1986). Recombinant retroviruses that transduce individual polyoma tumor antigens: effects on growth and differentiation. *Proc. Natl. Acad. Sci. USA.* **83**, 4307-4311.

Chomczynski, P. and Sacchi, N. (1987). Single-step method of RNA isolation by acid guanidinium thiocyanate-phenol-chloroform extraction. *Anal. Biochem.* **162**, 156-159.

Cole, M. D. (1986). The myc oncogene: its role in transformation and differentiation. *Ann. Rev. Genet.* **20**, 361-384.

Coppola, J. E., Lewis, B. A. and Cole, M. D. (1990). Increased retinblastoma gene expression is associted with late stages of differentiation in many different cell types. *Oncogene.* **5**, 1731-1733.

Courtneidge, S. A. and Smith, A. (1983). Polyoma virus transforming protein associates with the pruduct of the c-src cellular gene. *Nature.* **303**, 435-439.

Cowie, A., De villiers, J. and Kamen, R. (1986). Immmortalization of rat embryo fibroblasts by mutant polyomavirus large T antigens deficient in DNA binding. *Mol. Cell. Biol.* **6**, 4344-4352

Cowie, A. and Kamen, R. (1984). Multiple binding sites for polyomaviurs large T antigen within regulatory sequences of polyomavirus DNA. *J. Virol.* **52**, 750-760.

Davis, R. L., Weintraub, H. and Lassar, A. B. (1987). Expression of a single transfected cDNA converts fibroblasts to myoblasts. *Cell* **51**, 987-1000.

De Caprio, J. A., Ludlow, J. W., Figge, J., Shew, J., Huang, C., Lee, W., Marsilio, E., Paucha, E. and Livingston, D. M. (1988). SV40 large tumor antigen forms a specific complex with the product of the retinoblastoma susceptibility gene. *Cell* **54**, 275-283.

Dyson, N., Bernards, R., Friend, S. H., Gooding, L. R., Hassel, J. A., Major, E. O., Pipas, J. M., Vandyke, T. and Harlow, E. (1990). Larte T antigens of many polyomaviruses are able to form complexes with the retinoblastoma protein. *J. Virol.* **64**, 1353-1356.

Dyson, N., Howley, P. M., Munger, K. and Harlow, E. (1989). The human papilloma viurs 16 E7 oncoprotein is able to bind to the retinoblastoma gene product. *Science* **243**, 934-937.

Enkeman, S. A., Konieczny, S. F. and Taparowsky, E. J. (1990). Adenovirus 5 E1A represses muscle-specific enhancers and inhibits expression of the myogenic regulatory factor genes, Myo D and Myogenin. *Cell Growth and Diff.* **1**, 375-382.

Falcone, G., Alemà, S. and Tatò, F. (1991). Transcription of muscle-specific genes is repressed by reactivation of pp60 $^{v-src}$ in postmitotic quail myotubes. *Mol. Cell. Biol. Vol.* **11**, 3331-3338.

Falcone, G., Tatò, F. and Alemà, S. (1985). Distinctive effetcts of the viral oncogenes myc, erb, fps, and src on the differentiation program of quail myogenic cells. *Proc. Natl. Acad. Sci. USA.* **82**, 426-430.

Felsani, A., Maione, R., Ricci, L. and Amati, P. (1985). Coordinate expression of myogenic functions and polyoma virus replication. *Cold Spring Harbor Symp. Quant. Biol.* **50**, 753-757.

Fiszman, M. Y. and Fuchs, P. (1975). Temperature-sensitive expression of differentiation in transformed myoblasts. *Nature* **254**, 429-431.

Glenn, G. M. and Eckhart, W. (1990). Transcriptional regulation of early-response genes during polyomavirus infection. *J. Vir.* **64**, 2193-2201.

Gorman, C. H., Mofat, L. F. and Howard, B. N. (1982). Recombinant genomes which express chloramphenicol acetyltransferase in mammalian cells. *Mol. Cell. Biol.* **2**, 1044-1051.

Hall, C. V., Jacob, P. E., Ringold, G. M. and Lee, F. (1983). Expression and regulation of Escherichia coli lacZ gene fusions in mammalian cells. *J. Mol. Appl. Genet.* **2**, 101-109.

Herbomel, P., Bourachot, B., and Yaniv, M. (1984). Two Distinct Enhancers with different cell specificities coexist in the regulatory region of polyoma. *Cell* **39**, 653-662.

Holtzer, H., Biehl, J., Yeoh, G., Meganathan, R. and Keji, A. (1975). Effect of oncogenic virus on muscle differentiation. *Proc. Natl. Acad. Sci. USA.* **72**, 4051-4055.

Iujvidin, S., Fuchs, O., Nudel, U. and Yaffe, D. (1990). SV40 immortalizes myogenic cells: DNA synthesis and mitosis in differentiating myotubes. *Differentiation* **43**, 192-203.

Jaynes, J. B., Chamberlain, J. S., Buskin, J. N., Jonson, J. E. and Hauschka, S. D. (1986). Transcriptional regulation of the muscle creatine kinase gene and regulated expression in transfected mouse myoblasts. *Mol. Cell. Biol.* **6**, 2855-2864.

Kaplan, D. R., Whitman, M., Schaffausen, B., Raptis, L., Garcea, R. L., Pallas, D., Roberts, T. M. and Cantley, L. (1986). Phosphatidylinositol metabolism and polyoma-mediated transforamtion. *Proc. Natl. Acad. Sci. USA.* **83**, 3624-3628.

Kern, F. G., Pellegrini, S. and Basilico, C. (1986). Cis and trans-acting factors regulating gene expression from polyoma late promoter. In 'Cancer Cells' (Botchan, M., Grodzicker, T. and Sharp, P.A. eds.) Cold Spring Harbor Laboratory, Cold Spring Harbor, New York, **4**, 115-124.

Kingston, R. E., Cowie, A., Morimoto, R. I. and Gwinn, K. A. I. (1986). Binding of polyomavirus large T antigen to the human hsp70 promoter is not required for trans activation. *Mol. Cell. Biol.* **6**, 3180-3190.

Kornbluth, S., Sudol, M. and Hanafusa, H. (1987). Association of the polyomavirus middle-T antigen with c-yes protein. *Nature* **325**, 171-173.

Kypta, R. M., Hemming, A. and Courtneidge, S. A. (1988). Identification and characterization of p59^{fyn} (a src-like protein tyrosine kinase) in normal and polyoma virus transformed cells. *EMBO J.* **7**, 3837-3844.

Land, H., Parada, L. F. and Weinberg, R. A. (1983). Tumorigenic conversion of primary embryo fibroblasts requires at least two cooperating oncogenes. *Nature.* **304**, 596-601.

La Rocca, S. A., Grossi, M., Falcone, G., Alemà, S. and Tatò, F. (1989). Interaction with normal cells suppresses the transformed phenotpe of v-myc-transformed quail muscle cells. *Cell* **58**, 123-131.

Larose, A., Dyson, N., Sullivan, M., Harlow, E. and Bastin, M. (1991). Polyomavirus larg T mutants affected in retinoblstoma protein binding are defective in immortalization. *J. Vir. Vol.* **65**, 2308-2313.

Lassar, A. B., Buskin, N. J., Lockson, D., Davis, R. L., Apone, S., Hauschka, S. D. and

Weintraub, H. (1989a). MyoD is a sequence-specific DNA binding protein requiring a region of Myc homology to bind to the muscle creatine kinase enhancer. *Cell* **58**, 823-831.

Lassar, A. B., Thayer, M. J., Overell, R. W. and Weintraub, H. (1989b). Transformation by activated ras or fos prevents myogenesis by inhibiting expression of MyoD1. *Cell* **58**, 659-667.

Linckhart, T. A., Clegg, C. H. and Hauschka, S. D. (1981). Myogenic differentiation in permanent clonal myoblast cell lines: regulation by macromolecular growth factors in the culture medium. *Dev. Biol.* **86**, 19-30.

Maione, R., Felsani, A., Pozzi, L., Caruso, M. and Amati, P. (1989). Polyomavirus genome and polyomavirus enhancer-driven gene expression during myogenesis. *J. Virol.* **63**, 4890-4897.

Noda, T., Satake, M., Robins, T. and Ito, Y. (1986). Isolation and characterization of NIH 3T3 cells expressing polyomavirus small T antigen. *J. Virol.* **60**, 105-113.

Olson, E., Spizz, G. and Tainsky, M. A. (1987). The oncogenic forms of N-ras or H-ras prevent skeletal myogblast differentiation. *Mol. Cell. Biol.* **7**, 2104-2111.

Pallas, D. C., Shahrik, L. K., Martin, B. L., Jaspers, S., Miller, T. B., Brautigan, D. L. and Roberts, T. M. (1990). Polyoma small and middle T antigens and SV50 small t antigen form stable complexes with protein phosphatase 2A. *Cell* **60**, 167-176.

Payne, P., Olson, E., Hsiau, P., Roberts, R., Perryman, M. B. and Schneider, M. D. (1987). An activated c-Ha-ras allele blocks the induction of muscle-specific genes whose expression is contingent on mitogen withdrawal. *Proc. Natl. Acad. Sci. USA.* **84**, 8956- 8960.

Piechaczyk, M., Blanchard, J. M., Marty, L., Dani, C., Panabieres, S., El Sabouti, P., Fort, P. and Jeanteur, P. (1984). Post-transcriptional regulation of glyceraldehyde-3-phosphate dehydrogenase gene expression in rat tissues. *Nucleic Acids Res.* **12**, 6951-6953.

Rassoulzadegan, M., Cowie, A., Carr, A., Glaichenhous, N., Kamen, R. and Cuzin, F. (1982). The roles of individual polyoma virus early proteins in oncogenic transformation. *Nature.* **300**, 713-718.

Rassoulzadegan, M., Naghashfar, Z., Cowie, A., Carr, A., Grisoni, M., Kamen, R. and Cuzin, F. (1983). Expression of the large T protein of Polyoma virus promotes the establishment in culture of 'normal' rodent fibroblast cell lines. *Proc. Natl. Acad. Sci. USA* **80**, 4354-4358.

Ruley, E. H. and Fried, M. (1983). Sequence repeats in a polyomavirus DNA region important for gene expression. *J. Virol.* **47**, 233-237.

Sambrook, J., Fritsch, E. F. and Maniatis, T. (1989). Molecular cloning: a laboratory manual. Second Edition. Cold Spring Harbor Laboratory. Cold Spring Harbor New York.

Schneider, M. D., Perryman, M. B., Payne, P. A., Spizz, G., Roberts, R. and Olson, E. N. (1987). Autonomous expression of c-myc in BC3H1 cells partially inhibits but does not prevent myogenic differentiation. *Mol. Cell. Biol.* **7**, 1973-1977.

Spizz, G., Roman, D., Strauss, A. and Olson, E. N. (1986). Serum and fibroblast growth factor inhibit myogenic differentiation through a mechanism dependent on protein synthesis and indipendent of cell profiferation. *J. Biol. Chem.* **261**, 9483-9488.

Strauss, M., Hering, S., Lubbe, L. and Griffin, B. (1990). Immortalization and transformation of human fibroblasts by regulated expression of polyoma virus T antigens. *Oncogene* **5**, 1223-1229.

Tatò, F., Alemà, S., Dlugosz, A., Boettiger, D., Holtzer, H., Cossu, G. and Pacifici, M. (1983). Development of 'revertant' myotubes in cultures of Rous sarcoma virus transformed avian myogenic cells. *Differentiation* **24**, 131-139.

Tooze, J. (1981). Molecular biology of tumor viruses. DNA tumor viruses. (Cold Spring Harbor, N.Y.: Cold Spring Harbor Laboratory Press).

Treisman, R., Novak, V., Favaloro, J. and Kamen, R. (1981). Transformation of rat cellls by an altered polyoma virus genome expressing only the middle-T protein. *Nature* **292**, 595-600.

Vaidya, T. B., Rhodes, S. J., Taparowsky, E. J. and Konieczny, S. F. (1989). Fibroblast growth factor and transforming growth factor β repress transcription of the myogenic regulatory gene MyoD1. *Mol. Cell. Biol.* **9**, 3576-3579.

Webster, K. A., Muscat, G. and Kedes, L. (1988). Adenovirus E1A product suppress myogenic differentiation and inhibit transcription from muscle-specific promoters. *Nature* **332**, 553-557.

Weinberg, R. A. (1990). The retinoblastoma gene and cell growth control. *TIBS.* **15**, 199-202.

Weintraub, H., Tapscott, S. J., Davis, R. L., Thayer, M. J., Adam, M., Lassar, A. B. and Dusty Miller, A. (1989). Activation of muscle specific genes in pigment, nerve, fat, liver, and fibroblast cell lines by forced expression of MyoD. *Proc. Natl. Acad. Sci. USA.* **86**, 5434-5438.

Weydert, A., Daubas, P., Lazaridis, I., Barton, P., Garner, I., Leader, D. P., Bonhomme, F., Catalan, J., Simon, D., Guenet, J. L., Gros, F. and Buckingham, M. (1985). Genes for skeletal muscle myosin heavy chains are clustered and are not located on the same mouse chromosome as a cardiac myosin heavy chain gene. *Proc. Natl. Acad. Sci. USA.* **82**, 7183-7187.

Whyte, P., Buchkovich, K., Horowitz, J., Friend, S. H., Raybuck, M., Weinberg, R. A. and Harlow, E. (1988). Association between an oncogene and an anti-oncogene: the adenovirus E1A proteins bind to the retinoblastoma gene product. *Nature* **334**, 124-129.

Wigler, M., Silverstein, S., Lee, L., Pellicer, A., Cheg, Y. C. and Axel, R. (1977). Transfer of purified herpes virus thymidine kinase gene to cultured mouse cells. *Cell.* **11**, 223-232.

Wright, W. E., Sassoon, D. and Lin, V. K. (1989). Myogenin, a factor regulating myogenesis, has a domain homologous to MyoD. *Cell.* **56**, 607-617.

Yaffe, D. and Saxel, O. (1977). Serial passaging and differentiation of myogenic cells isolated from dystrophic mouse muscle. *Nature.* **270**, 725-727.

Yutzey, K. E., Rhodes, S. J. and Konieczny, S. F. (1990). Differential trans activation associated with the muscle regulatory factor MyoD1, myogenin, and MRF4. *Mol. Cell. Biol.* **10**, 3934-3944.

Zhu, Z. G., Veldman, G., Cowie, A., Carr, A., Schaffausen, B. and Kamen, R. (1984). Construction and functional characterization of polyomavirus genomes that separately encode the three early proteins. *J. Virol.* **51**, 170-180.

Printed in Great Britain © Society for Experimental Biology 1992   73

A CELL DIVISION COUNTER EXISTS IN THE CHICK EMBRYO MYOGENIC LINEAGE

M. NAMEROFF

Department of Biological Structure, University of Washington, Seattle, Washington 98195, USA

Summary

A computer assisted timelapse video microscopy system was developed to study the behaviour of individual chick embryo myogenic cells in culture. With this system it is possible to monitor and photograph the development of 50-100 myogenic clones simultaneously. Photos were made with a frame grabber and image compressor and were stored in digital form on the computer's hard disc. Colonies derived from 10-day chick embryo myogenic were followed over a period of 3 days. The cultures were then fixed, reacted first with a mouse monoclonal antibody against sarcomeric myosin, then with peroxidase-labelled goat anti-mouse IgG, and stained with diaminobenzidine to detect antibody binding. The dishes were then replaced into the video system and the computer was used to find the originally photographed colonies. Cells which bound the antibody against sarcomeric myosin were identified in the last image stored for each colony. The computer was then used to retrieve the images from each colony in reverse order and to determine the lineage ancestry of every antibody-positive cell in each clone. Several hundred colonies and many hundreds of cell divisions have been examined thus far. In all but one clone, every antibody-positive cell was produced by a symmetrical cell division in which its sister was also antibody-positive (or died). Antibody-positive cells were generated by a sequence of no more than 4 symmetrical cell divisions. Thus, most colonies consisted of 1, 2, 4, 8 or 16 nuclei in terminally differentiated muscle cells. Those colonies in which cell numbers was not a power of 2 were always produced by cell death or detachment. These results directly and unambiguously demonstrate a cell division counter in the end stages of myogenesis in the chick embryo.

Introduction

The experiments described here were addressed to the question: what causes a cell that is already in the skeletal myogenic lineage to withdraw from the cell cycle and initiate the synthesis of muscle-specific mRNAs and proteins. Though both external influences and internal programming must be involved in this decision, it is not clear how the final result is achieved. Moreover, there may not be a general answer to this question in the sense that different species may have devised different ways to achieve the same end. In previous studies of the behaviour of individual myogenic cells isolated from the pectoral muscles of chick embryos, we

Key words: lineage, cell division, stem cells, computers, timelapse, video microscopy.

found 2 general classes of cells (Quinn and Nameroff, 1983; Robinson, Quinn and Nameroff, 1984; Quinn, Holtzer and Nameroff, 1985; Quinn and Nameroff, 1986; Nameroff and Rhodes, 1989). One class produced small colonies containing 1 to 16 terminally differentiated muscle cells and no other cells; the second class produced large colonies containing hundreds of cells, only some of which were terminally differentiated muscle cells at any given time. Because the numbers of cells in the small colonies tended to be powers of 2 (1,2,4,8 and 16), we proposed that terminally differentiated myoblasts were generated by a series of 4 symmetrical cell divisions. Because the numbers of differentiated cells in the large colonies tended to be integer multiples of 16, we proposed that, in the chick embryo, there is a self-renewing myogenic stem cell with the following properties:

1. It can divide symmetrically to produce 2 more stem cells.
2. It can divide symmetrically to produce 2 committed precursor cells.
3. It can divide asymmetricaly to produce 1 stem cell and 1 committed replication.

In our terminology, a committed percursor is a myogenic cell with a highly restricted replication potential. The data indicated that the end result, for any committed cell whose immediate precursor was a stem cell, was the production of exactly 16 terminally differentiated, mononucleated skeletal muscle cells. These cells initiate synthesis of muscle-specific proteins including those required for fusion.

The major difficulty with the interpretation of our data was that we examined the end stage of the development of clones and could only infer the cellular dynamics which led to that state. Thus, for example, a clone which contained 8 nuclei in differentiated cells could have developed in any of the ways shown in Fig. 1. Moreover, small clones, in which all nuclei were in differentiated cells but the number of nuclei was not a power of 2, could have arisen without the cell death or detachment which we proposed. To determine unambiguously how clones develop from individual precursor cells requires following the fate of every cell produced by every division in the colony. One way to do this would be to use conventional timelapse cinematography. However, since we needed to follow a clone for at least 3 days, at best we could do 2 cells per week, perhaps 100 cells per year. And this does not allow for cell death in which experiments would be total losses. To circumvent the limitations of conventional timelapse, a new technology was developed: computer assisted timelapse video microscopy. This system allowed us to follow the development of many clones (up to 50-100) simultaneously. We report that small myogenic clones unambiguously arise by a series of symmetrical cell divisions and that deviations from this pattern are almost always a result of cell death or detachment within a colony.

Materials and methods
Cell culture
Myogenic cells were isolated from the pectoral muscles of 10-11 day chick

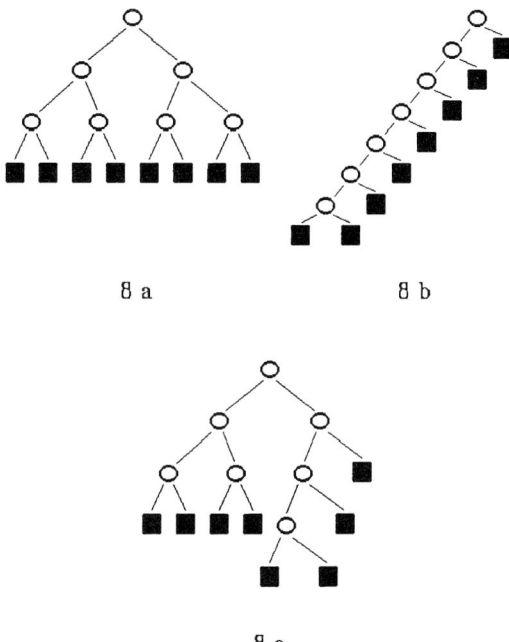

Fig. 1. Three ways by which 8 differentiated muscle cells (filled squares) could be generated from a single precursor cell. In 8a, only symmetric cell divisions are used. In 8b, all but the last division are asymmetric. 8c illustrates one of several ways in which a mixture of symmetric and asymmetric divisions could be used.

embryos as previously described (Nameroff and Rhodes, 1989). Cultures were set up by plating 1000-2000 cells into gelatin-coated 60mm dishes in 3 ml of medium (85 parts Eagle's MEM: 10 parts horse serum: 5 parts embryo extract). After an overnight incubation, the medium was changed and the culture was placed into the timelapse system.

Timelapse video microscopy

The timelapse system consisted of a Nikon inverted phase contrast Diaphot microscope equipped with an environmental chamber and a video camera. The stage of the microscope was operated by a Syn-Optics AFM-5000 XYZ stage controller. The stage controller was connected to the COM1 port of a DTK Keen-2530 computer with an 80386 processor operating at 25 Mhz and a 680 Mb hard disc. The output of the video camera was connected to an Image Technology Overlay frame grabber. The frame grabber output was to video monitor and to a Rapid Technology image compression board. The image compression board output was to the hard disc during initial photography and to the frame grabber when retreiving images from the hard disc. The details of the system will be presented elsewhere.

Up to 50 individual cells were located at the start of the experiment. Their X, Y and Z coordinates were stored by the computer and a programmable time delay was input. Thereafter, the computer moved the stage to the appropriate X, Y and Z coordinates of each clone in sequence and made a digital photo on the hard disc. Position and focus could be corrected during a run and provision was made for making montages if clone size was too large for one frame. A typical experiment lasted 3 days, took 300 images of each clone and, for 30 clones, used about 400 megabytes of disc space for the 9000 images made. For high quality images, compression ratios were about 6 or 7 to 1; i.e., a typical 250,000 byte image was compressed to 35-40,000 bytes. Compressed images were retrieved at a rate of about 2 per second for viewing results. The system was operated by software written specifically for this purpose by M.N.

Antibody staining

Cells were fixed and reacted with mouse monoclonal anti-myosin, MF 20 (Bader, Masaki and Fischman, 1982) followed by a secondary reaction with peroxidase-conjugated goat anti-mouse IgG and staining with diaminobenzidine as previously described (Quinn et al., 1985) The dishes were placed back on the stage of the microscope and the clones were found again by the computer. Sarcomeric myosin positive cells were identified and their ancestry traced in each clone. A lineage diagram was constructed for each clone in which positive cells were found.

Results and discussion

The results of examining 293 clones from 10 to 11-day chick embryo cells are shown in Table 1. About half (48.5%) of the cells isolated were myogenic. Of these, 89.4% produced small myogenic clones in which every cell was a differentiated muscle cell. About 10% of the myogenic cells produced large muscle colonies in which only some of the cells were positive for sarcomeric myosin. The remaining cells were fibroblasts. In this series, about 10% of the starting cells died before producing colonies. In every instance but one, the differentiated cells in small clones were produced by a series of 1 to 4 symmetrical cell divisions. In those small clones in which the number of nuclei was not a power of 2, cell death was observed at some point in the development of the clone. There was no apparent preference for detachment at any of the divisions and all cycles

Table 1.

Clone type	Number
Myogenic	142
Small (≤16 cells) 127	
Large (>16 cells) 15	
Fibroblast	121
Died/detached	30
TOTAL	293

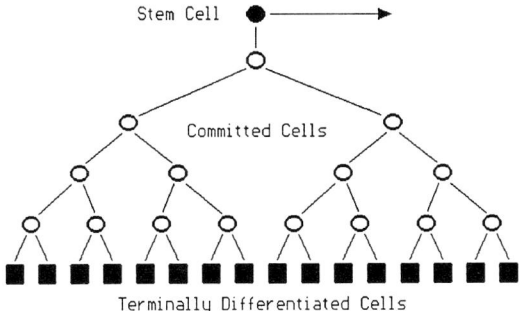

Fig. 2. Chick embryo skeletal muscle lineage. Cells at all stages of the lineage can be found in the muscle at most times during embryogenesis. At hatching and in the adult, only stem cells remain as muscle satellite cells. Direct evidence from computer assisted timelapse studies at present confirms the model only from the appearance of the first committed precursor cell.

were 8 to 11 hours in length. Less than 5% of the clones could not be analyzed for technical reasons; these were omitted from the analysis.

The results obtained so far are well described by a model in which committed myogenic cells are a heterogeneous population in the sense that each cell can only produce a predetermined number of progeny (see Fig. 2). However, it is not yet proven that a single committed cell which produces, for example, 4 differentiated muscle cells, *must always* have a direct lineal ancestor which produces 8 muscle cells; i.e., it is possible that stem cells divide to produce committed cells with randomly varying (though restricted) replication potential. Direct study of colonies produced by stem cells will settle this question.

The pattern of symmetrical divisions observed in the chick myogenic cells is not an artifact of the conditions employed. Myogenic cell lines did not display this behaviour. Neither is the 1, 2, 4, 8, 16 pattern an 'accident' of cell division. If that were the case, we should have found a 1, 2, 4, 6, 8, 10, 12, 14, 16 pattern because some pairs of cells would differentiate after any division.

Other studies (Quinn, Nameroff and Holtzer, 1984; Quinn, Norwood and Nameroff, 1988) have shown that the model depicted in Fig. 2 predicts reasonably well the kinetics of differentiation of both clonal and mass cultures under several types of environmental conditions. The data indicate that merely slowing or speeding up the cell cycle time can have dramatic effects on the differentiation kinetics without the necessity of postulating specific inducers of differentiation.

The computer assisted timelapse system offers the opportunity to investigate many other questions in regard to cell lineage. In particular, the following are presently under study:

Do growth factors act on a particular step in the lineage?

Do myogenic cells from other species behave like chick cells?

Do myogenic cells from other muscles and other developmental times obey the same lineage rules?

This new technology enables us to take a 'snap shot' of the developmental stages reached by myogenic cells at various times during embryogenesis.

References

Nameroff, M. and Rhodes, L. D. (1989). Differential response among cells in the chick embryo myogenic lineage to photosensitization by Merocyanine 540, *J. Cell. Physiol.* **141**, 475-482.

Quinn, L. S., Holtzer, H. and Nameroff, M. (1985). Generation of chick skeletal muscle cells in groups of 16 from stem cells. *Nature* **313**, 692-694.

Quinn, L. S. and Nameroff, M. (1983). Analysis of the myogenic lineage in chick embryos. III. Quantitative evidence for discrete compartments of precurser cells. *Differentiation* **24**, 111-123.

Quinn, L. S., Norwood, T. H. and Nameroff, M. (1988). Myogenic stem cell commitment probability remains constant as a function of organismal and mitotic age. *J. Cell. Physiol.* **134**, 324-336.

Robinson, M. M., Quinn, L. S. and Nameroff, M. (1984). BB Creatine Kinase and myogenic differentiation. Immunocytochemical identification of a distinct precursor compartment in the chick skeletal myogenic lineage. *Differentiation* **26**, 112-120.

MULTIMERIC STRUCTURES INFLUENCE THE BINDING ACTIVITY OF bHLH MUSCLE REGULATORY FACTORS

WOODRING E. WRIGHT[1], FRANÇOISE CATALA[2] and KAREN FARMER[1]

[1]Department of Cell Biology and Neuroscience, U.T. Southwestern Medical Center, Dallas, TX 75235
[2]Département de Biologie Moléculaire, Institut Pasteur, 7572 Paris, France

Summary

Sucrose gradients and molecular sieve chromatography were used to determine the native molecular weight of the basic HLH proteins myogenin, MyoD and E12. The muscle bHLH proteins not only formed dimers but also associated in a variety of higher order complexes. Although homodimers bind to DNA sequences such as the MEF-1 site in the creatine kinase enhancer, homotetramers and larger forms do not recognize this DNA sequence. Little evidence for complexes larger than dimers was found for the ubiquitous bHLH protein E12. Most of the myogenin remains in large complexes when myogenin and E12 are mixed. The same result was obtained in nuclear extracts from differentiated myotubes, in which most of the myogenin was found to be present in large complexes that do not bind to the creatine kinase enhancer. A fusion protein that contains only the myogenin HLH region fused to glutathione-S-transferase also forms large homomeric complexes. A model to explain these results is that each helix of the HLH motif can associate with a different subunit to form chains or ring structures. The presence of myogenin in nuclear extracts as both dimers that recognize known DNA sequences as well as higher order complexes that do not raises significant issues concerning the regulation of skeletal muscle bHLH protein activity during myogenesis.

Introduction

All of the known muscle determination/differentiation factors [MyoD (Davis, Weintraub and Lassar, 1987), myogenin (Wright, Sassoon and Lin, 1989; Edmonson and Olson, 1989), Myf5 (Braun, Buschhausen-Denker, Bober, Tannich and Arnold, 1989b) and MRF4 (Rhodes and Konieczny, 1989; Miner and Wold, 1990; Braun, Bober, Winter, Rosenthal and Arnold, 1990)] share the property of being able to transactivate the myogenic program: when any of these factors are transfected into C3H10T1/2 mesodermal cells colonies of myoblasts capable of forming myotubes are produced. This conversion is accomplished by single factors because they are all able to induce other regulatory factors, including

Key words: transcription factors, bHLH, myogenin, muscle differentiation, MyoD, E12.

the endogenous MyoD and myogenin genes (Rhodes and Konieczny, 1989; Miner and Wold, 1990; Thayer, Tapscott, Davis, Wright, Lassar, Weintraub, 1989; Braun, Bober, Buschhausen-Denker, Kotz, Grzeschik and Arnold, 1989a). The functional differences between these four factors remains largely unknown, although they show very different patterns of expression during development (Davis, et al. 1987; Wright et al., 1989; Rhodes and Konieczny, 1989; Miner and Wold, 1990; Braun et al., 1990; Sassoon, Lyons, Wright, Lin, Lassar, Weintraub and Buckingham, 1989; Ott, Bober, Arnold and Buckingham, 1991). It appears that at least two factors are expressed in actively differentiating cells (Montarras, Chelly, Bober, Arnold, Ott, Gros and Pinset, 1991).

These four skeletal muscle regulatory molecules are part of a larger family of DNA binding proteins that share a common structural domain, the basic helix-loop-helix (bHLH) motif (Murre, McCaw and Baltimore, 1989a). The basic region confers DNA binding specificity, while the helix-loop-helix region provides the amphipathic surfaces required for protein dimerization (Murre et al. 1989a; Murre, McCaw, Vassin, Caudy, Jan, Jan, Cabrera, Buskin, Hauschka, Lassar, Weintraub and Baltimore, 1989b; Voronova and Baltimore, 1990; Davis, Cheng, Lassar and Weintraub, 1990; Chakraborty, Brennan, Li Li and Olson, 1991). The MEF-1 site in the creatine kinase enhancer (Buskin and Hauschka, 1989) was the first binding site for these factors to be identified (Lassar, Buskin, Lockson, Davis, Apone, Hauschka and Weintraub, 1989), and similar sequences were later found in many other muscle specific enhancers/promoters. These sequences contain the general binding site for bHLH proteins, the E-box core sequence CANNTG (Murre et al. 1989a; Church, Ephrussi, Gilbert and Tonegawa, 1985).

E12 and E47, alternative splicing products of the E2A gene, are ubiquitous proteins not restricted to skeletal muscle (Murre et al. 1989a; Sun and Baltimore, 1991). A variety of evidence indicates that the bHLH family of muscle regulatory factors functions as heteromers with E2A proteins: 1) EMSA (electrophoretic mobility shift assay) activity increases significantly when *in vitro* translated MyoD (Murre et al. 1989b) or myogenin (Brennan and Olson, 1990; Braun, Gearing, Wright and Arnold, 1991) is mixed with E12/E47, with the activity being localized to a new mobility characteristic of the heterodimer; 2) Differentiation specific bands present in EMSA using nuclear extracts with the creatine kinase enhancer as probe are shifted by antibodies against both proteins (i.e., one band is shifted by either antimyogenin or antiE12 antibodies, while a second muscle-specific band is shifted by either antiMyoD or antiE12 antibodies) (Lassar, Davis, Wright, Kadesch, Murre, Voronova, Baltimore and Weintraub, 1991); 3) E2A antisense expression vectors (Lassar et al. 1991) can inhibit muscle differentiation; and 4) The protein Id ('Inhibitor of differentiation'), which lacks the DNA binding basic region but contains the HLH protein-protein interaction domain, is thought to function by forming inactive heterodimers with E2A proteins, thus sequestering them and preventing their interaction with the skeletal muscle bHLH factors (Benezra, Davis, Lockshon, Turner and Weintraub, 1990). These observations have led to the impression that the muscle bHLH family forms homomers inefficiently, and that the increased assembly into dimers that occurs with E2A proteins is responsible for the higher biological activity of heterodimers.

This investigation was designed to measure the native molecular weight of homomeric and heteromeric complexes involving myogenin. The results indicate that the small fraction of myogenin present as homodimers is not a consequence of inefficient self-assembly: most of the protein is present as tetramers or higher order complexes. Although myogenin homodimers are active in EMSA using the E-box containing MEF-1 site of the creatine kinase enhancer, tetramers fail to recognize this DNA sequence. Nuclear extracts from differentiated myotubes contain myogenin present as both heterodimers and in the form of higher order complexes. These observations raise significant questions about the partitioning of the bHLH regulatory factors between dimers versus higher order complexes and the roles of these complexes in the regulation of myogenesis.

Experimental procedures

Proteins

Fusion proteins to glutathione-S-transferase were purified on glutathione-agarose columns (Smith and Johnson, 1988). Nuclear extracts from differentiated C2C12 myotubes were prepared as described by Lassar et al. (1991).

Messenger RNA to be used for *in vitro* translations was transcribed from plasmids pBSM13⁻MGN#11 (myogenin; Wright et al., 1989), pEMSVα2-C11s (MyoD; Davis et al., 1987) and pBS-ATG-E12 (E12; Murre et al. 1989a). The messenger RNA was then translated in reticulocyte lysates according to the manufacturer's conditions (Promega, Madison, WI).

Sucrose gradients

Gradients were established by layering 0.5 ml aliquots of 20%, 15%, 10% and 5% sucrose in a 2.1 ml tube and incubating at 4°C for one hour before use (Bloom, Wagner, Pfister and Brady, 1988). In most cases, 100 μg GS-myogenin was run on gradients containing protein standards (200 μg thyroglobulin, S=19.4; 60 μg catalase, S=11.3; 30 μg BSA, S=4.3; 30 μg ovalbumin, S=3.7; in some cases hemoglobin [S=3.8] or myoglobin [S=2.0] were also used). Most gradients were centrifuged for 4 hours at 55,000 rpm in a TLS55 rotor in a Beckman TL100 centrifuge (acceleration=5 and deceleration=5), although these conditions were modified as needed to expand certain regions of the gradient.

Reticulocyte lysates containing *in vitro* translated proteins were too dense to analyze on 5%-20% gradients, and were examined on 20%-35% gradients centrifuged for 17 hours at 40,000 rpm. Protease inhibitors were included when nuclear extracts were analyzed.

HPLC

Samples were analyzed on a TSKG4000 PW molecular sieve column (Toso Haas, Phil., PA). Protein standards included thyroglobulin (D=2.65 $\times$ 10^{-7} cm^2

$\sec^{-1}$), catalase (D=4.1 × 10^{-7} cm^2 sec^{-1}), ovalbumin (D=7.8 × 10^{-7} cm^2 sec^{-1}) and myoglobin (D=11.3 × 10^{-7} cm^2 sec^{-1}).

Molecular weights

Molecular weights were calculated using the formula MW=RTS÷[D(1−$\bar{v}\rho$)], in which R is the ideal gas constant (8.31 × 10^7 erg deg^{-1} mol^{-1}), T=293 °K, S is the sedimentation coefficient, D is the diffusion coefficient, $\bar{v}$ is the protein partial specific volume (approximated as 0.725 cm^2 g^{-1}) and ρ is the density of water at 293 °K (0.9982 g cm^{-1}) [Bloom et al., 1988]. Sedimentation coefficients (S) were determined from plots of the S values of standards versus their position in sucrose gradients (Martin and Ames, 1961). Diffusion coefficients were determined by plotting 1/D versus K_{av} (K_{av}=[$V_{elution}$−V_{void}]÷[V_{total}−V_{void}]) (Cooper, 1977).

Results

Myogenin fusion proteins

The MEF-1 site in the creatine kinase enhancer (Buskin and Hauschka, 1989) contains a single E-box. However, two bands are found on EMSA when purified GS-myogenin is bound to an oligonucleotide containing the MEF-1 site. GS-myogenin was analyzed on both sucrose gradients and HPLC in order to investigate whether these two bands represented homodimers and homo-tetramers. Individual fractions were examined both for myogenin protein by SDS-PAGE and for activity against the MEF-1 site by EMSA. There was a clear discordance between the location of EMSA activity and protein on both sucrose gradients and HPLC.

Seven sucrose gradients and five HPLC analyses yielded calculated molecular weights of 136 Kda for the lower EMSA band, 156 Kda for the upper band, and 268 Kda for the GS-myogenin protein. A single subunit of glutathione-S-transferase fused to the 191 amino acids of myogenin included in this construct should have a molecular weight of approximately 55 Kda. The molecular weights determined from this analysis for both species active in EMSA are very close to what would be expected for dimers, whereas the protein is in a much larger complex close to that predicted for homotetramers.

The shape of the two dimer conformations is clearly different, since they migrated to slightly different positions on sucrose gradients. The explanation for their very different mobilities on EMSA remains unclear.

Both cleaved GS-myogenin and in vitro translated myogenin also associated into higher order complexes. The formation of tetramers is thus not an artifact of using a bacterially expressed fusion protein.

MyoD and E12

Two other bHLH proteins were examined for their ability to form higher order complexes. GS-MyoD behaved like myogenin, with most of the protein present as tetramers while the activity against the MEF-1 site was limited to homodimers.

Tetramers were also the predominant species using *in vitro* translated MyoD. In contrast, very little large complex formation was found using *in vitro* translated E12, in which the calculated molecular weight corresponded to homodimers (or a mixture of monomers and dimers). In mixtures of *in vitro* translated myogenin and E12, some of the myogenin associated with E12 to form heterodimers that were active against the MEF-1 site and some of the E12 became incorporated into higher order heteromeric complexes, although most of the myogenin remained large and most of the E12 remained small.

The HLH region is sufficient for tetramerization

Two different dimerization motifs might be able to cooperate to form tetramers. Evidence for a dimerization motif outside than the HLH domain was sought by examining deletion mutants in which either all of the amino acids prior to or after the bHLH domain had been removed. Neither mutant abolished large complex formation, suggesting that a second dimerization motif was not required for tetramerization. Mutants that only contained the bHLH (amino acids 71-138) or HLH (amino acids 94-138) motifs also formed tetramers, indicating that the HLH motif alone was sufficient to form higher order complexes.

Nuclear extracts

Nuclear extracts from cultured myotubes were examined to see if these observations using purified bacterially expressed or *in vitro* translated myogenin applied to 'authentic' myogenin. The myogenin in nuclear extracts should be both properly processed and complexed to its normal biological partner(s). The activity against the MEF-1 site in nuclear extracts migrated in sucrose gradients and HPLC at locations that yielded a calculated molecular weight of about 80 Kda, which is about the predicted size for myogenin/E2A protein heterodimers. Western blots of the different fractions showed a discordance with EMSA activity, and indicated that the myogenin protein was present in a much larger form of approximately 160 Kda. We have no information as to whether this larger complex represents homomers or myogenin associated with one or several other factors.

Discussion

These studies were undertaken to determine the size of the native complexes of myogenin and other bHLH proteins involved in regulating muscle specific differentiation. The results establish that most of the protein in purified myogenin is present as tetramers or larger complexes, with only a small fraction being in homodimers. Although MyoD behaves like myogenin, the ubiquitous factor E12 shows little large complex formation and exists primarily as monomers/dimers in homomeric solutions.

Both purified bacterially expressed fusion proteins and *in vitro* translated myogenin show this distribution of subunits into large complexes. Myotube nuclear extracts also contain myogenin complexes larger than homo- or heterodimers. The fact that these large complexes are present in a variety of different preparations, and particularly in nuclear extracts, argues strongly in favor of their representing a native form of the protein.

Non-specific aggregation is an unlikely explanation for large complex formation. The HLH domain is not only sufficient for the formation of large complexes, it is also required: a mutation eliminating the second helix (premature termination at myogenin amino acid 115) abolishes large complexes and yields monomers. The demonstration that the domain responsible for large complex formation must be structurally intact, and its localization to the same region implicated in specific protein-protein interactions, strongly argues that it represents a specific effect.

One model to explain this effect is that dimers would be formed by the association of each helix with a single adjacent subunit, whereas larger complexes would be formed if each helix interacted with a different subunit. The flexibility of the loop region would permit the molecule to twist to accommodate these different interactions. Although chains could form, the first and last subunits would then only have one of their two helices paired. The most likely structure would be if the helices at the ends interacted to form a ring, thus making a variety of complexes (dimers, tetramers, hexamers, octamers, etc) possible.

The overall affinity of DNA binding in dimers is generally produced by each subunit recognizing a half-site DNA sequence. We have confirmed this pattern for myogenin dimers by using the protein to retrieve its preferred binding sites from a mixture of random oligonucleotides. The consensus site that was obtained contained the expected E-box CANNTG within the palindromic sequence AACAGCTGTT (Wright, Bender and Funk, 1991). The present results indicate that myogenin complexes larger than dimers fail to exhibit significant binding to DNA sequences containing the CANNTG motif. It is likely that the basic domains responsible for DNA binding would be misaligned when each helix interacted with a different subunit to form the ring structures postulated above. The failure of these large complexes to show EMSA activity against the MEF-1 site would thus be a consequence of the inability of two basic domains to cooperate in the binding to two CTGTT-like half-sites due to their mispositioning in the ring structures.

Based on indirect measurements, two groups have recently reported that neither MyoD (Sun and Baltimore, 1991) nor myogenin (Chakraborty et al. 1991) oligomerize efficiently. Although we agree with the major conclusion of these authors, that MyoD or myogenin forms heterodimers with E2A gene products more efficiently than they form homodimers, our results indicate that both myogenin and MyoD oligomerize very well. It is the presence of most of the protein in higher order complexes that are inactive in EMSA that produces the low or absent EMSA activity observed in homomeric preparations. It will be necessary to reconsider the association constants calculated based upon MyoD monomer-dimer distributions (Sun and Baltimore, 1991) in light of the fact that most of the MyoD is present as tetramers or higher order complexes.

Although bacterially expressed myogenin is active against the MEF-1 site, *in vitro* translated myogenin homomers fail to exhibit detectable EMSA activity (Brennan and Olson, 1990; Braun et al. 1991). This may be a consequence of the increased efficiency of large complex formation found with *in vitro* translated myogenin. *In vitro* translated myogenin associated into very large complexes whose average size approximated octamers, whereas bacterial expressed myogenin fusion proteins primarily formed tetramers. Following *in vitro* translation, the fraction of myogenin present as homodimers may thus be too small to give detectable activity. The average size of *in vitro* translated MyoD is significantly smaller than that of *in vitro* translated myogenin, forming mainly tetramers like the bacterially expressed proteins. *In vitro* translated preparations of MyoD would thus contain enough homodimers to account for some EMSA activity.

In contrast to the skeletal muscle specific bHLH factors myogenin and MyoD, *in vitro* translated E12 was present primarily as either dimers or a mixture of monomers and dimers. This result is somewhat surprising given the fact that the HLH domain is sufficient to form higher order complexes and the similarity in the HLH domains between all three proteins. Subtle differences in the helices or the nature of the loop region may be responsible for the difference. However, we consider it more likely that regions outside the bHLH domain in E12 (such as that thought to inhibit DNA binding; Sun and Baltimore, 1991) are responsible for inhibiting tetramer formation.

We are reluctant to assign precise physical parameters to these complexes because of the variability, lack of resolution and simultaneous presence of multiple species (dimers, tetramers, octamers, etc.) in these experiments. The results nonetheless clearly indicate that myogenin forms highly asymmetric structures. The radii of ideal spherical proteins of the same MW as GS-myogenin dimers and tetramers would be 34Å and 42Å, whereas the values obtained for the Stokes Radii from our measurements are approximately 60Å and 80Å. A similar ratio between the Stokes Radii and that of comparable spherical proteins was observed for the myogenin protein and EMSA activity in myotube nuclear extracts. Purified bacterial expressed and crude myotube derived myogenin complexes thus both behave in a similar fashion and show large axial ratios.

The model described above does not explain the two conformations of dimers that are present in bacterially expressed myogenin. Both forms are present in all of our preparations (with the site of fusion at amino acid 1, 33 or 71), although the distribution between the two forms varies depending on the site of the fusion to glutathione-S-transferase. The two conformations are not dependent on the presence of glutathione-S-transferase, since both forms are found with the bHLH peptide released following cleavage of the myogenin aa.71-138 fusion protein. Only a single heterodimeric band is formed when GS-E12 and GS-myogenin are mixed (Wright et al. 1991). Similarly, only a single band shifted by antimyogenin monoclonal antibodies is observed in EMSA using nuclear extracts on the MEF-1 site (Lassar et al. 1991). Since neither *in vitro* translated myogenin nor bacculovirus expressed myogenin (Braun et al. 1991) exhibit EMSA activity and the two conformations are only detected by EMSA activity, we have been unable to determine whether the two conformations are a consequence of bacterial expression. The biological significance of these two conformations of myogenin homodimers remains unclear given the lack of evidence for their presence outside of bacterial expressed preparations.

The presence of large myogenin complexes in myotube nuclear extracts strongly suggests that these complexes are not an artifact of bacterially produced or *in vitro* translated preparations. At the present time we have no information as to their potential biological function. They might represent a reserve of myogenin protein sequestered in an inactive form and available for rapid shifts in concentration of the active species, or perhaps they recognize a DNA sequence different from that present in the MEF-1 site. It is possible that postranslational modifications might influence the partitioning of myogenin between dimeric and larger complexes. The demonstration that myogenin can form large complexes that don't bind to known DNA sequences indicates that the control of their formation could represent an important level of regulation of the function and activity of the bHLH muscle differentiation factors.

This work was supported by grants from the Muscular Dystrophy Association and the NIH (AG01228).

References

Benezra, R., Davis, R. L., Lockshon, D., Turner, D. L. and Weintraub, H. (1990). The protein Id: a negative regulator of helix-loop-helix DNA binding proteins. *Cell* **61**, 49-59.

Bloom, G. S., Wagner, M. C., Pfister, K. K. and Brady, S. T. (1988). Native structure and physical properties of bovine brain kinesin, and identification of the ATP-binding subunit polypeptide. *Biochemistry* **27**, 3409-3416.

Braun, T., Bober, E., Buschhausen-Denker, G., Kotz, S., Grzeschik, K.-H. and Arnold, H. H. (1989a). Differential expression of myogenic determination genes in muscle cells: possible autoactivation by the Myf gene products. *EMBO J.* **8**, 3617-3625.

Braun, T., Bober, E., Winter, B., Rosenthal, N. and Arnold, H. H. (1990). Myf-6, a new member of the human gene family of myogenic determination factors: evidence for a gene cluster on chromosome 12. *EMBO J.* **9**, 821-831.

Braun, T., Buschhausen-Denker, G., Bober, E., Tannich, E. and Arnold, H. H. (1989b). A novel human muscle factor related to but distinct from MyoD1 induces myogenic conversion in 10T1/2 fibroblasts. *EMBO J.* **8**, 701-709.

Braun, T., Gearing, K., Wright, W. E. and Arnold, H. H. (1991). Baculovirus-expressed myogenic determination factors require E12 complex formation for binding to the myosin-light-chain enhancer. *Euro. J. Biochem.* **198**, 187-193.

Brennan, T. J. and Olson, E. N. (1990). Myogenin resides in the nucleus and acquires high affinity for a conserved enhancer element on heterodimerization. *Genes Dev.* **4**, 582-595.

Buskin, J. N. and Hauschka, S. D. (1989). Identification of a myocyte nuclear factor which binds to the muscle-specific enhancer of the mouse creatine kinase gene. *Mol. Cell. Biol.* **9**, 2627-2640.

Chakraborty, T., Brennan, T. J., Li Li, D. E. and Olson, E. N. (1991). Inefficient homooligomerization contributes to the dependence of myogenin on E2A products for efficient DNA binding. *Mol. Cell. Biol.* **11**, 3633-3641.

Church, G. M., Ephrussi, A., Gilbert, W. and Tonegawa, S. (1985). Cell-type-specific contacts to immunoglobulin enhancers in nuclei. *Nature* **313**, 798-801.

Cooper, T. G. (1977). *The Tools of Biochemistry*, Wiley, New York.

Davis, R. L., Cheng, P-F., Lassar, A. B. and Weintraub, H. (1990). The MyoD DNA binding domain contains a recognition code for muscle-specific gene activation. *Cell* 60, 733-746.

Davis, R. L., Weintraub, H. and Lassar, A. B. (1987). Expression of a single cDNA converts fibroblast to myoblasts. *Cell* 37, 879-887.

Edmondson, D. G. and Olson, E. N. (1989). A gene with homology to the myc similarity region of MyoD1 is expressed during myogenesis and is sufficient to activate the muscle differentiation program. *Genes Dev.* 3, 628-640.

Lassar, A. B., Buskin, J. N., Lockson, D., Davis, R. L., Apone, S., Hauschka, S. D. and Weintraub, H. (1989). MyoD is a sequence-specific DNA binding protein requiring a myc homology to bind to the muscle creatine kinase enhancer. *Cell* 58, 823-831.

Lassar, A. B., Davis, R. L., Wright, W. E., Kadesch, T., Murre, C., Voronova, A., Baltimore, D. and Weintraub, H. (1991). Functional activity of myogenic HLH proteins requires hetero-oligomerization with E12/E47-like protein in vivo. *Cell* 66, 305-315.

Martin, R. G. and Ames, B. N. (1961). A method for determining the sedimentation behavior of enzymes: application to protein mixtures. *J. Biol. Chem.* 236, 1372-1379.

Miner, J. H. and Wold, B. (1990). Herculin, a fourth member of the MyoD family of myogenic regulatory genes. *Proc. Natl. Acad. Sci. U.S.A.* 87, 1089-1093.

Montarras, D., Chelly, J., Bober, E., Arnold, H., Ott, M.-O., Gros, F. and Pinset, C. (1991). Developmental patterns in the expression of Myf5, MyoD, myogenin and MRF4 during myogenesis. *The New Biologist* 3, 1-10.

Murre, C., McCaw, P. S. and Baltimore, D. (1989a). A new DNA binding and dimerization motif in the immunoglobulin enhancer binding, *daughterless*, MyoD1, and *myc* proteins. *Cell* 56, 777-783.

Murre, C., McCaw, P. S., Vassin, H., Caudy, M., Jan, L. Y., Jan, Y. N., Cabrera, C. V., Buskin, J. N., Hauschka, S. D., Lassar, A. B., Weintraub, H. and Baltimore, D. (1989b). Interactions between heterologous helix-loop-helix proteins generate complexes that bind specifically to a common DNA sequence. *Cell* 58, 537-544.

Ott, M. O., Bober, E., Arnold, H. and Buckingham, M. (1991). Early expression of the myogenic regulatory gene, Myf-5, in precursor cells of skeletal muscle in the mouse embryo. *Development* 111, 1097-1107.

Rhodes, S. J. and Konieczny, S. F. (1989). Identification of MRF4: a new member of the muscle regulatory factor gene family. *Genes Dev.* 3, 2050-2061.

Sassoon, D., Lyons, G., Wright, W. E., Lin, V. K., Lassar, A. B., Weintraub, H. and Buckingham, M. (1989). Expression of two myogenic regulatory factors myogenin and MyoD1 during mouse embryogenesis. *Nature* 341, 303-307.

Smith, D. B. and Johnson, K. S. (1988). Single-step purification of polypeptides expressed in Escherichia coli as fusions with glutathione S-transferase. *Gene* 67, 31-40.

Sun, X-H. and Baltimore, D. (1991). An inhibitory domain of E12 transcription factor prevents DNA binding in E12 homodimers but not in E12 heterodimers. *Cell* 64, 459-470.

Thayer, M. J., Tapscott, S. J., Davis, R. L., Wright, W. E., Lassar, A. B. and Weintraub, H. (1989). Positive autoregulation of the myogenic determination gene MyoD1. *Cell* 58, 241-248.

Voronova, A. and Baltimore, D. (1990). Mutations that disrupt DNA binding and dimer formation in the E47 helix-loop-helix protein map to distinct domains. *Proc. Natl. Acad. Sci. U.S.A.* 87, 4722-4726.

Wright, W. E., Bender, M. and Funk, W. (1991). CASTing for the myogenin consensus binding site. *Mol. Cell. Biol.* 11, 4104-4110.

Wright, W. E., Sassoon, D. A. and Lin, V. K. (1989). Myogenin, a factor regulating myogenesis, has a domain homologous to MyoD. *Cell* 56, 607-617.

Printed in Great Britain © *Society for Experimental Biology 1992*   89

ISOLATION AND FUNCTIONAL COMPARISON OF Dmyd, THE DROSOPHILA HOMOLOGUE OF THE VERTEBRATE MYOGENIC DETERMINATION GENES, WITH CMD1

BRUCE M. PATERSON[1,2], MASAKI SHIRAKATA[1],
SEIJI NAKAMURA[1], CLAUDE DECHESNE[1], UWE WALLDORF[2],
JUANITA ELDRIDGE[1], ANDREAS DÜBENDORFER[3],
MANFRED FRASCH[4] and WALTER J. GEHRING[2]

[1]Laboratory of Biochemistry, National Cancer Institute, National Institutes of Health, Bethesda, Maryland 20892
[2]Department of Cell Biology Biocenter, University of Basel, Klingelbergstr. 70, CH-4056 Basel, Switzerland
[3]Zoology Institute, University of Zürich, Künstlergasse 16, CH-8006 Zürich, Switzerland
[4]Max-Planck Institute for Developmental Biology, Abteilung III Genetik, D-7400 Tübingen, FRG

Abstract

We have isolated a cDNA clone, called Dmyd for *Drosophila* myogenic determination gene, from a 0-16 hour *Drosophila* embryo library that encodes a protein with structural and functional characteristics similar to the members of the vertebrate MyoD family (Paterson et al 1991). Dmyd encodes a polypeptide of 332 amino acids with 82% identity to MyoD in the 41 amino acids of the putative helix-loop-helix region and 100% identity in the 13 amino acids of the basic domain proposed to contain the essential recognition code for muscle specific gene activation. The gene is unique and maps to 95A/B on the right arm of the third chromosome. Low stringency hybridizations indicate Dmyd is not a member of a multigene family, similar to MyoD in vertebrates. Dmyd is a nuclear protein in *Drosophila*, consistent with its role as a nuclear gene regulatory factor, and is proposed to be a transiently expressed marker for a unique subset of muscle founder cells. We have used an 8kb promoter fragment from the gene, which contains the first 55 amino acids of the Dmyd protein, joined to lac Z to follow myogenic precursor cells into muscle fibers using antibodies to β-galactosidase and Dmyd. Unlike the myogenic factors in vertebrate muscle cells, Dmyd appears to be expressed at a much lower level in differentiated *Drosophila* muscles so it cannot be followed continuously as a muscle marker. This is reflected in the loss of expression of Dmyd RNA in 12-24 hour embryos, a major period of early myogenesis, as well as in the undetectable level of the nuclear antigen in primary cultures of embryonic and adult *Drosophila* muscle. Functional differences between Dmyd and CMD1 are described and explained in terms of a model which may give insight to the nature of homo and heterodimer formation in the bHLH family of proteins.

Key words: insect myogenesis, helix-loop-helix, invertebrate MyoD.

Introduction

The discovery of the myogenic determination genes (Davis et al 1987, Wright et al 1989, Braun et al 1989b, Rhodes and Konieczny 1989) in the vertebrates has helped to establish the idea that a small number of so-called "master regulatory genes" select and commit embryonic cells to a particular developmental pathway. The developmental role of the four different myogenic determination genes in the vertebrates has yet to be established but it is clear all members of this gene family, from amphibia to man, share a common structural motif with a very similar amino acid profile, a basic domain joined to a putative helix-loop-helix configuration, that is essential for function (Tapscott et al 1988). Furthermore, ectopic expression of any of the vertebrate myogenic factors in 10T1/2 mouse embryonic fibroblasts will not only activate the endogenous mouse determination genes but will also convert these cells to stable differentiating myoblast lines, confirming the similar function of these different vertebrate genes.

A disadvantage in studying these determination genes in vertebrate systems is reflected in the difficulties in carrying out a detailed genetic analysis of their function as well as in the lack of information on genes involved in the control of early vertebrate development. The isolation and characterization of a myogenic determination gene in *Drosophila* would provide a tremendous advantage in both genetic and early developmental analyses of myogenesis since so many of the genes controling early developmental events in *Drosophila* have now been isolated and studied in detail (Ingham 1988).

We have used a probe from the conserved basic helix-loop-helix domain of the avian myogenic factor, CMD1 (Lin et al 1989), to isolate a *Drosophila* cDNA clone, Dmyd, encoding a polypeptide very similar to the vertebrate myogenic determination genes (Paterson et al 1991). We demonstrated Dmyd is a transiently expressed nuclear protein that serves as a marker for myogenic precursor cells, or a subset thereof, in *Drosophila* development. Similar to the grasshoppers (Ho et al 1983, Ball and Goodman 1985a,b, Ball et al 1985a, 1985b), these cells may represent the muscle founder cells that are crucial in organizing and establishing the precise muscle pattern in each segment of the *Drosophila* body plan. Preliminary functional difference between Dmyd and CMD1 are described and explained based upon rules of homo and heterodimerization derived from structural models of CMD1 set forth in chimeric protein studies.

Materials and methods

Preparation of primary cultures of Drosophila *embryo cells*

Single drop cultures, using a cell suspension of 4-6 hour embryo cells, were prepared and treated with ecdysone as described earlier (Dubendorfer et al 1978). Cultures were fixed in formaldehyde:PBS (1:4) for 20-30 minutes, dehydrated with methanol then treated for antibody staining as for the whole mount embryos (Frasch et al 1987). !0T1/2 mouse fibroblasts were plated, transfected, assayed for CAT activity and stained with antibody as described (Lin et al 1989).

Gel shift assays

The gel shift assays with the MCK enhancer were performed as described (Davis et al 1990).

Isolation of the Dmyd cDNA and genomic clones

The probe used to isolate the *Drosophila* Dmyd cDNA clones was prepared by PCR amplification (Saiki et al 1988) of the basic helix-loop-helix domain in the avian CMD1 cDNA (7) (from nucleotides 413 to 637) using the following primers: 5′primer; TGGGCGTGCAAGCATATGAAGAGGAAGACC. 3′primer; TGCATCCTCCTGGAATTCTTACAGGGCCTGCAG. One ng of plasmid DNA containing the entire *Eco*R1 insert of the CMD1 cDNA clone was amplfied through 40 PCR cycles (94°C 1 min, 37°C 1 min, 72°C 1 min) and the amplified fragment was isolated by electroelution from a preparative 1.5% agarose gel. The insert was labeled by random priming. Roughly 5×10^5 clones were screened in a random primed lambda gt11 library prepared from 0-16 hour *Drosophila* embryos (prepared by Bernd Hovemann) using hybridization conditions of reduced stringency (Lin et al 1989, McGinnis et al 1984). A single positive clone was plaque purified and sequenced. The 1.3Kb *Eco*R1 insert from the clone, containing the entire open reading frame of Dmyd, was used to screen a second cDNA library, a 3-12 hour embryonic library in lambda gt 10 (from T. Kornberg, E-library) (Poole et al 1985), in order to obtain additional 5′ and 3′ sequence regions. Sequence analysis and protein consensus data were determined using the GCG programs as described previously (Lin et al 1989). For isolation of the genomic sequence, a lambda EMBL 3 *Drosophila* genomic library (Clos et al 1991) was screened with the 1.3kb EcoR1 Dmyd cDNA insert and a single clone was further characterized (to be presented elsewhere).

Construction and injection of p C.20 Dmyd-lac Z

A single genomic fragment containing 8kb of 5′ flanking sequences and the first 55 codons of the Dmyd protein was fused in frame with the lac Z gene of E. coli. This construct contained the polyadenylation signal and 3′ flanking sequences of the *Drosophila* hsp70 gene (Hiromi et al 1985) and was then cloned into the polylinker of the transformation vector Carneigie 20.1, a derivative of Carneigie 20, and injected into *Drosophila* embryos using standard procedures (Hiromi et al 1985). Five independent transformant lines were generated and balanced. Two of the lines gave the same patterns of β-galactosidase expression by activity staining and antibody detection (Ghysen and O'Kane 1989). The stronger staining line 14.1 (II) was analyzed in detail. A single P element containing the Dmyd-lacZ construct mapped to 47A on chromosome-2 (data not shown).

Preparation of Dmyd antibody

A *Rsa*1 fragment from the Dmyd cDNA (nucleotides 323 to 922) was inserted into the *Sma*1 site of the glutathione transferase fusion vector, expressed in JM109 and purified as described (Smith and Johnson 1988). New Zealand White Rabbits were injected subcutaneously with 200-300µg of fusion protein in complete Freund's adjuvant and boosted four times with the same amount of protein in incomplete adjuvant. Antibody was affinity purified as described (Lasko and Ashburner 1990).

β-galactosidase staining of whole mount Drosophila embryos

Embryos were collected, fixed and stained for enzyme activity as previously described (Hiromi et al 1985, Ghysen and O'Kane 1989).

Analysis of RNA and DNA

RNA was extracted from various developmental stages using the acid phenol proceedure (Chomczynski and Sacchi 1987). Both poly A^+ and total RNA were electrophoresed on formaldehyde gels and analyzed by Northern blot on nitrocellulose membranes with the probes indicated in the figure legends. Probes were labeled by random priming. Drosophila DNA was prepared (Walldorf et al 1985) and analyzed by Southern blotting onto nitrocellulose filters using 5µg of DNA per lane and the probes indicated. Chromosomal mapping was performed on salivary gland chromosomes using standard procedures (Pardue 1986).

Antibodies

Monoclonal antibody to E. coli β-galactosidase was purchased from Promega Laboratories and used at a dilution of 1:500 or 1:1000. Rabbit polyclonal antibodies to Drosophila muscle myosin heavy chain (MHC), were the generous gift of Dan Kiehart, Harvard University, and were used at a dilution of 1:500. Antibody stainings were carried out using the Vectastain elite kit from Vector Laboratories (Lin et al 1989, Frasch et al 1987).

Results

Isolation of Dmyd, a Drosophila homolog to the vertebrate myogenic determination genes

Using hybridization conditions of reduced stringency, we screened a Drosophila 0-16 hour random primed cDNA library in lambda gt11 with a PCR amplified region from the avian myogenic factor gene, CMD1, which contained the conserved basic helix-loop-helix domain. A single cDNA clone was plaque purified from approximately 5 × 105 recombinant phages. The 1.3kb EcoR1 insert did not contain the poly A tail region, therefore, an additional 3-12 hour embryonic cDNA library in lambda gt 10 was screened and a 3′ fragment containing a poly A sequence was obtained. The combined 1410 nucleotide sequence of the two clones, called Dmyd for Drosophila myogenic determination gene, is shown in figure 1. The AUG at position 263 fits the Drosophila translation initiation consensus sequence, C/AAAA/C AUG (Cavener 1987), and is followed by the longest open reading frame of 332 amino acids encoding a polypeptide of

36,208 daltons. Furthermore, this reading frame contains the consensus amino acid sequence of the basic helix-loop-helix region found in all the vertebrate myogenic factors (figure 2). Two AUG codons at positions 95 and 138 are immediately followed by stop codons and do not conform to the consensus sequence. Outside this region, the *Drosophila* polypeptide is highly divergent compared to the vertebrate factors (data not shown). Two conserved regions in the carboxy portion of the vertebrate factors belonging to the MyoD and Myf5 groups (Davis et al 1987, Braun et al 1989) (amino acids 176-186 (box 1) and 222-238 (box 2) in CMD1, and amino acids 199-209 (box 1) and 245-261 (box 2) in mouse MyoD (see figure 2 in (Lin et al 1989)), are greatly reduced in Dmyd with only the first six amino acids of box 2 (position 1091 to 1108, figure 1), SSLDCL, remaining. The particular role for these conserved sequences remains to be determined. Myogenin does not contain these regions and there is only partial homology in box 2 with the myf6 group (Rhodes and Konieczny 1989).

The amino acid sequences of the basic and helix-loop-helix domains are conserved between Dmyd and the vertebrate myogenic factors

Direct comparison of the amino acid sequences of the basic and helix-loop-helix regions of the vertebrate myogenic factors with the same region in Dmyd is given in figure 2. The lower case letters in any given sequence correspond either to a difference from the consensus sequence for the entire collection of sequences presented or the absence of a consensus. In the basic region, the consensus has been defined with reference to three clusters of basic amino acid residues previously used in the deletion analysis of mouse MyoD (Davis et al 1990). The amino acid sequence that includes basic clusters two and three has been conserved exactly from *Drosophila* to mouse and man in the MyoD and Myf5 groups (DRRKAATMRERR), whereas members of the myogenin and Myf 6 groups have replaced the amino acid sequence ATM with either ATL or ATV. It has been suggested that this region contains a recognition code for muscle specific gene activation. Taken together with the helix-loop-helix consensus for each group (figure 2), these comparisons would place Dmyd closer to the MyoD group of myogenic genes. The helix-loop-helix regions are more divergent within the consensus. Dmyd has two additional position changes in helix-2 out of nine changes in total; two glutamate residues (E) replace an arginine (R) and glutamine (Q) residue in the consensus, respectively. The remaining seven substitutions are positionally conserved. These observations are consistent with the results of the mutational studies on the mouse MyoD helix-loop-helix region which suggest that the exact sequence and spacing between the helicies can be varied within certain structural constraints with minimal effect on biological function,whereas the basic region is held practically invariant (Davis et al 1990). The spacing between the basic region and the helix-loop-helix domain, however, has been conserved in all the factors. The insertion of two alanines between arginine (R) 20 and leucine (L) 21 at the junction of these two regions (figure 2) completely inactivates mouse

```
          10                  30                  50
GGAAAAATCCCGAAAGTGAAACTATAACAAATTAAACTAAATAGAAACCACAGCCTAAAA
          70                  90                 110
CTTGTGTGTAACCATAACTCATAAATTTGTGTTAATGACCTAGCATACCTAGAAAAGGAG
         130                 150                 170
CTTAAACCACGTTAAAAATGGTAATAATCAACTGACTAAATTAACTGTGCCATCGTAATT
         190                 210                 230
AGTATCAGCAATTAAATTCAGCAACTTGTTGAACTACCAAAATTGTTTTGTAAAATATAA
         250                 270                 290
CAAAATCGTAGTGAAGTGAAAAATGACCAAGTATAATAGTGGCAGCAGTGAAATGCCTGC
                              M  T  K  Y  N  S  G  S  S  E  M  P  A
         310                 330                 350
GGCTCAAACCATCAAGCAGGAGTACCACAATGGCTATGGTCAGCCGACACATCCTGGATA
 A  Q  T  I  K  Q  E  Y  H  N  G  Y  G  Q  P  T  H  P  G  Y
         370                 390                 410
CGGATTTAGCGCCTATAGCCAACAGAATCCGATAGCCCATCCCGGCCAGAATCCACACCA
 G  F  S  A  Y  S  Q  Q  N  P  I  A  H  P  G  Q  N  P  H  Q
         430                 450                 470
GACACTGCAGAATTTCTTTAGCCGCTTCAATGCCGTCGGTGATGCGAGTGCGGGAAATGG
 T  L  Q  N  F  F  S  R  F  N  A  V  G  D  A  S  A  G  N  G
         490                 510                 530
TGGAGCGGCTTCCATCTCAGCCAACGGATCGGGTTCGTCTTGCAACTACAGTCATGCGAA
 G  A  A  S  I  S  A  N  G  S  G  S  S  C  N  Y  S  H  A  N
         550                 570                 590
TCATCATCCGGCGGAGCTGGACAAGCCGTTGGGCATGAATATGACACCGTCGCCCATCTA
 H  H  P  A  E  L  D  K  P  L  G  M  N  M  T  P  S  P  I  Y
         610                 630                 650
CACCACCGACTACGATGACGAGAACAGCAGTCTCAGCTCCGAGGAGCACGTCCATGCGCC
 T  T  D  Y  D  D  E  N  S  S  L  S  S  E  E  H  V  H  A  P
         670                 690                 710
CCTCGTCTGCTCCTCCGCCCAATCCTCCAGACCATGCCTCACCTGGGCCTGCAAGGCGTG
 L  V  C  S  S  A  Q  S  S  R  P  C  L  T  W  A  C  K  A  C
         730                 750                 770
CAAAAAGAAGAGCGTCACCGTGGACCGTCGAAAAGCGGCCACTATGAGGGAACGCCGGAG
```
K K K │S V T V D │R R K │A A T M R E │R R R
  1   BASIC   2   3
```
         790                 810                 830
ACTGCGAAAGGTTAACGAGGCCTTCGAGATCCTCAAGCGACGCACTTCATCCAATCCCAA
```
L R K V N E A F E I L K R R │T S S N P N
      HELIX 1
```
         850                 870                 890
CCAGCGCCTGCCGAAGGTTGAGATATTGCGCAATGCCATCGAGTATATCGAGAGCCTGGA
```
 Q R L P K V E │I L R N A I E Y I E S L E
      LOOP   HELIX 2
```
         910                 930                 950
GGATCTGCTACAGGAATCCAGTACCACACGCGATGGCGACAACCTGGCGCCCAGTTTGAG
```
 D L L Q E S S T T R D G D N L A P S L S
```
         970                 990                1010
CGGCAAAAGCTGCCAGTCCGATTATCTGAGCTCCTATGCTGGCGCTTATCTAGAAGATAA
 G  K  S  C  Q  S  D  Y  L  S  S  Y  A  G  A  Y  L  E  D  K
        1030                1050                1070
ACTTAGTTTTTACAACAAACATATGGAGAAATATGGTCAGTTTACAGACTTTGATGGCAA
 L  S  F  Y  N  K  H  M  E  K  Y  G  Q  F  T  D  F  D  G  N
        1090                1110                1130
TGCCAATGGCTCCAGTTTGGACTGTCTAAATCTGATTGTTCAGAGCATCAATAAGAGCAC
 A  N  G  S  S  L  D  C  L  N  L  I  V  Q  S  I  N  K  S  T
        1150                1170                1190
CACGAGTCCCATTCAAAATAAGGCCACGCCCTCCGCTTCAGATACCCAATCGCCGCCCTC
 T  S  P  I  Q  N  K  A  T  P  S  A  S  D  T  Q  S  P  P  S
        1210                1230                1250
ATCCGGAGCAACTGCACCCACTTCTCTGCACGTGAACTTCAAACGGAAGTGCAGCACTTA
 S  G  A  T  A  P  T  S  L  H  V  N  F  K  R  K  C  S  T  *
        1270                1290                1310
GCACTTAAGTATCAGCACCTTAGGCAATTGTAAAGCTATTTTTAAGAGGATACACGAGAT
        1330                1350                1370
ACCCAGTGACCCGAATAGGCCTTAAATTATTTGTATAGCATTAGAACTTAATTAAATGGT
        1390                1410
AATTCAAAACAGCAAAAAAAAAAAAAAAAA
```

Fig. 1. Nucleotide and predicted amino acid sequence of the longest ORF in the *Drosophila* Dmyd cDNA clone. The encoded polypeptide has 332 amino acids and a molecular weight of 36,208 daltons. The basic and helix-loop-helix regions are marked.

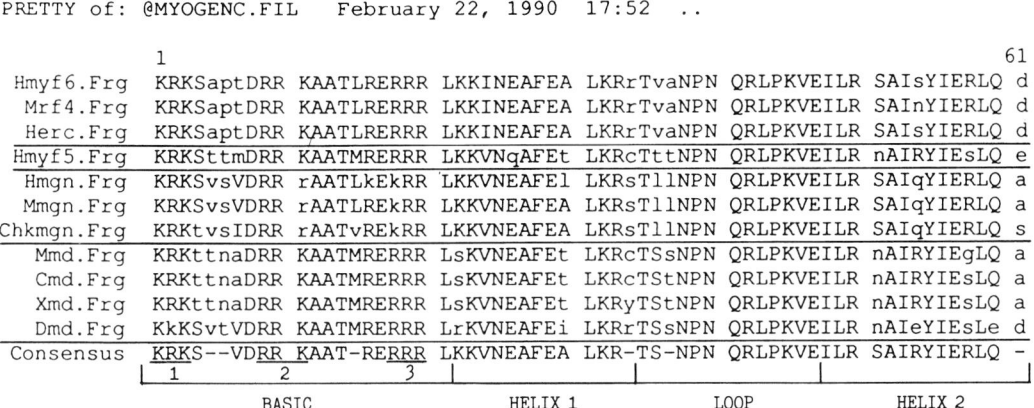

PRETTY of: @MYOGENC.FIL   February 22, 1990 17:52 ..

```
               1                                                            61
  Hmyf6.Frg   KRKSaptDRR KAATLRERRR LKKINEAFEA LKRrTvaNPN QRLPKVEILR SAIsYIERLQ d
  Mrf4.Frg    KRKSaptDRR KAATLRERRR LKKINEAFEA LKRrTvaNPN QRLPKVEILR SAInYIERLQ d
  Herc.Frg    KRKSaptDRR KAATLRERRR LKKINEAFEA LKRrTvaNPN QRLPKVEILR SAIsYIERLQ d
  Hmyf5.Frg   KRKSttmDRR KAATMRERRR LKKVNqAFEt LKRcTttNPN QRLPKVEILR nAIRYIEsLQ e
  Hmgn.Frg    KRKSvsVDRR rAATLkEkRR LKKVNEAFEl LKRsTllNPN QRLPKVEILR SAIqYIERLQ a
  Mmgn.Frg    KRKSvsVDRR rAATLREkRR LKKVNEAFEA LKRsTllNPN QRLPKVEILR SAIqYIERLQ a
  Chkmgn.Frg  KRKtvsIDRR rAATvREkRR LKKVNEAFEA LKRsTllNPN QRLPKVEILR SAIqYIERLQ s
  Mmd.Frg     KRKttnaDRR KAATMRERRR LsKVNEAFEt LKRcTSsNPN QRLPKVEILR nAIRYIEgLQ a
  Cmd.Frg     KRKttnaDRR KAATMRERRR LsKVNEAFEt LKRcTStNPN QRLPKVEILR nAIRYIEsLQ a
  Xmd.Frg     KRKttnaDRR KAATMRERRR LsKVNEAFEt LKRyTStNPN QRLPKVEILR nAIRYIEsLQ a
  Dmd.Frg     KkKSvtVDRR KAATMRERRR LrKVNEAFEi LKRrTSsNPN QRLPKVEILR nAIeYIEsLe d
  Consensus   KRKS--VDRR KAAT-RERRR LKKVNEAFEA LKR-TS-NPN QRLPKVEILR SAIRYIERLQ -
              |_____|_____|_____|_____|_____|
               1       2         3 |          |          |          |
                   BASIC               HELIX 1      LOOP         HELIX 2
```

Fig. 2. Dmyd is closely related to the MyoD group of myogenic factors. The sequence (single letter code) of the 61 amino acid core of the basic helix-loop-helix region from the four vertebrate myogenic gene families are compared to Dmyd. Lower case letters in each sequence represent a divergence from the consensus or the absence of a consensus. Upper case letters in a sequence that do not fit the consensus have equal probabilities. The similarity of the helix-loop-helix sequences and the invariance in the amino acid sequence of the 13 amino acids that include the basic 2 and 3 regions place Dmyd in the MyoD group. Sources for protein sequences are as follows: Myf6 (Braun et al 1990a), Mrf4 (Rhodes and Konieczny 1989), Herculin (Miner and Wold 1990), Myf5 (Braun et al 1989a), Myf4 (Hmgn) (Braun et al 1989), myogenin (Mmgn) (Wright et al 1989), Chicken myogenin (Chkmgn) (unpublished B.M.P.), mouse MyoD (Mmd) (Davis et al 1987), Chicken MyoD (Cmd) (Lin et al 1989), Xenopus MyoD (Xmd) (Hopwood et al 1989, Harvey 1990).

MyoD (Davis et al 1990). The spacing is conserved in Dmyd, underlining the importance of this configuration.

Dmyd is a unique gene and is not a member of a multigene family

Hybridization of the Dmyd cDNA insert to various restriction enzyme digests of *Drosophila* DNA indicates Dmyd is not a member of a multigene family since the same pattern of hybridization is seen under low and high stringency (figure 3 (Paterson et al 1991)). Hybridization of the same digests with the CMD1 PCR amplified basic helix-loop-helix domain gave the same major bands, however, the smaller *Bam*H1 fragment was the only hybridizing fragment in this digest (data not shown). In a complementary approach, degenerate primers with sequences representing the most strongly conserved regions of the basic and helix-loop-helix domains were used to PCR amplify *Drosophila* DNA. The only specific fragment obtained was a 950 bp piece corresponding to the basic region, an intron, and the helix-loop-helix region of Dmyd (data not shown). This again suggests there are few, if any, genes closely related to Dmyd in the *Drosophila* genome. The Dmyd

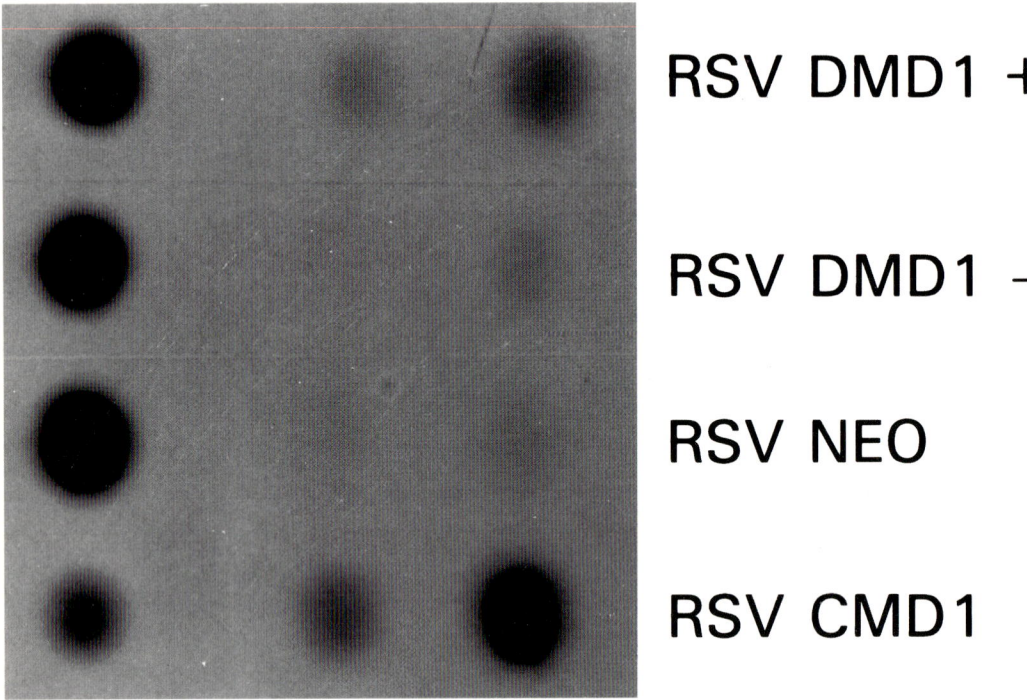

RSV DMD1 +

RSV DMD1 –

RSV NEO

RSV CMD1

Fig. 3. Activation of the chicken alpha cardiac actin promoter (Lin et al 1989) in response to cotransfected RSV CMD1 and RSV Dmyd (DMD1) in 10T1/2 mouse fibroblasts. Dmyd is at least 20-fold less active than CMD1.

cDNA was used to isolate the genomic clone containing the promoter region and this will be described elsewhere. Hybridization of the biotinylated genomic clone to polytene chromosomes mapped the gene to position 95A/B on the right arm of the third chromosome (data not shown).

Steady state levels of Dmyd RNA decrease during the period of larval myogenesis

Northern analysis of poly A^+ RNA from different developmental stages shows a single transcript of approximately 1.5kb when the Dmyd cDNA is used as a probe (figure 4 (Paterson et al 1991)). A weak signal is first seen at 3-6 hours of embryonic development using poly A^+ RNA and reaches a maximum level of expression around 9-12 hours. Thereafter, from approximately 12-24 hours of development, the signal is substantially reduced. By comparison, the 2.3kb transcript for the constitutively expressed gene, elongation factor1α (EF1α) (Walldorf et al 1985), is seen at all early stages. Analysis of total RNA throughout

Fig. 4. Dmyd (DMD1) complexes with E12 and binds poorly to the MCK enhancer compared to CMD1. E12, CMD1 and Dmyd were all synthesized in retic lysates and mixed appropriately for band shift assays as described previously (Davis et al 1990).

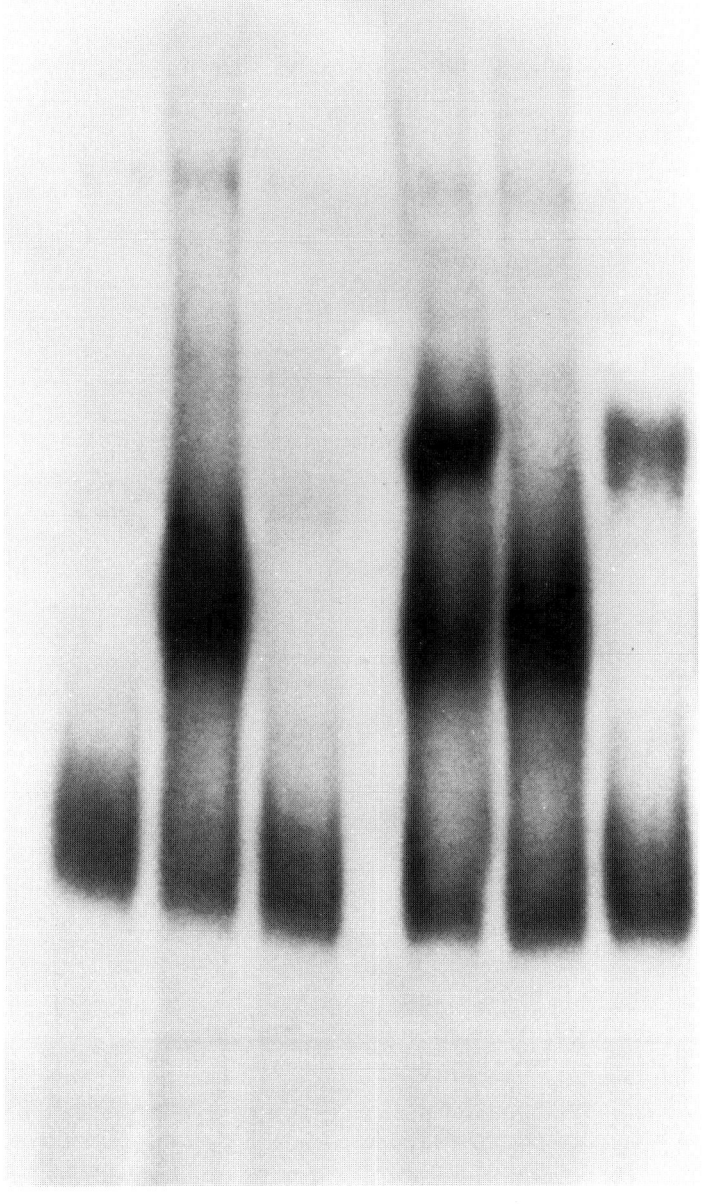

all major developmental stages, using both myosin heavy chain (MHC) and Dmyd probes, indicates Dmyd transcription preceeds that of MHC RNA and is again expressed coincident with MHC in the first through third instar larval stages and in late pupae. Expression is not detected for either transcript in early pupae or in adult flies. The MHC probe is derived from exon 17 (Collier et al 1990) and does not detect high levels of MHC RNA in adults (Paterson et al 1991). Larval muscle formation takes place between stages 13 (11.5 hours) and 16 (16 hours plus) (Leiss et al 1988) just as Dmyd RNA levels are decreasing.

The expression pattern of Dmyd during early embryogenesis

In order to establish the expression pattern of the Dmyd protein during development, we produced polyclonal antibodies against a glutathione transferase-Dmyd fusion protein synthesized in E. coli. Antibodies were specific for the fusion protein in total E. coli extracts and for Dmyd in extracts from transgenic flies expressing the protein under the hsp70 promoter (data not shown). Dmyd could not be detected in total extracts from any developmental stage by western blot analysis in the absence of heat shock induction. This is likely due to the small number of cells expressing the antigen at all stages (see below).

The earliest detection of Dmyd antigen is at stage11, 6 to 7 hours after fertilization (Fig. 5 (Paterson et al 1991)). This is a stage when the germ band is fully elongated, the tracheal pits have invaginated, and the proliferation of mesodermal cells is almost finished. Each of 14 segments, with the labium being the most anterior one, shows a cluster of stained nuclei on either side of the ventral midline and a smaller cluster more laterally. A third, small cluster of Dmyd-positive nuclei appears slightly later in a dorsal position in the segments from the labium through A7. During stage 11, the number of mesodermal nuclei containing Dmyd increases about 2 to 3 fold to a total of 30-40 per hemisegment, resulting in an additional ventral cluster and in a continuous band of nuclei in the dorso-lateral region of segments T1-A7. This pattern as well as the number of Dmyd-positive nuclei are basically maintained until completion of germ band retraction (stage 13, 10-11 h of development). However, differences in the number and pattern of stained nuclei between abdominal and more anterior segments become more obvious during this period. For example, the ventral clusters are reduced in the labium through T3 compared to A1-A7, and the dorso-lateral bands of labium and T1 appear to fuse to form a ring-shaped structure. As dorsal closure begins, the level of nuclear staining decreases rapidly. Only the nuclei of ventrally located syncitia in A1-A7, representing the precursors of muscles 26 and 27, and also some clusters in the head, continue to show high levels of Dmyd for and extended period. After dorsal closure, when the somatic muscles have differentiated, weak nuclear staining is still seen in several of these muscles: in the ventral external

Fig. 5. Dmyd is a nuclear antigen in mouse 10T1/2 fibroblasts. Cells were transfected with RSV Dmyd and stained with antibody to Dmyd (Paterson et al 1991). Even though there is good nuclear staining there is no myoblast conversion with or without cotransfection of MSV E12.

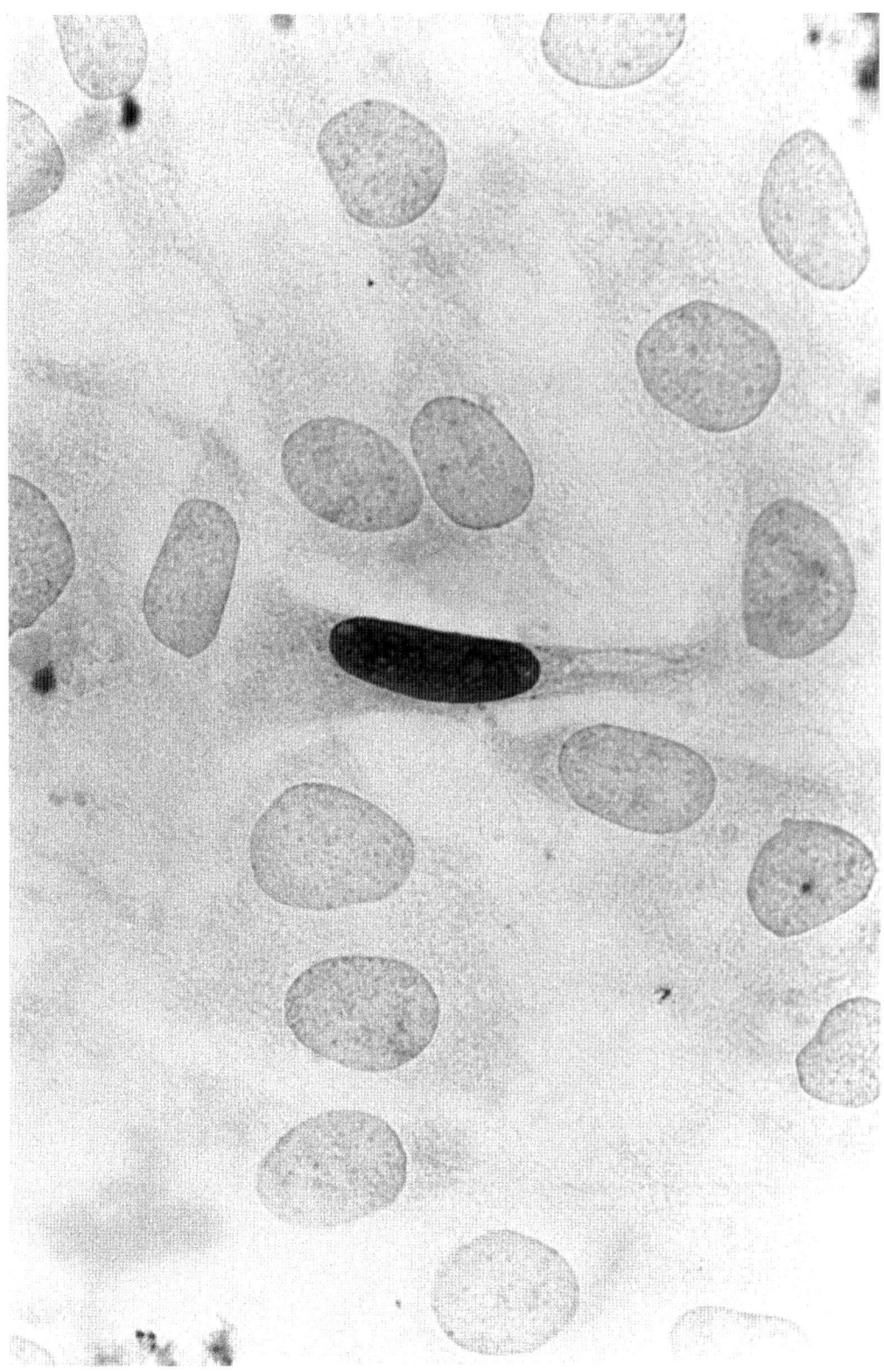

oblique muscles (muscles 15,16,17 (Crossley 1978, Hooper 1986), the ventro-lateral external oblique muscles (muscles 26,27) and the pleural external longitudinal muscle (muscle 12). Stronger staining persists in the pharyngeal muscles and in muscles of the anterior spiracles. These observations indicate that Dmyd is a marker for particular muscle precursor cells, where it is transiently expressed. Those precursor cells participate in the formation of a large number of muscle types in the somatic musculature of the larva. Furthermore, the decrease in the level of Dmyd detection with antibody is coincident with the decline in Dmyd RNA steady state levels mentioned above and the onset of larval myogenesis.

In our attempts to localize Dmyd antigen to mature muscle nuclei, we cultured cells from *Drosophila* 4-6 hour embryos. These cultures give rise to a variety of tissue types including larval muscle, nervous tissue, fatbody cells, hemocytes and larval epidermal cells. Cultures can be treated with ecdysone and will undergo metamorphosis, loosing embryonic muscles (Fig. 5 (Paterson et al 1991)), to give rise to adult tubular and fibrillar muscles. In this way three distinct muscle types can be analyzed for Dmyd antigen expression. In no instance was Dmyd detected in the nuclei of these muscle types grown in culture, yet antibody to horse radish peroxidase specifically stained all the neuronal cells (Jan and Jan 1982) (data not shown). We have not carried out a detailed study on the kinetics of in vitro myogenesis versus the pattern of Dmyd expression, so it is possible we have missed any early transient expression of Dmyd, however, it is clear Dmyd is not detectable in mature cultured *Drosophila* muscle.

It is important to note at this point that i) Dmyd antigen levels decrease at the onset of dorsal closure, coincident with the decrease in the steady state levels of Dmyd RNA and the onset of larval myogenesis, ii) the total number of Dmyd positive cells does not increase substantially enough to account for all myoblasts in the larval muscle and, iii) Dmyd positive cells are highly organized into specific abdominal and thoracic segment patterns, similar to cells that will give rise to the PNS (Ghysen and O'Kane 1989).

The developmental expression of a Dmyd-lac Z fusion protein under the control of the Dmyd promoter

Since we could not follow Dmyd expression as a conclusive marker for the terminal cell phenotype initially expressing the gene, we took a different approach. We used 8kb of 5′ genomic fragment that included the first 55 amino acid codons of Dmyd and fused it to lacZ to express a more stable cytoplasmic marker, β-galactosidase from E. coli, with the idea that enzyme activity and antigen could be followed into later developmental stages. This has worked succesfully with a variety of *Drosophila* genes (Hiromi et al 1985) and with various enhancer-trap *Drosophila* strains to follow developmentally regulated tissue specific genes (O'Kane and Gehring 1987). The Dmyd promoter lacZ construct was introduced into the P element vector Carnegie 20.1. Five independent transformants were characterized and two gave the same developmental patterns of β-galactosidase expression. The stronger staining line 14.1 (II) was chosen for detailed analysis. Both antibodies to Dmyd and β-galactosidase could be used to follow the activity of the Dmyd promoter. It should be noted that there is no nuclear localizing signal in this promoter construct so the staining pattern is now dependent upon the antibody used for detection; β-galactosidase antigen is cytoplasmic whereas Dmyd is both nuclear (endogenous) and cytoplasmic (fusion protein). Using antibodies to β-galactosidase, the fusion protein is first seen at full germ band extension in a pattern of expression similar to the Dmyd nuclear staining described above (Fig. 5 (Paterson et al 1991)). This expression continues to mimic the overall Dmyd nuclear pattern through full germ band retraction (data not shown) however the staining is more diffuse since the antigen is now located cytoplasmically. β-galactosidase activity is also clearly visible in whole mounts as a marker for the expression pattern of the fusion protein (data not shown) but the resolution is not as clear as with the antibodies to Dmyd or β-galactosidase.

When the Dmyd antibody is used for detection of the fusion protein the pattern is both nuclear and cytoplasmic in early stages, as explained above (data not shown). When germ band retraction is completed one can see the faint nuclear pattern in a background of cytoplasmically stained cells in each segment (data not shown). However, when nuclear staining is no longer detectable, cytoplasmic Dmyd staining now persists into later stages due to the stability of the fusion protein. Numerous muscle groups are clearly stained including the pharyngial muscles and muscles near the anterior spiracles. The patterns are identical in the horizontal views stained either with antibodies to myosin or Dmyd. The pattern is very reminscient of β-3 tubulin expression seen in somatic musculature and in the pharynx (Leiss et al 1988).

Based on our studies with the Dmyd promoter lacZ fusion, we conclude Dmyd is transiently expressed in a specific subset of highly organized muscle precursor cells and that one or a few of these precursors is associated with each of the muscle groups in a segment, as well as the muscles of the pharynx and the anterior spiracles.

The function of Dmyd in mouse 10T1/2 fibroblasts

Since Dmyd appeared to be virtually identical to the avian myogenic factor, CMD1, in the bHLH domain (Fig. 2), we decided to determine if Dmyd would behave in a fashion similar to CMD1 when expressed mouse 10T1/2 fibroblasts i.e. could Dmyd activate a muscle specific promoter in a cotransfection, could it convert 10T1/2 cells to muscle cells, is it a nuclear antigen, and can it form a heterodimer with members of the E2A family and bind to and E-box element? RSV-Dmyd only weakly activates a muscle specific promoter (Fig. 3), the alpha cardiac actin promoter, in cotransfections (20 fold less than CMD1), and Dmyd synthesized in retic lysates binds poorly to the MCK enhancer when combined with E12 (Fig. 4). There is no evidence for myoblast conversion with RSV Dmyd as judged by antibody staining with the monoclonal AB MF20 to skeletal muscle myosin (data not shown). Surprisingly, Dmyd is a good nuclear antigen in 10T1/2 cells as seen with antibody staining of transiently expressing cells (Fig. 5). So, in conclusion, it appears high levels of Dmyd expression in 10T1/2 cells will not convert these cells to muscle even though the essential portion of the basic domain is exactly conserved and there are only two major amino acid changes in the HLH region.

Dmyd and CMD1 chimeras reveal functional domains in CMD1

Since Dmyd was effectively a null functional mutant in 10T1/2 cells, we decided to construct chimeric proteins between CMD1 and Dmyd in order to identify conversion and activation domains in CMD1. The details of this study are to be presented elsewhere, but the following general conslusions emerge: 1) In conversion and activation assays the basic domains are equivalent between CMD1 and Dmyd, 2) the amino terminal portion of CMD1 5′ to the bHLH region contains the transcriptional activation domain, and 3) the carboxy terminal portions of Dmyd and CMD1 are essentially equivalent. The suprising result was the HLH domains are not equivalent since the Dmyd HLH will not work in a CMD1 background even though there are only minor differences in the amino acid sequences. These results have led to a model which suggests helix I and II on the same molecule interact through specific conserved hydrophobic residues in such a way as to favor the formation of conserved salt bridges which are thought to be critical in hetero and homo dimer formation in all the bHLH proteins.

Phosphorylation of the myogenic factors as a potential modulator of Dmyd/CMD1 DNA binding

The myogenic factors in vertebrates are phosphoproteins (Tapscott et al 1988). In order to study the biochemical role of this phosphorylation, we have used the Baculovirus system to overproduce phosphorylated CMD1 protein. The details of the study are to be presented elsewhere but the major results indicate Baculo-CMD1 complexes with E12 protein produced in E. coli and binds specifically to the MCK enhancer, as judged by the methylation interference pattern and binding competition assays (data not shown). Surprisingly, the binding affinity of CMD1 can be lowered by removal of the phosphate groups by treatment with calf intestinal phosphatase. Incubation of the dephosphorylated CMD1 in wheatgerm extracts rephosphorylates CMD1 and partially restores binding. At present, serine residues appear to be the sites of phosphorylation. We are in the process of mapping and sequencing the phosphorylated peptides.

Discussion

The Drosophila Dmyd gene is not a member of a multigene family

We have previously isolated the *Drosophila* myogenic determination gene (Dmyd) (Paterson et al 1991), a homolog to the myogenic determination factor genes of the vertebrates, due to its cross hybridization with the conserved basic helix-loop-helix domain of the avian myogenic factor CMD1. The amino acid sequence of this portion of the Dmyd polypeptide suggests it is closest to the vertebrate MyoD family since the essential basic region has been conserved exactly and homology is best with the helix-loop-helix region of the MyoD group of factors (figure 2). Remarkably, Dmyd appears to be a single copy gene with no closely related members similar to the four vertebrate myogenic factor genes. In the 207 nucleotides of the CMD1 basic helix-loop-helix domain used in the initial screen there are 38 nucleotide differences (18%) between *Drosophila* and chicken and 24 of these changes are in the third nucleotide position of the corresponding codon. The smaller genome size of *Drosophila* should favor the detection of related genes but none have been detected to date (unpublished observations). The Dmyd gene maps at 95A/B on the right arm of the third chromosome where there are currently no deficiencies or mutants known for this locus at the present time.

Dmyd is a transiently expressed nuclear antigen in a small number of highly organized muscle precursor cells

Unlike the vertebrate myogenic factor genes, which are expressed in both the dividing myoblast and in newly differentiated muscle fiber nuclei (Tapscott et al 1988, Lin et al 1989), Dmyd expression is transient and restricted to a small population of muscle precursor cells with characteristic thoracic and abdominal patterns. These Dmyd expressing cells cannot account for all the myoblasts required to produce the 25-30 distinct muscles in each segment of the developing embryo (Crossley 1978, Hooper 1986). Furthermore, Dmyd expression has never been detected in differentiated larval, tubular or fibrillar muscle fiber nuclei in vitro and is only weakly seen in some nascent myofibers in vivo, with the exception of the pharynx and anterior spiracles.

These observations are better understood when taken in the context of what is known about muscle formation in insects. The concept of the muscle pioneer or founder cell was first described in the developing grasshopper where one could identify large single cells of mesodermal origin attached to the ectoderm at sites that marked the insertion points of the future muscle fascicles (Ball et al 1985b). These pioneer cells serve as organizing centers for smaller mesodermal cells which collect nearby and subsequently fuse with the poineer cell to form a particular muscle. After their maturation, the muscles maintain the basic orientation and location of the original pioneer cells. It is thought that variations of this theme account for the organization and formation of all the muscles in the grasshopper.

More recent studies suggested that founder cells for individual muscles equivalent to the pioneer cells in grasshoppers also play a role during embryonic development of larval muscles in *Drosophila*. In a morphological study, Bate (Bate 1990) has concluded that each of the future larval muscles is prefigured by an appropriately located, syncitial precursor by the end of germ band shortening. These precursors seem to derive from doublets or triplets of fused mesodermal cells seen at even earlier stages which, in turn, are thought to be preceeded by individual founder cells. While in contrast to the grasshopper system, these putative founder cells cannot be recognized morphologically in *Drosophila*, the expression of the homeo box gene S59 (Dohrmann et al. 1990) in individual precursor cells of particular muscles also strongly argues for the existence of muscle founder cells in *Drosophila* embryos. Therefore, it is likely that mesodermal cells are divided into two classes: founder myoblasts that are committed to a particular fate and uncommitted, fusion competent cells.

The expression of Dmyd in defined mesodermal cells at a stage shortly before the occurrence of syncitial muscle precursors strongly suggests that those cells correspond to founder myoblasts. This idea is further supported by the fact that Dmyd expression starts at about the same time as that of S59, that Dmyd is expressed in a fixed number of cells per segment and in a highly organized, segment specific pattern, and by the observation that cells expressing the Dmyd-βGal fusion protein give rise to many of the larval muscles. While S59 is expressed only in a small subset of putative founder cells, Dmyd appears in a much larger number of cells. It remains to be shown if Dmyd expression defines the complete set of muscle founder cells or only a larger subset of them. At least in one precursor, that of muscle 27, the expression of Dmyd and S59 overlap (Dohrmann et al 1990). If our interpretation of Dmyd expressing cells being equivalent to muscle founder cells is correct, our results support the following model: Small numbers of pioneer cells are determined during germ band extension by upstream pattern formation genes. Segement polarity genes, homeotic genes, and genes exclusively expressed in the mesoderm like twist (Thisse et al 1987) and pox meso (Bopp et al 1989) are candidates for such determination genes. The expression of Dmyd in such cells may define their general identity as muscle founder cells, whereas the expression of homeo box genes like S59 in subsets of them may determine their particular developmental fates. These roughly 25-40 founder cells per hemisegment take up specific positions in relation to the ectoderm that will define particular embryonic muscles. Thus, the organization of the founder cells establishes the future muscle pattern in each segment. The founder cells recruit uncommitted mesodermal cells into a syncitium which then matures to establish the muscle in question. In most cases, Dmyd expression would cease or be reduced to very low levels just prior to or during syncitium formation. Our assumption that the recruited nascent myoblasts do not necessarily express Dmyd is in accordance with the observed absence or low level of this factor in differentiated muscle in vivo and in vitro.

Functional comparison between Dmyd and the vertebrate myogenic factors

It is not yet clear if Dmyd is a direct transcriptional activator of muscle specific genes in *Drosophila* as is thought to be the case for myogenic factors in vertebrates. while the similarities of the expression patterns of Dmyd and muscle specific tubulin would argue for such a possibility, the absence of Dmyd in most mature muscles suggests that different mechanisms are used for the maintenance of muscle specific gene expression.

All of the vertebrate myogenic factors, when introduced into the 10T1/2 mouse embryonic fibroblast, will convert these cells into stable muscle cells. Mouse

MyoD, expressed in a retroviral vector, has been shown to terminally convert primary dermal fibroblasts, chondroblasts, smooth muscle and retinal pigmented epithelial cells into striated muscle and can induce a variety of immortalized or transformed cell lines to express muscle genes (Choi et al 1990). In addition, over expression of CMD1 and MyoD in dividing cells or tumorgenic cells, respectively, appears to remove these cells from the dividing cell population (Lin et al 1989, Sorrentino et al 1990). Using various mutations of the MyoD protein these functions have been correlated with the conserved helix-loop-helix region. It was initially reported the complete basic and helix-loop-helix was sufficient in itself to convert 10T1/2 cells to muscle with nearly the same efficiency as the complete protein (Tapscott et al 1988). However, it has been shown recently using Gal4-Myf5 hybrid proteins that regions outside the basic helix-loop-helix domain are important for transcriptional activation (Braun et al 1990b). This makes it difficult to understand how the basic helix-loop-helix motif alone can convert cells in the absence of a transcriptional activation domain. One would have to assume the cellular partner for MyoD, another helix-loop-helix protein such as the E2A gene products E12 and E47 (Murre et al 1989), provides enough of a transcriptional activation domain to compensate for the deletion.

A comparison of the amino acid sequence homology of the Dmyd basic helix-loop-helix region with the MyoD family of vertebrate myogenic factors (figure 2) suggests Dmyd should behave similarly to MyoD in many of its functions. In our preliminary studies with Dmyd in 10T1/2 cells we have observed the following: Dmyd will activate weakly a vertebrate muscle specific promoter joined to the CAT reporter gene in cotransfection assays in 10T1/2 cells but the activation is 20 fold less compared to CMD1; Dmyd synthesized in reticulocyte lysates will also form a complex with E12 and bind specifically to the MCK enhancer sequence; Dmyd is a nuclear antigen in 10T1/2 cells and is clearly detected with antibodies. Surprisingly, Dmyd will not convert 10T1/2 cells to muscle, even when cotransfected with an expression vector that provides excess levels of E12 (to be presented elsewhere). At least in the case of Dmyd, and possibly the vertebrate myogenic factors, regions outside the basic helix-loop-helix domain are not only important for promoter activation (Braun et al 1990b) but play a role in conversion as well. The presence of the conserved MyoD-like domain is not sufficient to support 10T1/2 cell conversion and only weakly activates a muscle-specific promoter. In our hands, expression of just the basic helix-loop-helix region of Dmyd, alone or in combination with E12, does not convert 10T1/2 cells. In agreement with the original report (Tapscott et al 1988), we are not able to detect the presence of the basic helix-loop-helix polypeptide with polyclonal antibodies to Dmyd. However, nuclear staining is clear with Dmyd antibody when the entire protein is expressed in 10T1/2 cells yet conversion does not take place. Preliminary structural studies and modeling suggest a possible role for conserved charged amino acids in the HLH domain in the regulation of hetero and homodimerization. This would certainly regulate Dmyd function if dimerization with the vertebrate cellular partner was not as stable as with CMD1. It is not known if

Dmyd is phosphorylated. Our preliminary data would suggest this modification plays a role in modulating the binding affinity of the myogenic factors to their target sites, possibly through the regulation of dimerization, DNA binding or both. Failure to phosphorylate Dmyd in 10T1/2 cells may explain it lack of normal function.

Evolutionary considerations

The results presented here point to at least two distinct developmental pathways that have evolved to give rise to striated muscle. In the vertebrates early mesodermal cells in the somites become committed to particular cell lineages that will eventually give rise to essentially all of the muscle tissue in the organism. This lineage selection or determination is thought to involve the activation of one or more of the myogenic factor genes by earlier regulatory genes. The pattern of expression of the various myogenic factors suggests a temporal relationship that could be important in the regulation of the entire myogenic program, possibly in isoform switching, but this has not been established at present.

In grasshoppers, and likely in *Drosophila*, muscle formation does not depend upon the early establishment of mesodermal cell lineages where all cells contribute equally to the various muscle groups they form. Rather, the insects use a recruitment mechanism whereby a small number of muscle founder or pioneer cells establish the location and organization of a muscle through positional cues in the ectoderm and then proceed to recruit uncommitted mesodermal cells into a syncitium that gives rise to the muscle in question. Vertebrate evolution selected the lineage mechanism wheras arthropods evolved the recruitment scheme, possibly as a solution to the problems involved in forming the complex pattern repitition in the insect body plan. It has been shown that mutations in the homeotic gene Ubx alter the muscle pattern in the first two abdominal segments towards metathorax (Hooper 1986), suggesting a close evolutionary relationship between genes that establish segment identity and those genes involved in muscle formation.

Recently, a similar myogenic factor gene has been identified in *C. elegans* (Krause et al 1990), called CeMyoD, that also has conserved exactly the amino acid sequence containing basic clusters 2 and 3 (figure 2) but has 13 amino acid changes in the helix-loop-helix region when compared to MyoD. The gene appears to be unique by Southern blot analysis with only minor bands appearing under low stringency hybridization conditions. However, unlike Dmyd, CeMyoD continues to be expressed in the nuclei of the body wall muscles and appears restricted to these muscles. The muscles of the pharynx do not express CeMyoD whereas Dmyd is expressed in the *Drosophila* pharyngial muscles. It was not reported if CeMyoD functions at all in 10T1/2 cells.

Dmyd as a marker for muscle precursor cells will allow the idendification of upstream and downstream genes important in the development of *Drosophila* muscle.

We would like to thank Urs Kloter for carrying out the chromosomal in situ hybridizations and Dan Kiehart of Harvard University for the gift of the *Drosophila* muscle myosin heavy chain antibodies. This work was supported in part by the Laboratory of Biochemistry NCI-NIH while B.M.P was on sabbatical, by the Roche Research Foundation (grant No.), the Kantons Basel and the Swiss National Science Foundation, etc.

References

Ball, E. E. and Goodman, C. S. (1985a). Muscle development in the Grasshoper embryo. II Syncitial origin of the extensor tibiae muscle pioneers. *Dev. Biol.* **111**, 399-416.

Ball, E. E. and Goodman, C. S. (1985b). Muscle development in the grasshopper embryo.III Sequential origin of the flexor tibiae muscle pioneers. *Dev. Biol.* **111**, 417-424.

Ball, E. E., Ho, R. K. and Goodman, C. S. (1985a). Muscle development in the grasshopper embryo. I Muscles, nerves, and apodemes in the metathoracic leg. *Dev. Biol.* **111**, 383-393.

Ball, E. E., Ho, R. K. and Goodman, C. S. (1985b). Development of neuromuscular specificity in the grasshopper embryo: Guidance of motoneuron growth cones by muscle pioneers. *J. Neurosci.* **5**, 1808-1819.

Bate, M. (1990). The embryonic development of larval muscles in *Drosophila*. *Development* **110**, 791-804.

Bopp, D., Jamet, E., Baumgartner, S., Burri, M. and Noll, M. (1989). Isolation of two tissue-specific *Drosophila* paired box genes, Pox meso and Pox neuro *EMBO J.* **8**, 3447-3457.

Braun, T., Bober, E., Buschhausen-Denker, G., Kotz, S., Grzeschik, K.-H. and Arnold, H. H. (1989a). Differential expression of myogenic determination genes in muscle cells: possible autoactivation by Myf gene products. *EMBO J.* **8**, 3617-3625.

Braun, T., Buschhausen-Denker, G., Bober, E. and Arnold, H. H. (1989b). A novel human muscle factor related to but distinct from MyoD1 induces myogenic conversion in 10T1/2 fibroblasts. *EMBO J.* **8**, 701-709.

Braun, T., Bober, E., Winter, B., Rosenthal, N. and Arnold, H. H. (1990a). Myf-6, a new member of the human gene family of myogenic determination factors: evidence for a gene cluster on chromosome 12. *EMBO J.* **9**, 821-831.

Braun, T., Winter, B., Bober, E. and Arnold, H. H. (1990b). Transcriptional activation domain of the muscle-specific gene regulatory protein myf5. *Nature* **346**, 663-665.

Cavener, D. R. (1987). Comparison of the consensus sequence flanking translational start sites in *Drosophila* and vertebrates. *Nucleic Acids Res.* **15**, 1353-1361.

Choi, J., Costa, M. L., Mermelstein, C. S., Chagas, C., Holtzer, S. and Holtzer, H. (1990). MyoD converts primary dermal fibroblasts, chondroblasts, smooth muscle, and retinal pigmented epithelial cells into striated mononucleated myoblasts and multinucleated myotubes. *Proc. Nat. Acad. Sci. USA* **87**, 7988-7992.

Chomczynski, P. and Sacchi, N. (1987). Single-step method of RNA isolation by acid guanidium thiocyanate-phenol-chloroform extraction. *Anal. Biochem.* **162**, 156-159.

Clos, J., Westwood, T. J., Becker, P. B., Wilson, S., Lambert, K. and Wu, C. (1990). Molecular cloning and expression of a hexameric *Drosophila* heat shock factor subject to negative regulation. *Cell* **63**, 1085-1097.

Collier, V. L., Kronert, W. A., O'Donnell, P. T., Edwards, K. A. and Bernstein, S. I. (1990). Alternative myosin hinge regions are utilized in a tissue-specific fashion that correlates with muscle contraction speed. *Genes Dev.* **4**, 885-895.

Crossley, A. C. (1978). In The Genetics and Biology of *Drosophila* 2b (eds. M. Ashburner and T.R.F. Wright). 499-560. Academic Press.

Davis, R. L., Cheng, P.-F., Lassar, A. B. and Weintraub, H. (1990). The MyoD DNA binding domain contains a recognition code for muscle-specific gene activation. *Cell*, **60**, 733-746.

Davis, R. L., Weintraub, H. and Lassar, A. B. (1987). Expression of a single transfected cDNA converts fibroblasts to myoblasts. *Cell* **51**, 987-1000.

Dohrmann, C., Azpiazu, N. and Frasch, M. (1990). A new *Drosophila* homeo box gene is expressed in mesodermal precursor cells of distinct muscles during embryogenesis. *Genes Dev.* **4**, 2098-2111.

Dübendorfer, A., Blumer, A. and Deak, I. I. (1978). Differentiation in vitro of larval and adult muscles from embryonic cells of *Drosophila* Wilhelm Roux Arch. *devl Biol.* **184**, 233-249.

Frasch, M., Hoey, T., Rushlow, C., Doyle, H. and Levine, M. (1987). Characterization and localization of the *even-skipped* protein of *Drosophila. EMBO J.* **6**, 749-759.

Ghysen, A. and O'Kane, C. J. (1989). Neural enhancer-like elements as specific cell markers in *Drosophila. Development* **105**, 35-52.

Harvey, R. P. (1990). The Xenopus MyoD gene: an unlocalised maternal mRNA predates lineage-restricted expression in the early embryo. *Development* **108**, 669-680.

Hiromi, Y., Kuroiwa, A. and Gehring, W. J. (1985). Control elements of the segmentation gene fushi tarazu. *Cell* **43**, 603-613.

Ho, R. K., Ball, E. E. and Goodman, C. S. (1983). *Nature* **301**, 66-69.

Hooper, J. E. (1986). Homeotic gene function in the muscles of *Drosophila* larvae. *EMBO J.* **5**, 2321-2329.

Hopwood, N. D., Pluck, A. and Gurdon, J. B. (1989). MyoD expression in the forming somites is an early response to mesoderm induction in Xenopus embryos. *EMBO J.* **8**, 3409-3417.

Ingham, P. W. (1988). The molecular genetics of embryonic pattern formation in *Drosophila. Nature* **335**, 25-34.

Jan, L. Y. and Jan, N. J. (1982). Antibodies to horseradish peroxidase as specific neuronal markers in *Drosophila* and in grasshopper embryos. *Proc.Nat. Acad. Sci. USA* **79**, 2700-2704.

Krause, M., Fire, A., White Harrison, S., Priess, J. and Weintraub, H. (1990). CeMyoD accumulation defines the body wall muscle cell fate during C. elegans embryogenesis. *Cell* **83**, 907-919.

Lasko, P. F. and Ashburner, M. (1990). Posterior localization of Vasa protein correlates with, but is not sufficient for, pole cell development. *Genes Dev.* **4**, 905-921.

Leiss, D., Hinz, U., Gasch, A., Mertz, R. and Renkawitz-pohl, R. (1988). beta 3 tubulin expression characterizes the differentiating mesodermal germ layer during *Drosophila* embryogenesis. *Development* **104**, 525-531.

Lin, Z.-Y., Dechesne, C. A., Eldridge, J. and Paterson, B. M. (1989). An avian muscle factor related to MyoD1 activates muscle-specific promoters in nonmuscle cells of different germ layer origin and in BrdU-ttreated myoblasts. *Genes Dev.* **3**, 986-996.

McGinnis, W., Levin, M., Hafen, E., Kuroiwa, A. and Gehring, W. J. (1984). A conserved DNA sequence found in homeotic genes of *Drosophila* Antennapedia and bithorax complexes. *Nature* **308**, 428-433.

Miner, J. H. and Wold, B. (1990). Herculin, a fourth member of the MyoD family of myogenic regulatory genes. *Proc. Nat. Acad. Sci. USA* **87**, 1089-1093.

Murre, C., Schonleber-McCaw, P. and Baltimore, D. (1989). A new DNA binding and dimerization motif in immunoglobulin enhancer binding, *daughterless*, MyoD, and *myc* proteins. *Cell* **56**, 777-783.

O'Kane, C. J. and Gehring, W. J. (1987). Detection *in situ* of genomic regulatory elements in *Drosophila. Proc. Nat. Acad. Sci. USA* **84**, 9123-9127.

Pardue, M. L. (1986). in *Drosophila*, a practical approach, (Ed. D.B. Roberts). In situ hybridization to DNA of chromosomes and nuclei. 111-138. IRL Press.

Paterson, B. M., Walldorf, U., Eldridge, J., Dubendorfer, A., Frasch, M. and Gehring, W. J. (1991). The *Drosophila* homologue of vertebrate myogenic-determination genes encodes a transiently expressed nuclear protein marking primary myogenic cells. *Proc. Nat. Acad. Sci. USA* **88**, 3782-3786.

Poole, S., Kauvar, L. M., Drees, B. and Kornberg, T. (1985). *Cell* **40**, 37-43.

Rhodes, S. J. and Konieczny, S. F. (1989). Identification of MRF4: a new member of the muscle fregulatory factor gene family. *Genes Dev.* **3**, 2050-2061.

Saiki, R. K., Gelfand, D. H., Stoffel, S., Scharf, S. J., Higuchi, R., Horn, G. T., Mullis, K. B. and Erlich, H. A. (1988). Primer-directed enzymatic amplification of DNA with a thermostable DNA polymerase. *Science* **239**, 487-491.

Smith, D. B. and Johnson, K. S. (1988). Single-step purification of polypeptides expressed in E. coli as fusions with glutathione S-transferase. *Gene* **67**, 31-40.

Sorrentino, V., Pepperkok, R., Davis, R. L., Ansorge, W. and Phillipson, L. (1990). Cell proliferation inhibited by MyoD1 independently of myogenic differentiation. *Nature* **345**, 813-815.

Tapscott, S. J., Davis, R. L., Thayer, M. J., Cheng, P.-F., Weintraub, H. and Lassar, A. B. (1988). MyoD1: a nuclear phosphoprotein requiring a Myc homology region to convert fibroblasts to myoblasts. *Science* **242**, 405-411.

Thisse, B., El Messal, M. and Perrin-Schmitt, F. (1987). The twist gene: isolation of a *Drosophila* zygotic gene necessary for establishment of dorso-ventral pattern. *Nucleic Acids Res.* **15**, 3439-3453.

Walldorf, U., Hovemann, B. and Bautz, E. K. F. (1985). *Proc. Nat. Acad. Sci. USA* **82**, 5795-5799.

Wright, E. W., Sassoon, D. A. and Lin, V. K. (1989). Myogenin, a factor regulating myogenesis, has a domain homologous to MyoD. *Cell* **56**, 607-617.

Printed in Great Britain © *Society for Experimental Biology 1992*   111

DROSOPHILA ACTIN MUTANTS AND THE STUDY OF MYOFIBRILLAR ASSEMBLY AND FUNCTION

SPARROW, J. C., DRUMMOND, D. R., HENNESSEY, E. S., CLAYTON, J. D. and LINDEGAARD, F. B.

Department of Biology, University of York, York, YO1 5DD, U.K.

Abstract

The use of *Drosophila* mutations in the indirect flight muscle-specific actin gene, *Act88F*, to study actin structure/function and its assembly into thin filaments during myofibrillogenesis is described. Mutants with different phenotypic effects are discussed and attempts made to correlate the different properties of the mutants *in vivo* - myofibrillar structure, actin synthesis, accumulation and stability, heat shock response induction - with properties of the same mutations expressed by *in vitro* transcription/translation of the cloned actin genes - co-polymerisation, thermostability and protein conformation. Few of the properties show a complete correlation between the different classes of mutants. The nature of the diversity of the mutant effects is discussed. Questions as to how this will help in elucidating the molecular effects of the mutations and the assembly of thin filaments and myofibrils are considered. In addition, the efficacy of the co-polymerisation assay is examined.

The post-translational processing of this actin - by N-terminal processing, methylation and ubiquitination - are described. Data is presented that inhibition of the N-terminal processing of actin *in vitro* affects the ability of the actin to co-polymerise, and makes unprocessed actin behave as a capping protein. The possible *in vivo* importance of this phenomenon is discussed.

Introduction

The major proteins present in the myofibrils of striated muscle, their position within the sarcomere and their roles in producing or regulating tension development are well characterised. Despite this, the dynamics of myofibril assembly and the molecular details of how actin and myosin develop force remain largely unknown. To understand these processes we need information about the molecular structures of the proteins and of the dynamic interactions between the different molecules. These interactions are clearly complex and difficult to follow *in vivo*. However, genetical techniques can provide a powerful means by which to analyse these processes. Mutations introduce changes whose effects enable the molecular dissection of protein interactions by focussing on specific proteins or regions of proteins.

The genes for most of the major muscle proteins of *Drosophila* have been cloned

Key words: Drosophila, actin, muscle.

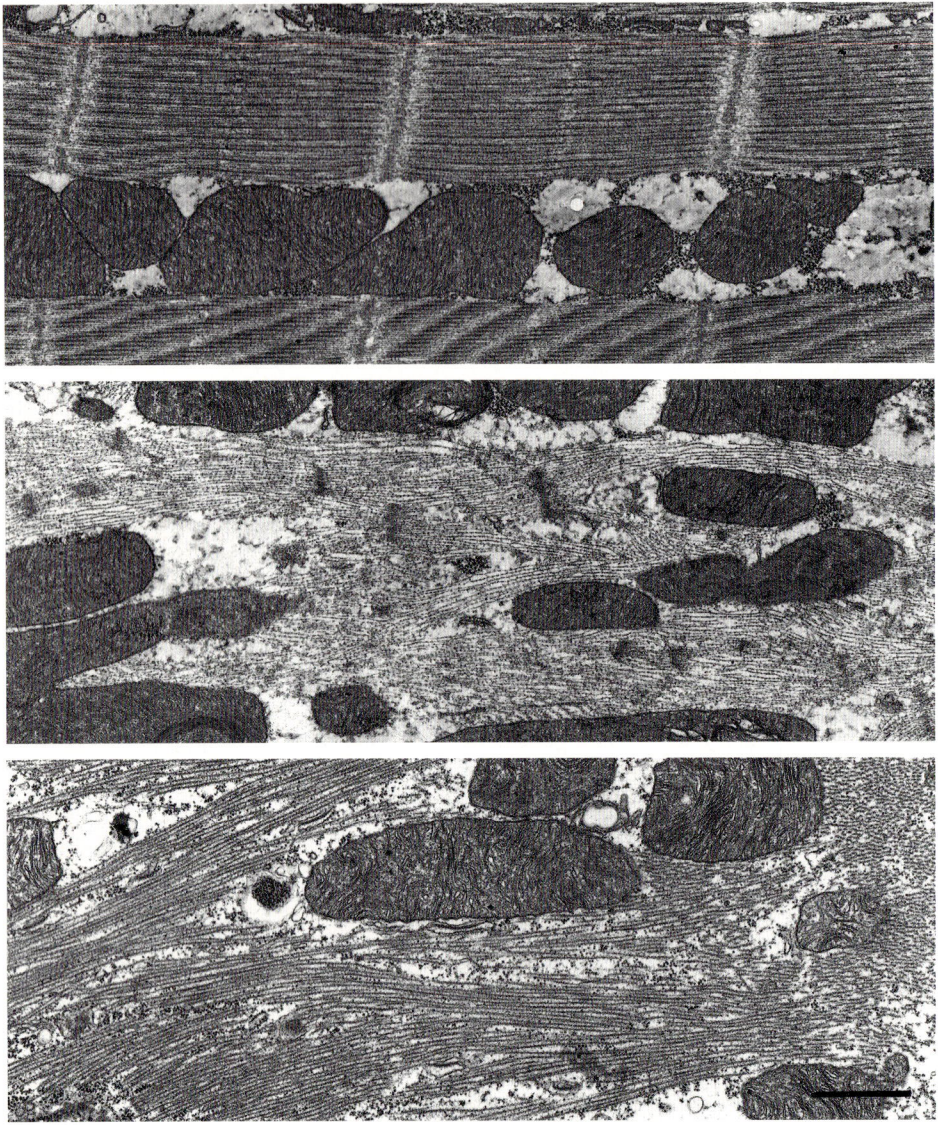

Fig. 1. Electron micrographs of indirect flight muscle myofibrils to show the effects of null mutations; (a) wild type (b), *Ifm(2)2* (c) and *KM88* (c). Bar=1 μm.

and sequenced and flightless mutations have been recovered in most of them. Most of the mutations produce structurally aberrant flight muscles.

The effects of null mutations for the single myosin heavy chain gene, *Mhc*, and the indirect flight muscle specific actin gene, *Act88F*, are particularly interesting in the analysis of myofibril assembly. Flies homozygous for *KM88*, a null mutation of the *Act88F* gene, lack actin in their indirect flight muscles (IFMs) (Hiromi and Hotta, 1985) but contain assemblies of bipolar thick filaments (see Fig. 1b) with discernible M-bands. However, thin filaments and Z-discs are absent (Beall,

Sepanski and Fyrberg, 1989). From the absence of Z-discs it can be argued that actin is required for Z-disc assembly or stability (Fyrberg, 1990). This is consistent with actin being a major Z-disc protein in insects (Bullard and Sainsbury, 1977). Null mutations of the single muscle myosin heavy chain gene, *Mhc*, are usually lethal but a small number of viable mutants, e.g. *Ifm(2)2* (now known as *Mhc⁷*), prevent synthesis of myosin in the IFMs. The IFMs of *Ifm(2)2* (Fig. 1c) lack thick filaments but do contain thin filament assemblies and transverse Z-disc-like structures are formed (Chun and Falkenthal, 1988; O'Donnell, Collier, Mogami and Bernstein, 1989; Beall et al., 1989). These observations suggest that the assembly of thick and thin filaments are relatively independent processes. Interactions between these filament systems are required to produce the very regular structure and periodicity of the sarcomere (Fyrberg, 1990; Epstein and Fischman, 1991). This is apparent, for example, in the absence of any periodicity in the spacing of the Z-disc-like structures in *Ifm(2)2* fibres.

The *Ifm(2)2* myosin null mutant has no effect on fibre morphology (Fig. 2). In contrast, the effects of *KM88* are dramatic; the normally cylindrical fibres assume a series of contorted shapes. Close examination reveals that structures which normally lie on the fibre surface, such as tracheae and nerves, become surrounded by the fibre body. Similar results are obtained with other *Act88F* mutants e.g. *raised*, (*rsd*) and *A138V* (Sparrow, unpublished). Clearly the presence of thin filaments alone generates normal fibre morphology but thick filaments on their own produce large changes in fibre shape. An explanation of this phenomenon is not obvious but may shed light on fibre development. These observations demonstrate the serendipitous results that are sometimes obtained when using mutants!

While all these observations demonstrate the power of mutants in analysing myofibril assembly, the use of null mutations is a limited and rather blunt tool. The next phase must be to refine this analysis by using more subtle mutations which affect specific protein interactions. This requires the identification of suitable mutants and an understanding of the effects of each mutation on the protein product. Since the relevant macromolecular interactions are likely to be too rapid, subtle or complex to be adequately analysed *in vivo*, it is inevitable that we will have to study many properties of the mutant proteins *in vitro*.

Our main interest is to obtain mutants which affect muscle contraction to study the molecular details of the actomyosin motor (Sparrow, Drummond, Peckham, Hennessey and White, 1991a). For this we have focussed on mutants of the *Act88F* gene to probe the relationships between actin structure and function. Actin is a highly conserved protein. We, like others, have found that the majority of actin mutants affect the ability of the myofibrils to assemble. It is with this aspect that we are concerned in this chapter.

Many mutations in the *Act88F* gene are known. They were first isolated from among flightless mutants induced by chemical mutagenesis of whole flies (Mogami and Hotta, 1981; Ball, 1987). More recently, *in vitro* mutagenesis of the cloned *Act88F* gene, followed by P-element transformation of the *KM88* strain, has been

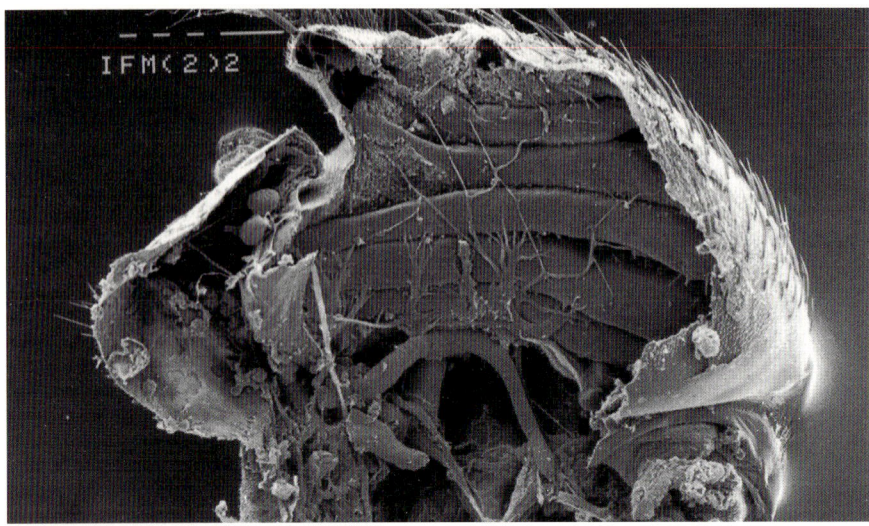

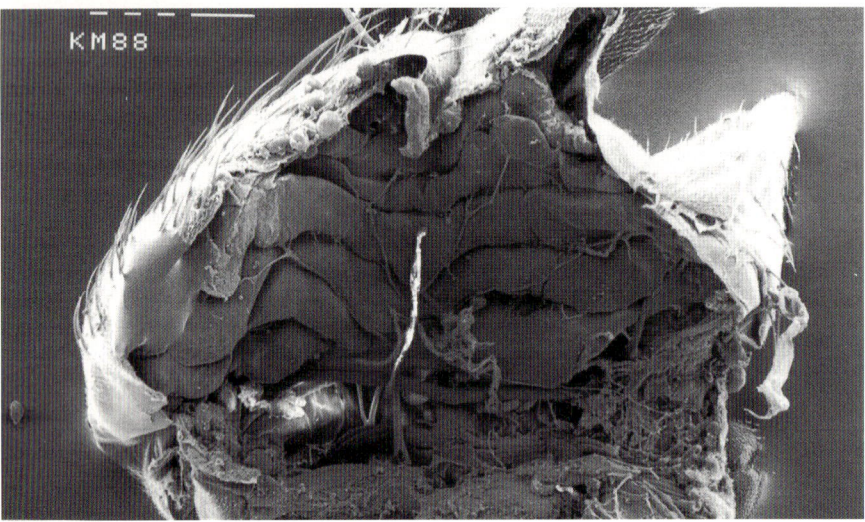

Fig. 2. Scanning electron micrographs of half thoraces to show the effects of null mutations on dorsolongitudinal indirect flight muscle (fibre) morphology; (a) *Ifm(2)2*, (b) *KM88*.

used to make mutants (Reedy, Beall and Fyrberg, 1989; Hiromi, Okamoto, Gehring and Hotta, 1986; Okamoto, Hiromi, Ishikawa, Yamada, Isoda, Maekawa and Hotta, 1986; Sakai, Okamoto, Mogami, Yamada and Hotta, 1990; Drummond, Hennessey and Sparrow, 1991a). Since the *Act88F* gene is expressed specifically in the IFMs of the fly (Fyrberg, Mahaffey, Bond and Davidson, 1983; Sanchez, Tobin, Rdest, Zulauf and McCarthy, 1983; Hiromi and Hotta, 1985) and is the only actin gene expressed in these muscles (Ball, Karlik, Beall, Saville, Sparrow, Bullard and Fyrberg, 1987) such mutations are both viable and fertile. With these techniques specific amino acid changes can be made in the protein and studied *in vivo*. Different *Act88F* mutants show a wide range of effects. For the purposes of discussion we have grouped them into five different groups based on their phenotypic effects. We discuss below specific examples of each class. These are taken largely from our own work.

1. Null mutations

KM88 is the only null mutation of the *Act88F* gene; it causes a termination codon at aminoacid position 79 (Hiromi and Hotta, 1985). The effects of this mutant on myofibrillar structure have been described above.

SDS-PAGE separation of myofibrillar proteins from *KM88* IFMs shows the absence, not only of actin but of all the other thin filament proteins as well. Radiolabelling of newly emerged flies shows that these proteins are synthesised but they do not accumulate sufficiently to be seen on Coomassie stained gels. This suggests that the stability of these proteins depends on their being assembled into thin filaments. In addition, it provides evidence that degradative processes are active during myogenesis and can recognise non-mutant proteins which have not assembled into myofilaments. These processes may be as important as the regulation of gene expression in determining the strict protein stoichiometries which must accompany the very regular structure of these myofibrils.

2. Mutants with low levels of actin accumulation

Very low levels of actin accumulate in the IFMs of the *raised*, (Mahaffey, Coutu, Fyrberg and Inwood, 1985), *V339I* (Drummond et al., 1991a) and *A138V* (Ball, 1987; Clayton, Lindegaard and Sparrow, unpublished) mutants. None produce myofibrils. Electron micrographs show that the IFMs of these mutants are similar to those of *KM88*, except that in each case Z-disc like structures are seen (Fig. 3; *rsd*, Mahaffey et al., 1985). Undoubtedly this difference can be explained by the small amounts of actin in these mutants. Whether the actin is preferentially assembled into these Z-disc-like structures is not known.

The *rsd* mutation causes a reduction in *Act88F* mRNA levels (Mahaffey et al., 1985) which may account for the low actin levels. In both *A138V* flies (Lindegaard, unpublished) and *V339I* (Drummond et al., 1991a) the ratio of ^{35}S-methionine incorporation into myosin heavy chain and actin is much higher than the myosin:actin ratio of accumulated protein from Coomassie stained gels of IFMs of 4-5 day old flies (see Table 1). However, the instability, at least for *V339I*, is not a simple reduction in the thermodynamic instability of the protein. In gels with a transverse urea gradient (Creighton, 1979), *V339I* actin is slightly less stable at lower urea concentrations than wild-type actin (Drummond, Hennessey and Sparrow, 1991b). This contrasts with its poor *in vivo* stability. Another mutant, E316K, has a greater reduction in thermodynamic stability yet is stable *in vivo*.

The phenotype of these mutants cannot be entirely explained by protein instability. It is possible, using P-element transformants to produce flies containing

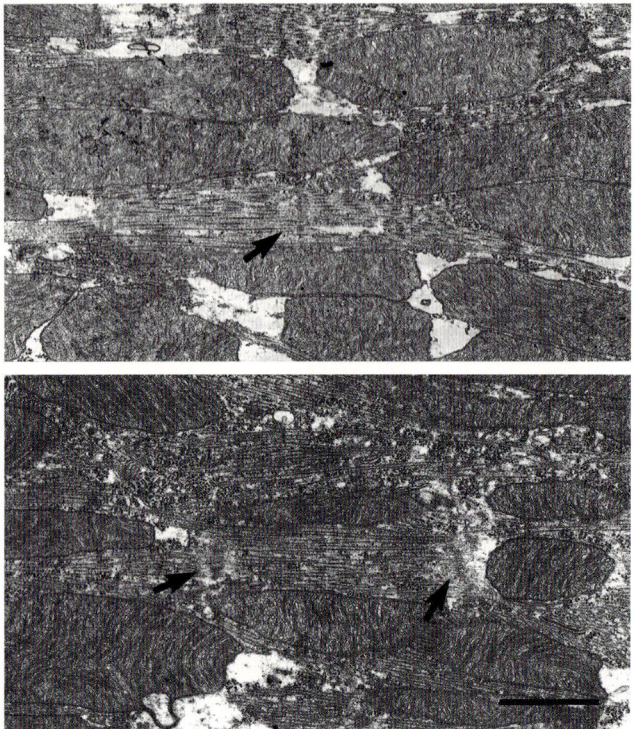

Fig. 3. Electron micrographs of mutant indirect flight myofibrils (a) *V339I* and (b) *A138V*. Arrows indicate Z-disc-like structures. Bar=1 µm.

3 or 4 wild-type copies of the *Act88F* gene. Such flies are flighted (Hiromi and Hotta, 1985; Clayton, unpublished). However, in flies with two wild-type and one or two mutant copies of *V339I* or *A138V* the flight ability is reduced. This type of effect is known as antimorphism. It indicates that the mutant actins are interfering with the assembly or function of the myofibrils despite the presence of wild-type actin monomers and their mutant effects are not simply due to reduced IFM actin content (note the flight of *KM88* in the presence of two wild-type copies, Fig. 4). It is not clear if this is also true for *rsd*. P-element transformation of this mutant with wild-type *Act88F* copies only partially rescued flight but no change in the coding region of the *rsd* gene has been detected (Mahaffey et al., 1985).

3. Mutants which accumulate actin but with no myofibrils

Light microscopy shows that *G366D* and *E364K* do not form myofibrils. The fibres of *E364K* and *G366D* have higher myosin:actin accumulation ratios than normal indicating a reduction in actin stability or synthesis (Table 1). Since the myosin:actin incorporation ratios for *G366D* and *E364K* are essentially the same as for *G368E*, which forms normal myofibrils, we interpret this as evidence that

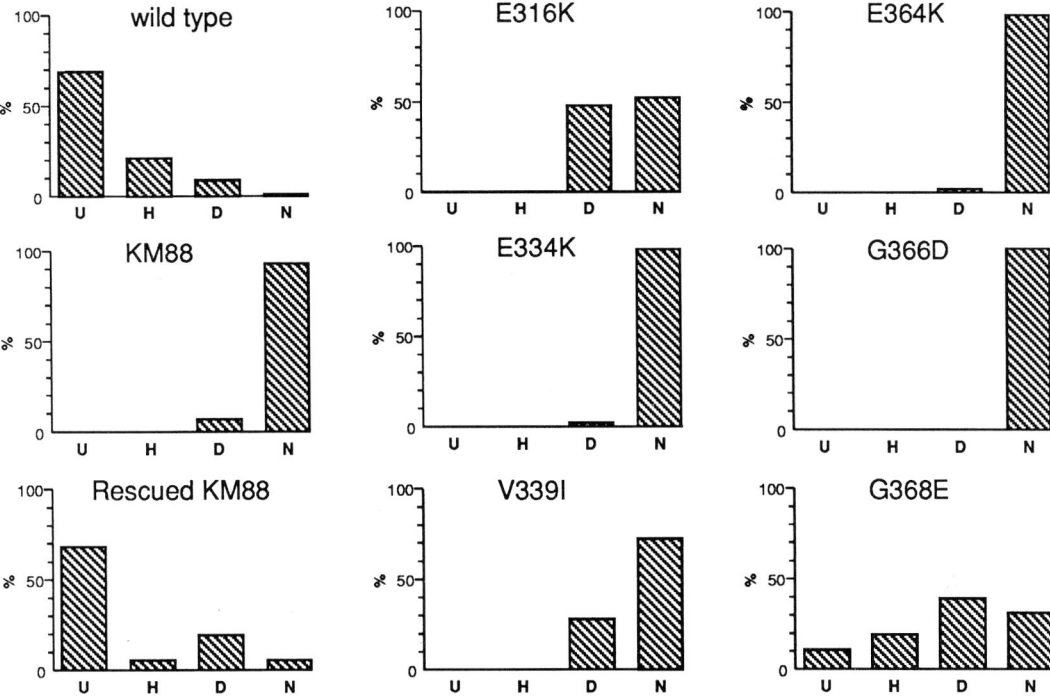

Fig. 4. Flight-testing of transformed flies. Flight-testing was performed in a flight box as described in Drummond et al., 1991a. Flies were scored for their flight behaviour on release; up (U), horizontal (H), down (D) and not (N). Each transformant genotype tested contained 2 copies of the wild-type *Act88F* gene and 2 copies of the transformed gene with the exception of the 'rescued KM88' which contain two wild-type gene copies transformed into a homozygous *KM88* strain. Data taken from Drummond et al. (1991a).

these two actins are synthesised at near normal levels but are not as stable *in vivo* as some of the other mutants (Drummond et al., 1991a).

These two mutants are typical of a large group of *Act88F* mutants which accumulate detectable levels of actin and induce a strong heat shock response in the IFMs (Hiromi and Hotta, 1985; Hiromi et al., 1986; Okamoto et al., 1986).

Both *E364K* and *G366D* are antimorphic for flight (Fig. 4) and myofibrillar structure, common features of *Act88F* mutants which cause a strong heat shock response. *E364K* and *G366D* form part of a cluster of missense and nonsense mutations. In the IFMs of *E364K* and *G366D* there is a large accumulation of hsp70 and hsp22. The hsp22 remains resistant to detergent extraction of the 'myofibrillar' pellet, suggesting a direct association of the two proteins which might account for the aggregates observed by Okamoto et al. (1986) in electron micrographs of similar *Act88F* mutants.

On non-dissociating gels *E364K* actin forms two distinct well separated bands; *G366D* formed two less well separated bands (Fig. 5). This suggests that each of these mutant actins are present as two distinct isomers with a $t_{1/2}$ for

Table 1. *Myosin:actin ratios*

Strain	Accumulation (m:a)	Incorporation of ^{35}S-methionine (m:a)
Wild-type	1.00±0.06	1.0±0.1
E316K	0.74±0.02	0.92±0.02
E334K	0.62±0.02	0.65±0.01
V339I	2.7±0.3	2.1±0.2
E364K	1.42±0.04	0.9±0.02
G366D	2.1±0.2	0.92±0.07
G368E	0.78±0.02	1.0±0.1

Accumulation ratio: The ratio of myosin heavy chain (m) to actin (a) in unskinned IFMs from 4-5 day old flies separated by 1-D SDS-PAGE, stained with Coomassie blue and quantified by densitometry.

Incorporation ratio: The ratio of incorporation of ^{35}S-methionine into myosin heavy chain (m) and actin (a) in newly emerged flies labelled for 2 hours. The proteins were separated by 1-D SDS-PAGE, the protein bands excised from gels and scintillation counted. The ratio is the average from 2 gels ± s.e.m. normalised to a wild-type ratio set at 1.0.

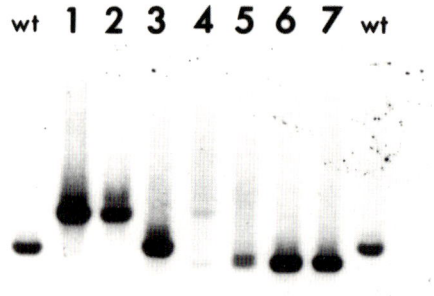

Fig. 5. Autoradiographs of wild-type and mutant *Act88F* actins transcribed and translated *in vitro* and separated by non-dissociating 1-D PAGE gels. wt, wild-type; lane 1, E316K; 2, E334K; 3, V339I; 4, E364K; 5, G366D; 6, G368E; 7, R372H (not described).

interconversion equivalent to at least the electrophoresis time of 45 minutes (Creighton, 1979) and contrasts with the other mutants which had a single major band (Drummond et al., 1991b).

4. Mutants with normal amounts of actin and myofibrils lacking Z- discs

E93K and *E334K* are the only members of a class of actin mutants in which myofibrils are present but Z-discs are absent (Sparrow, Reedy, Ball, Kyrtatas, Molloy, Hennessey, Durston and White, 1991b; Drummond et al., 1991a). Electron micrographs of *E93K* IFMs show the presence of interdigitating thick and thin filaments in myofibril-like assemblies but an absence, particularly in mature flies, of Z-discs. In pupae and young flies Z-disc like material is present but is lost from the myofibrils in the first few days following eclosion (Mary Reedy, personal communication). This is consistent with the absence from the myofibrils of high molecular weight proteins and α-actinin in 3-4 day old flies (Sparrow et al., 1991b). These proteins are Z-disc components and it seems likely that the *E93K* mutation affects the binding of a Z-disc protein to the actin filament. This aminoacid is located on the periphery of the small domain of actin (Kabsch, Mannherz, Suck, Pai and Holmes, 1990) in a region which contains an α-actinin binding site (Mimura and Asano, 1987). This site is on the outer surface of the F-actin filament (Holmes, Popp, Gebhard and Kabsch, 1990). Amino acid 334 is not close to amino acid 93 and is unlikely to form part of the same binding site.

E334K actin shows wild-type behaviour on both non-dissociating and urea gradient gels (Drummond et al., 1991b). Unlike *E93K*, it induces a weak heat shock response. This combination of an ability to assemble into myofibrils, albeit without Z-discs, and yet induce a weak heat shock response, is also a feature of *E316K* (see below) but *E334K* is more antimorphic for flight ability than *E316K* (Fig. 4).

An important feature of myofibrillogenesis is the assembly of the lattice of thick and thin filaments. Electron micrographs of transverse sections of wild-type flight muscle show a regular hexagonal lattice. This is also seen in skinned fibres no matter what physiological solutions they have been bathed in. However, incubation of *E93K* skinned fibres in a 'relaxing' solution before fixation for EM produces very little hexagonal lattice packing. The lattice can be largely restored if the fibres are taken from 'relaxing' solution and incubated in a 'rigor' solution before fixation (Sparrow et al., 1991b). Myosin heads are detached from the thin filaments in 'relaxing' solutions but form stable crossbridges in 'rigor' solutions. We interpret this as showing that crossbridges can re-establish the filament lattice, a process which it has been suggested (Fischman, 1972) might be important during myofibril assembly.

5. Actin mutants with almost normal myofibrillar structure

Only two *Act88F* mutants with near normal myofibrillar structure, *E316K* and *G368E*, are known. This is partly due to the fact that they were made by *in vitro* mutagenesis and P-element transformation (Drummond et al., 1991a) rather than by selection for flightlessness, although *E316K* is flightless. The highly conserved amino acid sequence of actin also predicts that mutations like these will be relatively rare, but these were recovered as two out of seven mutations with non-conservative amino acid changes.

Electron micrographs show that these mutants have almost completely normal myofibrillar structure; some lattice disorder is seen at the myofibrillar periphery (Fig. 6). Transformation with the *G368E* mutant partially rescues the flightlessness of the *KM88* strain and gives a wingbeat frequency of 178 ± 3 Hz compared to a

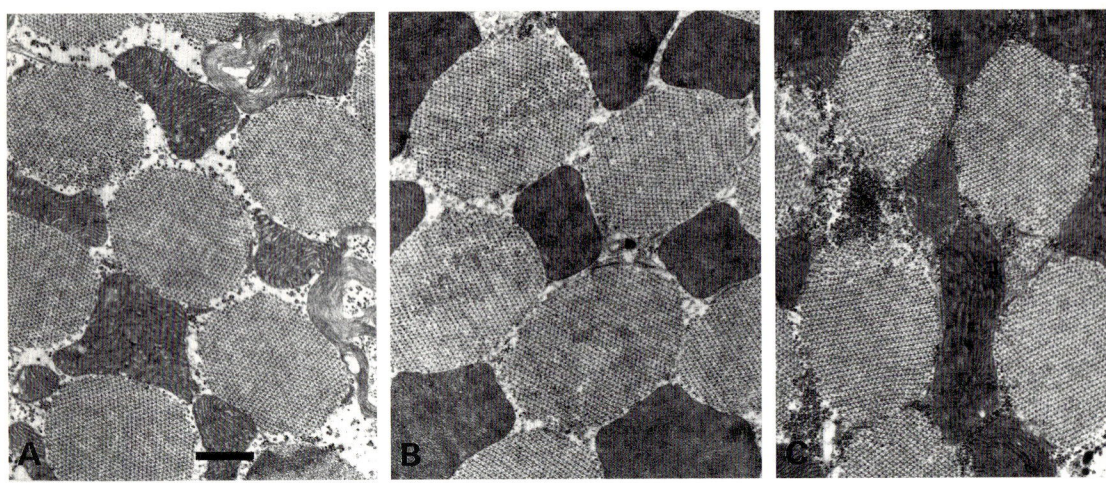

Fig. 6. Electron micrographs of indirect flight muscle myofibrils from (A) wild-type, (B) *E316K* and (C) *G368E* flies. Bar=1 μm.

wild-type frequency of 227 ± 4 Hz. Skinned muscle fibres from *G368E* and *E316K* produce normal muscle tension transients although with changes in the kinetics of different components of these responses (Drummond, Peckham, Sparrow and White, 1990).

The *E316K* mutation has a number of unique features. Despite the ability of skinned fibres to produce delayed tension responses, *E316K* flies are flightless. Electron micrographs of *E316K* fibres close to the tendon cells, which are the fibre attachment points to the cuticle, show myofibrillar abnormalities (Mary Reedy, personal communication) which probably explain the flightlessness. This suggests that despite the ability of *E316K* actin to form thin filaments there are subtle differences in the requirements for actin assembly in different parts of the fibre. This may explain its antimorphic effect on flight.

E316K induces a slight heat shock response. In non-dissociating gels *E316K* actin (like *E334K* actin) runs as single band with the mobility expected from its charge. In urea gradient gels *E316K* actin shows the greatest decrease in stability found with any of the six mutants studied (Drummond et al., 1991b) despite the fact that *E316K* actin accumulates normally *in vivo* (Table 1; Drummond et al., 1991a). The decreased thermostability of *E316K* actin does not correlate with its heat shock induction, since *E334K* actin shows normal thermostability yet gives a similar heat shock response. Similarly the unique instability of *E316K* actin in urea gradients and its normal accumulation *in vivo* contrasts with the lack of accumulation of *V339I* actin *in vivo*.

Antimorphism and heat shock induction

The combination of a strong heat shock response with strong antimorphism, e.g.

G366D and *E364K*, is shown by many *Act88F* mutants. (Mogami and Hotta, 1981; Karlik, Coutu and Fyrberg, 1984; Hiromi and Hotta, 1985; Hiromi et al., 1986; Okamoto et al., 1986). However, this is not always the case (Karlik, Saville and Fyrberg, 1987). Antimorphism must depend on the retention, in a mutant protein, of one or more of its normal binding sites. The observation that many actin mutants are antimorphic is undoubtedly related to the relatively large number of different binding sites on actin. In our investigation of actin mutants which induce the heat shock response the only common features of the strong inducers (*G366D* and *E364K*) was that from their accumulation in the IFMs they appeared fairly stable *in vivo*, yet on *in vitro* gel studies they alone showed evidence for multiple conformations (Drummond et al., 1991b).

Nearly all the strong inducers are clustered near the C-terminus of actin, perhaps indicating a region of the protein that on partial unfolding readily binds specific components of the heat shock system. We have observed a correlation between the accumulation of hsp70 and hsp22 and their insolubility during detergent (Triton X-100) extraction of *E364K* and *G366D* IFMs (Drummond et al., 1991a). The weak inducers, *E334K* and *E316K*, also induce a qualitatively different response, since they induce hsp22 but hsp70 is undetected.

The question remains unexplained (Parker-Thornburg and Bonner, 1987) as to why of all the mutations that have now been described in IFM muscle protein genes only *Act88F* mutations, and a significant fraction of them, produce a strong heat shock response. Throughout all living systems the heat shock response is induced by a wide variety of treatments with the common feature that they cause protein damage (Burdon, 1986). In IFMs the clustering of mutations within the *Act88F* gene which give a strong heat shock response suggests induction may be due to a rather localised unfolding of actin. This seems to correlate well with the presence of multiple conformers in non-dissociating *in vitro* conditions.

Polymerisation

Actin must polymerise to be biologically active. This is a major process of myofibrillogenesis. While the presence of thin filaments in the IFMs of an *Act88F* mutant indicates that the mutant actin can polymerise *in vivo*, the converse is not necessarily true. Actin polymerisation is necessary but not sufficient for thin filament assembly. The binding of myosin (Grazi, Magri and Rizzier, 1989) or tropomyosin (Hitchcock-deGregori, Sampath and Pollard, 1988; Weigt, Schoepper and Wegner, 1990) can stabilise actin filaments *in vitro*. *In vivo* a mutant actin, defective in binding an associated protein might not assemble into thin filaments.

Many assays are available for studying actin polymerisation *in vitro* (Pollard and Cooper, 1982) but few are appropriate for the small amounts of actin available from dissections of IFMs or *in vitro* translations. We have assessed two different assays for polymerisation ability with six actin mutants expressed *in vitro* (Hennessey, Harrison, Drummond and Sparrow, 1992). These are the co-polymerisation assay (Solomon and Rubenstein, 1987) and incorporation of

Table 2. *The ability of mutant and wild-type actins to co-polymerise and be incorporated into myofibrils* in vitro *and* in vivo

Actin	Distribution ratio[1]		Myofibril Appearance[4]
	Myofibril[2]	Co-polymerisation[3]	
Wild-type	1.0±0.1	1.00±0.01	N
E316K	0.9±0.2	1.8±0.3	N
E334K	0.8±0.2	1.2±0.2	D
V339I	1.2±0.2	1.5±0.2	A
E364K	1.1±0.2	2.0±0.2	A
G366D	0.8±0.1	1.9±0.3	A
G368E	1.4±0.3	1.7±0.3	N

Co-polymerisations were performed by method of Solomon and Rubenstein (1987). Wild-type and mutant *Act88F* actins were translated *in vitro* in the presence of ^{3}H-leucine and ^{35}S-methionine respectively. After polymerisation with bulk rabbit actin, the F-actin was pelletted by centrifugation and the pellet and supernatants scintillation counted. Incorporation of the *in vitro* actins into myofibrils was done by method of Bouche *et al.* (1988). After incubation the myofibrils were recovered by centrifugation, proteins separated by 1-D gels and autoradiographed. Actin bands were cut out and scintillation counted. Figures are mean ± s.e.m.

[1]Distribution ratio $= \dfrac{(^{35}\text{S-mutant}/^3\text{H-wild-type})\quad \text{(for supernatant)}}{(^{35}\text{S-mutant}/^3\text{H-wild-type})\quad \text{(for pellet)}}$

When ratio=1.0 both actin copolymerise equally well; ratio>1.0, wild-type polymerises better than mutant; ratio<1.0, mutant polymerises better than wild-type.
[2]data from Hennessey et al., 1992.
[3]data from Drummond et al., 1991a.
[4]Appearance of the myofibrils by phase contrast microscopy from IFMs of flies transformed with mutant actins, summarised from Drummond et al., 1991a. A=absent, D=disordered, N=normal.

labelled *in vitro* actins into purified myofibrils (Bouche, Goldfine and Fischman, 1988).

The distribution ratios obtained from wild-type and the six mutant actins (Table 2) are not significantly different from 1.0. They show no significant differences between wild-type and mutant actins and there is no correlation with the *in vivo* results on myofibril assembly. The binding of the short actin translation products (caused by internal initiation in our *in vitro* translations) was significant in both assays. This binding in the co-polymerisation assay was less than that found with the full length actin. In the myofbril binding assay the short products show a higher affinity than whole actin which raises questions about the site and specificity of actin binding in this complex binding substrate.

The low distribution ratios for the six mutant actins are in contrast to the much higher ratios determined for other mutants (Solomon, Solomon, Gay and Rubenstein, 1988) and the ratios we obtained with N-terminal processing-inhibited actins (Hennessey, Drummond and Sparrow, 1991). What does the polymerisation assay measure? The incubations contain a high concentration of bulk actin and a very low concentration of the *in vitro* translated actins in the ratio of bulk actin/*in vitro* actin of 10^5:1. A mutant or unprocessed actin may appear to polymerise when, in fact, it is only able to bind to the ends of filaments formed by the bulk actin. It may not be able to form a filament in the absence of other, normal actin (Hennessey, Harrison, Drummond and Sparrow, in press). This assay therefore measures co- polymerisation and capping ability of mutant actins, but cannot distinguish between them.

Post-translational processing

Actins are post-translationally modified in a number of ways, including N-terminal processing (Pollard and Cooper, 1986) and methylation of histidine-73 (Collins and Elzinga, 1975; Vandekerchove and Weber, 1979). The actin of *Drosophila* IFMs undergoes further post-translational modifications, including a charge change under the control of the mod^+ gene (Mahaffey et al., 1985) and ubiquitination of some of the actin to form a stable conjugate (Ball et al., 1987), known as arthrin (Bullard, Bell, Craig and Leonard, 1985). The *in vivo* functions of all these modifications are unknown. Flies homozygous for the mod^- mutation fly normally and have wild-type wingbeat frequencies (Drummond and Sparrow, unpublished).

The *Act88F* gene codes for N-Met-Cys-Asp-Actin and during post-translational processing the first two N-terminal amino acids are removed and the aspartate is acetylated (Rubenstein and Martin, 1983; Hennessey et al., 1991). Some actin mutants which cause amino acid substitutions close to the N-terminus prevent N-terminal processing *in vitro* (Solomon et al., 1988; Solomon and Rubenstein, 1989) but none of the changes affected the abilities of the mutants to polymerise or bind to DNase I. The authors concluded that N-terminal processing has no effect on actin function.

We have used a wild-type *Act88F* gene copy to translate actin *in vitro* with or without processing inhibitors to re-examine the above conclusion (Hennessey et al., 1991). Acetylation was inhibited with citrate synthase and oxaloacetate (Palmiter, 1977) and proteolysis with bestatin, an inhibitor of amino-peptidases, which inhibits the N-terminal processing of *Act88F* actin by rabbit reticulocyte lysate (Hennessey et al., 1991).

In the co-polymerisation assay (Solomon and Rubenstein, 1987) processed actin polymerises better than the unprocessed actins. The effect is much more pronounced with rabbit bulk actin rather than *Lethocerus* or *Calliphora* (a blow-fly) actin. Our ability to detect an effect of inhibiting actin processing on co-polymerisation may depend on using a heterologous actin as the bulk actin.

The unprocessed actins reduce the co-polymerisation of the processed actin translated *in vitro* with the bulk rabbit actin (Fig. 7). However, it does not affect the polymerisation of the bulk rabbit actin.

This is the first demonstration of an effect of N-terminal processing of actin. Since the unprocessed actins can prevent the processed actin from co-polymerising with the vast excess of rabbit actin ($10^5 \times$ the *in vitro* actin concentration) the *Drosophila* actins behave independently of the bulk actin.

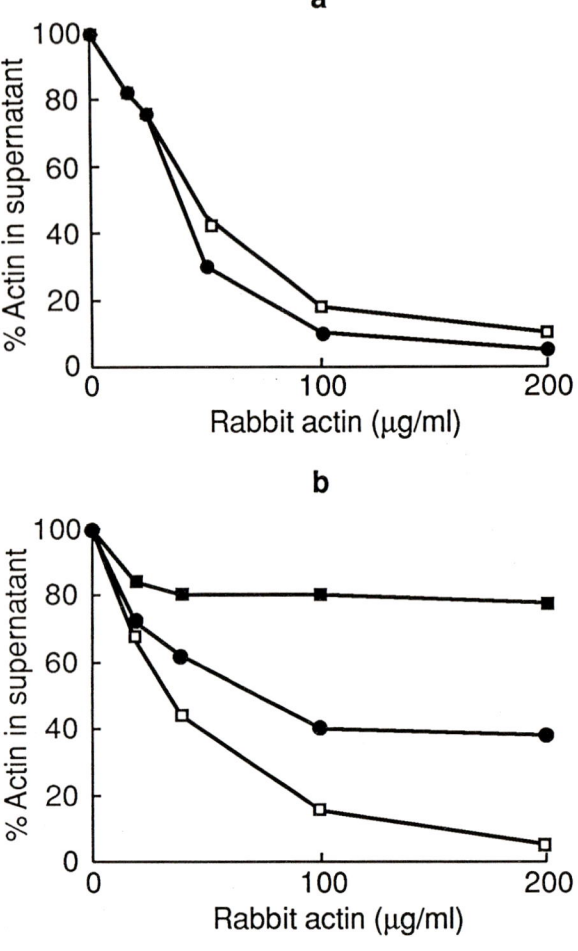

● Rabbit actin
● Processed actin
■ Unprocessed actin

Fig. 7. The effect of processing-inhibited actin on the ability of processed and bulk rabbit actin to polymerise. Radiolabelled wild-type *Act88F* actins were obtained by *in vitro* translation in rabbit reticulocyte lysate of RNAs synthesised *in vitro*. Processed actin was labelled with ^{3}H-leucine; processing inhibited actins with ^{35}S-methionine. After polymerisation the F-actin was pelleted by centrifugation and the percentage of each actin remaining in the supernatant determined. (a) co-polymerisation of processed actin with bulk rabbit actin; (b) co-polymerisation of processed actin with bulk rabbit actin in the presence of processing-inhibited actin. (●) processed actin; (■) unprocessed actin; (□) bulk actin. Data from Hennessey et al. (1991).

These experiments suggest that low concentration actin interactions can occur but possibly only between actins of the same or very similar sequence. This is supported by our observations that these effects are much reduced when an insect flight muscle actin is used as a bulk actin. This may explain why Solomon et al. (1988) found no effects of unprocessed human skeletal actin mutants on copolymerisation with bulk chicken breast muscle actin since these actins have identical amino acid sequences. The heterologous actin effect must be detecting subtle differences in actin.

The nature of these low concentration interactions is unclear, since the first event in the polymerisation is thought to be a conformational change induced in actin by the salt (Rich and Estes, 1976). Unprocessed actins can clearly interact with other actins prior to polymerisation. On this criterion unprocessed actins behave like capping proteins. There are several reports of modified actins acting as capping proteins (Ampe and Vandekerchove, 1987; Vandekerchove, Schering, Barmann and Aktories, 1987; Wegner and Aktories, 1988; Weigt, Just, Wegner and Aktories, 1989). As the N-terminus of actin has no direct role in polymerisation (DasGupta, White, Phillips, Bulinski and Reisler, 1990), processing is presumably required for a conformational change to give an actin structure which is fully polymerisation competent.

Our *in vitro* studies of N-terminal processing argue that it may have important functions. Is there a role for N-terminal processing *in vivo*? Does it have a role in regulating filament assembly in ways analogous to the functions of G-actin binding proteins? These are not easy questions to study *in vivo* but they are critical to our understanding of thin filament assembly. We are currently making N-terminal *Act88F* mutants for P-element transformations to investigate the effects of altering N-terminal processing *in vivo*.

Conclusions

The analysis of mutations in Drosophila flight muscles has already provided insights into the assembly (Fyrberg, 1990) and function (Sparrow et al., 1991a) of myofibrils in the IFMs. The studies on *Act88F* mutations reviewed here show a diverse range of effects on myofibrillar assembly. We have sought patterns and correlations in this diversity to determine further questions about the relationship between the amino acid sequence of actin, its molecular structure and its properties *in vitro* and *in vivo*.

The diversity of mutant effects has two obvious causes. First, the assembly of myofibrils is clearly complex. Second, actin has a number of binding sites for other actin monomers, thin filament associated proteins, Z-disc proteins and myosin. If actin mutations affect different binding sites this will lead to different phenotypes. In addition, mutations will, to differing degrees, affect the stability of actin. The accumulation of the protein *in vivo* will be a combination of the mutant protein's inherent stability under physiological conditions, interactions with other proteins which may serve to stabilise it and the ability of the protein degradative machinery to recognise aberrant molecular structure.

Examples of each of these considerations can be found in the five groups of mutants reviewed here. Some of these properties can be studied *in vivo*, others by *in vitro* assays.

Antimorphicity is a common feature of *Act88F* mutations *in vivo* and reveals the retention of binding sites, though these are difficult to characterise *in vivo* unless they lead to the loss of specific structures e.g. *E93K*, which led to the detection of a Z-disc protein binding site (Sparrow et al., 1991b). Co-polymerisation assays and affinity column binding assays (Hennessey, 1989) can be used to explore those binding sites which remain on mutant actins.

The instability of mutant actins can be detected *in vivo* but its cause is less easily determined. Most of the *Act88F* mutants described show decreased *in vivo* stability but those which are the least stable (e.g. *V339I*) behave essentially like wild-type in the *in vitro* assays. In fact, given the random generation of the six actin mutants (Drummond et al., 1991a) and the highly conserved amino acid sequence of actin, it was surprising that all of the mutant actins ran as discrete bands on non-dissociating gels, revealing folding into approximately the normal actin shape or discrete conformers.

The induction of the strong heat shock response appears to be diagnostic of localised unfolding and, at least in the *G366D* and *E364K*, mutants correlates with the detection of multiple conformers in non-dissociating gels. There is a clear correlation between the induction of the heat-shock response and the relative stability of the mutant actins *in vivo* (Drummond et al., 1991a).

The difficulties and promise of our approach are exemplified by the *E316K* mutation. The mutant actin assembles into almost normal myofibrils, which support the development of tension *in vitro* with subtle changes in rate constants, yet it produces specific structural defects at the ends of the myofibrils, induces a heat-shock response and on urea gradient gels is the least stable of all the mutations yet studied. Had we performed the *in vitro* assays on this actin before P-element transformation its thermodynamic instability might have persuaded us that it would be unstable *in vivo*, and of limited interest. This highlights the point that *in vivo* and *in vitro* properties are, not yet, easily correlated. This type of result runs counter to the belief that the application of protein engineering techniques to *in vitro* protein studies will allow us predict the properties and functions of the protein *in vivo*. The diverse, often unexpected, effects of actin mutations should provide useful insights of actin function *in vivo* and *in vitro*.

This work was supported by grants from the SERC and the EC Commission. We wish to thank Meg Stark and Anne Lawn for their excellent technical assistance.

References

Ampe, C. and Vanderkerchove, J. (1987). The F-actin capping proteins of *Physarum polycephalum*: cap42(a) is very similar, if not identical, to fragmin and is structurally and functionally very homologous to gelsolin: cap42(b) is *Physarum* actin. *EMBO J.* **6**, 4149-4157.

Ball, E. (1987). Isolation and characterization of *Drosophila melanogaster* dominant flightless mutations. DPhil. Thesis, University of York.

Ball, E., Karlik, C. C., Beall, C. J., Saville, D. L., Sparrow, J. C., Bullard, B. and Fyrberg, E. A. (1987). Arthrin, a myofibrillar protein of insect flight muscles, is an actin-ubiquitin conjugate. *Cell* **51**, 221-228.

Beall, C. J., Sepanski, M. A. and Fyrberg, E. A. (1989). Genetic dissection of *Drosophila* myofibril formation: effects of actin and myosin heavy chain null alleles. *Genes Dev.* **3**, 131-140.

Bouche, M., Goldfine, S. M. and Fischman, D. A. (1988). Post-translational incorporation of contractile proteins into a cell-free system. *J. Cell Biol.* **107**, 587-596.

Bullard, B. and Sainsbury, G. M. (1977). The proteins in the Z-line of insect flight muscle. *Biochem. J.* **161**, 399-403.

Bullard, B., Bell, J., Craig, R. and Leonard, K. (1985). Arthrin: a new actin-like protein in insect flight muscle. *J. Mol. Biol.* **182**, 443-454.

Burdon, R. H. (1986). Heat shock and the heat shock proteins. *Biochem. J.* **240**, 313-324.

Chun, M. and Falkenthal, S. (1988). *Ifm(2)2* is a myosin heavy chain allele that disrupts myofibrillar assembly only in the indirect flight muscles of *Drosophila melanogaster*. *J. Cell Biol.* **107**, 2613-2621.

Collins, J. H. and Elzinga, M. (1975). The primary structure of actin from rabbit skeletal muscle. *J. Biol. Chem.* **250**, 5915-5920.

Creighton, T. E. (1979). Electrophoretic analysis of the unfolding of proteins by urea. *J. Mol. Biol.* **129**, 235-264.

dasGupta, G., White, J., Phillips, M., Bulinski, J. C. and Reisler, E. (1990). *Biochemistry* **29**, 3319-3324.

Drummond, D. R., Peckham, M., Sparrow, J. C. and White, D. C. S. (1990). Alteration in crossbridge kinetics caused by mutations in actin. *Nature* **348**, 440-442.

Drummond, D. R., Hennessey, E. S. and Sparrow, J. C. (1991a). Characterisation of missense mutations in the *Act88F* gene of *Drosophila melanogaster*. *Mol. gen. Genet.* **226**, 70-80.

Drummond, D. R., Hennessey, E. S. and Sparrow, J. C. (1991b). Stability of mutant actins. *Biochem. J.* **274**, 301-303.

Epstein, H. F. and Fischman, D. F. (1991). Molecular analysis of protein assembly in muscle development. *Science* **251**, 1039-1044.

Fischman, D. F. (1972). Development of striated muscle. In 'Structure and Function of Muscle' (Bourne, G., ed.). Volume 1, 75-148, Academic Press, N.Y.

Fyrberg, E. A. (1990). Genetic approaches to myofibril form and function in *Drosophila*. *Trends in Genetics* **6**, 126-131.

Fyrberg, E. A., Mahaffey, J. W., Bond, B. J. and Davidson, N. (1983). Transcripts of the six *Drosophila* actin genes accumulate in a stage- and tissue-specific manner. *Cell* **33**, 115-123.

Grazi, E., Magri, E. and Rizzieri, L. (1989). The influence of substoichiometric concentrations of myosin subfragment 1 on the state of aggregation of actin under depolymerising conditions. *Eur. J. Biochem.* **182**, 277-282.

Hennessey, E. S. (1989). Biochemical studies on actin expressed *in vitro* from the *Drosophila Actin88F* gene. D.Phil. Thesis, University of York.

Hennessey, E. S., Drummond, D. R. and Sparrow, J. C. (1991). Post-translational processing of the amino terminus affects actin function. *Eur. J. Biochem.* **197**, 345-352.

Hennessey, E. S., Harrison, A., Drummond, D. R. and Sparrow, J. C. (1992). Mutant actin: a dead end? *J. Muscle Res and Cell Motil.*

Hiromi, Y. and Hotta, Y. (1985). Actin gene mutations in the *Drosophila*: heat shock activation in the indirect flight muscles. *EMBO J.* **4**, 1681-1687.

Hiromi, Y., Okamoto, H., Gehring, W. J. and Hotta, Y. (1986). Germline transformation with *Drosophila* mutant actin genes induces constitutive expression of the heat shock proteins. *Cell* **44**, 293-301.

Hitchcock-deGregori, S. E., Sampath, P. and Pollard, T. D. (1988). Tropomyosin inhibits the rate of actin polymerisation by stabilising actin filaments. *Biochemistry* **27**, 9182-9185.

Holmes, K. C., Popp, D., Gebhard, W. and Kabsch, W. (1990). Atomic model of the actin filament. *Nature* **347**, 44-49.

Kabsch, W., Mannherz, H. G., Suck, D., Pai, E. F. and Holmes, K. C. (1990). Atomic structure of the actin:DNase I complex. *Nature* **347**, 37-44.

Karlik, C. C., Coutu, M. D. and Fyrberg, E. A. (1984). A nonsense mutation within the *Act88F* actin gene disrupts myofibril formation in *Drosophila* indirect flight muscles. *Cell* **38**, 711-719.

Karlik, C. C., Saville, D. L. and Fyrberg, E. A. (1987). Two missense alleles of the *Drosophila melanogaster* Act88F actin gene are strongly antimorphic but only weakly induce synthesis of the heat shock proteins. *Mol. Cell Biol.* **7**, 3084-3091.

Mahaffey, J. W., Coutu, M. D., Fyrberg, E. A. and Inwood, W. (1985). The flightless Drosophila mutant *raised* has two distinct genetic lesions affecting accumulation of myofibrillar proteins in flight muscles. *Cell* **40**, 101-110.

Mimura, N. and Asano, A. (1987). Further characterisation of a conserved actin binding 27 KDa fragment of actinogelin and α-actinins and mapping of their binding sites on the actin molecule by chemical cross-linking. *J. Biol. Chem.* **262**, 4717-4723.

Mogami, K. and Hotta, Y. (1981). Isolation of *Drosophila* flightless mutants which affect myofibrillar proteins of indirect flight muscle. *Mol. gen. Genet.* **183**, 409-417.

O'Donnell, P. T., Collier, V. L., Mogami, K. and Bernstein, S. I. (1989). Ultrastructural and molecular analyses of homozygous viable *Drosophila melanogaster* muscle mutants indicates there is a complex pattern of myosin heavy-chain isoform distribution. *Genes Dev.* **3**, 1233-1246.

Okamoto, H., Hiromi, Y., Ishikawa, E., Yamada, T., Isoda, K., Maekawa, H. and Hotta, Y. (1986). Molecular characterisation of mutant actin genes which induce heat-shock proteins in *Drosophila* flight muscles. *EMBO J.* **5**, 589-596.

Palmiter, R. D. (1977). Prevention of NH_2-terminal acetylation of proteins synthesised *in vitro*. *J. Biol. Chem.* **252**, 8781-8783.

Parker-Thornburg, J. and Bonner, J. J. (1987). Mutations that induce the heat-shock response in *Drosophila*. *Cell* **51**, 763-772.

Pollard, T. D. and Cooper, J. A. (1982). Methods to measure actin polymerisation. *Methods Enzymol.* **85**, 182-210.

Pollard, T. D. and Cooper, J. A. (1986). Actin and actin-binding proteins. A critical evaluation of mechanisms and functions. *Ann. Rev. Biochem.* **55**, 987-1035.

Reedy, M. C., Beall, C. and Fyrberg, E. A. (1989). Formation of reverse rigor chevrons by myosin heads. *Nature* **339**, 481-483.

Rich, S. A. and Estes, J. E. (1976). Detection of conformational changes in actin proteolytic digestion: evidence for a monomeric species. *J. Mol. Biol.* **104**, 777-792.

Rubenstein, P. A. and Martin, D. J. (1983). NH_2-terminal processing of actin in mouse L-cells *in vivo*. *J. Biol. Chem.* **258**, 3961-3966.

Sakai, Y., Okamoto, H., Mogami, K., Yamada, T. and Hotta, Y. (1990). Actin with tumour-related mutation is antimorphic in *Drosophila* muscle: two distinct modes of myofibrillar disruption by antimorphic actins. *J. Biochem.* **107**, 499-505.

Sanchez, F., Tobin, S. L., Rdest, U., Zulauf, E. and McCarthy, B. J. (1983). Two *Drosophila* actin genes in detail. Gene structure, protein structure and transcripts during development. *J. Mol. Biol.* **163**, 533-551.

Solomon, L. R. and Rubenstein, P. A. (1987). Studies on the role of actin's N^7-methylhistidine using oligonucleotide-directed site specific mutagenesis. *J. Biol. Chem.* **262**, 11382-11388.

Solomon, L. R. and Rubenstein, P. A. (1989). Studies on the role of actin's aspartic acid 11 using oligonucleotide directed site specific mutagenesis (abstract). *J. Cell Biol.* **105**, 25a.

Solomon, T. L., Solomon, L. R., Gay, L. S. and Rubenstein, P. A. (1988). Studies on the role of actin's aspartic acid 3 and aspartic acid 11 using oligonucleotide-directed site specific mutagenesis. *J. Biol. Chem.* **263**, 19662-19669.

Sparrow, J., Drummond, D., Peckham, M., Hennessey, E. and White, D. (1991a). Protein engineering and the study of muscle contraction in *Drosophila* flight muscles. *J. Cell Sci. Suppl.* **14**, 73-78.

Sparrow, J., Reedy, M., Ball, E., Kyrtatas, V., Molloy, J., Durston, J., Hennessey, E. S. and White, D. (1991b). Functional and ultrastructural effects of a missense mutation in the indirect flight muscle-specific actin gene of *Drosophila melanogaster*. *J. Mol. Biol.* **222**, 963-982.

Vandekerchove, J. and Weber, K. (1979). The complete amino acid sequences of actins from bovine aorta, bovine heart, bovine fast skeletal muscle and rabbit slow skeletal muscle. *Differentiation* **14**, 123-133.

Vandekerchove, J., Schering, B., Barmann, M. and Aktories, K. (1987). *Clostridium perfringens* iota toxin ADP-ribosylates skeletal muscle actin in Arg-177. *FEBS Letts.* **225**, 48-52.

Wegner, A. and Aktories, K. (1988). ADP-ribosylated actin caps the barbed ends of actin filaments. *J. Biol. Chem.* **263**, 13739-13742.

Weigt, C., Just, I., Wegner, A. and Aktories, K. (1989). Non-muscle actin ADP-ribosylated by botulinum toxin C2 caps actin filaments. *FEBS Letts* **246**, 181-184.

Weigt, C., Schoepper, B. and Wegner, A. (1990). Tropomyosin-troponin complex stabilises the pointed ends of actin filaments against polymerisation and depolymerisation. *FEBS Lett.* **260**, 266-268.

Printed in Great Britain © *Society for Experimental Biology 1992*   131

EXPRESSION OF ACTIN ISOFORMS IN *ARTEMIA*

MARIA-ASUNCIÓN ORTEGA, MARIA-TERESA MACIAS, JOSE-LUIS MARTINEZ, IGNACIO PALMERO and LEANDRO SASTRE

Instituto de Investigaciones Biomédicas del C.S.I.C., C/Arturo Duperier, 4, 28029 - Madrid, Spain

Summary

Complementary DNA clones have been isolated from the crustacean *Artemia* that code for four different actin isoforms. The nucleotide sequence of these clones has been determined. The four clones are about 80% identical in their translated regions but unrelated in their untranslated regions. The cloned *Artemia* actins are very similar in their deduced amino acid sequences to other invertebrate actins, especially in the amino terminal region. The analyses of the steady-state levels of actin mRNAs during *Artemia* development has shown a parallel increase in the levels of all four mRNAs between five and ten hours of development. Whole-mount embryo hybridizations have shown that one of the clones codes for a muscular actin isoform while the other three clones code for cytoplasmic isoforms.

Introduction

Actin is encoded in most organisms by a gene family that consists of several highly homologous genes, coding for different actin isoforms. The expression of each isoform is differentially regulated according to a tissue- and developmental-specific pattern. The expression of actin isoforms is well established in vertebrates: there are four muscular isoforms that are expressed in the different types of muscles (skeletal and cardiac striated muscles and vascular and non-vascular smooth muscles) and two cytoplasmic isoforms that are expressed in non-muscular cells.

In invertebrates, the pattern of expression is more diverse. Some invertebrates, such as *Drosophila*, have a pattern of expression similar to vertebrates in the sense that there are four muscular isoforms that are expressed in different types of muscles and two cytoplasmic isoforms that are expressed in non-muscle cells (Fyrberg, Mahaffey, Bond and Davidson, 1983). In some other cases, such as sea urchin, there is just one muscular isoform and several cytoplasmic isoforms that are differentially expressed in other tissues, like oral or aboral ectoderm (Shott, Lee, Britten and Davidson, 1984). All invertebrate actin isoforms, including muscular isoforms, are more similar to vertebrate cytoplasmic isoforms than to vertebrate muscular isoforms (Vandekerckhove and Weber, 1984). Invertebrate muscular actins seem to have evolved independently from a cytoplasmic gene

Key words: actin, crustacean, gene expression.

which raises interesting questions about muscular actins evolution and the establishment of tissue-specific patterns of expression.

The knowledge of the actin gene family in different species is of great interest to study the evolution of these genes and the establishment of the mechanisms that regulate their specific patterns of expression. Because of this interest and due to the lack of studies on the actin gene family in many groups of invertebrates, we have undertaken the study of this gene family in a crustacean, the brine shrimp *Artemia* as representative of this group of arthropods.

Artemia has been used as a model system to study embryonic development. *Artemia* embryos can get under cryptobiosis at the gastrula stage in which case they get dehydrated, are surrounded by a hard shell and interrupt their metabolic activity. Cryptobiotic embryos (cysts) can be activated and the development of synchronous populations of embryos studied (reviewed by Marco, Garesse, Cruces and Renart, 1990). Our laboratory is interested in the regulation of gene expression during the activation of the cyst and its later development. Actin genes are a very interesting gene family to study in this system since the regulation of the expression of several highly related genes with different tissue specificities can be approached and compared to similar studies that are being carried out in other systems (see for example Chung and Keller, 1990; Hough-Evans, Franks, Zeller, Britten and Davidson, 1990). In this article we summarize some of the results that we have obtained on the isolation and characterization of cDNA clones coding for actin isoforms and the study of their temporal and tissue-specific patterns of expression.

Results and discussion

An *Artemia* cDNA library made from adults mRNA was screened for actin clones using *Drosophila* actin 5C (DmA2) clone as probe (Macias and Sastre, 1990). Two cDNA clones were isolated and one of them used as probe in two further screenings of *Artemia* cDNA libraries made from nauplii mRNA. A total of 17 clones were isolated that could be classified in four groups of overlapping clones according to their restriction maps and partial nucleotide sequences. The longest clone from each group was chosen for characterization, selected clones were: pArAct 205 (representative of a group of two clones), pArAct 211 (representative of 11 clones), pArAct 302 (representative of two clones) and pArAct 403 (representative of two clones). The nucleotide sequence of the selected clones was determined. The four clones contained a long open reading frame coding for proteins that were over 95% identical to insect actins. *Artemia* cDNA clones are about 80% identical to each other in the nucleotide sequence of their protein coding regions but are non homologous in their untranslated regions. Specific probes derived from these clones hybridize to mRNAs of different size (see below). These results strongly suggest that the clones characterized code for four different isoforms of actin in *Artemia* and are summarized in Table 1.

The actual number of actin isoforms expressed in *Artemia* could be higher than

Table 1. *Size of* Artemia *actin cDNA clones and homology of the amino acid sequence of their longest open reading frames with that of three insect actins and between them*

Clone	Size (bp)	% Amino acid identity				
		Insects	205	211	302	403
pArAct205	1458	95.8	100	98.4	94.5	97.1
pArAct211	1347	96.1	98.4	100	94.8	97.1
pArAct302	1033	95.3	94.5	94.8	100	96.6
pArAct403	1305	96.5	97.1	97.1	96.6	100

four since Southern blot experiments, hybridizing *Artemia* DNA digested with several enzymes to pArAct211 have shown than there might be as many as 8 to 10 actin genes in this organism (Macias and Sastre, 1990). The four cDNA clones isolated seem to represent the four major isoforms expressed in nauplii and adult animals.

The comparison of the amino acid sequences deduced from the isolated cDNA clones with those of other organisms has shown that *Artemia* actins are similar to other invertebrate actins. Of special significance is the homology at the amino terminus of the proteins. Clones pArAct 205, 211 and 403 code for a N-terminal methionine, followed by cystein and three acidic amino acids, valine at position 11 and methionine and cystein at positions 17 and 18, respectively, as described for other invertebrate actins (Fyrberg, Bond, Hershey, Mixter and Davidson, 1981). The clone pArAct 302 is a partial cDNA clone, lacking the sequences encoding the 49 N-terminal amino acids of the protein and can not be compared to other actins in this region.

The evolutionary relationship of the proteins deduced from *Artemia* actin cDNAs clones pArAct 205, 211 and 403 to some representative vertebrate and invertebrate actins was analyzed using the program Treealign described by Hein (1990). The phylogenetic tree deduced from these comparisons is represented in Fig. 1. The actin isoform encoded by pArAct 302 could not be analyzed with this program since we do not know its complete amino acid sequence. The phylogenetic tree shows that *Artemia* actins are closely related to invertebrate actins, in special to insect actins and more related to vertebrate cytoplasmic actins than to vertebrate muscular actins, as expected. The two *Artemia* isoforms more closely related are pArAct205 and 211, differing in just five amino acids although their untranslated regions do not conserve any significant homology. *Artemia* actins 205 and 211 are also very homologous to *Drosophila* actins 7 and 8, even more than to *Artemia* isoform 403. These results suggest than *Artemia* actins 205 and 211 and *Drosophila* actins 7 and 8 have a common ancestor different from that of *Artemia* actin 403 so that the animal from which insects and crustaceans were generated should have had at least two different actin genes.

As mentioned in the Introduction, we are interested in the study of the

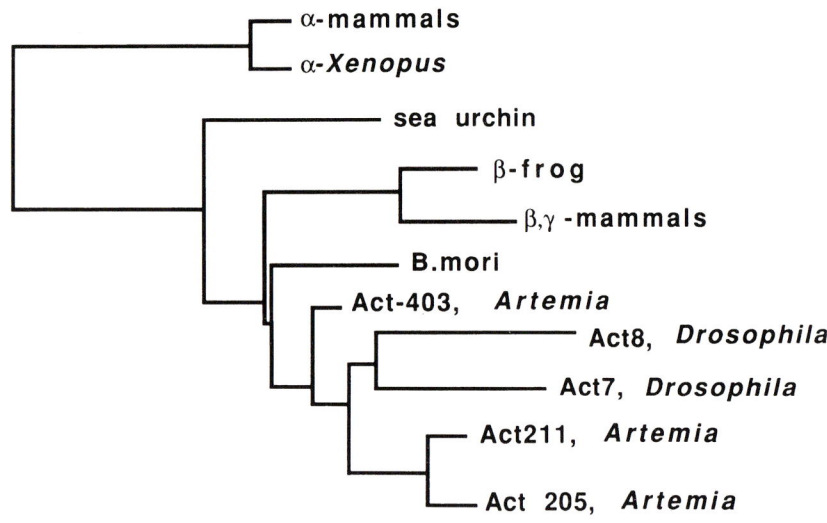

Fig. 1. Phylogenetic tree of *Artemia* actin isoforms. The phylogenetic tree has been derived from the amino acid sequences deduced from the nucleotide sequence of the clones pArAct205, 211 and 403 and the amino acid sequences of a group of representative actin isoforms obtained from available protein data banks. The amino acid sequences were compared and the phylogenetic tree constructed using the program described by Hein (1990) where the length of horizontal lines is proportional to the degree of divergence between each pair of proteins.

regulation of gene expression during *Artemia* cyst activation and embryonic development. We have studied the level of expression of the mRNA complementary to each cDNA clone in the cyst and at several stages of development by Nothern- and slot-blot experiments. Oligonucleotide probes complementary to the 3′ untranslated region of each cDNA clone were used in these experiments (Macias and Sastre, 1990). The more relevant results are summarized in Table 2. The mRNAs complementaries to clones pArAct205, 211 and 403 are present at low levels in the cyst. They must be present as stored mRNAs since there is no translation in the cyst (Tate and Marshall, 1990). When

Table 2. *Size of the mRNAs complementary to* Artemia *cDNA clones and their relative steady-state levels of expression in uncultured cysts (0 hr) and cyst cultured for different periods of time (5, 10, 15 and 20 hours)*

Clone	mRNA size	Relative level of mRNA expression				
		0 hr	5 hr	10 hr	15 hr	20 hr
pArAct205	5.2 kb	0.60	1.04	1.43	1.38	0.95
pArAct211	1.9 kb	0.30	0.30	1.77	1.70	1.19
pArAct302	1.6 kb	0.00	0.00	0.59	1.33	0.51
pArAct403	1.8 kb	0.14	0.24	1.28	1.14	0.78

the cyst is activated, mRNA levels increase markedly for these three actins, reaching maximal levels 10 hours after cyst activation, the levels decrease to half maximal levels by 30 hours of development and increase again there after. The mRNA complementary to pArAct302 is not present at detectable levels in the cyst, its synthesis is induced between 5 and 10 hours of development so that maximal levels are present by 15 hours of development. The level of pArAct 302 mRNA follows afterwards an oscillatory pattern, with relative maxima at 25 and 50 hours of development. One of the more interesting results of these experiments is that they show a coordinated induction of transcription of the four actin isoforms between 5 and 10 hours of development. A marked increase in mRNA levels has also been observed in the same period of development for other genes such as Ca-ATPase (Palmero and Sastre, 1989) and Na/K ATPase α subunit (Sastre, Palmero, Macias, Gil, Franco, Dominguez, Diaz-Guerra, Quintanilla, Cruces and Renart, 1989), suggesting that there is a general activation of transcription between 5 and 10 hours of cyst development.

At the present time, we are studying the tissue-specificity of expression of each actin isoform by whole-mount embryo hybridization using the specific oligonucleotide probes. Preliminary results using 65-hour-old nauplii are presented in Fig. 2. The more conclusive results are obtained with clone pArAct302 that seems to be expressed exclusively in the muscular fibers of the antennula, antenna and mandible that form the swimming and mandibular muscular systems of the nauplii (Fig. 2c). The other three isoforms seem to be expressed in non-muscular tissues and can be considered cytoplasmic actins. At least two actin isoforms seem to be expressed at the ectoderm of the nauplii, actins 211 and 403 (Fig. 2b and d). Actin 211 seems to be expressed as well in the mesoderm (Fig. 2b). Actin 205 seems to be expressed exclusively at the digestive tract, although the level of expression of actin 205 is relatively low and the digestive tract gives higher background hybridization so that more experiments would be necessary to ascertain the tissue specificity of expression of actin 205. We would like to enforce that the data shown in Fig. 2 are preliminary and that further experiments are being carried out in our laboratory to confirm these results and to analyze the expression of the actin isoforms at other stages of development.

The isolation of genomic clones coding for these actin isoforms is also in progress at our laboratory. Besides elucidating the intron/exon structure of these genes, we are interested in the study of their regulatory regions that determine the tissue-specificity and timing of expression of these genes during development. It is of particular interest in *Artemia* the study of cryptobiosis, when all metabolic and transcriptional activity is arrested, and the activation of the cyst. Experiments described in this article have shown that the transcription of actin genes is activated between 5 and 10 hours after resumption of development. The isolation and characterization of promotor and enhancer regulatory regions of the actin genes would allow the study of the mechanisms that block their transcription in the cyst and those that mediate their activation 5 to 10 hours after resumption of cyst development.

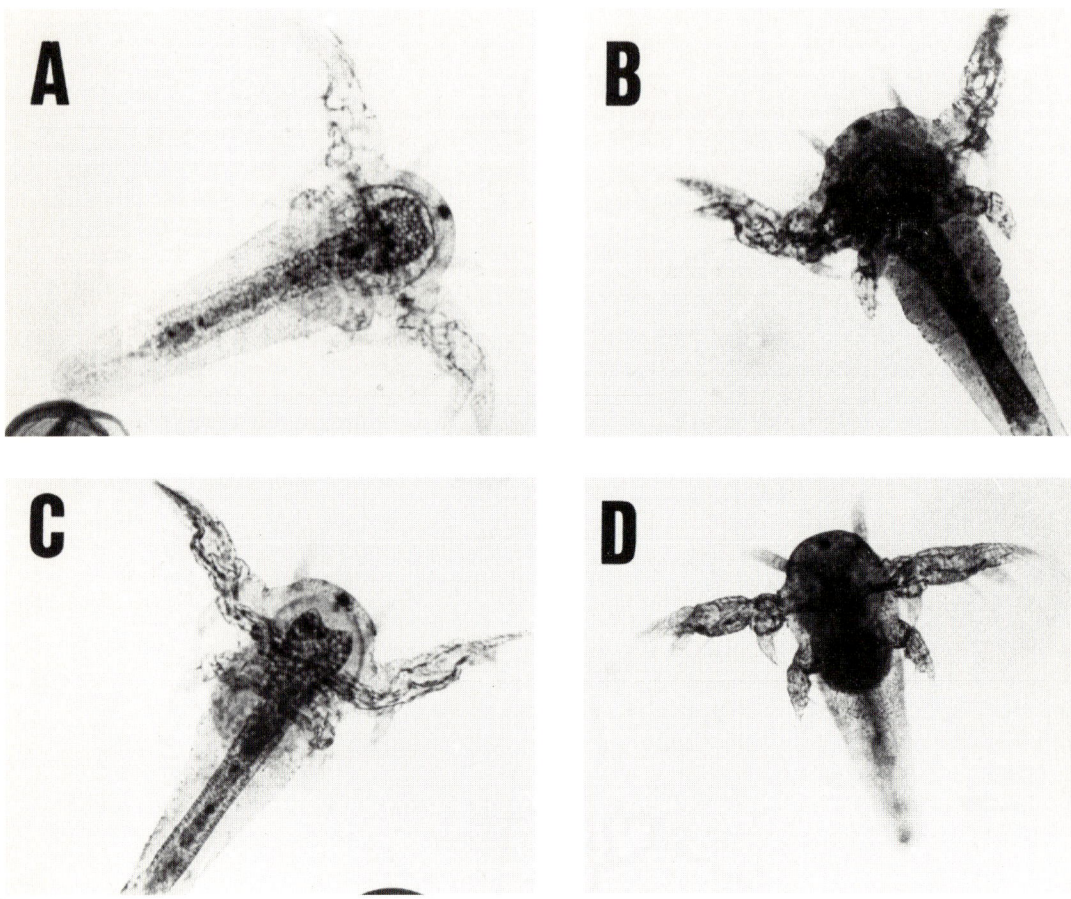

Fig. 2. Whole-mount embryo hybridization of *Artemia* nauplii to isoform specific actin probes. *Artemia* cysts were cultured for 65 hours, the resulting nauplii collected and fixed in 4% paraformaldehyde in phosphate buffered saline (PBS) (PBS is 137 mM NaCl, 2.7 mM KCl, 4.3 mM $Na_2HPO_4 \cdot 7H_2O$, 1.4 mM KH_2PO_4, pH 7.4) during at least two hours. Fixed nauplii were washed with methanol until most of *Artemia* pigments were lost. Permeabilization was performed by brief sonication of fixed nauplii, which were previously washed exhaustively with PBS. After permeabilization, the nauplii were fixed in 4% paraformaldehyde in PBS, digested with 100 μg/ml proteinase K in PBS for 10 minutes, blocked with 2 mg/ml glicine and fixed again in 4% paraformaldehyde in PBS. Oligonucleotides specific for each actin isoform were labelled with digoxigenin-dUTP (Boehringer Mannheim), using the enzyme terminal transferase. Hybridizations were made in 6×SSC, 5% Denhardt solution, 500 μg/ml sonicated boiled salmon sperm DNA, 50 μg/ml heparin, and 0.1% SDS at 45°C for 15 hours. Washing of nauplii was made at 45°C in mixtures of hybridization solution and PBT (PBT is PBS plus 0.1% of Tween 20) 4:1, 3:2, 2:3, 1:4 and PBT. Immunodetection of hybridization signals was performed according to the kits manufacturers' instructions. Panel (a) shows the results of the hybridization with the pArAct205 specific probe, panel (b) the results of the hybridization with the pArAct211 specific probe, panel (c) the results of the hybridization with the pArAct302 specific probe and panel (d) the results of the hybridization with the pArAct403 specific probe.

We would like to thank Antonio Fernandez, Silvia Kremenic and Javier Perez Diaz for their help in the preparation of the figures. The work presented in this article has been supported by grants from the Dirección General de Investigación Científica y Técnica (grant number PB87-0208) and Fondo de Investigaciones Sanitarias de la Seguridad Social (grant number 87/1561). M-A.O. is recipient of a fellowship from the Gobierno Autónomo Vasco.

References

Chung, Y-T. and Keller, E. B. (1990). Regulatory elements mediating transcription from the *Drosophila melanogaster* actin 5C proximal promoter. *Mol. Cell Biol.*, **10**, 6172-6180.

Fyrberg, E. A., Bond, B. J., Hershey, N. D., Mixter, K. S. and Davidson, N. (1981). The actin genes of *Drosophila*: Protein coding regions are highly conserved but intron positions are not. *Cell*, **24**, 107-116.

Fyrberg, E. A., Mahaffey, J. W., Bond, B. J. and Davidson, N. (1983). Transcripts of the six *Drosophila* actin genes accumulate in a stage- and tissue-specific manner. *Cell*, **33**, 115-123.

Hein, J. (1990). Unified approach to alignment and phylogenies. *Methods in Enzimol.* **183**, 626-644.

Hough-Evans, B. R., Franks, R. R., Zeller, R. W., Britten, R. J. and Davidson, E. H. (1990). Negative spatial regulation of the lineage specific CyIIIa actin gene in the sea urchin embryo. *Development*, **110**, 41-50.

Macias, M. T. and Sastre, L. (1990). Molecular cloning and expression of four actin isoforms during *Artemia* development. *Nucl. Acid Res.*, **18**, 5219-5225.

Marco, R., Garesse, R., Cruces, J. and Renart, J. (1990). Artemia Molecular Genetics. In *Hand book of Artemia biology* (Ed. R.A. Browne, P. Sorgeloos and C.N.A. Trotman). pp 1-19. CRC Press.

Palmero, I. and Sastre, L. (1989). Complementary DNA cloning of a protein highly homologous to mammalian sarcoplasmic reticulum Ca-ATPase from the crustacean *Artemia. J. Mol. Biol.*, **210**, 737-748.

Sastre, L., Palmero, I., Macias, M-T., Gil, I., Franco, E., Dominguez, E., Diaz-Guerra, M., Quintanilla, M., Cruces, J. and Renart, J. (1989). cDNA cloning of developmentally regulated *Artemia* genes. In *Cell and Molecular Biology of Artemia development*. (Ed. A.H. Warner, T.H. MacRae and J.C. Bagshaw). pp 319-328. Plenum Press, N.Y.

Shott, R. J., Lee, J. J., Britten, R. J. and Davidson, E. H. (1984). Differential expression of the actin gene family of *Strangylocentrotus purpuratus*. *Develop. Biol.*, **101**, 295-306.

Tate, W. P. and Marshall, C. J. (1990). Post-dormancy transcription and translation in the brine shrimp. In *Hand book of Artemia biology* (Ed. R.A. Browne, P. Sorgeloos and C.N.A. Trotman). pp 21-36. CRC Press..

Vandekerckhove, J. and Weber, K. (1984). Chordate muscle actins differ distinctly from invertebrate muscle actins. *J. Mol. Biol.*, **179**, 391-413.

Printed in Great Britain © *Society for Experimental Biology 1992*   139

SWITCHES IN FISH MYOSIN GENES INDUCED BY ENVIRONMENT TEMPERATURE IN MUSCLE OF THE CARP

GEOFFREY GOLDSPINK*, LUCIEN TURAY, EKKEHARD HANSEN, STEVEN ENNION* and GERALD GERLACH

Unit of Veterinary Molecular and Cellular Biology, The Royal Veterinary College, University of London, Royal College Street, London NW1 OTU

Summary

Fish are cold blooded animals and their muscle function is expected to be greatly affected by environmental temperature. Species that live in the Antarctic ocean have evolved a different contractile system to fish that live in the tropical waters. In the case of Antarctic fish they have a higher specific myofibrillar ATPase activity but 'the trade off' seems to be a lower thermal stability. They are thus capable of a greater muscle power output at low temperatures but the lower thermal stability means they are restricted to living at temperatures below +4°C. Some species, however, experience a wide range of seasonal variations in temperature. We found that these species adapt by changing their myofibrillar apparatus so that they have a higher specific ATPase which physiological studies indicate is due to a different type of myosin crossbridge for low temperature swimming. This is reversible and they develop a contractile system with a greater thermal stability and a commensurate loss of ATPase activity when their environment warms up again. There were several possibilities by which this may be achieved including expression of different isoform genes or the post- translational processing of existing proteins. To elucidate the mechanism we made a carp genomic library and screened this for myosin heavy chain gene using mammalian cDNA sequences under moderate stringency conditions. The clones were restriction mapped which resulted in 28 non overlapping sequences. This indicated that the carp had a reasonably large family of myosin heavy chain genes that is about twice the size of that in mammals. Rather fortuitously the first sequence to be identified was from the gene that is predominantly expressed in white muscle at warm temperatures. This was done by extracting the RNA from red and white muscle of fish acclimated to different 25°C, 18°C or 8°C and carrying out Northern analysis using the gene fragment as the probe. The time course for the expression of this gene when carp maintained at a low temperature were acclimated to a warm temperature was slightly in advance of the change in myofibrillar ATPase which suggested that this strategy for adaptation is regulated at the transcriptional level. Hence these

*Present address: Department of Anatomy and Developmental Biology, The Royal Free Medical School, London University, London NW3 2PF.

Key words: fish myosin, temperature, crossbridges, myosin heavy chain genes, myosin isoforms.

species of fish can adapt to seasonal changes in temperature by expressing different myosin heavy chain isoform genes and rebuilding their myofibrils for either warm or cold temperature swimming. At the present time we are characterising the 5′ regulatory (promoter) sequence of this gene to see how a temperature switch may operate. We also feel that the myosin heavy chain promoter of value for producing transgenic fish as mammalian gene constructs under the control of mammalian promoters not expressed or are weakly expressed. Also it is of interest to compare the coding sequences for the different isoforms particularly the S1 (crossbridge region) to determine how subtle differences can result in marked changes in contractility of red and white and warm and cold adapted fish muscle.

Design of contractile apparatus for cold or warm temperature swimming

Some years ago we became interested in differences in design of the contractile apparatus in fish adapted to function at widely differing environmental temperatures. We therefore studied the properties of myofibrils isolated from ranging from Antarctic species which live within a range of -1 to $+4°C$ and fish found in hot springs in the Rift Valley in Africa which experience temperatures of up to 40°C (Johnston, Frearson and Goldspink, 1973). We found that the myofibrillar system in the Antarctic fish had a much higher specific myofibrillar ATPase but their contractile systems was much more rapidly denatured by heat than was the case in fish adapted to warmer environments (Johnston, Davison and Goldspink, 1975a; Johnston, Davison, Walesby and Goldspink, 1975b). Myofibrils were used for the ATPase assays because the myosin from cold water fish was very unstable. There seemed to be a 'trade off' between high ATPase activity which is necessary to give the higher cross bridge cycling time and hence the higher power output needed at low environmental temperatures and the thermal stability of the myofibrillar proteins that is required for functioning at the warmer environmental temperatures (Fig 1 top). We then asked the question, what about fish such as carp (e.g. gold fish) which may be required to function at cold temperatures of just above 0°C in a pond in temperate country to say 25°C in an ornamental pool in the tropics. Rather to our surprise we found that some species of fish including carp (Fig 2) could adapt to a different environmental temperature by producing a different myofibrillar system with a high specific ATPase for cold water swimming and another myofibrillar system with a higher thermal stability for warm water swimming (Johnston and Goldspink, 1975c; Penney and Goldspink, 1979; Heap, Watt and Goldspink, 1985). This strategy means that the power output of the musculature is not reduced so drastically as it would be if the temperature was to drop by 10°C or even as much as 20°C during the seasonal change from summer to winter. Swimming performance experiments including EMG measurements have shown that this molecular reconstruction of the contractile apparatus does indeed improve the power output at low environment temperature (Heap and Goldspink, 1985, Rome, Loughna and Goldspink, 1985). This has been shown to be related to

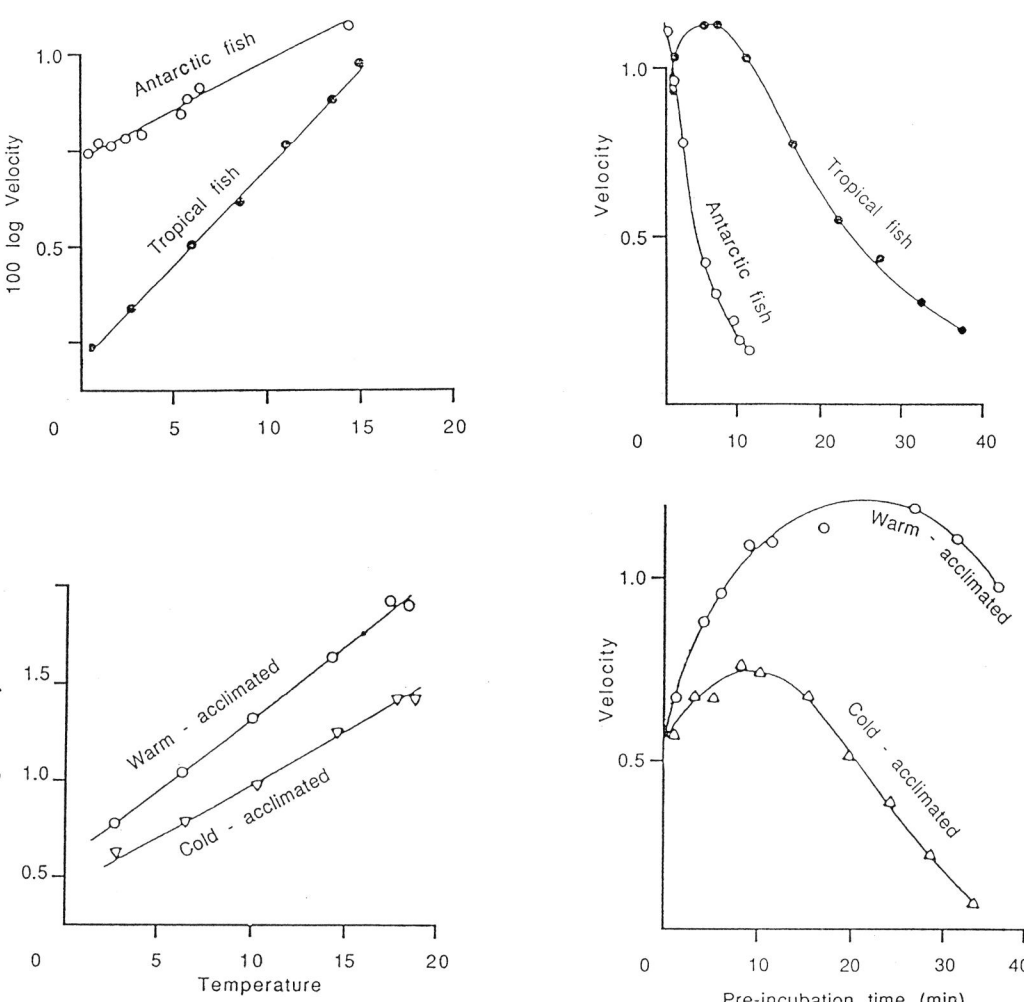

Fig. 1. *Top Left*: The effect of temperature on the specific activity of white muscle myofibrils of an Antarctic fish, *Notothernia* (○) and a tropical fish, *Amphirion* (●). Note that the vertical axis is a log scale so that the differences in the ATPase activity are considerable. *Top Right*: Shows the thermal stability of the contractile system as followed by the loss of ATPase activity when the myofibrils from the Antarctic fish (○) and the tropical fish (●) were maintained at 37°C. Note that the contractile proteins in the Antarctic fish are much less stable at warmer temperatures than those of the tropical fish. *Bottom Left*: The effect of adaptation of carp to different temperature on the specific activity of white muscle myofibrils of warm adapted (○) and cold adapted (▽) fish. Note that the scale is again a log scale so that the increase in myofibrillar ATPase specific activity associated with acclimation to a cold temperature are considerable. From Johnston *et al.* (1975a). *Bottom Right*: Shows the decrease in thermal stability of the myofibrillar ATPase system when carp are acclimated to 25°C and 8°C. The acclimatory temperature was not reduced below 8°C as the carp stopped eating at 5°C. Note that the 'trade off' in increased ATPase on acclimation to a cold temperature seems to be the loss of thermal stability which is possibly because the more enzymatically active myosin molecule has a more open 3 dimensional configuration (From Johnston *et al.*, 1975b). Velocity of the reaction is measured as ATPase specific activity in μmol Pi mg^{-1} protein min^{-1}.

changes in the contractile properties of the muscle fibres (Heap, Watt and Goldspink, 1987).

Switching of transcription of different genes or post-translational processing?

Examining the evidence we postulated that this type of adaptation to a different environmental temperature must involve switching of expression from one set of genes to another set of genes. There were other possibilities such as post-translational modification of the myosin heavy chain (myosin hc). However, this process was completely reversible (Fig 2 left), (Heap *et al.*, 1985), also the thermal stability and energy of activation measurements indicated different proteins. It was also found that adaptation did not occur if the fish were starved before subjecting them to different temperatures. If the fish were starved after they had adapted to 25°C or to 8°C then the myofibrillar ATPase reverted to a default state which was intermediate (Fig 2 right). Hence there seemed to be three quantal

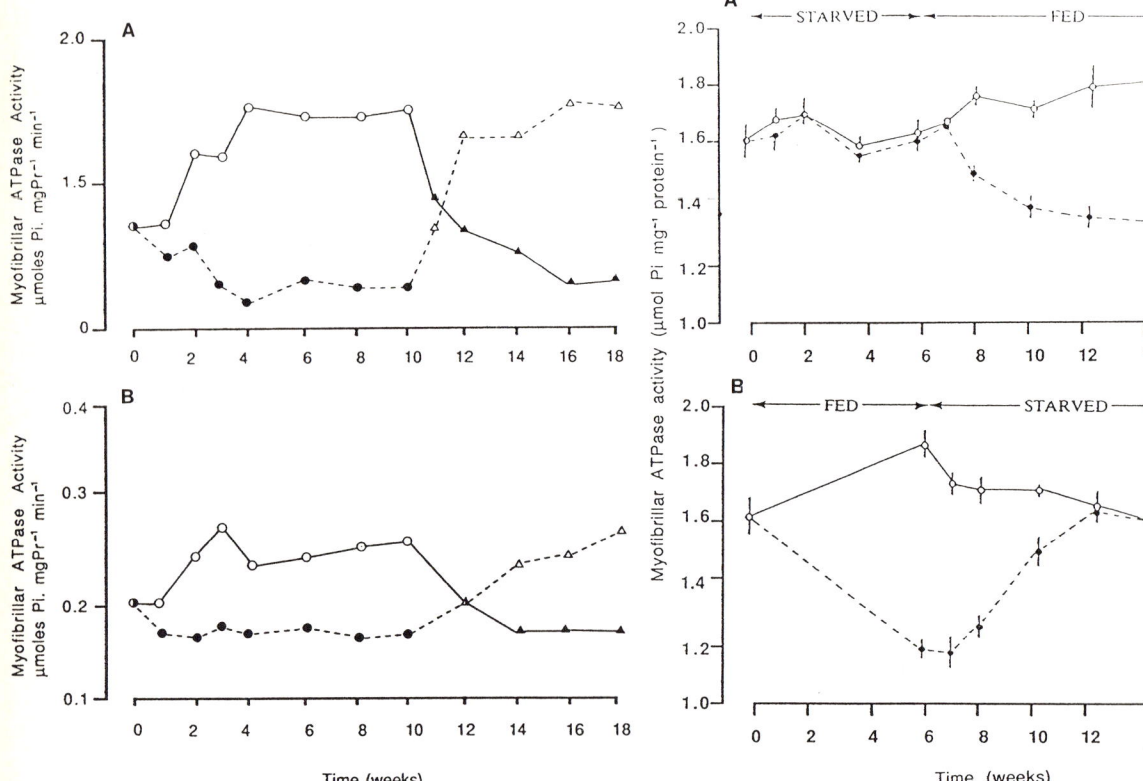

Fig. 2. *Left*: The reversibility of the change in the myofibrillar ATPase activity in white A and red B muscle in warm (28°C) and cold (10°C) acclimated carp. *Right*: Shows how the acclimatory changes in carp muscle are prevented when food is withheld from the fish either before A or after B subjecting them to a warm (28°C) or a cold temperature (8°C). From Heap *et al.* (1986).

states rather than a continuum (Heap *et al.*, 1986). As the information for mammalian muscle genes became available this demonstrated the existence of different myosin heavy chain isoforms (Whalen *et al.*, 1981; Weydert *et al.*, 1985; Wieczorek *et al.*, 1985) coded for by separate genes and it was plausible that different muscle genes may exist in fish for adaptation to warm or to cold environments. As the myosin heavy chain genes codes for the cross bridge including the actin binding and ATPase sites these were the obvious myofibrillar protein genes to focus on. At the time protein separation methods were not particularly sensitive for large proteins and different isoforms of myosin hc might have more or less the same change and mass. Changes at the protein level of myosin hc isoforms in warm and cold acclimated muscle have however, just been achieved by a Japanese group (Hwang *et al.*, 1990). Using protein separation methods (Crockford and Johnston, 1990) also showed that there was a change in myosin light chain 3. Therefore as part of the 'fine tuning' it seems that other contractile protein isoforms are also being produced during temperature acclimation.

Quite subtle changes in sequence may be involved in altering the contractile properties of the myofibrils. It therefore seemed that only a molecular approach could answer if there are different genes involved and if only small sequences differences are responsible for the marked changes in contractility and thermal stability indicated by the physiological and biochemical measurements. Also it was of considerable interest to see by what mechanism certain genes might be regulated by changes in a purely physical parameter such as environmental temperature, if indeed separate myosin heavy chain genes were involved.

Molecular biology

We constructed a carp lambda 2001 library and screened this for myosin hc using mammalian myosin hc cDNA obtained from Dr Whittenhofer's (Maeda *et al.*, 1987) and Dr Zak, Shina, and Umeda's laboratories (Sinha *et al.*, 1982). We also isolated a fish actin clone using a mouse sequence. The clones were then restriction mapped (Fig 3) and 28 non overlapping myosin heavy chain genes were identified indicating that there might be as many as twice the number of myosin hc genes in the carp as in mammals (Gerlach *et al.*, 1990). This was worrisome until we learnt that the bird may have as many as 35 different myosin hc genes (Robbins *et al.*, 1986). The next stage was to find out which fragments belonged to which genes. For this purpose we extracted RNA from red and white epaxial muscle from warm (25°C) intermediate (18°C) and cold adapted fish (8°C) and carried out Northern analysis (Fig 4) using the labelled myosin hc, DNA fragment as the probe (Gerlach *et al.*, 1990). The actin DNA fragment was also hybridized to the same nitrocellulose membranes to provide an indication as to whether the same amount of message had been loaded in each lane. It was possible that although the total RNA loaded was the same the percentage of message might differ as the mRNA only constitutes 5% of total RNA. In Northern blots of RNA from spleen, red and

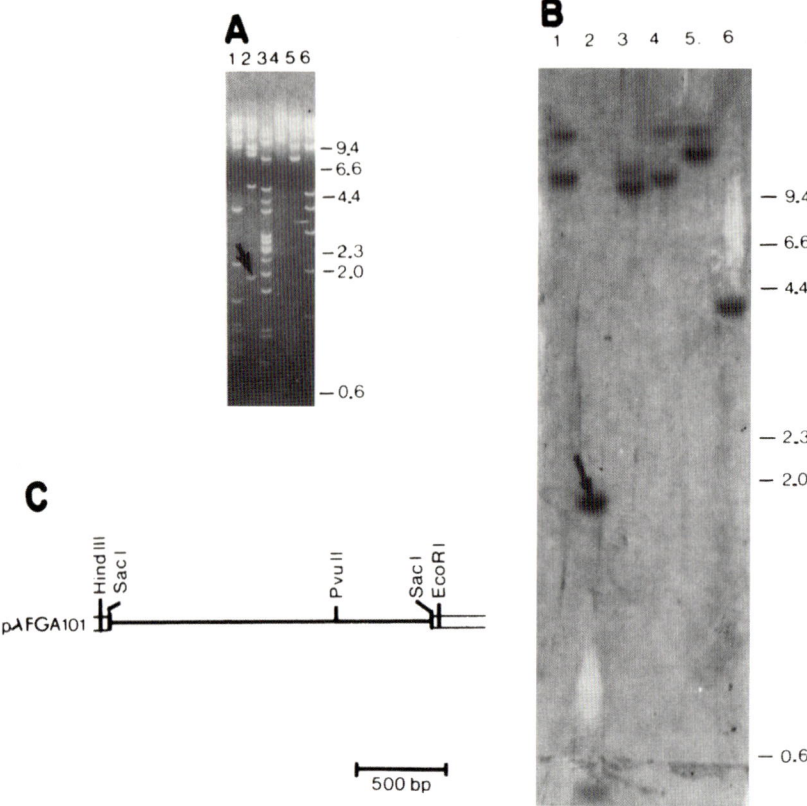

Fig. 3. Restriction enzyme profiles, A and hybridization profile, B and the physical map, C of the FG2 fragment indicated by arrow on B which was characterised as a sequence belonging to the warm white myosin h c isoform gene. The probe used in B to detect the fish myosin hc fragments was the rabbit pMHCB174 (Umeda *et al.*, 1981) encoding the heavy meromyosin region of the rabbit beta cardiac (slow skeletal) using moderate to low stringency conditions. From Gerlach *et al.* (1990).

white muscle our fragment FG2 (which was an overlapping sequence which was a more specific probe FG 1706) was found to hybridize strongly only to the white muscle RNA. The actin probe was found to hybridize spleen as well as red and white muscle RNA but the band was in a slightly different position as this actin would be cytoskeletal β actin DNA. Thus the Northern analysis showed quite distinctly that the fragment belonged to an adult white muscle myosin hc gene. The next step was to see if this was a fragment of a myosin hc gene that is expressed at either warm or at cold environmental temperatures. The Northerns using RNA extracted from white muscle were carried out at different stringencies to check that the amount of mRNA loaded with equal amounts of RNA from fish adapted to different temperatures was more or less the same. At low stringency, because there was sufficient homology between the gDNA and all the myosin hc RNAs the FG2 probe hybridized more or less equally to all lanes. However, when the

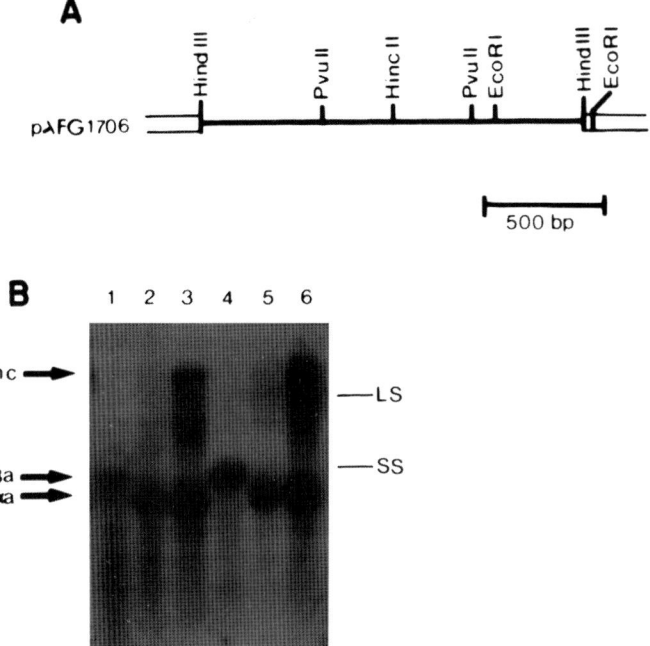

Fig. 4. Restriction map of the FG1706 probe which was derived from the FG2 fragment A as used in Northern analysis B of carp RNA from spleen (lanes 1 and 4) red muscle (2 and 5) and white muscle (3 and 6). A carp actin DNA was also hybridized to the nitrocellulose membrane to establish that approximately the same amount of message had been loaded into each lane. The autoradiograph shows that the ^{32}P labelled myosin hc probe (hc) hybridized only to the white muscle RNA whereas the actin probe bound equally to all lanes including the spleen RNA. The positions of the large and small ribosomal RNAs (LS and SS) are given as a size markers. From Gerlach *et al.* (1990). This study showed that the clone was part of the isoform gene that is expressed in adult white muscle.

stringency was increased the FG 1706 still hybridized strongly to RNA from 25°C acclimated fish (Fig 5). There was some detectable hybridization to the 18°C but not to the 8°C acclimated fish. Thus the first clone to be characterised appears to be from the adult, white, warm skeletal muscle hc gene (Gerlach *et al.*, 1990). The time course for the expression of this gene (Turay *et al.*, 1991) was then established by acclimating carp cold adapted fish to 25°C for periods of up to two months and by measuring the change in the myofibrillar ATPase with the increased expression of the white, warm, myosin hc gene. The expression was appreciable after 3 days (Fig 6) and reached a maximum by about 20 days. This time sequence is slightly quicker than the change in protein which indicates that the control of this type of adaptation is probably at the transcriptional level. Our molecular biology studies have shown there are 28 non overlapping clones and these must presumably represent enough separate myosin hc genes to code for cytoskeletal, cardiac, smooth, embryonic, red, pink and white muscle skeletal myosin as well as the

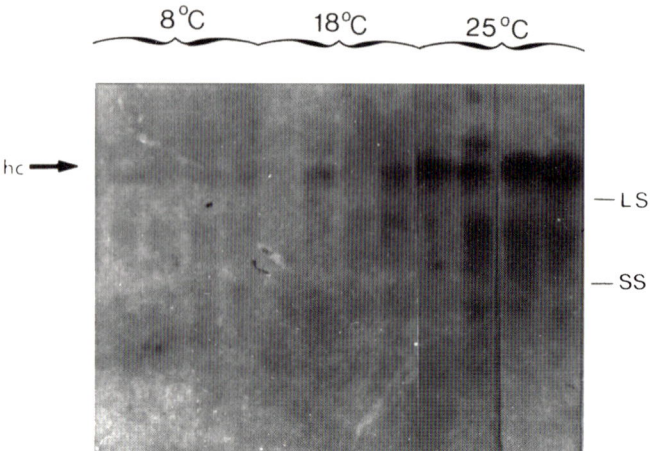

Fig. 5. Northern analysis of RNA extracted from carp acclimated for 4 weeks to 8°C (lanes 1 to 4) 18°C (5 to 8) and 25°C (9 to 12) using the p FG1706 probe (position denoted by hc). The positions of the large and small ribosomal RNAs (LS and SS) are given as a size markers. This showed that the sequences FG 2 and its derivative FG 1706 belonged to the carp adult white warm myosin hc isoform gene.

warm, cold and default skeletal isoforms. There appears to be about 13 different mammalian myosin hc genes and as the carp is tetraploid it appears to have had twice the opportunity to evolve genes for warm and for cold temperature adaptation. There must presumably have been strong selection pressure for the ability to move economically and sometimes quickly at low temperatures in order to locomote efficiently and to predate or escape predators. Therefore in the tetraploid species that are subjected to wide seasonal variations in temperature, the possibility existed for the retension and expression of genes that were beneficial in warm or cold environments.

The next phase of the work is to look at the regulatory sequences to see how the promoter differ for genes that are switched on by increased or decreased temperature. It is possible that differential expression of warm and cold isoform genes is not a direct temperature effect. The switches in transcription may be bought about by metabolic effects, changed hormone levels or even mechanical activity. Myosin isoform switching has been shown to occur in mammalian muscle in that both stretch and force development cause fast muscle in the rabbit to commence to express the slow isoform genes (Williams *et al.*, 1986; Loughna *et al.*, 1990; Goldspink *et al.*, 1991). It is also possible that changes in temperature may differentially effect trans-activators or -repressors.

This question of adaptation to temperature is relevant to global warming. Although the carp which is a eurythermal species, can adapt to changed environmental temperatures, stenothermal fish such as the trout and salmon, that have to live within a narrow temperature range. The question as to why these fish cannot adapt should be answered - do they not have the hot and cold myosin genes or have these isoform genes become non functional?

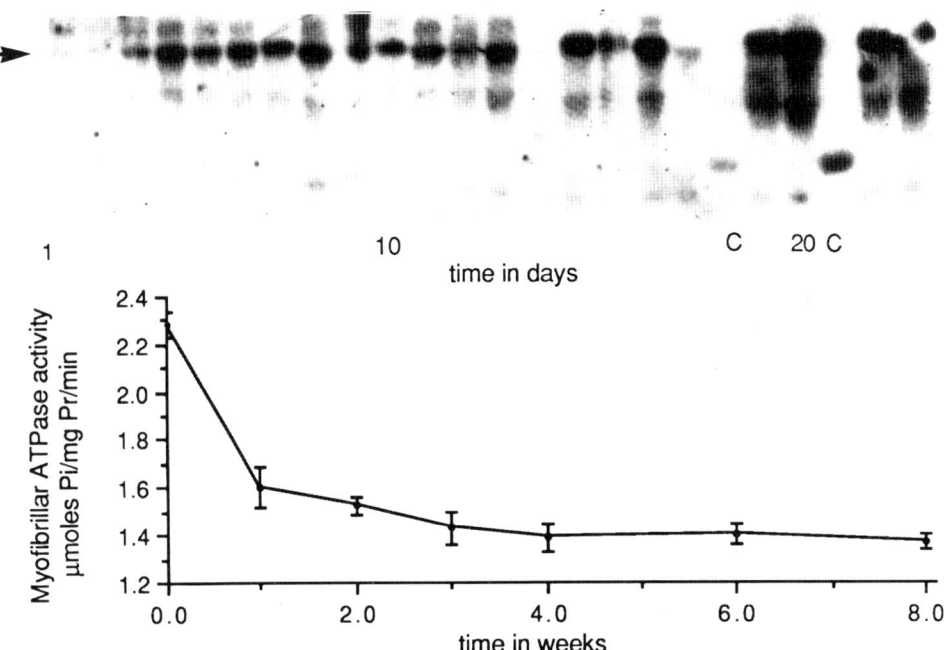

Fig. 6. Top: The time course of expression of the white warm isoform gene as shown by Northern analysis of the RNA extracted from fish that have been acclimated to 28°C following maintenance at 10°C, using the pFG1706 probe. Bottom: The change in the myosin hc protein was also measured using the myofibrillar ATPase method. The new mRNA become apparent at three days but was very abundant by two weeks. This appropriately just precedes the change at the protein level which takes two to four weeks. Hence it seems that this adaptation mechanism is controlled at the level of transcription.

We intend to study the coding region of different isoforms particularly the S_1 fragment ATPase site and hinge regions to see what sequence differences are associated with the different cross bridge cycle properties between say red and white as well as warm and cold adapted muscle. In this way the studies on fish myosin isoform genes could shed light on the basic mechanisms of muscular contraction. To return to the regulatory sequences it seems that the promoters for these myosin genes may well be of value for producing transgenic fish. Attempts thus far to produce transgenic fish using mammalian type gene constructs have not been successful because although the construct is integrated, it is not expressed. It seems that the fish cells do not recognise the mammalian promoters so we plan to use constructs under the control of a fish myosin hc promoter. In addition to introducing genes into the germline we are currently carrying out somatic gene introduction fusion genes. A recent finding by our group (Hansen *et al.*, 1991) has shown that a single direct injection of DNA into muscle tissue as described by Wolff *et al.* (1990) works well for gene constructs under the control of certain muscle specific or general promoters. We reason that the fish myosin hc promoter would be a good sequence to drive different coding sequences including growth factors hormones and DNA vaccines as this would be very cost effective as it requires only a single injection.

Therefore as well as understanding muscle growth and adaptation the study of fish genes could have 'applied spin offs'. Certainly from a fundamental point of view it offers a system for studying a system in a eukaryotic animal that is switched on and off by a simple physical signal which can be accurately controlled as it does not involve the use of hormones or chemicals the levels of which are difficult to control at the cellular level and for which it is often difficult to distinguish between direct and indirect (toxic) effects.

This work was carried out with the aid of research grants from the Natural Research Council and the Wellcome Trust of the UK.

References

Crockford, T. and Johnston, I. A. (1990). Temperature acclimation and the expression of contractile protein isoforms in the skeletal muscle of the common carp (*Cyprinus carpio L*). *J. Comp Physiol (B)*. **160**, 23-30.

Gerlach, G., Turay, L., Malik, T. A., Lida, J., Scutt, A. and Goldspink, G. (1990). Mechanism of temperature acclimation in the carp: a molecular approach. *Am. J. Physiol*. **259**, R237-244.

Goldspink, G., Scutt, A., Martindale, J., Jaenicke, T., Turay, L. and Gerlach, G-F. (1991). Stretch and force generation induce rapid hypertrophy and isoform gene switching in adult skeletal muscle. *Biochem. Trans*. **19**, 368-373.

Hansen, E., Fernandes, K., Goldspink, G., Butterworth, P., Umeda, P.K. and Chang, K-C. (1991). Strong expression of foreign genes following direct gene injection into fish muscle. *FEBS lett*. **290**, 73-76.

Heap, S. P. and Goldspink, G. (1985). Alterations to the swimming performance of carp, *Cyprinus carpio*, as a result of temperature acclimation. *J.Fish Biol*. **29**, 747-753.

Heap, S. P., Watt, P. W. and Goldspink, G. (1985). Consequences of thermal change on the myofibrillar ATPase of five freshwater teleosts. *J.Fish Biol*. **26**, 733-738.

Heap, S. P., Watt, P. W. and Goldspink, G. (1986). Myofibrillar ATPase activity in the carp, *Cyprinus carpio*: interactions between starvation and environmental temperature. *J. Exp. Biol*. **123**, 378-382.

Heap, S. P., Watt, P. W. and Goldspink, G. (1987). Contractile properties of goldfish fin muscles following temperature acclimation. *J. Comp. Physiol. B* **157**, 219-225.

Hwang, G. C., Watabe, S. and Hashimoto, K. (1990). Changes in carp myosin ATPase induced by temperature acclimation. *J.Comp.Physiol. B*. **160**, 233-239.

Johnston, I. A., Davison, W. and Goldspink, G. (1975a). Adaptations in magnesium-activated myofibrillar ATPase activity induced by environmental temperature. *FEBS Lett*. **50**, 293-295.

Johnston, I. A., Davison, W., Walesby, N. J. and Goldspink, G. (1975b). Temperature adaptation in myosin of Antarctic fish. *Nature*, **254**, 74-75.

Johnston, I. A., Frearson, N. and Goldspink, G. (1973). The effect of environmental temperature on the properties of myofibrillar ATPase from various species of fish. *Biochem. J*. **133**, 735-738.

Johnston, I. A. and Goldspink, G. (1975c). Thermodynamic activation parameters of fish myofibrillar ATPase enzyme and evolutionary adaptations to temperature. *Nature*, **257**, 620-622.

Loughna, P. T., Izumo, S., Goldspink, G. and Nadal-Ginard, B. (1990). Rapid changes in sarcomeric myosin heavy chain gene and alpha-actin expression in response to disuse and stretch. *Development*, **109**, 217-223.

Maeda, K., Sczakiel, G. and Wittinghofer, A. (1987). Characterization of cDNA coding for the complete light meromyosin portion of rabbit fast skeletal muscle myosin heavy chain. *Eur. J. Biochem.* **167**, 97-102.

Penney, R. K. and Goldspink, G. (1979). Compensation limits of fish muscle myofibrillar ATPase enzyme to environmental temperature. *J. Therm. Biol.* **4**, 269-272.

Penney, R. K. and Goldspink, G. (1981). Temperature adaptation by the myotomal muscle of fish. *J.Therm.Biol.* **6**, 297-306.

Robbins, J., Horan, T., Gulick, J. and Kropp, K. (1986). The chicken myosin heavy chain family. *J.Biol.Chem.* **261**, 6606-6612.

Rome, L. C., Loughna, P. T. and Goldspink, G. (1985). Temperature acclimation: improved sustained swimming performance in carp at low temperatures. *Science.* **228**, 194-196.

Sinha, S. M., Umeda, P. K., Kavinsky, C. J., Rajamanickam, C., Hsu, H-J., Jakovcic, S. and Rabinowitz, M. (1982). Molecular cloning of mRNA sequences for cardiac α- and β-form myosin heavy chains: expression in ventricles of normal, hypothyroid, and thyrotoxic rabbits. *Proc.Natl.Acad.Sci.* USA **79**, 5847-5851.

Turay, L., Gerlach, G.-F. and Goldspink, G. (1991). Changes in myosin h.c. gene expression in the common carp during acclimation to warm environmental temperatures. *J. Physiol.* **435**, 102P.

Umeda, P. K., Sinha, A. M., Jakovcic, S., Meren, S., Hsu, H-J., Subramanian, K. N. and Zak, R. (1981). Molecular cloning of two fast myosin heavy chain cDNAs from chicken embryo skeletal muscle. *Proc.Natl.Acad.Sci.* USA. **78**, 2843-2847.

Weydert, A., Daubas, P., Lazaridis, I., Barton, P., Garner, I., Leader, D., Bonhomme, F., Catalan, J., Simon, D., Guenet, J., Gros, F. and Buckingham, M. (1985). Genes for skeletal muscle myosin heavy chains cluster not located on the same mouse chromosome as a cardiac myosin heavy gene. *Proc.Natl.Acad.Sci.* USA. **82**, 7183-7187.

Whalen, R. G., Sell, S. M., Butler-Brown, G. S., Schwarty, K., Bouveret, P. and Pinset-Harstrom, I. (1981). Three myosin heavy-chain isozymes appear sequentially in rat muscle development. *Nature Lond.* **292**, 805-809.

Wieczorek, D. F., Periasamy, M., Butler-Brown, Whalen, G. and Nadal-Ginard, B. (1985). Co-expression of multiple myosin heavy chain genes in addition to a tissue specific one in extraocular musculature. *J.Cell Biol.* **101**, 618-629.

Williams, P. E., Watt, P., Vicik, V. and Goldspink, G. (1986). Effect of stretch combined with electrical stimulation on type of sarcomeres produced at the ends of muscle fibres. *Exp.Neurol.* **93**, 500-509.

Wolff, J. A., Malone, R. W., Williams, P., Chong, W., Ascadi, G., Jani, A. and Felgner, P. L. (1990). Direct gene transfer into mouse muscle in vivo. *Science.* **247**, 1465-1468.

Printed in Great Britain © Society for Experimental Biology 1992   151

REGULATION OF MUSCLE GENE EXPRESSION IN CRUSTACEA OVER THE MOULT CYCLE

ALICIA J. EL HAJ, PAUL HARRISON and NIA M. WHITELEY

School of Biological Sciences, University of Birmingham, P.O. Box 363, Birmingham, B15 2TT

Summary

Muscle growth in Crustacea may occur during specific stages of the moult cycle, focused around ecdysis when the old cuticle is shed and the new cuticle expands. In order to determine the moult stages in which sarcomeric proteins are synthesized and the regulatory factors involved, actin mRNA levels have been measured in the muscles of two crustaceans. These levels have been followed throughout the moult cycle and in response to passive stretch of walking leg muscle *in vivo* and to exogenous ecydsteroids applied to muscle preparations *in vitro*. Actin mRNA levels in both claw and leg muscles were elevated during the pre and postmoult stages of the moult cycle. However, varying patterns of expression are found in claw and leg muscle at specific stages of pre and postmoult. There was no increase in actin mRNA expression in extensor leg muscles *in vitro* after 6 hours exposure to elevated premoult levels of ecdysteroids. Immobilization of intermoult walking legs to maintain the extensor muscle in continuous passive stretch did not result in increased levels of actin mRNA after 5 days. These results are discussed in relation to the regulation of muscle growth over the moult cycle and to the molecular processes which may be responsible for controlling muscle protein synthesis.

Introduction

Crustacean growth is an intermittent process centred around the moult, or ecdysis, when the old, calcified, exoskeleton is shed. Immediately following ecdysis, body volume increases greatly due to high rates of water uptake, which in the American lobster (*Homarus americanus*) occurs across the lining of the gut (Mykles, 1980). In the swimming crab, *Callinectes sapidus*, water uptake into the haemolymph results in a 5 fold increase in hydrostatic pressure, enlarging the space available for new tissue growth (Defur, Mangum and McMahon, 1985). The events surrounding ecdysis, in preparation for (premoult; stages D1 to D4), in recovery from (postmoult; stages A, B, C1 to C3) and the intermediate stage (intermoult; stage C4), comprise the crustacean moult cycle. Ecdysis is the physical culmination of a series of preparatory physiological and cellular events which occur throughout the body of the animal. Many of these premoult changes appear to be regulated by ecdysteroid hormones, principally 20-hydroxyecdysone and ponasterone A. During premoult, haemolymph ecdysteroid titres increase

Key words: Crustacea, muscle, gene, actin, moult cycle.

markedly through stages D1 to D2 and then fall in D3 to D4, immediateley prior to ecdysis (Snyder and Chang, 1991). These fluctuations in haemolymph ecdysteroid titre have been measured in a variety of crustaceans and are essential for a successful moult (Chang, 1989).

Early studies on growth in Crustacea were on intermoult animals of different sizes dealing mainly with moult frequencies and increments in the linear dimensions of the carapace (Hartnoll, 1965; Crothers, 1967). The specific stages of the moult cycle when growth of most soft tissues occurs have been poorly studied. However, growth of the claw and leg muscles has been investigated by observations of morphological characteristics and measurements of biochemical changes in the muscle fibres over the moult cycle (Houlihan and El Haj, 1985; Skinner, 1966; Mykles and Skinner, 1985a; Govind, She and Lang, 1977; El Haj, Govind and Houlihan, 1984). Longitudinal growth of crustacean muscle fibres occurs by addition of new sarcomeres or by lengthening of existing sarcomeres (Houlihan and El Haj, 1985). Both mechanisms may be operating at different phases of growth with sarcomere lengthening occuring during larval stages until the first adult stage when addition of new sarcomeres occurs (Govind et al. 1977). Increases in myofibrillar surface area is due to an increase in the size of the myofibres as a result of the addition of thick and thin filaments by longitudinal myofibrillar splitting as in vertebrate muscle (El Haj et al. 1984; Goldspink, 1980).

Morphological measurments during muscle lengthening of the muscle fibres in the legs of the American lobster, *Homarus americanus*, and the shore crab, *Carcinus maenas*, have shown that increases in fibre length occur over ecdysis and during the immediate postmoult period (El Haj et al. 1984; Houlihan and El Haj, 1985; El Haj and Houlihan, 1987). In the lobster extensor muscle, there is a shift in frequency of myofibril area over the moult from a greater percentage of large myofibres in premoult fibres to predominantly smaller myofibrils in the postmoult fibre (El Haj et al. 1984); this suggests that myofibril splitting occurs over the moult possibly in response to stretching of the new cuticle. Subsequently actin and myosin filaments may be added during the intermoult, causing enlargement of the myofibrils and therefore the fibre to maximal intermoult size.

Measurements of protein synthesis rates in the leg muscles of *C. maenas* using radiolabeled amino acid uptake experiments both *in vivo* and *in vitro* have shown that there is a 10 fold increase in the rates of protein synthesis during the premoult period (D3 to D4) (El Haj and Houlihan, 1987). These *in vivo* experiments employed an adaptation of a flooding dose technique devised by Garlick, McNurlan and Preedy (1980) which enabled whole body and muscle protein synthesis rates to be measured. Unfortunately due to the intermittent pattern of growth in crustaceans, protein degradation rates could not be determined with this method. Thus, there is no data available on total protein turnover and the relative contribution of changes in protein degradation. These results do suggest, however, that the leg muscles are synthesizing new sarcomeric proteins during the premoult period and possibly assembling new sarcomeric units during ecdysis and the immediate postmoult stages.

In contrast to leg muscles, claw muscles undergo massive atrophy during premoult to allow the remaining claw muscle to be pulled through the narrow basi-ischium joint during ecdysis (Mykles and Skinner, 1982). The chelae muscle of the Bermuda land crab, *Gecarcinus lateralis* Freminville, atrophy by approx 30-60% during premoult (Skinner, 1966). After ecdysis, the muscle grows to fill the newly expanded exoskeleton of the chelae. Studies on the rates of incorporation of pulse-labelled amino acids into claw muscles in the Bermuda land crab, *Geocarcinus lateralis* (Skinnner, 1965; Mykles and Skinner, 1985a), however, revealed two peaks in protein synthesis rate, one in premoult and the other in postmoult. These two peaks can be explained as a premoult increase due to the production of muscle degrading proteinase enzymes and the postmoult increase due to muscle growth (Mykles and Skinner, 1985b). Leg muscles do not undergo atrophy prior to ecdysis as they are sufficiently narrow to pass through the joints at the base of the legs and the growth pattern appears to be slightly different. Muscle growth in crustaceans therefore takes place at different stages of the moult cycle in different somatic muscle groups. This location dependant variation in muscle synthesis rates suggests that differential regulation of sarcomeric proteins may be involved. The regulatory mechanisms and the factors responsible for controlling moult related muscle growth have not yet been determined although we propose that both hormonal and mechanical stimulation are involved.

Ecdysteroids have been shown to directly effect protein synthesis in the exoskeleton, hypodermis and hepatopancreas (Traub, Gellissen and Spindler, 1987; Paulson and Skinner, 1991). They may also be involved in the regulation of protein synthesis rates in crustacean muscle tissue as separate studies on *C. maenas* have shown that increased levels of protein synthesis coincide with the elevation in haemolymph ecdysteroid titre (Lachaise, Lageuex, Leyerasen and Hoffman, 1976). Ecdysteroids may also vary in their relative importance in different species of Crustacea, e.g. in *Carcinus*, Ponasterone A is the predominant ecdysteroid (Lachaise, Carpentier, Somme, Colardeau and Beydon 1986; Lachaise, Meister, Hetni and Lafont, 1989). It is likely that these hormones are working in a co-ordinated fashion. There is very little direct *in vitro* or *in vivo* evidence to link ecdysteroids with protein production in muscle tissue. *In vitro* experiments on the direct action of ecdysteroids on protein synthesis in integumentary and hepatopancreas tissue isolated from the crayfish, *Astacus leptodactylus*, and the land crab, *G. lateralis*, showed changes which were related to events *in vivo* (Traub et al. 1987; Paulson and Skinner, 1991). Preliminary experiments by Whiteley, Taylor and El Haj (1992) have shown that leg extensor muscles from *Carcinus maenas* exposed to 20 hydroxyecdysone (10^{-6} M; 10^{-7} M), *in vitro,* resulted in a significant increase in total RNA synthesis rates compared to control muscles incubated in hormone free media after 24 hours incubation. These results suggest a direct effect of ecdysteroids on muscle tissue but do not necessarily indicate an increase in mRNA transcription. If ecdysteroids are involved in modulating muscle growth, these hormones must also be responsible for promoting synthesis and degradation in different muscle groups in the body.

A prerequisite for the effect of steroid hormones at the gene level is the presence of nuclear hormone receptors in target tissues. In insects, nuclear ecdysteroid receptors which have the ability to bind to DNA at specific target sites have been localized and characterised (Segraves and Richards, 1990). The ecdysone response elements found in insects are relatively small fragments, 23 bp, which are necessary and sufficient for the ecdysone response of a target promoter in transfection assays (Riddihough and Pelham, 1986). Cherbas, Lee and Cherbas (1990) have identified 3 binding sites and derived a putative consensus sequence for ecdysone binding. The sequence derived has homologies with response elements of a number of vertebrate receptors, i.,e. those of thyroid hormone and retinoic acid. In Crustacea, intracellular high affinity saturable ecdysteroid binding components have been found in the hypodermis and mid gut gland of the crayfish *Orconectes limosus* (Kuppert, Wilhelm and Spindler, 1978) and *Astacus leptodactylus* (Spindler, Dinan and Londershausen, 1984; Londershausen and Spindler, 1985). Ecdysteroid hormone receptors have also been found in *Carcinus* and *Artemia* (Spindler et al. 1984). Although the cytoplasm of several target tissues in the crayfish are able to bind to 20-hydroxyecdysone, the capacity to bind to this hormone was much less in the muscle tissue compared to the hypodermis and male gonads (Kuppert et al. 1978). However, it has been shown that the affinities of the ecdysteroid receptors are very similar in different crustacean species and that affinities in different tissue in the same species are similar with the receptor having the highest affinity for Ponasterone A, followed by 20 hydroxyecdysone and then ecdysone (Spindler et al. 1984). Membrane bound ecdysteroid binding sites may also be present but this carrier mediated response has only been shown in the crayfish hypodermis (Spindler and Spindler-Barth, 1989). It is also possible that receptor expression may be differentially regulated throughout the moult cycle in different tissues. But to date, the presence of intracellular ecdysteroid receptor proteins has not as yet been shown in intermoult muscle tissue examined.

In addition to these endocrine factors, mechanical influences may also play a role in the control of muscle protein synthesis especially after ecdysis when the muscles are stretched (Fig. 1). Muscle lengthening in intermoult animals has been induced by maintaining the MC joint at 45° angle thereby maintaining the extensor muscle in a stretched position. After three weeks of stretch, there was an maximal increase in the length of the muscle fibres by 10% of control fibre length (Houlihan and El Haj, 1985). Similar experiments on increases of muscle fibre length in *C. maenas* over the moult cycle demonstrated an increase of 29% in length between one day prior to ecdysis and five days after (Houlihan and El Haj, 1985). Mammalian muscle has been shown to be responsive to passive stretch (Loughna, Izumo, Goldspink and Nadal Ginard, 1990) and crustacean muscle may be responding in a similar manner.

In order to define the factors regulating muscle growth, we have addressed the question at the molecular level by measuring mRNA levels of a sarcomeric protein, actin throughout the moult cycle in claw and leg muscle. By using a

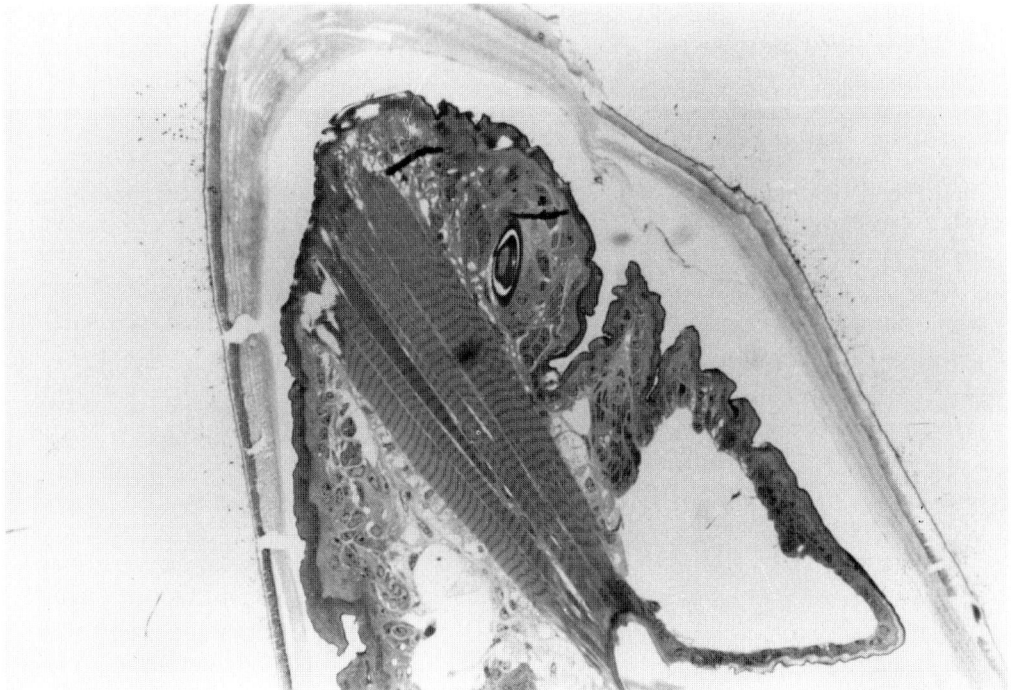

Fig. 1. Light micrograph longitudinal section of the muscle and overlying cuticle in the M-C joint in the walking leg of *Homarus americanus* 6 hours prior to ecdysis. The section illustrates the folding of the new cuticle with the attached muscle which will expand during increased haemostatic pressure in the leg of the animal over ecdysis resulting in stretch of muscle fibres. (×120)

heterologous cDNA for actin, we can take advantage of the strong homologies that exist throughout the animal kingdom from animals as diverse as mammals, insects and echinoderms (Engel, Gunning and Kedes, 1982; Vaudekerckhoe and Weber, 1984). Recently, Sastre and colleagues (see Chapter x; Macias and Sastre, 1990) have isolated actin mRNA for *Artemia sp.* which has been used for this study. Initially, a comparison has been made of the mRNA species identifiable in the different crustacean species used in these experiments by Northern analysis using mammalian and *Artemia* actin probes.

Our previous studies of moult related growth in *C. maenas* walking leg muscles determined that levels of actin mRNA were elevated during premoult and postmoult period compared with the intermoult stage (Whiteley et al. 1992). However, these studies were carried out on animals selected to represent general moult cycle stages and specific pre and post moult stages of the cycle were not identified. As part of an ongoing study investigating the factors responsible for regulating the seasonal moult cycle and associated muscle growth of the crayfish, *Austropotamobius pallipes*, animals have been induced to moult in the laboratory by changing environmental conditions. This has conveniently provided muscle samples which can be collected at specific stages of pre and postmoult. In this way we can monitor actin mRNA levels in both leg and claw muscles at specific points in the moult cycle. Studies have also been extended to the lobster, *H. gammarus*. Immobilization experiments similar to those carried out on *Carcinus* (Houlihan and El Haj, 1985) have been conducted on walking legs of *H. gammarus* to investigate the effects of passive stretch on mRNA synthesis. Isolated lobster extensor muscles have also been used to determine the direct effect of ecdysteroids on muscle actin mRNA synthesis.

In summary, levels of actin gene expression using 2 actin cDNA probes have been followed in the leg and claw muscle of 2 decapod crusaceans *in vivo* and *in vitro*; in response to exogenous ecdysteroids and experimentally applied mechanical stretch, to determine the possible role of these factors on the regulation of protein synthesis at the level of transcription.

Materials and methods

In vivo *experiments*

Freshwater crayfish, *A. pallipes*, (license to collect animals granted by English Nature) were collected by SCUBA divers from a pool, Litchfield, Staffordshire and maintained in tanks of aerated dechlorinated tapwater at 15°C and in constant light. Animals were moult staged according to Stevenson (1975). Leg and claw muscles were dissected from animals at the desired moult stage, frozen in liquid nitrogen and stored at −70°C for analysis.

European lobsters, *Homarus gammarus*, were purchased from the wholesale market, Birmingham, and maintained in a recirculated aerated sea water system, 15°C, 35‰S. For experiments on passive stretch, second and third walking legs were fully flexed between the meropodite and carpodite segments and tied with rubber bands to maintain the carpopdite extensor muscle in a stretched position. Legs were maintained in this position for periods up to 5 days after which they were removed together with the contralateral untreated legs and the carpopodite extensor and flexor muscles were isolated and stored in liquid nitrogen.

In vitro *experiments*

Fourth and fifth walking legs of intermoult *H. gammarus* were removed by autotomy. The anterior portion of the carapace of the merus segment was removed and the carpopodite extensor and accessory flexor muscles dissected free. The exposed flexor muscle within the remaining exoskeleton of the merus segment was placed in a sterile culture dish and incubated at 15°C in a recirculated modified version of Cole's (1941) lobster saline which contained amino acids at concentrations similar to those of the haemolymph of *H. gammarus* (Camien, Garlet, Duchateau and Flordin, 1951).

Control and experimental preparations were established using left and right legs. After one hour of incubation 20 hydroxyecdysone was added to the experimental media at a final concentration of 1500 ng/ml, equivalent to that measured by Snyder and Chang (1991) in the haemolymph of premoult American lobsters. Incubations were continued for a further six hours and terminated by freezing the tissue in liquid nitrogen.

Northern analysis and slot blots

Total RNA was isolated from the muscle tissue by acid guanidine thiocyanate/phenol/chloroform technique described by Chomcznski and Sacchi (1987).

For the Northern blots, 10 μg samples of total RNA were incubated for 15 minutes in a solution containing 50% formamide, 6.1% formaldehyde and 1× gel running buffer (see below). Following snap cooling in ice the RNA was separated by electrophoresis under denaturing conditions on a 1.5% agarose gel containing 6.1% formaldehyde and 1× gel running buffer. The running buffer was 20 mM MOPS (3(N-Morphlino)-propane-sulphonic acid), 5 mM sodium acetate (pH 7.0) and 1 mM disodium EDTA. Following electrophoresis, the gels were soaked for 20 min in 0.05 N NaOH, rinsed in several changes of water and the RNA transfered to nitrocellulose membranes by capillary blotting (Dabre, 1988).

Samples of total RNA between 0.25 and 1 μM were slot blotted onto nitrocellulose membranes with a Hybri-slot manifold (BRL, Maryland, U.S.A.) using the standard protocol described by Sambrook, Fritsch and Maniatis (1989). After transfer, the nitrocellulose membranes from both Northern and slot blots were dried at room temperature for 30 minutes and baked in a vacuum oven for 2 hours at 80°C.

Probe labelling and hybridization

In our initial studies, we used a mammalian cDNA probe to skeletal α actin to identify actin mRNA levels in *Carcinus* leg muscles over the moult cycle. The mammalian cDNA (Minty, Caravetti, Robert, Cohen, Duabas, Weydert, Gros and Buckingham, 1981) is 1150 bp in length coding for approx 90% of the α actin mRNA and approx 200 nucleotides of 3' non coding sequence and has been used in this study for Northern analysis of crustacean samples. In addition for Northern and slot blot anlysis, an actin cDNA insert from the Crustacean *Artemia sp.* contained in plasmid pArAct205 (Macias and Sastre, 1990) was used. The EcoRI insert of this plasmid was isolated from a low melting point agarose gel and 25 ng samples were labelled by random primer labelling with (aP^{32})dCTP to a specific activity of approximateley 10e^9 dpm/μg.

Northern and slot blots were prehybridized at 65°C for 3 hours in 6×SSC (1×SSC is 150 mM sodium chloride and 15 mM sodium citrate pH 7.0) 0.5% SDS, 5× Denhardt's solution (1× is 0.2% ficoll, 0.02% polyvinylpyrolidone and 0.02% bovine serum albumin) with 100 μg/ml denatured sonicated salmon sperm DNA. They were hybridized overnight at 65°C in 6×SSC, 0.5% SDS, 2× Denhardts solution, 100 μg/ml denatured sonicated salmon sperm DNA and the probe at a concentration of approx 2×10^6 dpm/ml. Nonspecifically bound probe was removed from the filters by washing in 4×SSC and 0.1% SDS; 2×SSC and 0.1%

SDS; and 1×SSC and 0.1% SDS at 65°C for varying amounts of time depending of the activity of each filter. After washing, the filters were exposed to x-ray film (Hyperfilm-MP, Amersham) in the presence of intensifying screens and left at −70°C for 4-16 h. Hybridization of the slot blots was quantified by scanning densitometry (LKB 2202 Ultrascan Laser).

Results and discussion

The results of Northern analysis of total RNA extracted from the walking leg muscles of the two crustaceans and also from mouse skeletal muscle and *Artemia* is shown in Fig. 2. Northern blot A was hybridized with mouse α actin cDNA, and Northern B with Artemia actin cDNA(pArAct205). In both cases the actin cDNA probes hybridized to mRNAs of approximately the same size (approx 1.6 kb) in each of the species studied. The actin probe isolated from mammalian muscle was originaly used to identify actin mRNAs in mouse cell lines which were found to be 1.65 and 2.1 kb in size (Minty et al. 1981). This probe has also been used to identify

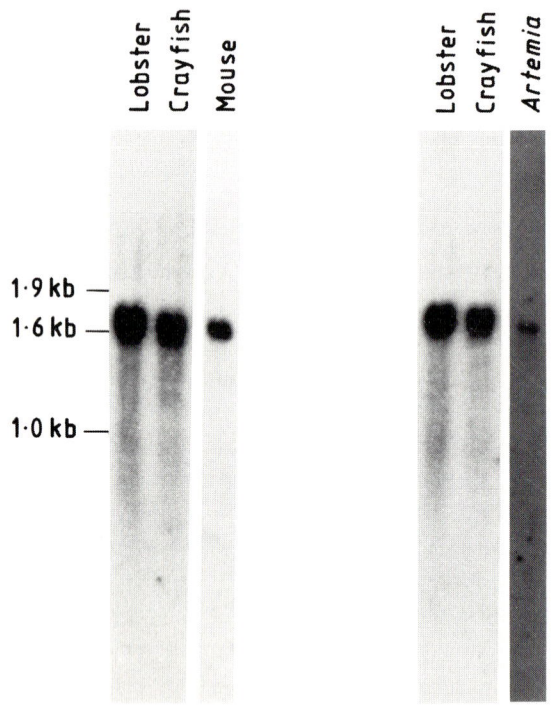

A. Mouse actin probe   B. *Artemia* actin probe

Fig. 2. Northern blot of total RNA extracted from the leg muscles of the European lobster, *Homarus gammarus* and the freshwater crayfish, *Austropotambius pallipes*, hybridized with a mouse actin cDNA (A) and Artemia actin cDNA probe (B). Hybridization conditions are as described in materials and methods.

2 bands of 1.6 and 1.8 kb in size in the leg muscles of *Carcinus* (Whiteley et al. 1992). This evidence combined with the present results shown in Fig. 2 clearly support the use of these heterologous actin probes for further determination of actin mRNA levels in crayfish using slot blot analysis. This is not surprising as actin has been shown to be a highly conserved protein throughout the animal kingdom and is found in various isoforms in many eukaryotic cells. The four muscle isoforms present in mammals including skeletal muscle α actin, cardiac muscle actin and two smooth muscle actins are closely related to each other. The amino acid sequence of these actins is well conserved in the majority of species studied so far. In contrast, the number of genes encoding the different actin proteins varies in number quite considerably throughout the animal kingdom, from a single gene for yeast (Gallwitz and Sures, 1980) up to 25-30 genes for humans, some of which are linked on a chromosome (Engel et al. 1982).

Previous work on *Carcinus* leg muscles by Whiteley et al. (1992) followed changes in actin mRNA levels over the moult cycle using the mammalian mouse α actin probe and found large increases in actin mRNA levels in pre and postmoult leg muscle. These changes have now been quantified in the crayfish using slot blot analysis and the actin cDNA probe isolated from *Artemia sp*. In the leg muscle a large elevation in actin mRNA levels occurred in stage D1 and D3 prior to the moult with slightly elevated levels remaining into postmoult stage B (Fig. 3a). Intermoult levels (C) were taken as control levels of actin expression and comparisons are given as a percentage change from these values. In the crayfish claw muscle (Fig. 3a), a different pattern of actin mRNA expression was measured over the moult cycle with actin mRNA levels slightly elevated in premoult stage D1, a marked decrease immediateley prior to the moult during the D3 phase and maximal levels during the postmoult B stage. The accompanying haemolymph ecdysteroid concentrations are given in Fig. 3b which shows that hormone titres were inceased 4 fold to reach peak values during the premoult phase D1, fell below intermoult levels during D3 just prior to the moult remaining slightly elevated during the immediate postmoult phase B.

Levels of mRNA expression for the sarcomeric protein actin vary between claw and leg muscles in individual crayfish. Although elevation in actin mRNA occurs in both premoult and postmoult stages, there are differences in magnitude of the response. In the leg muscle, the increase in actin mRNA levels observed in the crayfish agree with previous data on leg muscle showing elevated rates of protein synthesis (El Haj and Houlihan, 1985) and elevated actin mRNA levels in *Carcinus* (Whiteley et al. 1992) during pre and postmoult. This indicates that sarcomeric actin may be synthesized in the leg muscles predominantly during D1 and D3 stages. The elevation in actin mRNA levels in postmoult may be achieved by continued transcription of new mRNA in response to a maintained stimulus or by stabilizing the actin mRNA through the late premoult and the early postmoult stage. Preliminary histological studies of the ecdysial attachment site of leg extensor fibres noted the presence of small assembling sarcomeric units present immediateley following the moult (El Haj et al. 1984). Under the electron

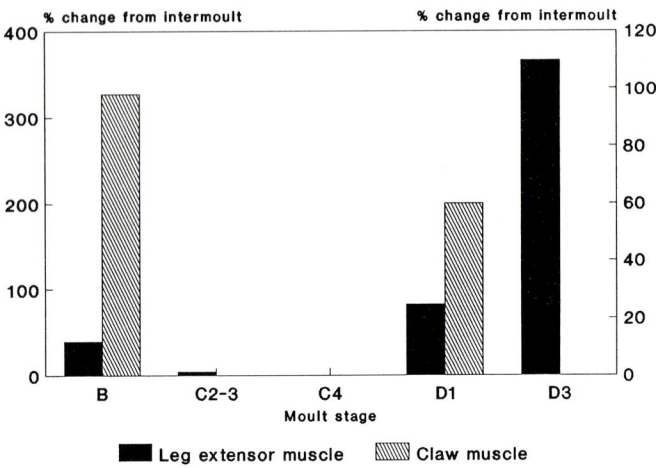

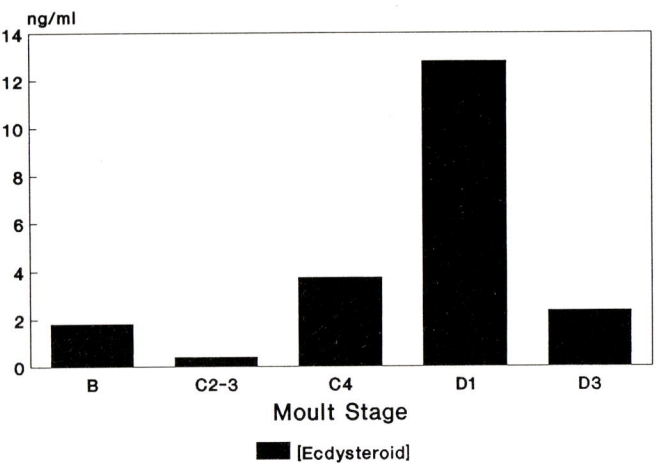

Fig. 3. A) Actin mRNA levels over the moult cycle of the crayfish, *A. pallipes*, in the leg and claw muscle. Values are expressed as a percentage of intermoult levels, in the case of leg, stage C4 and in the case of the claw, stage C2- 3. Leg and claw muscles are isolated from the same individual. B) Haemolymph ecdysteroid concentrations in the crayfish used in (a) expressed as ng/ml. Ecdysteroids were measured using a standard radioimmunoassay using primary antibodies to active ecdysteroids (kindly donated by J. Koolman, Marburg).

microscope these regions showed disassembled actin and myosin filaments. By three to four days following ecdysis, these sites were no longer found. Autoradiographic evidence has shown that in postmoult muscle fibres, proteins are being added to the exoskeletal region of the fibres to a greater extent than the rest of the muscle fibre (El Haj et al. 1987). It may be that the mRNA for actin and myosin are transcribed and translated during premoult in preparation for ecdysis with sarcomeric assembly and further translation and production of proteins occurring during the postmoult period at the exoskeletal end of the fibre.

In contrast, the greatest elevation in actin mRNA levels in claw muscle occurs during D1 and postmoult with levels falling immediateley prior to the moult. This agrees with the findings of Mykles and Skinner (1985a) who showed that atrophy occurs prior to the moult and the majority of muscle growth occurs during the late postmoult phase. Mykles and Skinner (1985a) have also measured protein synthesis rates in the claw and shown that increased rates of protein synthesis occur during both the pre and postmoult stages. This increase in total protein synthesis rates may include synthesis of degradative enzymes resulting in atrophy of the claw muscle. Mykles and Skinner (1985b) have demonstrated the importance of calcium dependant proteinases in moult induced claw muscle atrophy. Although rates of protein synthesis may be increased in all muscles during premoult in response to a systemic stimulus, the elevated degradation rates in the claw muscle and associated proteinase activity results in localized catabolism of proteins and a net reduction in protein turnover.

Ecdysteroid titre is elevated in the haemolymph prior to the moult which coincides with the elevation in actin mRNA levels in both leg and claw muscle during this stage. Ecdysteroids may be involved in the regulation of these changes in expression. Future studies on the presence of ecdysteroid response regions on sarcomeric protein genes could provide further support. However, changes in actin mRNA levels in both claws and leg muscles cannot be explained solely in terms of hormonal control as ecdysteroid titres fell prior to ecdysis, e.g. in the leg muscle during postmoult, when the ecdysteroid titre has fallen, mRNA levels remain elevated. In the cheliped muscle, actin mRNA levels are at their highest during postmoult. As previously outlined, this could be due to the stabilization of actin mRNA through ecdysis to prevent degradation. The stimulus for actin mRNA synthesis could be the premoult peak in hormones which exert an influence on the transcription of actin genes. During ecdysis, mechanical stimulation may be involved in the regulation of translation and protein modifications for the assembly of sarcomeric proteins.

If ecdysteroids are acting on the muscle to promote sarcomeric protein synthesis, it should be possible to promote actin mRNA transcription *in vitro*. After incubation for six hours in an ecdysteroid concentration similar to that found in premoult, there was no significant elevation in levels of actin gene expression between treated and control intermoult extensor leg muscle maintained *in vitro* ($P>0.01$). The results could be due to the tissue or stage related absence of receptors for ecdysteroids expressed in the muscle or due to the time frame of the experiment. Whiteley et al. (1992) showed that total RNA synthesis increased in *Carcinus* extensor muscle after 24 hours *in vitro* in response to premoult concentrations of ecdysteroids. This indicates that the leg muscles are responsive to ecdysteroid treatment but does not determine which RNA population is being synthesized. It may be possible that ecdysteroids take longer than 6 hours to have an effect on transcriptional rates of the actin gene. Mammalian studies have

Actin mRNA levels in stretched extensor
muscles compared to unstretched controls

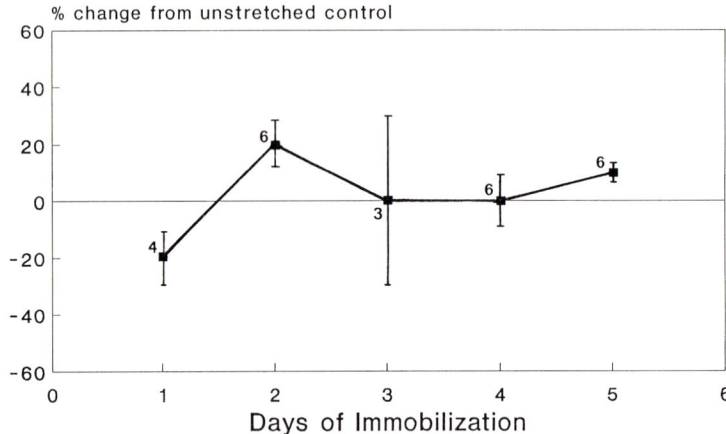

Fig. 4. The effect of passive stretch on the relative abundance of actin mRNA in carpopodite extensor muscles compared to unstretched controls. The numbers of treatments are given beside each point and the values are means ± standard errors.

demonstrated that some steroids, such as oestrogen, may act at the level of DNA accessibility, i.e. steroid hormones promote hypersensitive sites and hypomethylation of specific genes which then result in increased access for stage and tissue specific transcription factors and elevated transcription rates (Burch and Weintraub, 1983). The ecdysteroid receptors may not be accessible in intermoult lobster muscle until premoult. Different ecdysteroids may be acting synergistically to regulate gene transcription. Clearly, we need to monitor the levels of ecdysteroid in the muscle and if possible identify the presence of nuclear ecdysteroid receptors in different muscles throughout the moult cycle. Currently, we are isolating a clone for lobster actin and are making a genomic library which should also allow us to investigate the control sequences on the actin gene to identify the presence of the ecdysteroid response region which has been located in other arthropod species (Lepesant and Richards, 1990).

In order to determine if the postmoult elevation in synthesis rates may be due to mechanical stimulation, the effects of passive stretch on the muscles in the walking leg have been assessed. Immobilization experiments similar to those carried out in *Carcinus* (Houlihan and El Haj, 1985) were carried out in *H. gammarus* with the extensor muscle held in an elongated position. No variation in actin mRNA levels were measured after 5 days immobilization versus the opposing control extensor muscle (Fig. 4). There was also no significant change at the three time points analysed. Houlihan and El Haj (1985) demonstrated a 10% increase in length of certain muscle fibres in response to immobilization over a period of 3 weeks, however no significant increase in muscle length was measured after 1 week. Passive stretch may require a longer period to act on the muscle before changes in mRNA levels are observed which are then translated into addition of sarcomeric units resulting in elongated fibres. By 5 days following ecdysis, muscle fibres have reached their maximal length and protein synthesis rates have fallen to intermoult levels (El Haj et al. 1984; El Haj et al. 1987). It is also necesary to determine the amount of stretch imposed during the moult. Using this model *in vivo*, the extensor muscle may only be maintained in the static fully extended position (El Haj and Houlihan, 1985), however, we may require dynamic *in vitro* force transducer experiments to fully mimic and characterize the full magnitude of the effects of stretch imposed during the moult *in vivo*.

In conclusion, mRNA levels for the sarcomeric muscle protein, actin, fluctuate throughout the moult cycle and correlate with changes in rates of protein synthesis. Muscle growth may occur at different stages surrounding ecdysis in different parts of the body. The controlling influences are yet to be determined but may involve a complex interaction between hormonal and mechanical factors.

References

Burch, J. B. and Weintraub, H. (1983). Temporal order of chromatin structural changes associated with activation of the major chicken vitellogenin gene. *Cell* **33**, 65-76.

Camien, M. N., Garlet, H., Duchateau, G. and Flordin, M. (1951). Nonprotein amino acids in muscle and blood of some marine and freshwater Crustacea. *J. Biol. Chem.* **193**, 881-885.

Chang, E. S. (1989). Endocrine regulation of moulting in Crustacea. *Rev. Aquatic Sci.* **1**, 131-157.

Cherbas, L., Lee, K. and Cherbas, P. (1990). The Induction of Eip 28/29 by ecdysone in Drosophila cell lines, *Invert. Repro. Devel.* **18(1-2)**, 108-115.

Chomczynski, P. and Saachi, N. (1987). Single step method of RNA isolation by acid guanidium thiocycnate - phenol - chloroform extraction. *Analyt. Biochem.* **162**, 156-159.

Cole, W. H. (1941). A perfusing solution for the lobster (Homarus sp.) heart and the effects of its constituent ions on the heart. *J. Gen Physiol.* **25**, 1-6.

Crothers, J. H. (1967). The biology of the shore crab, *Carcinus maenas* (L.) *Fld. Stud.* **2**, 407-434.

Dabre, P. D. (1988). Introduction to Practical Molecular Biology. Chichester:Wiley.

DeFur, P. L., Mangum, C. P. and McMahon, B. R. (1985). Cardiovascular and ventilatory changes during ecdysis in the blue crab, *Callinectes sapidus* Rathburn. *J. Crust. Biol.* **5(2)**, 207-215.

El Haj, A. J., Govind, C. K. and Houlihan, D. F. (1984). Growth of lobster leg muscle fibres over intermoult and moult. *J. Crust. Biol.* **4(4)**, 536-545.

El Haj, A. J. and Houlihan, D. F. (1987). In vitro and in vivo protein synthesis rates in a crustacean muscle during the moult cycle. *J. exp. Biol.* **127**, 413-426.

Engel, J., Gunning, P. and Kedes, L. (1982). Human actin proteins are encoded by a multigene family. In Muscle Development, Molecular and Cellular Control. (Eds. M.L. Pearson and H.F. Epstein), pp. 107-117. New York: Cold Spring Harbour.

Gallwitz, D. and Sures, I. (1980). Structure of split yeast gene: Complete nucleotide sequence of the actin gene in *Saccharomyces cerevisiae*. *Proc. Natl. Acad. Sci.* **77**, 2246-2254.

Garlick, P. J., McNurlan, M. A. and Preedy, V. R. (1980). A rapid and convenient technique for measuring the rate of protein synthesis in tissues by injection of ^{3}H phenylalanine. *Biochem. J.* **192**, 719-723.

Goldspink, G. (1980). Growth of muscle. In Development and Specialization of Skeletal Muscle. SEB Seminar Series 7. (Ed. D.F. Goldspink) pp. xxxx. Cambridge: Cambridge University Press.

Govind, C. K., She, J. and Lang, F. (1977). Lengthening of lobster muscle fibres by two age dependant mechanisms. *Experientia.* **33**, 35-36.

Hartnoll, R. G. (1965). Notes on the marine grapsid crabs of Jamaica. *Proc. Linn. Soc. Lond.* **175**, 113-47.

Houlihan, D. F. and El Haj, A. J. (1985). An analysis of muscle growth. In Factors in Adult Growth (Ed. A.M., Wenner) A.A. Blaskema Press:Amsterdam.

Kuppert, P., Willhelm, S. and Spindler, K-D. (1978). Demonstration of cytoplasmic receptors for the moulting hormones in crayfish. *J. Comp. Physiol.* **128**, 95-100.

Lachaise, F., Carpentier, G., Somme, G., Colardeau, J. and Beydon, P. (1989). Ecdysteroid synthesis by crab Y organs. *J. exp. Zool.* **252**, 283-292.

Lachaise, F., Lageuex, M., Leyerasen, R. and Hoffman, J. A. (1976). Metabolisme de l'ecdysone au cours due development de *Carcinus maenas* (Brachyura, Dacapoda). *C.R. Acad. Sc. Paris.* **283(D)**, 943-946.

Lachaise, F., Meister, M., Hetni, C. and Lafont, R. (1986). Studies on the biosynthesis of ecdsyone by the Y organs of *Carcinus maenas*. *Molec. Cell Endo.* **45**, 253-261.

Lepesant, J-A. and Richards, G. (1989). Ecdysteroid regulated genes. In Ecdysone from Chemistry to Mode of Action (ed. J. Koolman), p.p. 355-367. Stuttgart: Georg Thiem Verlag.

Londershausen, M. and Spindler, K-D. (1985). Uptake and binding of moulting hormones in crayfish. *Amer. Zool.* **25**, 187-196.

Loughna, P. T., Izumo, S., Goldspink, G. and Nadal-Ginard, B. (1990). Disuse and passive stretch cause rapid alterations in expression of developmental and adult contractile protein genes in skeletal muscle. *Development.* **109**, 217-223.

Macias, M. T. and Sastre, L. (1990). Molecular cloning and expression of four actin isoforms during Artemia development. *Nucleic Acids Res.* **18(17)**, 5219-5225.

Minty, A. J., Caravatti, M., Robert, B., Cohen, A., Daubas, P., Weydert, A., Gros, F. and Buckingham, M. (1981). Mouse actin messenger RNAs. *J. Biol. Chem.* **256(2)**, 1008-1014.

Mykles, D. L. (1980). The mechanism of fluid absorption at ecdysis in the American lobster, *Homarus americanus*. *J. Exp. Biol.* **84**, 89-101.

Mykles, D. L. and Skinner, D. M. (1982). Crustacean muscle: atrophy and regeneration during moulting. In Basic Biology of Muscles: A comparative Approach (ed. B.M. Twarog, R.J.C. Levine and M.M. Dewey). pp. 337-357. New York: Raven Press.

Mykles, D. L. and Skinner, D. M. (1985a). Muscle atrophy and restoration during moulting. In Crustacean Issues, Vol II Crustacean growth (Ed. A.M., Wenner) pp. 3-14. New York: Balkema Press.

Mykles, D. L. and Skinner, D. M. (1985b). The role of calcium dependant proteinases in moult induced claw muscle atrophy. In Intracellular Protein Catabolism (Eds. Khairallah, E., Bow, J.S. and Bird, J.W.) pp. 141-150. New York: Alan R Liss.

Paulson, C. R. and Skinner, D. M. (1991). Effects of 20-hydroxyecdysone on protein synthesis in tissues of the land crab *Gecarcinus lateralis*. *J. exp. Zool.* **257**, 70-79.

Riddihough, G. and Pelham, H. R. B. (1986). Activation of the Drosophila hsp27 promoter by heat shock and by ecdysone involves independant and remote regulatory sequences. *EMBO J.* **5(7)**, 1653-1658.

Sambrook, J., Fritsch, E. F. and Maniatis, T. (1989). Molecular cloning: a laboratory manual. 2nd Editon. New York: Cold Spring Harbor Press.

Segraves, W. A. and Richards, G. (1990). Regulatory and developmental aspects of ecdysone-regulated gene expression. *Invert. Reprod. Dev.* **18(1-2)**, 67-76.

Skinner, D. M. (1965). Amino acid incorporation into protein during the moult cycle of the land crab, *Gecarcinus lateralis*. *J. exp. Zool.*, 226-233.

Skinner, D. M. (1966). Breakdown and reformation of somatic muscle during the moult cycle of the land crab, *Gecarcinus lateralis*. *J. Exp. Zool.* **163**, 115-124.

Snyder, M. J. and Chang, E. S. (1991). Ecdysteroids in relation to the moult cycle of the American lobster, *Homarus americanus* 1. Haemolymph titres and metabolites. *Gen Comp. Endocrinol.* **81**, 133-145.

Spindler, K-D., Dinan, L. and Londershausen, M. (1984). On the mode of action of ecdysteroids in crustaceans In Biosynthesis and Mode of Action of Invertebrate Hormones (eds. J.A. Hoffman and M. Porchet), pp. 255-264. Berlin: Springer-Verlag.

Spindler, K-D. and Spindler Barth, M. (1989). Uptake of ecdysteroids. In Ecdysone from Chemistry to Mode of ACtiuon. (Ed Jan Koolman), pp. 245-249. Stuttgart. Georg Thieme Verlag.

Stevenson, J. R. (1975). The molting cycle in the crayfish. Recognising the molting stagea effects of ecdysone and changes during the cycle. *Freshwater crayfish* **2**, 255-269.

Traub, M., Gellissen, G. and Spindler, K-D. (1987). 21(OH) ecdysone induced transition from intermolt to premolt protein biosynthesis patterns in the hypodermis patterns in the crayfish, *Astacus leptodactylus, in vitro. Gen Comp. Endocrinol.* **65**, 469-477.

Vaudekerckhoe, J. and Weber, K. (1984). Chordate muscle actins differ distinctly from invertebrate muscle actins. *J. Mol. Biol.* **179**, 310-413.

Whiteley, N. M., Taylor, E. W. and El Haj, A. J. (1992). Actin gene expression during muscle growth in *Carcinus maenas. J. exp. Biol.* in Press.

Printed in Great Britain © Society for Experimental Biology 1992   167

cDNA CLONING AND SEQUENCE COMPARISONS OF HUMAN AND CHICKEN MUSCLE C-PROTEIN AND 86kD PROTEIN

VAUGHAN, K. T., WEBER, F. E. and FISCHMAN, D. A.

Department of Cell Biology and Anatomy, Cornell University Medical College, 1300 York Avenue, New York, NY 10021

Summary

Thick filaments in vertebrate striated muscles are composed of myosin heavy chain (MHC) and myosin light chains (MLCs) plus at least eight other proteins: C-protein, 86kD protein (birds) or H-protein (mammals), M-protein, myomesin, titin, MM-creatine kinase, skelemin, and AMP-deaminase. Except for CPK and AMP deaminase, none have well defined functions. Analysis of cDNA clones encoding chicken C-protein and 86kD protein has revealed a high degree of shared amino acid identity, particularly in the C-terminal 40kD. To identify functionally significant regions, the human counterpart of each protein was cloned, sequenced and analysed. Two human C-protein cDNAs were isolated with significant homology to chicken fast C-protein. Clone H75, with 69% identity to chicken fast C-protein, shows the same pattern of hybridization as the chicken fast C-protein in chicken muscles. The other clone, H8 with 60% identity, shows a pattern of hybridization in chicken muscles which is consistant with the expression of chicken slow C-protein. The human 86kD protein shares 66% DNA sequence identity with the chicken 86kD protein. Assuming that essential sequences would be conserved during evolution, we compared the chicken and human proteins using PALIGN. Chicken and human fast C-proteins possess 66% peptide identity over their deduced length plus 10% conservative substitutions. Human slow C-protein and chicken fast C-protein share 44% peptide sequence identity, plus 16% conservative substitutions. Chicken and human 86kD proteins are also very similar: 54% peptide identity plus 20% conservative substitutions. This high degree of sequence identity between chicken and human C- and 86kD proteins suggests selective pressure on the primary sequence. Recent primary sequence analyses of projectin and mini-titins from Drosophila, twitchin from C. elegans, C-protein, smMLCK, 86kD protein, and M-protein from the chicken, titin from the rabbit, and skelemin from the mouse reveals that all these proteins possess multiple internal repeats of approximately 100 amino acids. These repeating domains are of two types: one is homologous to the internal repeats which define the C-2 subset of the immunoglobulin superfamily, the other is related to the fibronectin type III repeat. Both human C-proteins possess comparable internal repeats and preliminary evidence suggests the the presence of the same repeats in human 86kD. This duality of repeat structure is found in many extracellular

Key words: myofibrillar, cloning, muscle, repeats.

proteins and is typified by the N-CAMs. To our knowledge, this family of myosin-binding proteins is the first example of intracellular, nonmembrane-associated proteins which share significant primary structural features with cell surface/extra-cellular adhesion molecules. The genomic organization of the C-proteins and 86kD protein was analyzed in somatic cell hybrids. Human fast C-protein was localized to chromosome 19 and human slow C-protein to chromosome 12. The human 86kD protein has been mapped to chromosome 1. We conclude that genes of the C-protein family are genetically unlinked.

Introduction

The presence of myosin-associated proteins in striated muscle thick filaments has been demonstrated in both vertebrates and invertebrates. This group of proteins includes: MCK (Walliman and Eppenberger, 1985), AMP Deaminase (Cooper and Trinick, 1984), C-protein (Dennis, Shimizu, Reinach and Fischman, 1984; Bennett, Craig, Starr and Offer, 1986), 86kD protein (Bahler, Eppenberger and Walliman, 1985a), M-protein (Grove, Holmbom and Thornell, 1987), H-Protein (Bennett et al. 1986), X-protein (Bennett et al., 1986), myomesin (Grove et al. 1987), titin (Wang, McClure and Tu, 1979), twitchin (Benian, Kiff, Neckelman, Moerman and Waterston, 1989), and skelemin (Price, 1987).

C-protein (M_r-140kD), 86kD (M_r-86kD) protein and H-protein (M_r-69kD) have been localized to an area of the thick filament called the C-region (Squire, 1981). This region, covering the inner two-thirds of the crossbridge zone, is so-called because it is the location of C-protein, one of the more abundant of the thick filament-associated proteins. This protein, identified by Offer (Offer, Moos and Starr, 1973), as a contaminant of myosin preparations from rabbit muscle, is found in a 1:8 molar ratio with myosin. Studies using polyclonal and monoclonal antibodies to C-protein have localized C-protein to a series of transverse stripes in the crossbridge zone of the thick filament in rabbit (Craig and Offer, 1976) and chicken skeletal muscle (Dennis et al. 1984). C-protein has been shown to bind myosin in two places, on LMM and S-2 (Moos, Offer, Starr and Bennett, 1975; Starr and Offer, 1978) and has also been shown to bind F-actin (Moos, Mason, Besterman, Feng and Dubin, 1978), and regulated actin filaments (Yamamoto, 1986).

Fast, slow and cardiac isoforms of C-protein have been identified (Reinach, Masaki and Fischman, 1982; Yamamoto and Moos, 1983; Dhoot, Hales, Grail and Perry, 1985; Kawashima, Kitani, Tanaka and Obinata, 1986). The posterior latissimus dorsi (PLD) has been shown to express two isoforms, the fast and slow, in all sarcomeres (Reinach, Masaki and Fischman, 1983). The cardiac isoform is phosphorylated in a catacholamine-dependant pathway (Hartzell and Glass, 1984).

Rabbit H-protein (Starr and Offer, 1982) and chicken 86kD protein (Bahler et al. 1985a) seem to be evolutionary homologues. They behave similarly during purification and separation on hydroxyapatite (Starr and Offer, 1982; Bahler et al., 1985a). They are found in comparable abundance to C-protein and they have similar localizations on the thick filament (Bennett et al. 1986; Bahler, Eppenberger and Walliman, 1985b). Binding experiments have not been published for H-protein, but chicken 86kD protein has been shown to bind myosin using myosin affinity chromatography (Bahler et al., 1985a).

Recent work on the physiology of rabbit C-protein suggests that C-protein may function in the fine-tuning of muscle contraction. Partial extraction of C-protein results in a sensitization of rabbit skeletal muscle skinned fibers to calcium, and this can be reversed by readdition of purified C-protein (Hoffman, Hartzell and Moss, 1991a). Extraction of C-protein accelerates the rate of contraction under certain conditions (Hoffman, Greaser and Moss, 1991b). Normal rates can be restored by readdition of the purified C-protein.

The cDNA cloning of chicken fast C-protein allowed analysis of the deduced primary sequence (Einheber and Fischman, 1990). Notable features of the primary sequence include the finding of internal repeats approximately 100 amino acids in length. These repeats could be divided into two groups based on similarity to known protein sequences. The first group, encompassing 6 of the 9 repeats, bears resemblance to the C-2 subset of the immunoglobulin superfamily (Williams and Barclay, 1988). The second group, covering the remaining three repeats is more similar to the fibronectin III repeat (Petersen, Thogersen, Skorstengaard, Vibe-Pedersen, Sahl, Sottrup-Jensen and Magnusson, 1983; Kornblihtt, Umezawa, Vibe-Pedersen and Baralle, 1985).

The C-2 repeats were originally described for a series of cell-surface molecules (Williams, 1987). These proteins shared some features with the constant domain and some with the variable domains of the IgG molecule. The fibronectin molecule has three types of internal homology. The type III repeat, in the center of the molecule, possesses the cell-binding activity of the protein (Kornblihtt et al., 1985).

A large family of cell-surface associated proteins, typified by N-CAM (Cunningham, Hemperly, Murray, Prediger, Brackenbury and Edelman, 1987) possesses both C-2 and type III repeats. cDNA cloning of the 86kD protein revealed the presence of the same repeat duality (Fischman, Vaughan, Weber and Einheber, 1991). While the 86kD protein is smaller and has only two copies of each repeat, the pattern of repeats is the same as the C-terminal 40kD of C-protein: III,C-2,III,C-2.

Materials and methods

Using cDNAs for the chicken fast skeletal muscle C-protein and 86kD protein, we screened a human fetal muscle cDNA library. This lambda gt10 library was oligo-dT primed and a generous gift of Dr. Louis Kunkel. 1×10^6 pfu were plated at 50,000 pfu per plate. cDNA probes for either C-protein or the 86kD protein were labelled using the random-priming method with α^{32}P-dCTP. Duplicate filter plaque- lifts were processed as per the manufacturer (Nylon-N, Amersham).

Probes were separated from unincorporated counts using G-50 spin columns (BM), and plaque-lift hybridization was carried out according to the manufacturers recommendations (Rapid Hybrization Solution, Amersham). After hybridization, the filters were washed twice for 10 minutes each with $2\times$ SSC, 0.1% SDS at room temperature, then twice for 15 minutes each with $1\times$ SSC, 0.1% SDS at 65°C, and finally twice for 15 minutes each with $0.1\times$ SSC at 65°C. Filters were exposed overnight at -70°C with Kodak X-OMAT film. Each clone was taken through three rounds of screening until pure. The insert of each was characterized by obtaining lambda DNA using the plate lysate method and purifying the lambda DNA using the mini-lambda columns (Qiagen, Inc.). The inserts were liberated by EcoR1 digestion and the size was analysed by horizontal agarose gel electrophoresis in $1\times$ TAE buffer.

The cDNA inserts subcloned into M13 were sequenced by the Sanger method as modified by USB for the Sequenase and Taquence sequencing kits (USB), using α^{35}S-dATP. Sequencing products were resolved on a 6% acrylamide/6M urea/TBE sequencing gel, using the 0.4mm-0.8mm wedge spacers available with the IBI STS-45 sequencing apparatus. After fixing and drying, the gels were exposed to Kodak X-OMAT film overnight at room temperature.

The insert sequences were tranferred to the PCGENE sequence analysis package (Intelligenetics, CA) using a digitized gel reader. These sequences were assembled into a single meld using ASSEMGEL and compared to other sequences using NALIGN also contained in the PCGENE package. DNA sequences were compared to the current releases of Genbank using the FASTA program (Pearson and Lipman, 1988).

Northern blots were performed by standard methods. Chicken total RNA samples were electrophoresed in formaldehyde/MOPS/agarose gels and transferred to Nylon membranes by capillary transfer in $10\times$ SSC. Blots were washed with $2\times$ SSC and air dried. After UV crosslinking, the blots were hybridized as described above with labelled human cDNAs and washed. These blots were then autoradiographed as above.

The human chromosomal localization of the cDNAs was performed on human-hamster somatic cell hybrid blots obtained from the BIOS Corp., New Haven, CT. Hybrid panels were restriction-digested, electrophoresed and transferred by the company. Human cDNA probes labelled as above were used in Southern blots of these panels according to the manufacturers instructions. The pattern of positive and negative hybrids allowed identification of the chromosomal localization of the gene in question.

Results

Screening of the fetal human library with the chicken fast C-protein probe resulted in the isolation of two distinct groups of cDNAs. Since the library was oligo-dT primed, all clones within a group possessed the same 3' terminus, yet differed in the degree of 5' extension. Sequence analysis of these contiguous

Table 1. *Hybridization of human C-protein cDNAs to chicken muscle RNA*

RNA	H75	H8
Pectoralis major (fast C-protein)*	X	
Posterior latissimus dorsi (fast and slow C-protein)*	X	X
Anterior latissimus dorsi (slow C-protein)*		X

*(Reinach et al., 1983).

RNA was purified from chicken Pectoralis Major (PM), chicken Posterior Latissimus Dorsi (PLD), or chicken Anterior Latissimus Dorsi (ALD). These RNAs were probed with ^{32}P-labelled human cDNA H75 or H8. X's signify a positive hybridization signal.

groups showed that group H75 had the highest sequence identity with chicken fast C-protein (69%). Translation of this cDNA revealed 66% peptide sequence identity with chicken C-protein, with an additional 10% conservative substitutions.

The second group, H8, showed slightly lower sequence identity with chicken fast C-protein (60%). Translation of this cDNA revealed a deduced protein which shared 44% peptide sequence identity with chicken fast C-protein, plus 16% conservative substitutions.

Screening of the library with the chicken 86kD protein probe resulted in the isolation of overlapping clones for the human 86kD protein. The largest of these clones, 1.9 kb, had 66% sequence identity with the chicken 86kD cDNA. The deduced protein showed 54% peptide identity, with 20% conservative substitutions.

Although we suspected that clone H75 represented human fast C-protein based on its high similarity to chicken fast C-protein, we sought additional evidence. Since human muscle groups are of mixed fiber-type, thereby making northern analysis by muscle group difficult, we probed RNA from chicken muscles, many of which are of almost pure fiber-type with respect to C-protein expression. Northern analysis of Pectoralis Major (PM) (pure fast C-protein), Posterior Latisimus Dorsi (PLD) (fast and slow C-protein), and Anterior Latisimus Dorsi (ALD) (pure slow C-protein) (Reinach et al., 1983), showed different muscle specificities for the two human C-protein probes (Table 1). H75 hybridized with RNA from PM and PLD, consistent with its encoding a fast isoform of C-protein. H8 recognized RNA from PLD and ALD muscle, making it a candidate for a slow C-protein cDNA. Interestingly, neither probe labelled cardiac muscle (results not shown), consistent with unpublished northern blots using the fast chicken C-protein probe and chicken ventricular RNA.

Alignment of the new human C-protein peptide sequences with the known chicken sequences reveals domains of very high identity interspersed with areas of lower identity (Fig. 1). The high degree of similarity between the chicken and human fast C-proteins indicates that that these domains of high identity are abundant. The slow human C-protein exhibits fewer of these domains, and they are spread further apart. The human 86kD protein has large pockets of total

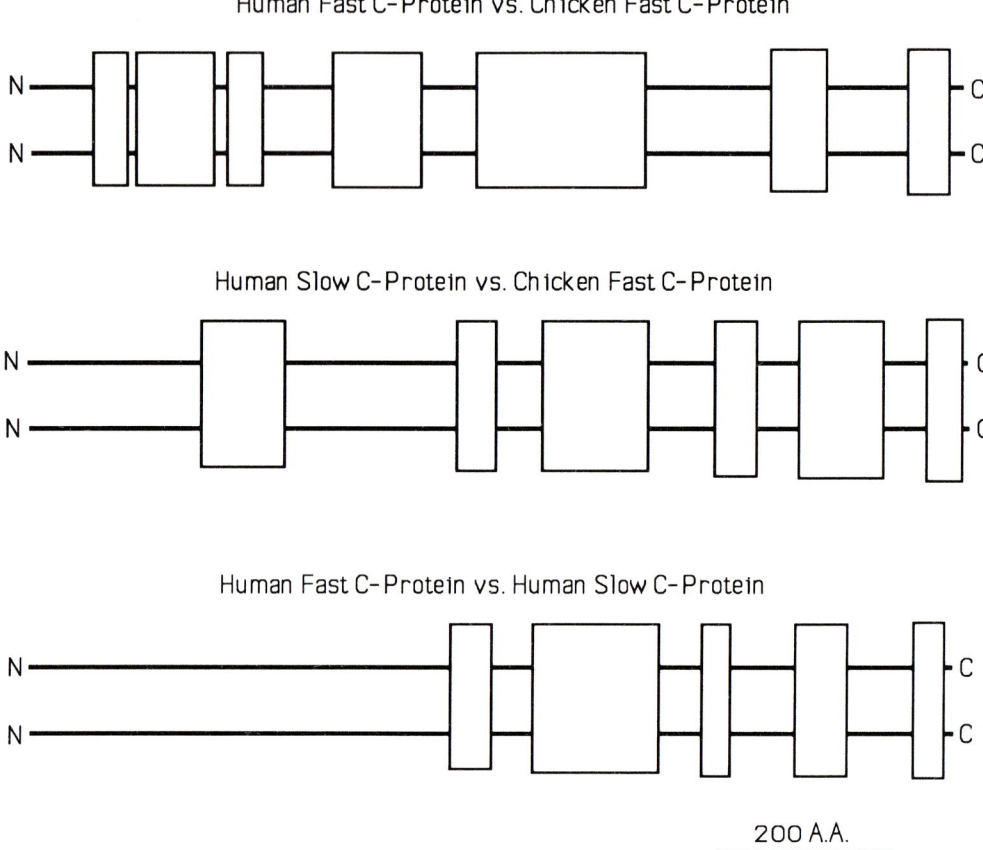

Fig. 1. Alignment of Human C-Proteins with Chicken Fast C-Protein. Deduced amino acid sequences of chicken fast C-protein, human fast C-protein, and human slow C-protein were aligned with each other using PALIGN. Lines represent linear protein sequences with the N-termini to the left and the C-termini to the right. Open boxes over the pair of sequences represent stretches of amino acid sequence where there is perfect identity or one residue mismatch.

identity in the middle of the protein (data not shown). We had hoped to identify discrete domains of high identity which could help us pinpoint functionally significant sequences. If such sequences exist, they are restricted to the C-termini of the three proteins; the N-termini of all three proteins are quite divergent.

Both human C-proteins were found to possess the same pattern of repeats seen in chicken fast C-protein: C-2,C-2,C- 2,C-2,III,III,C-2,III,C-2 (Fig. 2). Preliminary evidence indicates that the human and chicken 86kD proteins also possess the same pattern of internal repeats: III, C-2, III, C-2.

To identify the chromosomal location of the C-proteins and 86kD proteins, we probed Southern blots of human-hamster somatic cell hybrids. Fast human C-protein was localized to human chromosome 19 with two separate panels of

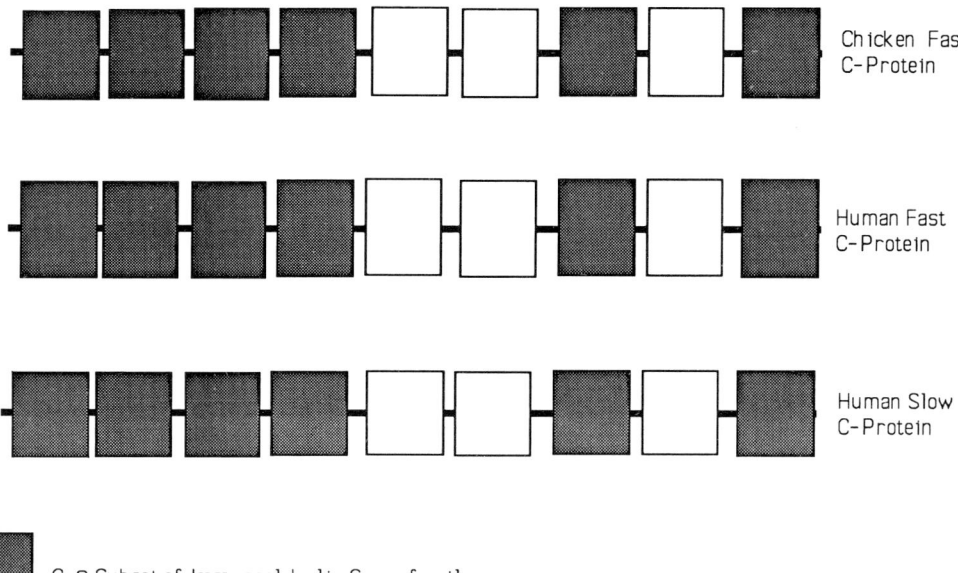

Fig. 2. Repeat structure of human C-proteins. Lines represent linear amino acid sequences. Hatched boxes indicate repeats of the C-2 subset of the immunoglobulin superfamily. Open boxes indicate repeats of the fibronectin III family. N-terminus is to the left, C-terminus is to the right.

hybrids (Fig. 3). Slow human C-protein was localized to chromosome 12 using the same methods and preliminary evidence suggest that human 86kD protein is found on chromosome 1.

Discussion

Screening a human fetal cDNA library, we were able to isolate clones for three human myofibrillar proteins: fast C-protein, slow C-protein, and 86kD protein. The identity of these cDNAs was ascertained using hybridization and sequence analysis. In addition, for the human C-proteins, northern analysis showed that the two cDNAs actually reflect different isoforms of C-protein. This is the first cDNA cloning of a slow isoform of C-protein. If H8 encodes X-protein (Bennett et al., 1986), this would be strong evidence that X-protein is in reality a slow isoform of C-protein.

Primary sequence analysis of the human C-proteins reveals that pockets of high identity are found throughout the molecule rather than in discrete domains. However, the N-termini of both proteins are quite divergent. Nucleotide sequence differences are dispersed throughout the two cDNAs making it highly unlikely that

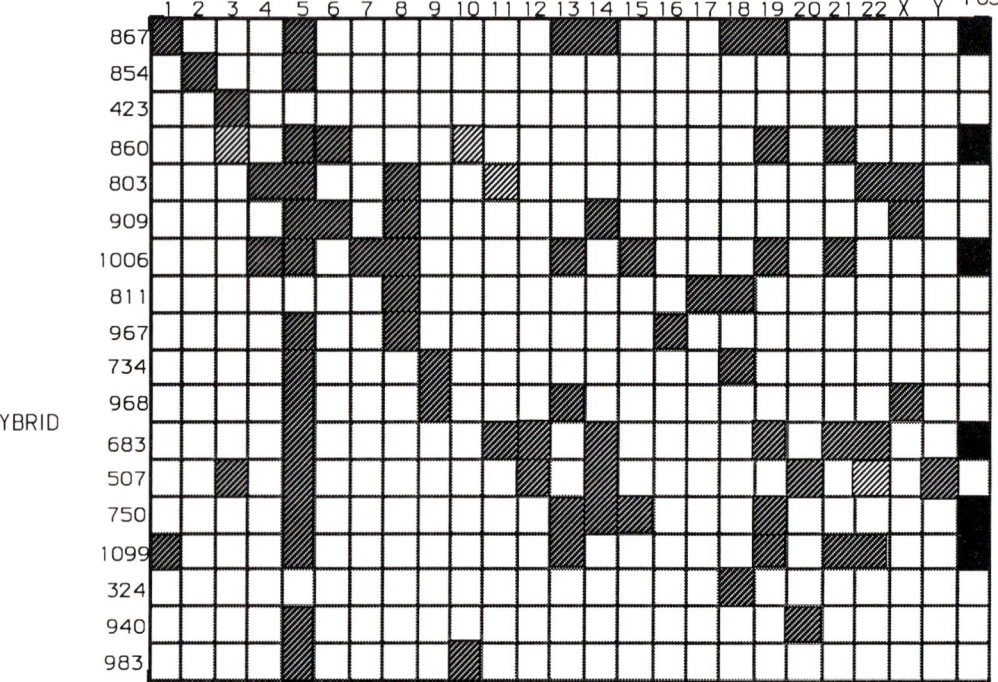

Fig. 3. Summary of hybrids for H75. This grid represents a summary of the results of probing a human-hamster somatic cell hybrid panel. The X-axis shows each of the human chromosomes which are labelled above. The Y-axis shows each of the hybrids. By crossing the two axes, one sees the human chromosome complement in each hybrid. The far right column shows which hybrids were positive with the H75 probe, and only human chromosome 19 fits the pattern of positive and negative results.

the two isoforms are encoded by one gene. The genomic studies conclusively prove this point (see below).

All three proteins exhibit C-2 and fibronectin III internal repeats found in the respective chicken homologues. The functional significance of these repeats is unknown, however, in extracellular proteins these repeats are thought to function in protein-protein interactions. Knowing the complexity of thick filaments in vertebrate skeletal muscle, a role in assembly for these proteins is worth considering. During myogenesis, immunofluorescent detection of C-protein is first apparent when definitive sarcomeres assemble (Schultheiss, Lin, Lu, Murray, Fischman, Weber, Masaki, Inamura and Holtzer, 1990).

While it appears that these proteins have the internal repeats described for many recently cloned myosin-binding proteins (Benian et al., 1989; Einheber and Fischman, 1990; Labeit, Barlow, Gautel, Gibson, Holt, Hsieh, Francke, Leonard, Wardale, Whiting and Trinick, 1990; Olson, Pearson, Neddleman, Hurwitz, Kemp and Means, 1990; Fischman et al., 1991; Price, 1991; Masaki, T., personal communication), these repeats do not appear to be more identical than surrounding amino acid sequences. This observation is especially true for fast human C-protein.

The chromosomal location of each human protein has been determined. Fast C-protein is found on chromosome 19, slow C-protein on chromosome 12, and 86kD protein is tenatively assigned to chromosome 1. Knowing that several muscle genes such as MMCK and Myotonic Dystrophy are located on chromosome 19, we are using high resolution in situ hybridization to more precisely localize the fast C-protein gene.

In summary, three vertebrate C-protein cDNAs have now been isolated, one from the chicken and two from man. Two cDNAs encoding 86kD protein have also been obtained, one from man and the other from chickens. The cardiac-type isoform of C-protein remains to be identified.

The authors express their gratitude to F. Reinach, S. Einheber and T. Mikawa for suggestions and discussion during the course of this work. The research has been supported by grants form NIH (AR32147) and MDA. F. Weber has been a fellow of the Revson-Winston Foundation.

References

Ayme-Southgate, A., Vigoreaux, J., Benian, G. and Pardue, M. L. (1991). Drosophila has a twitchin/titin-related gene that appears to encode projectin. *P.N.A.S.* **88**, in press.

Bahler, M., Eppenberger, H. and Walliman, T. (1985a). Novel Thick Filament Protein of Chicken Pectoralis Muscle: the 86kD Protein. I. Purification and Characterization. *J. Mol. Biol.* **186**, 381-391.

Bahler, M., Eppenberger, H. and Walliman, T. (1985b). Novel Thick Filament Protein of Chicken Pectoralis Muscle: the 86kD Protein. II. Distribution and Localization. *J. Mol. Biol.* **186**, 393-401.

Benian, G., Kiff, J., Neckelman, N., Moerman, D. and Waterston, R. (1989). The sequence of twitchin: an unusually large protein implicated in the regulation of myosin activity in C. elegans. *Nature* **342**, 45-50.

Bennett, P., Craig, R., Starr, R. and Offer, G. (1986). The ultrastructural location of C-protein, X-protein and H-protein in rabbit muscle. *J. Muscle Res. Cell Motil.* **7**, 550-567.

Cooper, J. and Trinick, J. (1984). Binding and location of AMP deaminase in rabbit psoas muscle myofibrils. *J. Mol. Biol.* **177**, 137-152.

Craig, R. and Offer, G. (1976). The location of C-protein in rabbit skeletal muscle. *Proc. R. Soc. Lond. B. Biol.* **192**, 451-461.

Cunningham, B., Hemperly, J., Murray, B., Prediger, E., Brackenbury, R. and Edelman, G. (1987). Neural cell adhesion molecule: structure, immunoglobulin-like domains, cell surface modulation, and alternative RNA splicing. *Science* **236**, 799-806.

Dennis, J., Shimizu, T., Reinach, F. and Fischman, D. A. (1984). Localization of C-protein isoforms in chicken skeletal muscle: ultrastructural detection using monoclonal antibodies. *J. Cell Biol.* **98**, 1514-1522.

Dhoot, G., Hales, M., Grail, B. and Perry, S. (1985). The isoforms of C-protein and their distribution in mammalian skeletal muscle. *J. Muscle Res. Cell Motil.* **6**, 487-505.

Einheber, S. and Fischman, D. A. (1990). Isolation and initial characterization of a cDNA clone encoding avian skeletal muscle C-protein: an intracellular member of the immunoglobulin superfamily. *P.N.A.S.* **87**, 2157-2161.

Fischman, D. A., Vaughan, K. T., Weber, F. E. and Einheber, S. (1991). Myosin binding proteins: intracellular members of the immunoglobulin superfamily. In *Frontiers of Muscle Research*, ed. Ozawa, E. (Amsterdam: Elsevier). in press.

Grove, B., Holmbom, B. and Thornell, L. (1987). Myomesin and M-Protein: Differential expression in embryonic fibers during pectoral muscle development. *Differentiation* **34**, 106-114.

Hatzell, H. and Glass, D. (1984). Phosphorylation of purified cardiac muscle C-protein by purified cAMP-dependant and endogenous Ca^{2+}-calmodulin-dependant protein kinases. *J. Biol. Chem.* **259**, 15587-15596.

Hoffman, P., Greaser, M. and Moss, R. (1991b). C-protein limits shortening velocity of rabbit skeletal muscle fibres at low levels of Ca^{2+} activation. *J. Physiol.* **439**, 710-715.

Hoffman, P., Hartzell, H. and Moss, R. (1991a). Alterations in Ca^{2+} sensitive tension due to partial extraction of C-protein from rat skinned cardiac myocytes and rabbit skeletal muscle fibers. *J. Gen. Physiol.* **97**, 1141-1163.

Kawashima, M., Kitani, S., Tanaka, T. and Obinata, T. (1986). The earliest form of C-protein expressed during striated muscle development is immunologically the same as cardiac-type C-protein. *J. Biochem* **99**, 1037-1047.

Kornblihtt, A. R., Umezawa, K., Vibe-Pedersen, K. and Baralle, F. (1985). Primary structure of human fibronectins: differential splicing can generate at least 10 polypeptides from a single gene. *EMBO J.* **4**(7), 1755-1759.

Labeit, S., Barlow, D., Gautel, M., Gibson, T., Holt, J., Hsieh, C., Francke, U., Leonard, K., Wardale, J., Whiting, A. and Trinick, J. (1990). A regular pattern of two types of 100-residue motif in the sequence of titin. *Nature.* **345**, 273-276.

Moos, C., Mason, C., Besterman, J., Feng, I. and Dubin, J. (1978). The binding of skeletal muscle C-protein to F-actin, and its relation to the interaction of actin with myosin subfragment-1. *J. Mol. Biol.* **124**, 571-586.

Moos, C., Offer, G., Starr, R. and Bennett, P. (1975). Interaction of C-protein with myosin, myosin rod and light meromyosin. *J. Mol. Biol.* **97**, 1-9.

Nave, R., Furst, D., Vinkemeier, U. and Weber, K. (1991). Purification and physical properties of nematode mini-titins and their relation to twitchin. *J. Cell Science* **98**, 491-496.

Offer, G., Moos, C. and Starr, R. (1973). A new protein of the thick filaments of vertebrate skeletal myofibrils. Extraction, purification and characterization. *J. Mol. Biol.* **74**, 653-676.

Olson, N. J., Pearson, R. B., Neddleman, D. S., Hurwitz, M. Y., Kemp, B. E. and Means, A. R. (1990). Regulatory and structural motifs of chicken gizzard myosin light chain kinase. *P.N.A.S.* **7**, 2284-2288.

Pearson, W. R. and Lipman, D. J. (1988). Improved tools for biological sequence comparison. *P.N.A.S.* **85**, 2444-2448.

Petersen, T., Thogersen, H., Skorstengaard, K., Vibe-Pedersen, K., Sahl, P., Sottrup-Jensen, L. and Magnusson, S. (1983). Partial primary structure of bovine plasma fibronectin: three type of internal homology. *P.N.A.S.* **80**, 137-141.

Price, M. (1987). Skelemins: cytoskeletal proteins located at the periphery of M-discs in mammalian striated muscle. *J. Cell Biol.* **104**, 1325-1336.

Price, M., Brooks, C. and Gomer, R. (1991). Skelemins are members of a family of myosin-associated proteins with immunoglobulin superfamily C-2 and fibronectin III domains. *J. Cell. Biochem* **15C**, 63.

Reinach, F., Masaki, T. and Fischman, D. A. (1982). Isoforms of C-protein in adult chicken skeletal muscle: Detection with monoclonal antibodies. *J. Cell Biol.* **95**, 78-84.

Reinach, F., Masaki, T. and Fischman, D. A. (1983). Characterization of the C-protein from posterior latissimus dorsi muscle of the adult chicken: heterogeneity within a single sarcomere. *J. Cell Biol.* **96**, 297-300.

Schultheiss, T., Lin, Z., Lu, M., Murray, J., Fischman, D. A., Weber, K., Masaki, T., Imamura, M. and Holtzer, H. (1990). Differential distribution of subsets of myofibrillar proteins in cardiac nonstriated and striated myofibrils. *J. Cell Biol.* **110**, 1159-1172.

Squire, J. (1981). in *The Structural Basis of Muscle Contraction*. p. 344. (New York: Plenum Press).

Starr, R. and Offer, G. (1978). The interaction of C-protein with heavy meromyosin and subfragment-2. *Biochem. J.* **171**, 813-816.

Starr, R. and Offer, G. (1982). Preparation of C-protein, H-protein, X-protein, and phosphofructokinase. *Meth. Enzymol.* **85**, 130-138.

Wallimann, T. and Eppenberger, H. (1985). Localization and function of M-line-bound creatine kinase. *Cell Muscle Motil.* **6**, 239-285.

Wang, K., McClure, J. and Tu, A. (1979). Titin: major myofibrillar components of striated muscle. *P.N.A.S.* **76**, 3698-3702.

Williams, A. (1987). A year in the life of the immunoglobulin superfamily. *Immunol. Today* **8**, 298-303.

Williams, A. and Barclay, A. (1988). The immunoglobulin superfamily - domains for cell surface recognition. *Ann. Rev. Immunol.* **6**, 381-405.

Yamamoto, K. (1986). The binding of skeletal muscle C-protein to regulated actin. *FEBS Lett.* **208**, 123-127.

Yamamoto, K. and Moos, C. (1983). The C-proteins of rabbit red, white, and cardiac muscles. *J. Biol. Chem.* **258**, 8395-8401.

Printed in Great Britain © *Society for Experimental Biology 1992*   179

MULTIPLE PRODUCTS OF THE DUCHENNE MUSCULAR DYSTROPHY GENE

DAVID YAFFE, ADINA MAKOVER, DORON LEDERFEIN, DEBORA RAPAPORT, SARA BAR, EFRAT BARNEA and URI NUDEL

Department of Cell Biology, The Weizmann Institute of Science, 76100 Rehovot, Israel

Summary

The gene which is defective in Duchenne Muscular Dystrophy (DMD) extends over 2300 kb of the X chromosome. Its product in the muscle is a 14 kb mRNA encoding a 427 kd rod-shaped protein called dystrophin. A 14 kb transcript encoding a very similar isoform of dystrophin is produced in the brain. The brain 14 kb mRNA is transcribed from the same gene but controlled by a different promoter, located at least 75 kb upstream from the muscle dystrophin promoter. The regulation of these promoters is very stringently controlled. The muscle-type but not the brain-type dystrophin mRNA is found in cloned skeletal muscle cells and its presence is correlated with the appearance of multinucleated fibers. The brain type is expressed in neurons, while in glia cells the muscle-type promoter is active.

A third DMD gene transcript which is only 6.5 kb long has been identified. It contains the sequence coding for the C-terminal domain and the cysteine-rich domain of dystrophin but not the large region encoding the spectrin-like repeats and the N-terminal domain. The cell type distribution of this transcript is also very different from that of the two 14 kb mRNA isoforms. It is the major product of the DMD gene in many nonmuscle tissues including brain.

Using monoclonal antibodies we have identified a 77 Kd protein which seems to be the translation product of this mRNA. As expected from the distribution of the 6.5 Kb mRNA, this protein is the major DMD gene product detectable in brain and many other nonmuscle tissues; it is undetectable in skeletal muscle but is present in the heart and stomach (as is the 6.5 Kb mRNA).

Introduction

Duchenne Muscular Dystrophy (DMD) is an X-linked recessive disease, which results in progressive degeneration of the muscle and death in the second or third decade of life. In addition to the severe damage in the muscle, some 30% of the affected patients also suffer various degrees of mental retardation (reviewed in Moser, 1984). Becker Muscular Dystrophy (BMD), a less severe disease, is allelic to DMD.

The gene which is defective in Duchenne Muscular Dystrophy (DMD) spans over 2300 kb and is the largest gene known to date. The transcription products of

Key words: muscular dystrophy, Duchenne, gene product muscle.

the gene in the muscle is a 14 kb mRNA encoding a 427 kd membrane associated protein called dystrophin. The predicted amino acid sequences suggest that dystrophin is a rod shaped protein consisting of 4 domains (Koenig, Monaco and Kunkel, 1988):

(a) An N-terminal domain which is partially homologous to the actin binding domain of α-actinin.

(b) A large, rod-shaped domain consisting of 24 triple helical, 109 amino acid repeats similar to the repeat domains of spectrin.

(c) A cysteine rich segment, partially homologous to the Ca^{++} binding domain of Dictyostelium α-actinin.

(d) A unique C-terminal domain which is not homologous to any known protein.

Early studies suggested that the gene is transcribed specifically in muscles. However, it was later demonstrated that the gene is also transcribed in the brain and several other nonmuscle tissues (Nudel, Robzyk and Yaffe, 1988; Chamberlain, Pearlman, Muzny, Gibbs, Ranier, Reeves and Caskey, 1988). The 14 kb mRNA expressed in the brain codes for a protein very similar to the muscle dystrophin isoform. However, its 5′ end is derived from a different exon and its expression is regulated by a different promoter (Nudel, Zuk, Einat, Zeelon, Levy, Neuman and Yaffe, 1989; Feener, Koenig and Kunkel, 1989; Yaffe, Zuk, Einat, Shinar, Levy, Neuman, Fuchs and Nudel, 1989; Makover, Zuk, Breakstone, Yaffe and Nudel, 1991) located between 75-300 kb upstream from the muscle dystrophin promoter (den Dunnen, Casula, Makover, Bakker, Yaffe, Nudel and van Ommen, 1991; Boyce, Beggs, Feener and Kunkel, 1991). In this communication we review some of our investigations related to the control of expression of muscle-type and brain-type dystrophins. We also describe a novel 6.5 kb nonmuscle DMD gene product which differs greatly from the two known dystrophin mRNAs.

Results and discussion

The dystrophin mRNA isoforms in muscle and brain cells

PCR analysis, using primers specific for the muscle or the brain dystrophin mRNAs, showed that the expression of the two dystrophin promoters is very stringently controlled: in cloned skeletal muscle cell colonies consisting only of proliferating mononucleated myoblasts no dystrophin mRNA was detectable. In colonies containing multinucleated fibers the muscle promoter transcript but not the brain promoter transcript was detectable.

In neuronal cell cultures only the mRNA of the brain-type dystrophin was detected. However, in glia cell cultures the mRNA product transcribed from the muscle-type promoter but not from the brain-type promoter was found (Barnea, Zuk, Simantov, Nudel and Yaffe, 1990, and Table 1). The expression of the muscle-type promoter in glia cells was also reported by Chelly, Hamard, Koulakoff, Kaplan, Kahn and Berwald-Netter, (1990).

Table 1. *Cell type distribution of DMD gene transcripts*

	Muscle type (14 kb)	Brain type (14 kb)	'Liver type' (6.5 kb)
Skeletal muscle			
Myoblasts	−	−	±
Fibers	+++	−	(−)
Brain			
Neurons	−	+	++
Glia	+	−	+++
G4	+	−	++

Quantitative values are rough estimates based on several Northern blots, RNAase protection analyses and PCR, as described in Bar et al. (1990), Barnea et al. (1991) and Rapaport et al. (1992). The signals of the 14 kb mRNA in skeletal muscle are represented by +++. −, undetectable.

Both promoters have been isolated and characterized. The muscle-type promoter contains a TATA consensus sequence and several elements which are common in the promoter enhancer region of several other muscle specific genes. When inserted 5′ to a reporter gene, it confers developmentally regulated expression in transfected muscle cells (Yaffe, Barnea, Bar, Makover and Nudel, 1991; Klamut, Gangopadhyay, Worton and Ray, 1990).

The promoter of the mouse brain-type dystrophin is very different (Makover et al., 1991 and Fig. 1). The sequence upstream of the putative transcription initiation site does not contain a TATA box at about 30 bp upstream from the cap site, nor does it contain other motifs commonly found in promoters of eukaryotic genes. A TATA like sequence (ATAATAAAT) is found 33 bp downstream from the suggested transcription start site. However, its function is not clear, as a DNA fragment of the promoter region which does not contain this sequence, is nevertheless very active in driving the transcription of a reporter gene in transfected cells (detailed below).

The transcription initiation site of the brain-type mRNA was determined by sequencing of 5′ extended cDNA clones and by the RNAase protection technique. It is located 265 nucleotides upstream from the first AUG which is in frame with the coding sequence of dystrophin. However 9 additional AUG triplets, which are not in frame with the open reading frame of the dystrophin mRNA, or which are followed by stop codons, are found in the first exon of the mouse brain-type dystrophin mRNA, 5′ to the initiator AUG (Fig. 1). Such upstream AUG codons may be involved in the regulation of translation of the mRNA.

In view of the lack of consensus sequences usually found in the relevant positions in eukaryotic promoters, we checked the functional activity of the putative promoter of the 14 kb brain-type transcript. A fragment extending from nucleotide −1200 to nucleotide + 11 of the brain-type first exon (the putative cap site numbered 1) was inserted into the pSVO-CAT vector containing a promoter-

```
-377  TCATTTGAAAAAGTAAGTCTTGTATTTTATACTAAAACAGAGGCATAGAACTGAGCACTCA

-316  ACCCCCTGATTTCAAAAGTAGAATCAAAGAGAGAGTTTTACCCTTAAAGAGCAGAGAAAAT

-255  AAGTTAAAAACAGTCTAATTTTCTCTAAAAATGATATGATGTGTGGCTAAAAACTCTCCTT

-194  TTAGTATAATTAGGAAAAGAATCTAATCCAAAGGCTTTGTATCTGTACCGGAGCAAGCAGA

-133  TACTGAAAGAGATTCGCAGATCCTCTGTTTTTTTTTTTAGGCAGGAAGAATGCTTGTTAAA

-72   TGCAGAACGCTGCTCCGGCTCATGTGTTTGCTCCGAGGTGGAGGTTTTGTTCGACTGACGT
                 ▾
-11   ATCAGATAGTCAGAGTGGTTACCACACCGACGTTGTGGCAGCTGCATAATAAATGACTGAG

51    AGAATCATGTTAGGCATGCCCACCTAACCTAACTTGAATCATGCGAAAGGGGAGCTGTTGG

112   AATTCAAATAGACTTTCTGGTTCCCAGCAGTCGGCAGTAATAGAATGCTTTCAGGAAGATG

173   ACAGAATCAGGAGAAAGATGCTGTTTTGCGCTATCTTGATTTGTTACAGCAGCCAACTTAT

234   TGGCATGATGGAGTGACAGGAAAAACAGCTGGCATGGAAG/gtaggattattaaagctat
```

Fig. 1. The promoter region of the mouse brain-type 14 kb mRNA. The arrow indicates the transcriptional start site. The ATG triplets in the first exon are underline. Only the last ATG (in bold letters) is in frame with the dystrophin coding sequence. The sequence from the first intron is given in italics. (From Makover et al., 1991.)

less bacterial chloramphenicol acetyl transferase (CAT) structural gene. The construct was assayed in N18TG-2 neuroblastoma transfected cells. The pSVO-CAT construct and the construct pβ-CAT, which contains the β-actin gene promoter (Melloul, Aloni, Calvo, Yaffe and Nudel, 1984), and is known to be expressed in many cell types, were used as controls. Background CAT activity only was found in extracts of the neuroblastoma cells stably transfected with the promoter-less construct pSVO-CAT. In contrast a relatively high CAT activity was observed in extracts of N18TG-2 cells transfected with pBD1200-CAT, containing the brain-type promoter of dystrophin. Very similar results were obtained in the transient transfection experiments (Makover et al., 1991). Thus, the 1.2 kb DNA fragment derived from the putative promoter region of the brain-type dystrophin mRNA can activate a promoter-less CAT construct in transfected neuroblastoma cells. These results strongly support our conclusions regarding the cap site, based on the RNAase protection assays and PCR mediated primer extension.

It is of interest that while the level of dystrophin mRNA in neurogenic cells is much lower than that of β-actin mRNA, the cloned dystrophin gene fragment confers comparable levels of CAT activity in transfected cells. This suggests that the low level of dystrophin mRNA in neuronal cells may depend on additional cis-elements, which are not present in the dystrophin gene fragment tested, on the location of the gene in the chromatin and its conformation, or on post-transcriptional events (which may be related to the large size of the gene and of the primary transcript).

The nucleotide sequence of the promoter and first exon region in the mouse and human brain-type dystrophin mRNA is almost identical. However, Boyce et al. (1991), using a different approach to determine the cap site of the human brain-type dystrophin mRNA, identified several transcription start sites, all of them located downstream from that determined by us for the mRNA of mouse brain dystrophin. It is not clear whether the difference is due to genetic differences or the methods of determination of the transcription initiation sites used in the two investigations.

A novel nonmuscle product of the DMD gene

Probes for the first several kilobases of dystrophin mRNA commonly used for the analysis of dystrophin gene expression did not detect transcripts of this gene in the liver. Therefore, liver mRNA was routinely used as a negative control. However, when we used probes for the 3′ untranslated region (3′UTR) of dystrophin mRNA in RNAase protection assays, significant levels of these sequences were detected in the liver. Further analysis using the RNAase protection assay, Northern blot analysis, PCR analysis and cDNA cloning showed that the liver contains an mRNA, transcribed from the DMD gene, which is very different from the known 14 kb mRNA isoforms which encode the muscle and brain-type dystrophins. It does not contain the sequence coding for the large domain of spectrin like repeats and the N-terminal domain (Fig. 2). Northern blot analysis showed that it migrates as a 6.5 kb mRNA. Sequencing of cloned cDNA isolated from a liver cDNA library has confirmed that the sequence of the 3′ untranslated region of the novel mRNA, as well as the sequences encoding the C-terminal domain and cysteine-rich domain of dystrophin, are identical to those of the 14 kb dystrophin mRNA, except for a few differences caused by alternative splicing. However, sequences 5′ to the region encoding the cysteine-rich domain are completely different and are not found anywhere else in the muscle or brain type dystrophin mRNAs (Bar, Barnea, Levy, Neuman, Yaffe and Nudel, 1990, and unpublished). Genomic DNA cloning and sequencing has shown that the sequence 5′ to the point of divergence is derived from a different exons located upstream to the common sequence. These results show that the 'liver type' 6.5 kb mRNA and the 14 kb dystrophin mRNAs are transcribed from the same gene. The big differences in structure are due to alternative splicing and most probably alternative initiation of transcription.

Knowing part of the sequence unique for the 6.5 kb mRNA, we were able to construct probes for RNAase protection assay which could distinguish between the 6.5 kb and the two 14 kb mRNAs. Using such probes in RNAase protection assays we found that the tissue distribution of the 6.5 kb mRNA is very different from that of dystrophin mRNA: It is the main DMD gene transcript in many nonmuscle tissues. In brain cell cultures it is present in relatively high quantities in neuronal cells and is definitely the major DMD gene product found in glia cells (Fig. 3, Table 1 and Rapaport et al., 1992).

As shown in Fig. 3, relatively very small amounts of the mRNA sequence diagnostic for the 6.5 kb mRNA are detected in skeletal muscle mRNA. This could be due to low expression of this mRNA in muscle fibers or due to the expression of this mRNA in mononucleated cells which are in the skeletal muscle. We therefore analyzed the presence of these sequences in RNA from a cloned myogenic cell line, L185, a subline of the L8 myogenic cell line which expresses dystrophin mRNA after cell fusion (Barnea et al., 1990 and manuscript in preparation). RNA was extracted from L185 cultures containing proliferating mononucleated cells and from differentiated cultures containing multinucleated fibers. As expected from previous studies (Nudel et al., 1988; Barnea et al., 1990) the strong signal diagnostic for the 14 kb mRNA was obtained only with RNA extracted from the differentiated cultures. However, RNA from both differentiated and undifferentiated cultures protected, to a small extent, a fragment which is diagnostic for the 6.5 kb mRNA, demonstrating that myogenic cells contain this mRNA. In contrast to the 14 kb mRNA, the levels of this mRNA did not increase during differentiation, but somewhat decreased. Furthermore, in RNA extracted from differentiated cultures exposed for 2 days to cytosine arabinoside, which selectively kills proliferating cells, the diagnostic band for the 6.5 kb mRNA was undetectable. It thus seems that mononucleated myoblasts contain small amounts of the novel 6.5 kb mRNA, but the expression of this DMD gene transcript is down-regulated during cell fusion and formation of skeletal muscle fibers.

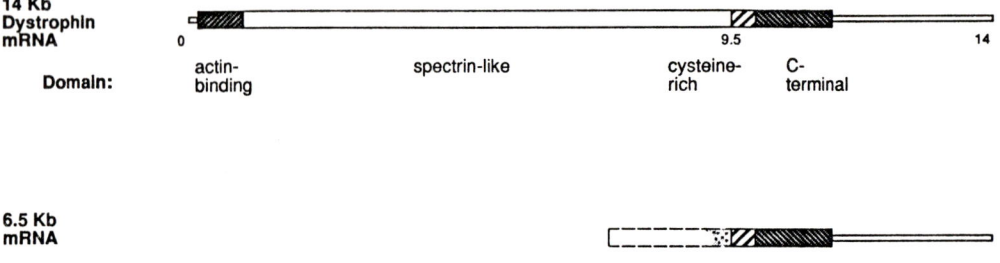

Fig. 2. Schematic description of the dystrophin 14 kb mRNA and the novel 6.5 kb mRNA. The 6.5 kb mRNA and the 14 kb mRNA were aligned according to the sequence homology in the 3′ untranslated region and the regions coding for the C-terminal domain and cysteine-rich domain.
TOP: The 14 kb brain and muscle dystrophin mRNAs. Thin bars: Untranslated regions. Wide bar: Coding region. The regions coding for the various domains are indicated on the bar and described below it.
BOTTOM: The postulated structure of the 6.5 kb mRNA. The dashed line borders the sequence unique to the 6.5 kb mRNA. The divergence between the sequence common to the dystrophin mRNA and the novel 6.5 kb mRNA is at the 5′ end of the sequence coding for the cysteine rich domain. The dotted area indicates the region of the unique sequence for which the sequence is already known.

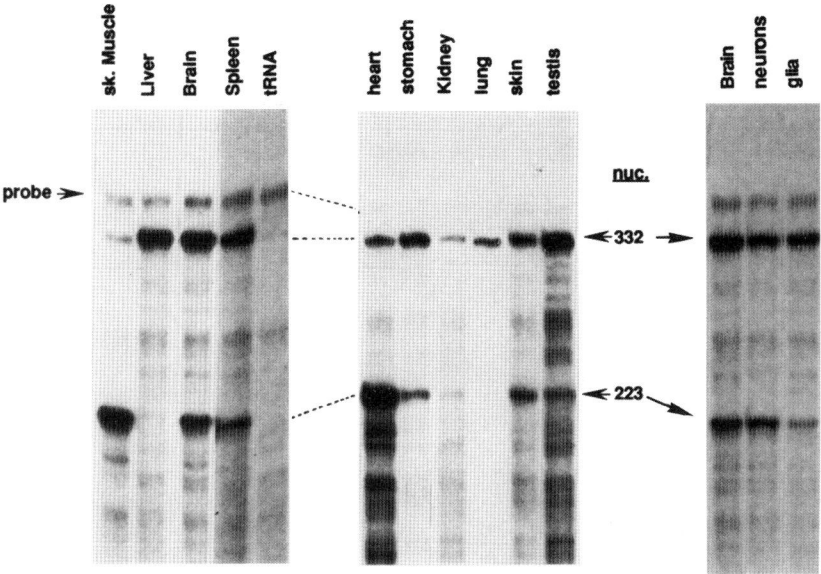

Fig. 3. The 6.5 kb mRNA is a major product of the DMD gene in many tissues. RNAase protection assay was performed using 15 μg of total RNA from the indicated rat tissues. Hybridization and digestion by RNAase were done as described Bar et al. (1990). After size-fractionation of the protected RNA fragments on the polyacrylamide/urea sequencing gels, the gels were fluorographed for 18 h at −70°. The probe used in this experiment was complementary to 223 nucleotides shared by the 14 kb mRNA and the 6.5 kb mRNA plus 109 nucleotides unique to the 6.5 kb mRNA. Accordingly, the 6.5 kb protected a 332 nucleotide fragment of the probe, while the muscle and brain 14 kb mRNAs protected a fragment of 223 nucleotides (from Rapaport et al., 1992).

The protein product of the 6.5 kb mRNA

As described above, the 6.5 kb mRNA encodes the C-terminal domain and the cysteine-rich domain of dystrophin, but not the large spectrin-like repeat domain and the N-terminal domain. In collaboration with N. Augier, J. Léger and J.J. Léger (Montpellier) and G. Morris (Wales), we used monoclonal antibodies raised against epitopes in the C-terminal domain of dystrophin to identify this product on Western blots. Protein samples extracted from various rat tissues and the human hepatoma cell line HepG2 were reacted with MANDRA 1 (Fig. 4). The antibody reacts specifically with the X-linked dystrophin but not with the dystrophin related protein (DRP), which is the product of a gene located on human chromosome 6. This antibody strongly stained dystrophin in muscle extracts (400 kd band). In brain extract it weakly stained dystrophin but strongly stained a polypeptide migrating as a 77 kd protein. In liver extracts only a 77 kd protein was detectable. The cell type and tissue distribution of this 77 kd polypeptide was very similar to the tissue distribution of the 6.5 kb mRNA, as determined by Northern blots and RNAase protection assays. The same results were obtained with another antibody raised against the C-terminal domain of chicken dystrophin.

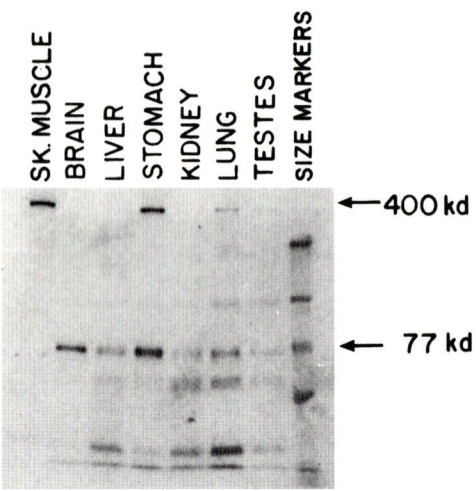

Fig. 4. Identification of a 77 kd protein that reacts with monoclonal antibodies against the C-terminal domain of dystrophin. Proteins from the indicated rat tissues were extracted and size fractionated on 3-7% gradient polyacrylamide/SDS gels. After transfer onto nitrocellulose filter the blots were stained using Mab MANDRA1 which was raised against epitopes in the C-terminal domain of dystrophin (from Lederfein et al., 1992).

Antibodies raised against the spectrin-like repeats produced a distinct signal with extracts of the relevant tissues, in the 400 kd region but not in the 77 kd region, as expected from the structure of the 6.5 kb mRNA.

Mdx mice do not synthesize dystrophin due to a point mutation at position 3185 of the dystrophin mRNA, resulting in a stop codon (Sicinski, 1989). Since our previous data indicated that the 6.5 kb mRNA does not contain this region of the dystrophin mRNA (Fig. 2), we anticipated that if the 77 kd protein is a genuine product of the 6.5 kb mRNA its production will not be affected by the mutation which abolishes the production of dystrophin in the *mdx* mice. Indeed, when tissue extracts of *mdx* mice were subjected to Western blot analysis, the 77 kd polypeptide but not dystrophin was observed in the appropriate tissue extracts. These results show that the 77 kd polypeptide is not translated from the 14 kb dystrophin mRNA. They also rule out the possibility that the 77 kd protein is a degradation or cleavage product of dystrophin (Lederfein et al., 1992).

In conclusion, it seems that the giant DMD gene is highly complex and codes for multiple products, which are expressed in a wide range of tissues, of which the muscle is only one. Our studies clearly demonstrate the existence of at least three products. Two of them, the muscle- and brain-type dystrophins, are very similar; however, their expression is regulated by two different promoters. The third product of the gene greatly differs from dystrophin. It is a 77 kd protein, encoded by a 6.5 kb mRNA. It contains only the cysteine-rich and C-terminal domains of dystrophin. The latter seems to be responsible for the association of dystrophin with the cell membrane by forming a complex with glycoproteins (Ervasti and Campbell, 1991). The 77 kd protein lacks, however, the sequence encoding the large domain of spectrin-like repeats and the N-terminal domain, which shares homology with the actin-binding domain of α-actinin. It is the major known DMD gene product in nonmuscle tissues. In many of them, the amounts of this mRNA is comparable to the amount of dystrophin mRNA in the muscle. The abundancy of this protein in brain cells is of special interest, since it raises the possibility that its absence or misfunction in these cells is causally related to the mental retardation found in a significant proportion of DMD affected children.

Since the 77 kd protein is encoded by mRNA which is significantly smaller than dystrophin, it is anticipated that the synthesis of this DMD gene product will not be affected in a significant proportion of DMD affected children (as in the mdx mice described above). Indeed, in a preliminary study, we have found that in several DMD affected human fetuses the production of the 77 kd protein was unaffected (in collaboration with Hagit Prigojin and with Ruth Shomrat of Ichilov Hospital, Tel Aviv). A study of the presence or absence of the normal product of this mRNA in DMD affected children and its correlation with the pathology of the disease should be very informative.

The valuable technical assistance of Zehava Levy, Sara Neuman and Ora Fuchs is gratefully acknowledged. This work was supported by research grants from the Muscular Dystrophy Association of the USA, the Association Française contre les Myopathies, the Israel Academy of Sciences, the Israel Ministry of Health, and the Leo and Julia Forchheimer Center for Molecular Genetics. We wish to thank Mali Baer and Dorit Zuk for editorial assistance.

References

Bar, S., Barnea, E., Levy, Z., Neuman, S., Yaffe, D. and Nudel, U. (1990). A novel product of the Duchenne muscular dystrophy gene which greatly differs from the known isoforms in its structure and tissue distribution. *Biochem J* **272**, 557-560.

Barnea, E., Zuk, D., Simantov, R., Nudel, U. and Yaffe, D. (1990). Specificity of expression of the muscle and brain dystrophin gene promoters in muscle and brain cells. *Neuron* **15**, 881-888.

Boyce, F. M., Beggs, A. H., Feener, C. and Kunkel, L. M. (1991). Dystrophin is transcribed in brain from a distant upstream promoter. *Proc Natl Acad Sci USA* **88**, 1276-1280.

Chamberlain, J. S., Pearlman, J. A., Muzny, D. M., Gibbs, R. A., Ranier, J. E., Reeves, A. A. and Caskey, C. T. (1988). Expression of the murine Duchenne muscular dystrophy gene in muscle and brain. *Science* **239**, 1416-1418.

Chelly, J., Hamard, G., Koulakoff, A., Kaplan, J.-C., Kahn, A. and Berwald-Netter, Y. (1990). Dystrophin gene transcribed from different promoters in neuronal and glial cells. *Nature* **344**, 64-65.

den Dunnen, J. T., Casula, L., Makover, A., Bakker, B., Yaffe, D., Nudel, U. and van Ommen, G.-J. B. (1991). Mapping of the dystrophin brain promoter: a deletion of this region is compatible with normal intellect. *Neuromusc. Disord.* (in press).

Ervasti, J. M. and Campbell, K. P. (1991). Membrane organization of the dystrophin-glycoprotein complex. *Cell* **66**, 1121-1131.

Feener, C. A., Koenig, M. and Kunkel, L. M. (1989). Alternative splicing of the human dystrophin mRNA generates isoform at the carboxy terminus. *Nature* **338**, 509-511.

Klamut, H. J., Gangopadhyay, S. B., Worton, R. G. and Ray, P. N. (1990). Molecular and functional analysis of the muscle-specific promoter region of the Duchenne muscular dystrophy gene. *Mol. Cell. Biol.* **10**, 193-205.

Koenig, M., Monaco, A. P. and Kunkel, L. M. (1988). The complete sequence of dystrophin predicts a rod-shaped cytoskeletal protein. *Cell* **53**, 219-228.

Lederfein, D., Levy, Z., Augier, N., Mornet, D., Morris, G., Fuchs, O., Yaffe, D. and Nudel, O. (1992). A 71 kd protein is a major product of the Duchenne Muscular Dystrophy gene in brain and other nonmuscle tissues. *Proc. Natl. Acad. Sci. USA* **89**, 5346-5350.

Makover, A., Zuk, D., Breakstone, J., Yaffe, D. and Nudel, U. (1991). Brain-type and muscle-type promoters of the dystrophin gene differ greatly in structure. *Neuromusc Disord* **1**, 39-45.

Melloul, D., Aloni, B., Calvo, J., Yaffe, D. and Nudel, U. (1984). Developmentally regulated expression of chimeric genes containing muscle actin DNA sequences in transfected myogenic cells. *EMBO J.* **3**, 983-990

Moser, H. (1984). Duchenne muscular dystrophy: Pathogenic aspects and genetic prevention. *Human Genet* **66**, 17-40.

Nudel, U., Robzyk, K. and Yaffe, D. (1988). Expression of the putative Duchenne muscular dystrophy gene in differentiated myogenic cell cultures and in the brain. *Nature* **331**, 635-638.

Nudel, U., Zuk, D., Einat, P., Zeelon, E., Levy, Z., Neuman, S. and Yaffe, D. (1989). Duchenne muscular dystrophy gene product in brain is not identical to its product in muscle. *Nature* **337**, 76-78.

Rapaport, D., Lederfein, D., Den Dunnen, J. T., Grootscholten, P. M., Van Ommen, G. J. B., Fuchs, O., Nudel, U. and Yaffe, D. (1992). Characterization and cell type distribution of mRNA sequences of a novel, major product of the Duchenne Muscular Dystrophy gene. *Differentiation* **49**, 187-194.

Sicinski, P., Geng, Y., Ryder-Cook, A. S., Barnard, E. A., Darlison, M. G. and Barnard, P. J. (1989). The molecular basis of muscular dystrophy in the *mdx* mouse: A point mutation. *Science* **244**, 1578-1580.

Yaffe, D., Barnea, E., Bar, S., Makover, A. and Nudel, U. (1991). Promoters and isoforms of transcripts of the DMD gene. In: *Frontiers in Muscle Research* (eds. Ozawa, E., Masaki, T. and Nabeshima, Y.), pp. 351-394. New York: Elsevier Sci. Publ. (Biomed. Div.).

Yaffe, D., Zuk, D., Einat, P., Shinar, D., Levy, Z., Neuman, S., Fuchs, O. and Nudel, N. (1989). Tissue- and stage-specificity of expression of the Duchenne muscular dystrophy gene In: *Cellular and Molecular Biology of Muscle Development*, (eds. Kedes L.H. and Stockdale, F.E.), pp. 963-961. New York: Alan R. Liss.

Printed in Great Britain © *Society for Experimental Biology 1992*   189

THE ROLE OF THE SKELETAL MUSCLE RYANODINE RECEPTOR GENE IN MALIGNANT HYPERTHERMIA

DAVID H. MACLENNAN[1], *KINYA OTSU*[1], *JUNICHI FUJII*[1], *FRANCESCO ZORZATO*[1], *MICHAEL S. PHILLIPS*[1], *PETER J. O'BRIEN*[2], *ALAN L. ARCHIBALD*[3], *BEVERLEY A. BRITT*[4], *ELIZABETH F. GILLARD*[5] *and RONALD G. WORTON*[5]

[1]Banting and Best Department of Medical Research, C.H. Best Institute, University of Toronto, Toronto, Ontario, Canada, [2]Department of Pathology, University of Guelph, Guelph, Ontario, Canada, [3]AFRC Institute of Animal Physiology and Genetics Research Edinburgh, Scotland, [4]Department of Anaesthesia, Toronto General Hospital, Toronto, Ontario, Canada, [5]Department of Genetics, Hospital for Sick Children, Toronto, Ontario, Canada

Summary

Malignant hyperthermia (MH) is an inherited, potentially lethal condition in which sustained muscle contracture with attendant hypermetabolism and hyperthermia is triggered in humans, heterozygous for the gene defect, by inhalational anaesthetics and skeletal muscle relaxants, and in pigs, homozygous for the defect, by stress. Because muscle contracture could result from a defective Ca^{2+} release channel, we have focussed our attention on the linkage of MH to defects in the gene (*RYR1*) encoding the skeletal muscle Ca^{2+} release channel. We have cloned and sequenced human *RYR1* cDNA and found restriction fragment length polymorphisms (RFLPs) in the human gene. We also localized *RYR1* to human chromosome 19q13.1. Studies of the cosegregation of MH with these RFLPs established *RYR1*/MH linkage on human chromosome 19q13.1 (lod score of 4.2; recombinant fraction 0.0). We then sequenced MH and normal porcine *RYR1* cDNAs. Mutation of C1843 to T, leading to substitution of Cys for Arg^{615}, was the sole amino acid change noted between MH and normal animals. Linkage of this mutation to MH was established in a study of 338 informative meioses (lod score of 102; recombinant fraction 0.0). We identified the corresponding mutation in 1 of 35 human MH families studied and found cosegregation of the mutation and MH. The combination of a high lod score with crossing of a species barrier supports the causal nature of this mutation. Future studies are aimed at finding the major human MH mutations and establishing assays for their accurate diagnosis.

Introduction

As new concepts in basic biological science emerge, they frequently become directly relevant to problems in disease. This has been the case for the inherited neuromuscular disease, malignant hyperthermia (MH), which is manifested most

Key words: malignant hyperthermia, ryanodine receptor gene, calcium release channel, anaesthetics, skeletal muscle relaxants.

commonly in humans as anaesthetic-induced muscle contracture, accompanied by high fever. The syndrome became widely recognized in the mid 1950s, following the introduction of halothane and succinylcholine as the most popular combination of anaesthetic and muscle relaxant. In certain families, the combination was fatal to individuals genetically predisposed to the disease (Denborough and Lovel, 1960).

At the same time, studies of the mechanisms controlling muscle contraction revealed that the interaction of actin and myosin was regulated by Ca^{2+}, mediated through the Ca^{2+} binding protein troponin (Ebashi, 1963). Moreover, muscle Ca^{2+} concentration, in turn, was shown to be regulated by a membrane system, the sarcoplasmic reticulum, which is responsible for initiating muscle relaxation by pumping Ca^{2+} to the lumenal space of the membrane, storing it, and later releasing it to initiate another round of muscle contraction (Ebashi, Endo and Ohtsuki,1969). Ca^{2+} was also shown to control glycolysis in muscle through its activation of phosphorylase kinase (Brostrom, Hunkeler and Krebs, 1971). From such studies, it became apparent that a defect in Ca^{2+} regulation, leading to the continued presence of Ca^{2+} within the sarcoplasm, could induce both muscle contracture and extensive glycolysis, leading to the high turnover of ATP which could be responsible for the elevated temperature associated with MH episodes (Kalow, Britt, Terreau and Haist, 1970; Endo, 1977).

Defects in Ca^{2+} regulation could result from defects in the Ca^{2+} pump, responsible for Ca^{2+} uptake, or from defects in the Ca^{2+} release channel. No primary defect in Ca^{2+} uptake could be confirmed, even in the presence of halothane (O'Brien, 1986a; Nelson, 1988). With the development of three different assays for Ca^{2+} release (Endo, 1977; Ohnishi, 1979; Meissner, 1984), it became possible to observe a defect in Ca^{2+} release associated with MH muscle (Endo, Yagi, Ishizuka, Horiuti, Koga and Omaha, 1983; Ohnishi, Taylor and Gronert, 1983; O'Brien, 1986b). Within a few years, the Ca^{2+} release channel itself was identified through its high affinity binding of the modulator ryanodine, hence the name 'ryanodine receptor' (Inui, Saito and Fleischer, 1987; Lai, Erickson, Block and Meissner, 1987; Campbell, Knudson, Imagawa, Leung and Sutko, 1987) and soon thereafter, cDNA encoding it was cloned (Takeshima, Nishimura, Matsumoto, Ishida, Kangawa, Minamino, Matsuo, Ueda, Hanaoka, Hirose and Numa,1989; Zorzato, Fujii, Otsu, Phillips, Green, Lai, Meissner and Maclennan, 1990). Comparative studies were then possible for both the protein (Mickelson, Gallant, Litterer, Johnson, Rempel and Louis, 1988; Knudson, Mickelson, Louis and Campbell, 1990; Fill, Coronado, Mickelson, Vilven, Ma, Jacobson and Louis, 1990; Fill, Stefani and Nelson, 1991; Carrier, Villaz and Dupont, 1991) and for the DNA encoding it (Fujii, Otsu, Zorzato, de Leon, Khanna, Weiler, O'Brien and MacLennan, 1991; Otsu, Khanna, Archibald and Maclennan, 1991; Gillard, Otsu, Fujii, Khanna, de Leon, Derdemezi, Britt, Duff, Worton and MacLennan, 1991) making it possible to demonstrate the association with MH of defects in the *RYR1* gene and the Ca^{2+} release channel expressed from it.

In this paper we review the studies that have led to the identification of defects in the Ca^{2+} release channel gene that are associated with MH under the headings *Malignant hyperthermia*; *Ryanodine receptor*; *Ryanodine receptor/MH linkage*; and *Future prospects*.

Malignant hyperthermia

Malignant hyperthermia is an inherited myopathy in which skeletal muscle contracture with attendant hypermetabolic and hyperthermic reactions is triggered most commonly by a combination of inhalational anaesthetics and skeletal muscle relaxants in humans heterozygous for the defect and by stress or a halothane challenge in swine homozygous for the defect (O'Brien, 1987; Britt, 1991). In humans, the syndrome appears in about 1 in 15,000 anaesthetics administered to children and in about 1 in 50,000 adult anaesthetics (Britt, 1991).

Since MH episodes, if untreated, can be fatal or can cause neurological, liver or kidney damage in surviving patients, the key to management of the disease is early detection, coupled with measures to reverse its course. In the case of human MH, infusion of dantrolene (Harrison, 1975; Britt, 1984) to lower intracellular Ca^{2+} has proven to be exceptionally efficacious in combat of the disease. The death rate has been lowered from 90% to about 7% using appropriate monitoring and therapy (Britt, 1991).

A major goal of MH research has been to identify MH susceptible individuals in advance of anaesthesia. Kalow et al. (1970) approached this problem through an *in vitro* caffeine/halothane contracture test in which fibers from muscle biopsies are attached to a force displacement transducer and then exposed to incremental doses of caffeine, to caffeine in the presence of halothane, or to single doses of halothane. Experience with variants of the test over a 20 year period has given rise to arbitrary cut off points defining normality and susceptibility (Britt, 1989; Ording, 1987). The test is invasive, expensive and, because of the potentially catastrophic results of a false negative diagnosis, errs on the side of false positives. Nevertheless, the test has been widely used to distinguish MH individuals among members of families in which an MH reaction has occurred. Results of the test support autosomal dominant inheritance of human MH.

MH in swine is associated with lean, heavily muscled breeds. Indeed, the MH gene appears to contribute 2 to 3% to dressed carcass weight (Webb and Simpson, 1986; Simpson and Webb, 1989). MH is a serious economic problem in the pork industry because it leads to sudden, stress-induced deaths and to pale, soft, exudative meat, probably resulting from contracture, hypermetabolism and hyperthermia in the carcass itself (Ludvigsen, 1953; Harrison, 1979).

Swine can be tested for MH susceptibility through a halothane challenge test which will induce rapid contracture in live homozygous (n/n) pigs but not in heterozygous (N/n) or normal (N/N) pigs (Eikelenboom and Minkema, 1974). If carried out carefully, few pigs are lost to this test, but most heterozygotes escape detection and may breed in the next generation. Polymorphisms in marker genes flanking the halothane sensitivity gene (*HAL* locus) have also been used in attempts to identify *N/n* and *n/n* individuals (Andresen and Jensen, 1977; Archibald and Imlah, 1985). Analysis of inheritance of linked markers, *GPI* and *PGD*, lying in the *HAL* linkage group, leads to an indirect test for the heterozygote which is 95% accurate but is expensive (Gahne and Juneja, 1985). At this time, therefore, there is still a need for an inexpensive accurate, non invasive test for both human and porcine MH.

Ryanodine receptor

During the 1980s, assays for the function of the Ca^{2+} release channel were developed (Morii and Tonomura, 1983; Miyamoto and Racker, 1982; Meissner, 1984; Meissner, 1986; Meissner, Darling and Eveleth, 1986; Smith, Coronado and Meissner, 1985; Smith, Imagawa, Ma, Fill, Campbell and Coronado, 1988). Rapid Ca^{2+} release from isolated muscles was shown to be activated by micromolar Ca^{2+} and millimolar adenine nucleotides and inhibited by millimolar Mg^{2+}. Calmodulin at micromolar concentrations was shown to inhibit Ca^{2+} release, apparently by direct protein-protein interactions with the Ca^{2+} release channel. Single channel measurements in planar bilayers showed that Ca^{2+} release is mediated by a ligand gated channel with a conductance greater than 100 pS in 50 mM Ca^{2+}. Accordingly, it became possible to evaluate the Ca^{2+} release function of skeletal muscle sarcoplasmic reticulum in MH sensitive (MHS) and normal (MHN) individuals (Endo et al., 1983; Ohnishi et al., 1983; Nelson, 1983; Kim, Streter, Ohnishi, Ryan, Roberts, Allen, Meszaros, Antoniu and Ikemoto, 1984; Mickelson, Ross, Reed and Louis, 1986; O'Brien, 1986; Ohnishi, Waring, Fang, Horiuchi, Flick, Sadanaga and Ohnishi, 1986; Nelson, 1988; Mickelson, Gallant, Rempel, Johnson, Litterer, Jacobson and Louis, 1989; O'Brien, Klip, Britt and Kalow, 1990; Fill et al., 1990; Fill et al., 1991; Carrier et al., 1991). Such studies demonstrated that the rate and extent of Ca^{2+} release from MHS sarcoplasmic reticulum exceeds the rate of Ca^{2+} release from MHN sarcoplasmic reticulum and that the threshold of activation of Ca^{2+} release by such agents as caffeine, ATP and even Ca^{2+} itself is lowered in MHS sarcoplasmic reticulum. Dantrolene, the antidote for MH in the operating room (Britt, 1984), inhibits halothane induced Ca^{2+} release in isolated sarcoplasmic reticulum (Ohnishi et al., 1986), and Ca^{2+}-induced Ca^{2+} release at 38°, but not at 22° (Ohta, Ito and Ohga, 1990).

The Ca^{2+} release channel has been isolated (Inui et al., 1987; Lai and Meissner, 1987; Lai, Erickson, Rousseau, Liu and Meissner, 1988; Campbell et al., 1987). It is a tetrameric complex made from 4 identical 565,000 Da subunits. Three dimensional image reconstruction of the functional tetramer (Wagenkneckt et al., 1989) shows a quatrefoil structure with hydrophobic segments of the four identical subunits forming a membrane spanning baseplate and hydrophilic segments forming a cytoplasmic domain that surrounds and decorates the central baseplate. Four internal channels radiate from a common origin above this baseplate and open into vestibules in 4 quadrants. The radiating channels may be formed from the faces of interactions among the 4 subunits.

In addition to the hypersensitivity of the Ca^{2+} release channel to various triggering agents (O'Brien, 1986b), ryanodine receptors from MH pigs were shown to have altered ryanodine binding when binding was measured as a function of free Ca^{2+} (Mickelson, Gallant, Litterer, Johnson, Rempel and Louis, 1988). Since channel opening is altered by alterations in Ca^{2+} concentration and since ryanodine binding is a function of channel opening, it is possible that Ca^{2+} regulation of channel opening was being measured. Moreover, Ca^{2+} induced channel closing was found to be altered in the MH channel (Fill et al., 1990). The tryptic digestion pattern for the MH channel was also found to be altered, consistent with the loss of a single tryptic digestion site in the MH pig (Knudson et al., 1990).

In the last few years, cDNAs encoding rabbit, human and porcine ryanodine receptors have been cloned and sequenced (Takeshima et al., 1989; Zorzato et al., 1990; Fujii et al., 1991). The cDNAs are over 15,000 bp long, encoding proteins of some 5035 amino acids, and the human gene is over 200,000 bp long (M.S. Phillips et al., in preparation). Searching for mutants in a gene of this size is a daunting task.

Ryanodine receptor/MH linkage

Our cloning of the human ryanodine receptor cDNA (Zorzato et al., 1990) allowed us to demonstrate several restriction fragment length polymorphisms (RFLPs) in the human *RYR1* gene (MacLennan, Duff, Zorzato, Fujii, Phillips, Korneluk, Frodis, Britt and Worton, 1990). In a study of linkage between inheritance of MH and one or more of these polymorphisms, we found cosegregation in 23 meioses in 9 families, leading to a lod score of 4.2 favoring linkage with a recombinant fraction of 0.0. The probability of linkage of more than 10,000 to 1 allowed us to identify *RYR1* as a candidate gene for MH in humans.

This finding was expected on the basis of earlier physiological observations. It was also consistent with a variety of other observations relevant to linkage in both pigs and humans. Andresen and Jensen (1977) demonstrated linkage between inheritance of MH and polymorphisms in *GPI*. Later studies (reviewed in Archibald and Imlah, 1985) established a linkage group for the porcine *HAL*, *GPI* and *PGD*, localized near the centromere of pig chromosome 6 (Davies, Harbitz, Fries, Stranzinger and Hauge, 1988; Chowdhary, Harbitz, Makinen, Davies and Gustavsson, 1989; Harbitz, Chowdhary, Thomsen, Davies, Kaufman, Kran, Gustavsson, Christensen and Hauge, 1990). The homologous region around the human *GPI* locus was known to be on the long arm of chromosome 19 (Lusis, Heinzmann, Sparkes, Scott, Knott, Geller, Sparkes and Mohandas, 1986), making this a candidate region for human MH localization. We localized *RYR1* to human chromosome 19q13.1 in the same region as human *GPI* (MacKenzie, Korneluk, Zorzato, Fujii, Phillips, Iles, Wieringa, Le Blond, Bailly, Willard, Duff, Worton

Fig. 1. Amino acid sequence of the porcine ryanodine receptor. Normal Yorkshire animals have Arginine (R) at position 615, while MH Pietrain have cysteine (C) at position 615. This substitution is believed to be causative of porcine MH and the corresponding human mutation to be causative of some forms of human MH.

and MacLennan, 1990). McCarthy, Healy, Heffron, Lehane, Deufel, Lehmann-Horn, Farrall and Johnson, (1990), using a series of linked chromosome 19q markers, established linkage of the MH locus to the region of human chromosome 19 where we had colocalized *RYR1* and the MH gene (MacLennan et al., 1990).

Linkage of *RYR1* to MH provided the incentive to search for sequence differences in the *RYR1* gene between MH and normal individuals. We initiated parallel programs to search for such mutations in both swine and humans (Fujii et al., 1991; Otsu et al., 1991; Gillard et al., 1991, 1992).

In a comparison of the cDNA sequences of MH (Pietrain) and normal (Yorkshire) pigs, we found 18 nucleotide polymorphisms, but only a single amino acid sequence change (Fujii et al., 1991). The substitution of T for C at nucleotide 1843 leads to the substitution of Cys for Arg at amino acid residue 615 (Fig. 1). Since this mutation resulted in the loss of a *Hin*PI restriction endonuclease site and the gain of a *Hgi*AI site, it was possible to analyse the mutation either by

Table 1. *Haplotype/genotype analysis using polymorphic restriction endonuclease sites within the* RYR1 *gene in six swine breeds*

Genotype		Number of animals		
		T/T	T/C	CC
HinPI	$-/-$	32/32		
HinPI	$+/-$		60/60	
HinPI	$+/+$			49/49
BanII	$+/+$	32/32	32/60	21/49
BanII	$+/-$		28/60	14/49
BanII	$-/-$			14/49
RsaI	$-/-$	32/32	33/60	23/49
RsaI	$+/-$		27/60	10/49
RsaI	$+/+$			16/49

PCR-amplified products that surround the polymorphic sites at nucleotides 1,843 (HinPI) 4,332 (BanII) and 13,878 (RsaI) were analyzed (in rows) for the presence ($+$) or absence ($-$) of appropriate restriction endonuclease sites. Since the polymorphism at nucleotide 1,843 alters the amino acid sequence, it is potentially causal of MH. The other two polymorphisms do not alter amino acid sequence and are not considered causal of the disease. On the basis of nucleotide 1843, animals were grouped (in columns) as: T/T (proven n/n); C/T (probable N/n); and C/C (probable N/N). The number of animals in each class were tabulated from studies of six breeds: Pietrain, Yorkshire, Poland China, Duroc, Landrace and Hampshire. No n/n or N/n Hampshire animals and no N/N or N/n Pietrain animals were observed in our study. Note that all T/T (n/n) animals have the haplotype HinPI$^-$ BanII$^+$ RsaI$^-$ and that each T/C (N/n) animal has the potential for this haplotype (no T/C animals with the BanII$^{-/-}$ or RsaI$^{+/+}$ genotype were detected, although these genotypes are common in C/C (N/N) animals).

restriction endonuclease digestion or by differential oligonucleotide probe analyses. Our initial studies showed an association between inheritance of the mutation and MH in some 80 animals from 5 different breeds. We then went on to analyse linkage in backcrosses between British Landrace N/n and n/n animals (Otsu et al., 1991). In this study, 376 animals were tested, including 338 representing informative meioses. Phenotypic diagnoses were based on the halothane challenge test and confirmed by *GPI* and *PGD* haplotype analysis. Cosegregation of the phenotype with the Cys for Arg[615] substitution was complete, leading to a lod score favoring linkage of 102 with a recombinant fraction of 0.0.

The fact that the identical mutation appeared in 5 lean, heavily muscled pig breeds led to the possibility that the mutation in all breeds arose in a founder animal. Haplotype and genotype analysis using three markers covering over 100 kb within the *RYR1* gene showed the inheritance of a *HinPI$^-$ BanII$^+$ RsaI$^-$* haplotype in every n/n animal examined and the potential for the inheritance of this chromosome in every N/n animal (Table 1). By contrast, the *HinPI$^+$ BanII$^-$ RsaI$^+$* haplotype was common in several animals in all breeds in which normal animals were found. Association of this haplotype with the MH phenotype suggests that the disease did originate in a founder animal and was selected for in breeding stock.

Why should a disease gene be selected? There is evidence that the MH gene contributes 2 to 3% to dressed carcass weight (Webb and Simpson, 1986; Simpson and Webb, 1989). Moreover, the animals are lean and heavily muscled. We speculate that a hypersensitive Ca^{2+} release channel could give rise to spontaneous muscle contraction and the continual toning of the muscle may lead to muscle hypertrophy. The utilization of ATP for spontaneous contraction would limit the deposition of fat. Alternatively, the MH mutation may be very closely linked to gene(s) responsible for the desirable carcass traits and maintained by linkage disequilibrium.

Having demonstrated linkage between MH and a substitution of Cys for Arg^{615}, we searched for the corresponding mutation in 35 human MH families. We found the equivalent mutation in a single family of 5 members in which the mutation segregated with MH. A lod score of 102 favoring linkage at θ max=0.0 in swine, combined with the crossing of a species barrier between swine and humans, strongly supports the proposal that this mutation is causative of MH in most pigs and in at least some human families, establishing *RYR1* as a very strong candidate gene for MH.

Linkage between *RYR1* and human MH has not been found in all human families studied. In a large French Canadian family, we found no linkage if criteria for diagnosis of MH in the halothane/caffeine contracture tests were rigorously applied (MacKenzie, Allen, Lakey, Crossan, Nalan, Mettler, Worton, MacLennan and Korneluk, 1991). On the other hand, by using caffeine/halothane contracture test limits below those currently recommended, complete linkage between *RYR1* and MH genes was demonstrated in this same family (lod score of 3.84; θ max=0.0. The logical conclusion is that criteria for diagnosis are set very conservatively, albeit with good reason, and that this creates an ongoing problem of inaccurate diagnosis in linkage studies.

Nevertheless, there are cases where diagnosis appears to be accurate and where no linkage can be discerned (Levitt, Nouri, Jedlicka, McKusick, Marks, Shutack, Fletcher, Rosenberg and Meyers, 1991; Deufel, Golla, Iles, Meindl, Meitinger, Schindelhauer, DeVries, Pongratz, MacLennan, Johnson and Lehmann-Horn, 1992). In at least some of these cases, both false positives and false negatives would have to be invoked to prove linkage. The finding of genetic heterogeneity is perhaps not surprising in light of the fact that patients with other muscle diseases are subject to MH episodes (Brownell, Paasuke, Elash, et al., 1983; Brownell, 1988; Heiman-Patterson, Martino, Rosenberg, et al., 1988; O'Brien, et al., 1990). Thus individuals with central core disease, myotonia congenita, myotonia dystrophica, limb-girdle muscular dystrophy, Brody's disease or Duchenne muscular dystrophy have had MH reactions. Such a reaction could be visualized as resulting from a normal Ca^{2+} release channel which is triggered to open by a rise in Ca^{2+} in a muscle cell in which poor Ca^{2+} regulation exists due to defects in the Ca^{2+} pump, calsequestrin or dihydropyridine receptor subunits, for example, or from increased membrane permeability due to secondary causes (O'Brien et al., 1990). Thus other defective proteins leading to poor Ca^{2+} regulation within the cell may eventually be shown to give rise to other forms of MH susceptibility.

Future directions

To date, only Arg^{614} to Cys has been shown to be a candidate mutation for human MH. Since this occurs in only a few percent of MH families, we must conclude that the major mutations are still to be discovered. Since there will probably be many mutations in human MH, more rapid screening methods for the identification of such mutants must be utilized. To this end, we have analysed some 80 exon/intron boundaries of an estimated 100 in the *RYR1* gene (M.S. Phillips et al., in preparation). We have also identified most of the polymorphisms that are likely to exist in the coding sequence of *RYR1* (Gillard et al., 1992). We are, therefore, in a position to amplify each exon by the polymerase chain reaction, using intron primers, and to analyse each exon sequence for the presence of such polymorphisms through the measurement of single strand conformational polymorphism (SSCP) mobility shifts in acrylamide gels (Orita, Suzuki, Sekiya and Hayashi, 1989). This work, which is currently underway, should yield a variety of candidate MH mutations in human *RYR1*.

Once most of the mutations are known, it should be possible to follow inheritance of these mutations and to designate those members of MH families who are MH susceptible. This will assure that MH susceptible individuals will receive a regimen of 'safe' anaesthetics, while normal family members can safely utilize more conventional anaesthetics. Furthermore, if it should turn out that the number of independent mutations is small and if most can be detected in a simple test, it may be possible to develop population screening for MH susceptibility. Although we think the likelihood of this is small, it is, nevertheless, a worthy goal if all cases of MH reactions are to be prevented.

For the swine industry, the outlook is more straightforward. With a single refined, restriction endonuclease test for the disease, (Otsu et al., 1992) muscle or blood samples as small as 50 μl can be used to obtain rapid, accurate diagnosis. Accordingly, it will be feasible to identify and remove the MH gene from the porcine population within a very short time if that should prove desirable. On the other hand, if the gene is truly beneficial in terms of food conversion, leanness and muscularity, breeding programs utilizing $n/n \times N/N$ lines may be set up that will produce N/n slaughter animals. Research is currently underway to evaluate the usefulness of the gene to the pork industry.

In this paper we have outlined how basic science has illuminated our understanding of the cause of a disease that is important both to human medicine and to the agricultural industry and provided the basis for its genetic dissection and diagnosis. Understanding of the genetic basis for the disease will provide not only a secure diagnosis for human carrriers of the defective gene but will also be of tremendous economic benefit to the pork industry.

This research was supported by grants from the Medical Research Council of Canada (D.H.M., and R.G.W.); the Muscular Dystrophy Association of Canada (D.H.M., J.F., and R.G.W.); the Heart and Stroke Foundation of Ontario (D.H.M., F.Z., M.S.P., and P.J.O.); the National Science and Engineering Research Council of Canada (P.J.O.); the Malignant Hyperthermia Association of Canada (B.A.B.) and the UK Ministry of Agriculture, Fisheries and Food (A.L.A.).

References

Andresen, E. and Jensen, P. (1977). Close linkage established between the *HAL* locus for halothane sensitivity and the PHI (phosphohexose isomerase) locus in pigs of the Danish Landrace breed. *Nord. Vet. Med.* **29**, 502-504.

Archibald, A. L. and Imlah, P. (1985). The halothane sensitivity locus and its linkage relationships. *Anim. Blood Groups Biochem. Genet.* **16**, 253-263.

Britt, B. A. (1984). Dantrolene. *Can Anaesth. Soc. J.* **31**, 61-75.

Britt, B. A. (1989). The North American caffeine halothane contracture test. In *Malignant Hyperthermia: Current concepts.* (M.A. Nalda Felipe, S. Gottmann and H.J. Khambatta, eds.), Normed Verlag: Madrid, pp. 53-69.

Britt, B. A. (1991). Malignant hyperthermia: a review. In *Thermoregulation: Pathology, Pharmacology and Therapy* (E. Schonbaum and P. Lomax eds.) Pergamon Press Inc. New York, pp. 179-292.

Brostrom, C. O., Hunkeler, F. L. and Krebs, E. G. (1971). The regulation of skeletal muscle phosphorylase kinase by Ca^{2+}. *J. Biol. Chem.* **246**, 1961-1967.

Brownell, A. K. W. (1988). Malignant hyperthermia: relationship to other disease. *Brit. J. Anaesth.* **60**, 303-308.

Brownell, A. K. W., Paasuke, R. T., Elash, A., Fowlow, S. B., Seagram, C. G. F., Diewald, R. J. and Friesen, C. (1983). Malignant hyperthermia in Duchenne muscular dystrophy. *Anaesthesiology* **58**, 180-182.

Campbell, K. P., Knudson, C. M., Imagawa, T., Leung, A. T., Sutko, J. L., Kahl, S. D., Raah, C. R. and Madson, L. (1987). Identification and characterization of the high affinity [^{3}H]ryanodine receptor of the junctional sarcoplasmic reticulum Ca^{2+} release channel. *J. Biol. Chem.* **262**, 6460-6463.

Carrier, L., Villaz, M. and Dupont, Y. (1991). Abnormal rapid Ca^{2+} release from sarcoplasmic reticulum of malignant hyperthermia susceptible pigs. *Biochim. Biophys. Acta* **1064**, 175-183.

Chowdhary, B. P., Harbitz, I., Makinen, A., Davies, W. and Gustavsson, I. (1989). Localization of the glucose phosphate isomerase gene to p12-q21 segment of chromosome 6 in pig by *in situ* hybridization. *Hereditas* **111**, 73-78.

Davies, W., Harbitz, I., Fries, R., Stranzinger, G. and Hauge, J. G. (1988). Porcine malignant hyperthermia carrier detection and chromosomal assignment using a linked probe. *Anim. Genet.* **19**, 203-212.

Denborough, M. A. and Lovel, R. R. H. (1960). Anaesthetic deaths in a family. *Lancet* **ii**, 45.

Deufel, T., Golla, A., Iles, D., Meindl, A., Meitinger, T., Schindelhauer, D., DeVries, A., Pongratz, D., MacLennan, D. H., Johnson, K. J. and Lehmann-Horn, F. (1992). Evidence for genetic heterogeneity of malignant hyperthermia susceptibility. *Am. J. Hum. Genet.* **50**, 1151-1161.

Ebashi, S. (1963). Third component participating in the superprecipitation of 'natural actomyosin'. *Nature* **200**, 1010.

Ebashi, S., Endo, M. and Ohtsuki, I. (1969). Control of muscle contraction. *Quart. Rev. Biophys.* **2**, 351-384.

Eikelenboom, G. and Minkema, D. (1974). Prediction of pale, soft, exudative muscle with a non-lethal test for halothane induced porcine malignant hyperthermia syndrome. *Netherlands J. Vet. Sci* **99**, 421-426.

Endo, M. (1977). Calcium release from the sarcoplasmic reticulum. *Physiol. Rev.* **57**, 71-108.

Endo, M., Yagi, S., Ishizuka, T., Horiuti, K., Koga, Y. and Amaha, K. (1983). Changes in the Ca-induced Ca release mechanism in sarcoplasmic reticulum from a patient with malignant hyperthermia. *Biomed. Res.* **4**, 83-92.

Fill, M., Coronado, R., Mickelson, J. R., Vilven, J., Ma, J., Jocobson, B. A. and Louis, C. F. (1990). Abnormal ryanodine receptor channels in malignant hyperthermia. *Biophys. J.* **50**, 471-475.

Fill, M., Stefani, E. and Nelson, T. E. (1991). Abnormal human sarcoplasmic reticulum Ca^{2+} release channels in malignant hyperthermia skeletal muscle. *Biophys. J.* **59**, 1085-1090.

Fujii, J., Otsu, K., Zorzato, F., De Leon, S., Khanna, V. K., Weiler, J., O'Brien, P. J. and MacLennan, D. H. (1991). Identification of a mutation in the porcine ryanodine receptor associated with malignant hyperthermia. *Science* **253**, 448-451.

Gahne, B. and Juneja, R. K. (1985). Prediction of the halothane (Hal) genotypes of pigs by deducing Hal, Phi, Po2, Pgd haplotypes of parents and offspring: Results from a large-scale practice in Swedish breed. *Anim. Blood Groups Biochem. Genet.* **16**, 265-283.

Gillard, E. F., Otsu, K., Fujii, J., Khanna, V. K., De Leon, S., Derdemezi, J., Britt, B. A., Duff, C. L., Worton, R. G. and MacLennan, D. H. (1991). A substitution of Cysteine for Arginine 614 in the ryanodine receptor is potentially causative of human malignant hyperthermia. *Genomics* **11**, 751-755.

Gillard, E. F., Otsu, K., Fujii, J., Duff, C. L., de Leon, S., Khanna, V. K., Britt, B. A., Worton, R. G. and MacLennan, D. H. (1992). Polymorphisms and deduced amino acid substitutions in the coding sequence of the ryanodine receptor (RYR1) gene in individuals with malignant hyperthermia. *Genomics* **13**, 1247-1254.

Harbitz, L., Chowdhary, B., Thomsen, P., Davies, W., Kaufman, U., Kran, S., Gustavsson, I., Christensen, K. and Hauge, J. (1990). Assignment of the porcine calcium release channel gene, a candidate for the malignant hyperthermia locus, to the 6p11-q21 segment of chromosome 6. *Genomics* **9**, 243-248.

Harrison, G. G. (1975). Control of the malignant hyperpyrexic syndrome in MHS swine by dantrolene sodium. *Br. J. Anaesth.* **47**, 62-65.

Harrison, G. G. (1979). Porcine malignant hyperthermia. *Int. Anesthesiol. Clin.* **17**, 25-62.

Heiman-Patterson, T., Rosenberg, H., Fletcher, J. E. and Tahmoush, A. J. (1988). Nalothane-caffeine contracture testing in neuromuscular disease. *Muscle and Nerve* **11**, 453-457.

Inui, M., Saito, A. and Fleischer, S. (1987). Purification of the ryanodine receptor and identity with feet structures of junctional terminal cisterae of sarcoplasmic reticulum from fast skeletal muscle. *J. Biol. Chem.* **262**, 1740-1747.

Kalow, W., Britt, B. A., Terreau, M. E. and Haist, C. (1970). Metabolic error of muscle metabolism after recovery from malignant hyperthermia. *Lancet* **ii**, 895-898.

Kim, D. H., Sreter, F. A., Ohnishi, S. T., Ryan, J. F., Roberts, J., Allen, P. D., Meszaros, L. G., Antoniu, B. and Ikemoto, N. (1984). Kinetic studies of Ca^{2+} release from sarcoplasmic reticulum of normal and malignant hyperthermia susceptible pig muscles. *Biochim. Biophys. Acta* **775**, 320-327.

Knudson, C. M., Mickelson, J. R., Louis, C. F. and Campbell, K. P. (1990). Distinct immunopeptide maps of the sarcoplasmic reticulum Ca^{2+} release channel in malignant hyperthermia. *J. Biol. Chem.* **265**, 2421-2424.

Lai, F. A., Erickson, H., Block, B. A. and Meissner, G. (1987). Evidence for a junctional feet-ryanodine receptor complex from sarcoplasmic reticulum. *Biochem. Biophys. Res. Commun.* **143**, 704-709.

Lai, F. A., Erickson, H. P., Rousseau, E., Liu, Q.-Y. and Meissner, G. (1988). Purification and reconstitution of the calcium release channel from skeletal muscle. *Nature* **331**, 315-319.

Levitt, R. C., Nouri, N., Jedlicka, A. E., McKusick, V. A., Marks, A. R., Shutack, J. G., Fletcher, J. E., Rosenberg, H. and Meyers, D. A. (1991). Evidence for genetic heterogeneity in malignant hyperthermia susceptibility. *Genomics* **11**, 543-547.

Ludvigsen, J. (1953). Muscular degeneration in hogs. *15th Internat. Vet. Cong., Stockholm* **1**, 602-606.

Lusis, A. J., Heinzmann, C., Sparkes, R. S., Scott, J., Knott, T. J., Geller, R., Sparkes, M. C. and Mohandas, T. (1986). Regional mapping of human chromosome 19: Organization of genes for plasma lipid transport (APOC1, -C2 and -E and LDLR) and the genes C3, PEPD and GPI. *Proc. Natl. Acad. Sci. USA* **83**, 3929-3933.

MacKenzie, A. E., Allen, G., Lahey, D., Crossan, M. L., Nolan, K., Mettler, G., Worton, R. G., MacLennan, D. H. and Korneluk, R. (1991). A comparison of the caffeine halothane muscle contracture test with the molecular genetic diagnosis of malignant hyperthermia. *Anaesthesiology* **75**, 4-8.

MacKenzie, A. E., Korneluk, R. G., Zorzato, F., Fujii, J., Phillips, M., Iles, D., Wieringa, B., Le Blond, S., Bailly, J., Willard, H. F., Duff, C., Worton, R. G. and MacLennan, D. H. (1990). The human ryanodine receptor gene: its mapping to 19q13.1, placement in a chromosome 19 linkage group and exclusion as the gene causing myotonic dystrophy. *Am. J. Hum. Genet.* **46**, 1082-1089.

MacLennan, D. H., Duff, C., Zorzato, F., Fujii, J., Phillips, M., Korneluk, R. G., Frodis, W., Britt, B. A. and Worton, R. G. (1990). Ryanodine receptor gene is a candidate for predisposition to malignant hyperthermia. *Nature* **343**, 559-561.

McCarthy, T. V., Healy, J. M. S., Heffron, J. J. A., Lehane, M., Deufel, T., Lehmann-Horn, F., Farrall, M. and Johnson, K. (1990). Localization of the malignant hyperthermia susceptibility locus to human chromosome 19q12-13.2. *Nature* **343**, 562-564.

Meissner, G. (1984). Adenine nucleotide stimulation of Ca^{2+} induced Ca^{2+} release in sarcoplasmic reticulum. *J. Biol. Chem.* **259**, 2365-2374.

Meissner, G. (1986). Evidence of a role for calmodulin in the regulation of calcium release from skeletal muscle sarcoplasmic reticulum. *Biochemistry* **25**, 244-251.

Meissner, G. (1986). Ryanodine activation and inhibition of the Ca^{2+} release channel of sarcoplasmic reticulum. *J. Biol. Chem.* **261**, 6300-6306.

Meissner, G., Darling, E. and Eveleth, J. (1986). Kinetics of rapid Ca^{2+} release by sarcoplasmic reticulum. Effects of Ca^{2+}, Mg^{2+}, and adenine nucleotides. *Biochemistry* **25**, 236-244.

Mickelson, J. R., Ross, J. A., Reed, B. K. and Louis, C. F. (1986). Enhanced Ca^{2+}-induced calcium release by isolated sarcoplasmic reticulum vesicles from malignant hyperthermia susceptible pig muscle. *Biochim. Biophys. Acta* **862**, 318-328.

Mickelson, J. R., Gallant, E. M., Litterer, L. A., Johnson, K. M., Rempel, W. E. and Louis, C. F. (1988). Abnormal sarcoplasmic reticulum ryanodine receptor in malignant hyperthermia. *J. Biol. Chem.* **263**, 9310-9315.

Mickelson, J. R., Gallant, E. M., Rempel, W. E., Johnson, K. M., Litterer, L. A., Jacobson, B. A. and Louis, C. F. (1989). Effects of the halothane-sensitivity gene on sarcoplasmic reticulum function. *Am. J. Physiol.* **257**, C787-C794.

Miyamoto, H. and Racker, E. (1982). Mechanism of calcium release from skeletal sarcoplasmic reticulum. *J. Membr. Biol.* **66**, 193-201.

Morii, H. and Tonomura, Y. (1983). The gating behavior of a channel for Ca^{2+}-induced Ca^{2+} release in fragmented sarcoplasmic reticulum. *J. Biochem.* **93**, 1271-1285.

Nelson, T. E. (1983). Abnormality in calcium release from skeletal sarcoplasmic reticulum of pigs susceptible to malignant hyperthermia. *J. Clin. Invest.* **72**, 862-870.

Nelson, T. E. (1988). SR function in malignant hyperthermia. *Cell* **9**, 257-265.

O'Brien, P. J. (1986a). Porcine malignant hyperthermia susceptibility: Increased calcium sequestering activity of skeletal muscle sarcoplasmic reticulum. *Can. J. Vet. Res.* **50**, 329-337.

O'Brien, P. J. (1986b). Porcine malignant hyperthermia susceptibility: Hypersensitive calcium-release mechanism of skeletal muscle sarcoplasmic reticulum. *Can. J. Vet. Res.* **50**, 318-328.

O'Brien, P. J. (1987). Etiopathogenetic defect of malignant hyperthermia: Hypersensitive calcium-release channel of skeletal muscle sarcoplasmic reticulum. *Vet. Res. Comm.* **11**, 527-559.

O'Brien, P. J. (1990). Microassay for malignant hyperthermia susceptibility: hypersensitive ligand-gating of the Ca channel in muscle sarcoplasmic reticulum causes increased amounts and rates of Ca-release. *Mol. Cell. Biochem.* **93**, 53-59.

O'Brien, P. J., Klip, A., Britt, B. A. and Kalow, B. I. (1990). Malignant hyperthermia susceptibility: Biochemical basis for pathogenesis and diagnosis. *Can. J. Vet. Res.* **54**, 83-92.

Ohnishi, S. T. (1979). Calcium-induced calcium release from fragmented sarcoplasmic reticulum. *J. Biochem.* **86**, 1147-1150.

Ohnishi, S. T., Taylor, S. and Gronert, G. A. (1983). Calcium-induced Ca^{2+} release from sarcoplasmic reticulum of pigs susceptible to malignant hyperthermia. The effects of nalothane and dantrolene. *FEBS Lett.* **161**, 103-107.

Ohnishi, S. T., Waring, A. J., Fang, S-R. G., Horiuchi, K., Flick, J. L., Sadanaga, K. K. and

Ohnishi, T. (1986). Abnormal membrane properties of the sarcoplasmic reticulum of pigs susceptible to malignant hyperthermia: modes of action of halothane, caffeine, dantrolene, and two other drugs. *Arch. Biochem. Biophys.* **247**, 294-301.

Ohta, T., Ito, S. and Ohga, A. (1990). Inhibitory action of dantrolene on Ca-induced Ca^{2+} release from sarcoplasmic reticulum in guinea pig skeletal muscle. *Eur. J. Pharmacol.* **178**, 11-19.

Ollivier, L., Sellier, P. and Monin, J. G. (1975). Determinisme genetique du syndrome d'hyperthermie maligne chez le porc de Pietrain. *Ann. Genet. Sel. Anim.* **7**, 159-166.

Ording, H. (1987). The European MH group: protocol for *in vitro* diagnosis of susceptibility to MH and preliminary results. In (Britt B.A. ed.) Malignant Hyperthermia. *Martinus Nijhoff Boston* pp. 269-

Orita, M., Suzuki, Y., Sekiya, T. and Hayashi, K. (1989). Rapid and sensitive detection of point mutations and DNA polymorphisms using the polymerase chain reaction. *Genomics* **5**, 874-879.

Otsu, K., Khanna, V. K., Archibald, A. L. and MacLennan, D. H. (1991). Co-Segregation of porcine malignant hyperthermia and a probable causal mutation in the skeletal muscle ryanodine receptor gene in backcross families. *Genomics* **11**, 744-750.

Otsu, K., Phillips, M. S., Khanna, V. K., deLeon, S. and MacLennan, D. H. (1992). Refinement of diagnostic assays for a probable causal mutation for porcine and human malignant hyperthermia. *Genomics* **13**, 835-837.

Simpson, S. P. and Webb, A. J. (1989). Growth and carcass performance of British Landrace pigs heterozygous at the halothane locus. *Anim. Prod.* **49**, 503-509.

Smith, J. S., Coronado, R. and Meissner, G. (1985). Sarcoplasmic reticulum contains adenine nucleotide activated calcium channels. *Nature* **316**, 446-449.

Smith, J. S., Imagawa, T., Ma, J., Fill, M., Campbell, K. P. and Coronado, R. (1988). Purified ryanodine receptor from rabbit skeletal muscle is the Ca^{2+} release channel of sarcoplasmic reticulum. *J. Gen. Physiol.* **92**, 1-26.

Takeshima, H., Nishimura, S., Matsumoto, T., Ishida, H., Kangawa, K., Minamino, N., Matsuo, H., Ueda, M., Hanaoka, M., Hirose, T. and Numa, S. (1989). Primary structure and expression from complementary DNA of skeletal muscle ryanodine receptor. *Nature* **339**, 439- 445.

Wagenknecht, T., Grassucci, R., Frank, J., Saito, A., Inui, M. and Fleischer, S. (1989). Three-dimensional architecture of the calcium channel/foot structure of sarcoplasmic reticulum. *Nature* **338**, 167-170.

Webb, A. J. and Simpson, S. P. (1986). Performance of British Landrace pigs selected for high and low incidence of halothane sensitivity. 2. Growth and carcass traits. *Anim. Prod.* **43**, 493-503.

Zorzato, F., Fujii, J., Otsu, K., Phillips, M., Green, N. M., Lai, F. A., Meissner, G. and MacLennan, D. H. (1990). Molecular cloning of cDNA encoding human and rabbit forms of the Ca^{2+} release channel (ryanodine receptor) of skeletal muscle sarcoplasmic reticulum. *J. Biol. Chem.* **265**, 2244-2256.

EXPRESSION OF MUSCLE GENES IN THE MOUSE EMBRYO

MARGARET BUCKINGHAM, DENIS HOUZELSTEIN, GARY LYONS*, MARCIA ONTELL**, MARIE-ODILE OTT and DAVID SASSOON***

Department of Molecular Biology, CNRS URA 1148, Pasteur Institute, 25 rue du Dr. Roux, 75015 Paris, France

* Department of Anatomy, University of Wisconsin Medical School, 1300 University Avenue, Madison, WI 53706 USA

** Department of Anatomy and Cell Biology, University of Pittsburgh School of Medicine, 3550 Terrace Street, Pittsburgh, PA 15261 USA

*** Department of Biochemistry, Boston University School of Medicine, 80 East Concord Street, Boston Mass, USA

Summary

Using isogene specific probes and *in situ* hybridization on sections, we have examined the expression of structural and regulatory genes in the mouse embryo during the formation of cardiac and skeletal muscle. The temporal and spatial information thus obtained about the onset of expression of muscle genes provides insight into the regulation of myogenesis *in vivo*.

Actin and myosin sequences present in different compartments of the adult heart are initially all co-expressed in the cardiac tube (between 7-8 days). The process of spatial restriction to atrial or ventricular compartments of the heart takes place asynchronously later. In contrast, the onset of expression of actin and myosin genes in the first skeletal muscle, the myotome, which corresponds to the central compartment of the somite, as well as their subsequent down-regulation in different skeletal muscle masses, takes place very asynchronously. One might predict that factor(s) responsible for the transcriptional activation of these genes are present in sufficient quantity in the cardiac tube, whereas in skeletal muscle individual genes are responding to variable levels of factor(s).

In fact the four myogenic regulatory sequences present in the mouse - MyoD1, myogenin, myf-5 and myf-6 - do show distinct patterns of expression during the development of skeletal muscle. None of these sequences have been detected in the heart. In the myotome there is no general correlation between the appearance of a particular myogenic sequence and the activation of a particular structural gene. A striking example of this is provided by the muscle isoform of creatine phosphokinase. We would propose that each muscle structural gene has a different threshold of activation, depending on the quantity and nature of the myogenic factor present.

Key words: Myogenesis *in vivo* - Myogenic regulatory sequences - Actin and myosin transcripts - Dystrophin transcripts - *In situ* hybridization - Mouse embryo - Myotomes - Cardiac tube - Limb bud.

We have also examined the onset of expression of the X-linked dystrophin gene known to be expressed in adult heart and skeletal muscle. In the myotome dystrophin transcripts are first detected at the time when myosin heavy chains first accumulate and muscular contraction is initiated. In contrast in the cardiac tube dystrophin transcripts are not detected initially, at a time (from 8 days) when the heart contracts. This observation can be correlated with the pathology of the disease which points to a more essential role of dystrophin in skeletal muscle.

No muscle structural gene examined is expressed in the somite prior to myotome formation. If the myogenic regulatory sequences are implicated in muscle cell determination then they should be expressed in the dermomyotome of the immature somite which gives rise to muscle precursor cells. Only myf-5 is present as early as this, in the region from which myotomal cells are derived, and myf-5 may therefore be implicated in an earlier determination step. Muscle masses which form elsewhere in the embryo result from the migration of cells present in a different region of the dermomyotome. For these cells, the role of myogenic factors in the early stages of myogenesis is less clear. The subsequent profile of expression of myogenic factors as muscle structural genes are activated is also different from that in the myotome, perhaps indicating a difference in myogenic lineage for cells from this region of the somite.

Introduction

One of the major attractions of skeletal muscle as a model for the study of differentiation has been the availability of cell culture systems (see Buckingham, 1985). This has greatly facilitated our understanding of the cell biology, and has permitted detailed functional analysis of the regulation of genes encoding proteins which are markers of differentiated muscle. Isolation of the myogenic regulatory sequences represented a major step forward in understanding the molecular machinery for tissue specific regulation of transcription. This was achieved by substractive hybridization with RNA from cultured cells at different stages of myogenesis (Davis, Weintraub and Lassar, 1987; Wright, Sassoon, and Lin, 1989) and depended on the availability of appropriate cell lines. Our current concepts of how the four mammalian myogenic factors - MyoD1, myogenin, myf-5 and myf-6 - function are mainly based on experiments with cultured cells (Olson, 1990; Weintraub, Davis, Tapscott, Thayer, Krause, Benezra, Blackwell, Turner, Rupp, Hollenberg, Zhuang, and Lassar, 1991). They are clearly transcriptional factors, capable of activating many muscle genes. In addition all four sequences have the striking property of converting non-muscle cells to myogenesis, the extent of the transformation depending on the cell line. On this basis they have been described as myogenic determination factors. No equivalent sequences have been isolated in the heart. Slower progress in characterizing the molecular regulation of cardiac muscle genes no doubt in part reflects the lack of good cardiac cell lines. As a first step towards elucidating how the activation of muscle genes takes place as striated muscles form during mammalian embryogenesis *in vivo*, we have used the *in situ* hybridization technique (see Sassoon, Garner and Buckingham, 1988) to describe when and where muscle specific genes are expressed in the mouse embryo. Firstly the expression of a number of different structural genes is compared with that of the myogenic sequences and the role of the latter as transcription factors is discussed. Secondly their expression is considered in the context of a potential role in the commitment of cells to the myogenic lineage prior to muscle formation.

Actin and myosin genes

The first striated muscle to form in the mammalian embryo is the heart (Rugh, 1990) which is evident in the mouse as a cardiac tube between 7-8 days p.c. (post coitum), resulting from the fusion of the two cardiac primordia, which are formed from lateral mesoderm. From 8 days p.c., the first contractions are detectable. By 9 days p.c. the heart acquires a more adult morphology with distinguishable atrial and ventricular compartments. In the cardiac tube from the earliest stages that we have examined (c. 7.5 days p.c.) all the major myosin sequences are already accumulated (Lyons, Schiaffino, Sassoon, Barton and Buckingham, 1990a). Cardiac actin, and to a lesser estent, skeletal actin transcripts are also present (Sassoon et al. 1988). These genes are among the first characteristic of an adult phenotype to be expressed during embryogenesis, and cardiac actin, for example, is a potentially interesting early marker for tracing cardiac cells as they emerge from the precardiac mesoderm. Subsequently, as the heart acquires its mature morphology, different myosin sequences become restricted to atrial or ventricular compartments. This process takes place asynchronously; the β-MHC (myosin heavy chain), for example, is only present in the ventricle by 10.5 days p.c., whereas the ventricular MLC1V (myosin alkali light chain) form is still present in the atria at this time, and is confined to the ventricular compartment only by 17.5 days p.c. (Table 1).

The first skeletal muscles to appear in the embryo are the myotomes which form in the central compartment of the somites as they mature, following a rostral-caudal gradient, from about 8.5 days p.c. in the most anterior somites. In contrast to the situation in the cardiac tube, the onset of expression of the different actin and myosin genes is very asynchronous (Table 2), (Sassoon et al. 1988; Lyons, Ontell, Cox, Sassoon and Buckingham, 1990b). Cardiac actin, for example, is present as a transcript and protein (Lyons, Buckingham, and Mannherz, 1991) a whole day before transcripts of the first myosin heavy chain, MHCemb, are detectable. The other myosin heavy chain gene to be transcribed early is the β-MHC. The first myosin light chain transcripts to accumulate at a high level are those encoding MLC1A, also a cardiac isoform. It is striking that actin and myosin genes expressed in the heart are major transcripts in early skeletal muscle, to be replaced in the case of cardiac actin and MLC1A by skeletal muscle isoforms as development proceeds. One surprising result was that the so-called perinatal or neonatal myosin heavy chain gene is already transcribed in the embryo by 10.5 days p.c., although in this case it is not clear that the protein is actually present so

Table 1. *Expression of actin and myosin genes in developing mouse cardiac muscle*

Days p.c.	8.0	9.5	10.5	11.5	12.5	14.5	15.5	17.5	0.0	7.0 p.p.
Atrial muscle										
MLC1A	+++	+++	+++	+++	+++	+++	+++	+++	+++	+++
MLC1V	+++	+++	+++	++	+	+	+	+/-	-	-
MHCα	+++	+++	+++	+++	+++	+++	+++	+++	+++	+++
MHCβ	+++	+	-	-	-	-	-	-	-	-
α-cardiac actin	+++	+++	+++	+++	+++	+++	+++	+++	+++	+++
Ventricular muscle										
MLC1A	+++	+++	+++	+++	+++	+++	++	+	-	-
MLC1V	+++	+++	+++	+++	+++	+++	+++	+++	+++	+++
MHCα	+++	+++	++	++	++	+	+	++	++	+++
MHCβ	+++	+++	+++	+++	+++	+++	+++	+++	++	-
α-cardiac actin	+++	+++	+++	+++	+++	+++	+++	+++	+++	+++

This table summarizes results from *in situ* hybridization experiments, carried out in collaboration with S. Schiaffino and reported in Lyons et al. (1990a). Abbreviations: MLC, myosin alkali light chain; A, atrial; V, ventricular; MHC, myosin heavy chain; p.c., post coitum; p.p. post partum.

Table 2. *Expression of actin and myosin genes in the myotomes and during subsequent skeletal muscle development*

Days p.c.	8.5	9.5	10.5	11.5	12.5	14.5	15.5
α-cardiac actin	+	++	+++	+++	+++	+++	+++
α-skeletal actin	±	+	++	++	+++	+++	+++
MLC1A	−	+	+++	+++	+++	+++	+++
MLC1F	−	±	++	++	+++	+++	+++
MHCemb	−	±	++	+++	+++	+++	++
MHCβ	−	±	+	+	+	+	+
MHCpn	−	−	+	++	+++	+++	+++
MLC1V	−	−	−	−	−	−	+
MLC3F	−	−	−	−	−	(+)	+

This table summarizes results from *in situ* hybridization experiments reported in Lyons et al. (1990b). Abbreviations: MLC, myosin alkali light chain; MHC, myosin heavy chain; A, atrial; V, ventricular; F, fast skeletal; emb, embryonic; p.m., perinatal; p.c. post coitum.

early (see Lyons et al. 1990b). As the skeletal muscle matures different actin and myosin genes either cease to be expressed (e.g. MLC1A) or become restricted to a particular fibre type (e.g. MHCβ) (Table 2). Depending on the muscle, myosin genes cease to be expressed uniformly throughout the tissue, as they are in the embryonic muscle masses, from about 15 days. Again this process takes place very asynchronously. Clearly each myosin or actin gene follows a distinct pattern of expression responding to different regulatory signals or to different quantitative levels of the same signal. One might predict that in the cardiac tube these signals or transcriptional factors required for the different cardiac actin and myosin genes are present initially together at a high level, whereas in skeletal muscle they accumulate asynchronously. The same gene in the two types of striated muscle is probably responding to different molecular signals, since in the cases examined the DNA sequences required for tissue specific expression of the gene differ (see Ordahl, 1991).

Myogenic regulatory sequences in differentiating muscle

From the earliest time point that we have examined (7.5 days p.c.), there is no detectable expression of any of the myogenic regulatory sequences in the cardiac tube or heart during embryogenesis in the mouse. With the exception of myf-5 which is expressed in the somite prior to myotome formation these sequences are restricted to skeletal muscle. As in the case of the structural genes described previously, each sequence has a distinct temporal pattern of expression (Sassoon, Lyons, Wright, Lin, Lassar, Weintraub and Buckingham, 1989; Ott, Bober, Lyons, Arnold and Buckingham, 1991; Bober, Lyons, Braun, Cossu, Buckingham, and Arnold, 1991). In the myotome, myogenin transcripts accumulate rapidly together with those of myf-5 which become restricted to this compartment of the somite. Myf-6 transcripts also accumulate slightly later, followed by

MyoD1. As indicated in Table 3 there is a two day time lag between myotome formation and MyoD1 expression. These observations, based on *in situ* hybridization experiments, are confirmed by analysis of transcripts present, using the very sensitive polymerase chain reaction technique (Montarras, Chelly, Bober, Arnold, Ott, Gros, and Pinset, 1991). In the muscle masses of the limb, myf-6 is not detected initially (Bober et al. 1991). Transitory expression of myf-5 (Ott et al. 1991) is followed by co-expression of myogenin and MyoD1 (Sassoon et al. 1989) (Table 3). This is in contrast to what happens in the myotome, and may reflect the fact that the cells which migrate out from the somite correspond to a different myogenic lineage present in the dermomyotome (see later). Alternatively different environmental cues may be responsible for this striking difference in the myogenic regulatory programme. In the myotome myf-6 expression decreases rapidly from 10.5 days. Myf-5 expression also decreases and transcripts are no longer detectable by 14 days. Myf-6, unlike myf-5, begins to be re-expressed in all skeletal muscles from about 16 days and is a major transcript post-natally when MyoD1 and myogenin decline (Bober et al. 1991; Rhodes and Konieczny, 1989). There is no indication that these myogenic factors are preferentially associated with a particular type of muscle fibre, as foetal muscles mature. At earlier stages there is no indication that secondary versus primary fibres have a distinct myogenic factor composition. Muscle masses appear uniformly labelled.

The data summarized here are based on levels of transcripts. Antibodies for myf-5 and myf-6 which work on mouse sections are not yet available. However in the case of MyoD1 and myogenin it has been possible to look at the equivalent proteins (Cusella-De Angelis, Lyons, Sonnino, De Angelis, Vivarelli, Farmer, Wright, Molinaro, Bouché, Buckingham and Cossu, 1992). In the limb the proteins are detectable together at the same time as the corresponding mRNAs. In the myotome, on the other hand, the MyoD1 protein is present from the same time as its mRNA but myogenin protein is not detectable prior to this time, while the transcript is clearly present earlier. At the protein level, therefore, MyoD1 and myogenin appear to be co-expressed in the somite only from a later stage. Prior to this time in the myotome, myf-5 and myf-6 are the candidate myogenic factors. Indeed, the expression of this pair of genes during the early period of myotome maturation, and not in the differentiating muscle masses of the limb, is probably related to the absence of myogenin and MyoD1 proteins in the former.

A key question currently is the extent to which the myogenic factors are interchangeable. It is noteworthy that in the chick the temporal pattern of expression of apparently equivalent sequences is different. CMD1, the MyoD1 homologue, for example, is already expressed in the myotome of the chick at an earlier stage than that at which it first appears in the mouse (Lyons, Mühlebach, Moser, Masood, Paterson, Buckingham and Perriard, 1991). The fact that each sequence has a distinct pattern of expression during muscle development strongly suggests that each gene responds differently to regulatory signals; *in vivo*, autoactivation between family members, if it occurs, is not simple. The temporal

Table 3. *Expression of myogenic sequences during skeletal muscle development*

Days p.c.	Rostral myotomes								Forelimb buds and visceral arches							Late foetal muscle
	8.5	9.0	10.5	11.0	11.5	12.0	13.0	14.0	9.5	10.5	11.0	11.5	12.0	13.0	14.0	17.5
myf-5	+	++	++	++	+	+	−	−	−	+	++	+	+	−	−	−
myogenin	(++)	(+++)	+++	+++	+++	+++	+++	+++	−	−	++	++	+++	+++	+++	+++
myf-6	−	+	++	+	−	−	−	−	−	−	−	−	−	−	−	+++
MyoD1	−	−	+	++	+++	+++	+++	+++	−	++	++	++	+++	+++	+++	+++

This table summarizes results from *in situ* hybridization experiments carried out in collaboration with W. Wright (myogenin), A. Lassar and H. Weintraub (MyoD1), E. Bober and H. Arnold (myf-5 and myf-6), as reported in Sassoon et al. (1989); Ott et al. (1991); Bober et al. (1991). The positive results obtained for myogenin transcripts at 8.5 and 9.0 days are placed in parentheses because the protein is not detectable at these stages (see Cusella-De Angelis et al. 1991).

differences between sequences also strongly suggest that each gene plays a different role during myogenesis *in vivo*.

The myogenic regulatory genes code for DNA binding proteins which have been shown to act as positive regulators of the tissue specific transcription of a number of structural genes expressed in skeletal muscle (see Olson, 1990; Weintraub et al. 1991). It is likely that many such genes are regulated by this family, although this is only demonstrated at present for certain myosin light chain and actin genes, of those listed in Table 2. Taking these genes as examples, one can say that transcription of cardiac actin or MLC1A for example (see Buckingham, Biben, Catala, Lyons and Ott, 1991) is activated when myf-5 and myf-6 are probably the main myogenic factors present. This is also the case for MLC1F whose transcription depends on a proximal promoter and 3' enhancer element (Donoghue, Ernst, Wentworth, Nadal-Ginard and Rosenthal, 1988). This gene undergoes an interesting developmental regulation since MLC3F, whose transcription depends on a second promoter but the same enhancer element, is expressed later when a different set of myogenic factors is present. MLC1V which is detectable as an RNA transcript from 15.5 days is activated at a time when myf-5 is no longer present and MyoD1 and myogenin are the major myogenic sequences, whereas genes such as those encoding the adult fast myosin heavy chains are activated in the perinatal period when myf-6 is present (see Cox and Buckingham, 1991). Although the onset of expression of different muscle genes takes place when a subset of myogenic factors are present, this does not mean that only this subset can activate the gene *in vivo*. This is clearly demonstrated by a comparison between limb muscle and myotome. Similar muscle structural genes are expressed in the limb with an approximately similar temporal sequence (Ontell, Buckingham et al., in preparation) in a situation where MyoD1 and myogenin, instead of myf-5 and myf-6, are the myogenic sequences present in the differentiating muscle mass. Does this mean that the myogenic factors or pairs of factors are redundant? Qualitatively, some interchange is clearly possible, but there may be quantitative restrictions. Different quantities of different myogenic factors may be necessary for activation of a given muscle gene. The levels of MyoD1 and myogenin in the limb required for activation of the cardiac actin gene, for example, may be considerably higher than those of myf-5 and myf-6 in the myotome. Their relative efficiencies may also depend on other transcriptional factors present. This kind of proposal is testable *in vitro* and there is already some evidence that transactivation for certain genes is more efficient with one factor rather than another (e.g. Yutzey, Rhodes and Konieczny, 1990). The idea of activation thresholds for muscle genes is also potentially important when the onset of expression of structural genes is compared with that of the different myogenic factors. In fact there is no correlation between the two, except perhaps in a few cases. A striking example of this is provided by the gene encoding the muscle isoform of creatine phosphokinase (M-CPK) (Lyons et al. 1991), which has been extensively used as a model to look at the mechanism by which a myogenic factor such as MyoD1 activates transcription. Transcripts of this gene only begin to be detectable in the myotomes at 13 days p.c., two days after the onset of MyoD1 expression and at a time when there is no obvious change in the subset of myogenic factors present. If the M-CPK gene requires a certain threshold concentration of MyoD1/myogenin then this may be only attained 2-3 days after these myogenic factors first appear. Another gene such as the MHCpn, for example, may require a much lower level of these factors and is therefore activated immediately. Other factors may also be required at different threshold levels to activate muscle structural genes. Similar considerations may also apply to the autoactivation of the myogenic factor family *in vivo*.

In addition to the question of muscle gene activation, there is also that of the maintenance of muscle gene expression. MLC1F continues to be expressed in adult muscle where myf-6 is the main myogenic factor present. If myogenic factors are involved in the maintenance of transcription - and this should be testable in antisense experiments on differentiated muscle cells - then different genes may again have different threshold requirements. The modulation of muscle gene expression in perinatal and adult muscle also depends on well defined physiological factors such as innervation or thyroid hormone (see Kelly, 1987), although how these effectors relate to regulatory circuits involving the myogenic factors is not yet clear. In embryonic muscle environmental signals remain ill-defined.

Yet another important point concerns the down-regulation of structural genes and indeed of those for myogenic factors too, as development proceeds. One can again invoke levels of factors required for maintenance of expression which fall below a threshold value. Alternatively negative regulators may be involved (see Blau and Baltimore, 1991), and indeed there begin to be examples of positive/negative regulation operating on the same myogenic factor consensus sequences in certain muscle genes (e.g. Numberger, Dürr, Kues, Koenen and Witzemann, 1991).

Dystrophin expression in the mouse embryo

Another muscle structural gene whose expression we have examined during mouse embryogenesis is that of the X-linked dystrophin gene, lesions in which are responsible for the myopathies of Duchenne and Becker (see Hoffman and Kunkel, 1989). This gene is known to be expressed in the adult in muscle, brain and other tissues, depending on the promoter used, and transcript generated (see Barnea, Zuk, Simentov, Nudel and Yaffé, 1990). In collaboration with Jeff Chamberlain we have used a 5′ coding probe covering exons 16 to 20 of the mouse gene (Chamberlain, Pearlman, Muzny, Gibbs, Ranier, Reeves, and Caskey, 1988). We were interested in seeing when and where the gene is expressed in the mouse embryo (Houzelstein, D., Lyons, G. E., Chamberlain, J. and Buckingham, M. E., 1992) compared to the other structural and regulatory sequences we had analysed during striated muscle development.

Dystrophin transcripts are first detectable in the heart from 9-9.5 days p.c. The signal reaches maximum intensity by about 14 days and is considerably lower after birth. The onset of expression is therefore considerably later than that of the actins and myosins, and occurs at least one day after the tubular heart begins to beat. This suggests that in normal development the initial absence of dystrophin is not deleterious, perhaps because other spectrin-related proteins are present which can serve as anchors between the muscle sarcomere and the cell membrane in the way in which dystrophin is thought to function (see Ervasti and Campbell, 1991).

In contrast, in skeletal muscle in the myotome, dystrophin transcripts begin to be detectable from 9.5 days p.c. at the same time as the first myosin heavy chain sequences appear and therefore prior to the formation of a functional contractile apparatus. Myf-5/myf-6 are the myogenic factors present at this time, and may therefore be involved in activating the gene in this tissue. Dystrophin messenger RNA continues to be present in all skeletal muscles, irrespective of fibre type, throughout embryonic and foetal development. After birth the level of transcripts decreases. The fact that muscle contractility is not observed prior to the onset of expression of the dystrophin gene points to the importance of the protein for the correct functioning of skeletal muscle. This can be correlated with the gravity of Duchenne muscular dystrophy in this tissue in humans, once the initial phase of muscle regeneration has been exhausted in the postnatal period.

The myogenic regulatory sequences as muscle determination factors prior to muscle formation during embryogenesis

All four myogenic sequences when transfected into 10T1/2 mouse embryonic fibroblasts will convert these cells to myogenesis (see Olson, 1990; Weintraub et al. 1991). Many other cell types of different embryonic origin, which are apparently less poised for myogenesis than the 10T1/2 cell, will also show varying degrees of conversion. This phenomenon probably occurs because the regulatory genes of this family autoactivate each other, together with other regulatory and structural muscle genes resulting in a 'cascade' of differentiation. The question is to what extent the myogenic regulatory sequences play a critical role *in vivo* at an earlier stage in the myogenic pathway. Are they responsible for committing precursor cells to myogenesis? In order to address this question in descriptive terms initially, it is important to review briefly the origin of skeletal muscle in mammals.

In higher vertebrates our understanding of the origin of skeletal muscle is mostly based on experiments with chick/quail chimaeras which demonstrated that muscle precursor cells are present in the somites (e.g. Chevallier, Kieny and Mauger, 1977; Christ, Jacob and Jacob, 1977. See Bellairs, Ede, and Lash, 1986; Ott, Robert and Buckingham, 1990) or in the case of certain head muscles in the pre-chordal plate which precedes the first somite (Wachtler and Jacob, 1986). With the possible exception of the latter where an additional neural crest component has been suggested (Le Douarin, 1982), no other source of skeletal muscle cells has been identified. Initially somites consist of balls of epithelial-like cells which form in a rostro-caudal gradient by segmentation of paraxial mesoderm on either side of the neural tube (see Theiler, 1989; Ott et al. 1990). The somite rapidly differentiates into two cellular compartments, the dermomyotome which contains precursor cells for muscle or skin, and the sclerotome which contributes to the formation of skeletal structures. Cells migrate out from the dermomyotome mainly from the ventral lateral edge (Jacob, Christ and Jacob, 1978) to form muscle masses elsewhere in the embryo; somites at the level of the limbs contribute the cells for formation of limb musculature (Wachtler, Christ and Jacob, 1982). The first skeletal muscle to form is the myotome, in the central region of the somite, as a result of migration of precursor cells from the cranio-medial edge of the dermomyotome (Kaehn, Jacob, Christ, Hinricksen and Poelmann, 1988). At later stages myotomes contribute to the formation of trunk musculature, such as the intercostal muscles. In mammals, the results of experiments on the transplantation and ablation of somites, strongly suggest that the basic strategy of muscle formation is similar to that described for avian systems (Milaire, 1976; Ede and El-Gadi, 1986), although the chronology and the location within the dermomyotome of migrating muscle precursor cells are less well defined.

If the myogenic regulatory sequences play a role in muscle determination *in vivo* one might expect them to be expressed in the dermomyotome before myotome formation, and also in precursor muscle cells before and during their migration to other sites of muscle formation in the embryo. In fact the only myogenic regulatory gene to be expressed as early as this is myf-5 (Ott et al. 1991) (Fig. 1). Myf-5 transcripts are not detectable in the embryo prior to somite formation but are present at low levels in the first somites as they form (from 8 days). Even at this stage the distribution of transcripts does not appear to be uniform, but is rather higher in cells adjacent to the neural tube. This would suggest that there may already be some distinction between muscle precursors and other cells. At the next stage of somite maturation this becomes evident. There is no labelling of cells in the sclerotome, whereas cells in the dermomyotome particularly in the dorsal lip region adjacent to the neural tube are strongly positive when hybridized to a myf-5 specific probe. In transverse section this corresponds to the region where myotomal precursor cells would be expected to be present. Interestingly, at least in the chick, the cells in this region are dividing rapidly before moving into the myotomal compartment (Kaehn et al. 1988). These observations demonstrate that the myf-5 gene is activated prior to muscle formation in the mouse embryo, and the location of the transcripts would suggest that myf-5 may be implicated in the commitment of muscle precursor cells which will differentiate in the myotome.

The question of the migratory muscle precursor cells is unresolved by these *in situ* hybridization experiments. Some cells throughout the dermomyotome are myf-5 positive, although we have not detected any concentration of label in the ventral lateral region. Myf-5 transcripts are detectable at other sites in the embryo, in the visceral arches or the limb buds, for example, before any other muscle marker examined, including the other myogenic sequences (Ott et al. 1991) (Table 3). However the earliest time at which we obtain a positive signal in the limbs (at

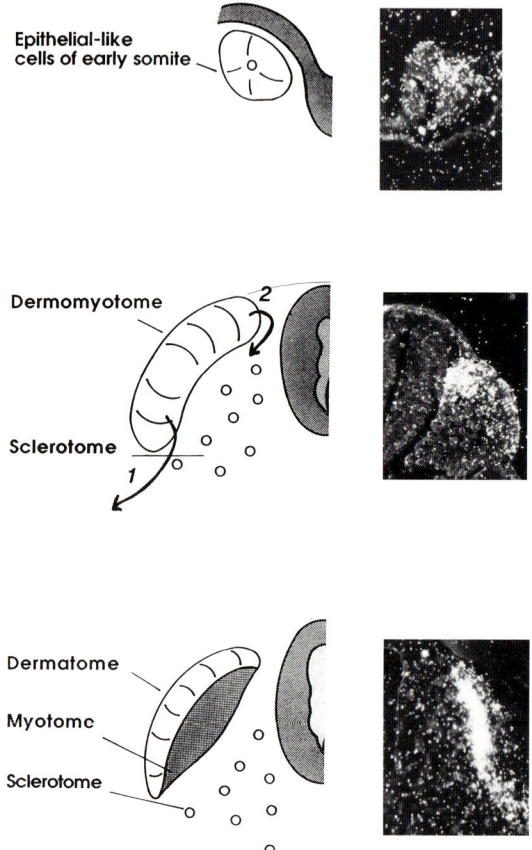

Fig. 1. Expression of myf-5 transcripts in the developing somite. A diagrammatic representation of the somite is shown on the left with a dark field photomicrograph of a section of a somite at a corresponding stage hybridized to a myf-5 specific probe (see Ott et al. 1991) on the right. The somite is shown initially as a ball of epithelial-like cells, with myf-5 transcripts already detectable. At the next stage dermomyotome and sclerotome are present with myf-5 transcripts concentrated in the region of the dermomyotome from which cells will migrate to form the myotome (indicated as the arrow (2)). Cells which migrate to form muscle masses elsewhere in the embryo are indicated by the arrow (1). At the third stage shown, the myotome has formed and myf-5 transcripts are concentrated here. Abbreviations: nt, neural tube; s, somite; dm, dermomyotome; sc, sclerotome.

10 days in the forelimbs, or 10.5 days in the hindlimbs) is later than the time when myogenic precursor cells would be expected to have migrated to sites in the limb, based on chick/quail experiments. The *in situ* hybridization technique with ^{35}S-labelled probes does not permit resolution at the single cell level and it is perhaps not surprising that signal is not detectable in cells as they migrate out from the somite. However we would expect to detect positive cells concentrated in the pre-muscle masses of the limb buds, as soon as they have migrated from the somites,

given the numbers of cells probably involved (e.g. see Wachtler et al. 1982; Milaire, 1976). It is also possible that cells which have been myf-5 positive in the dermomyotome cease to express this gene, and then re-express it again at a slightly later stage. Alternatively the muscle precursor cells, which form extra-myotomal muscle masses, and which are clearly committed to myogenesis in that they never contribute to another tissue type, may not be initially programmed by myf-5, but by another type of determination factor, acting upstream of the MyoD1 family. These cells located in the lateral part of the dermomyotome may correspond to a different myogenic precursor population from that which will form the myotome. As discussed previously, the sequence of myogenic factors expressed in muscle masses founded in the embryo as a result of migration of these cells is different. In addition, Selleck and Stern (1991) have shown recently in the chick that cells in the lateral part of the somite are derived from a region of the primitive streak which is distinct from that giving rise to cells in the medial part of the somite, from which the myotome is probably derived.

Perspective

It is only by interfering with the endogenous genes encoding the myogenic factors that their role in muscle formation can be more precisely appreciated. This is also true where the perplexing diversity of isoforms of muscle structural proteins is concerned. The homologous recombination approach, where a simple 'knock out' of a structural or myogenic gene is the first target, is certainly now possible for the mouse. Experiments of this kind are underway in our own and other laboratories. By 'knocking out' myf-5, for example, it should be possible to assess whether the gene plays a determining role in the subsequent formation of some or all skeletal muscles. As a result of experiments of this kind it should also be possible to determine whether the myogenic sequences are redundant and therefore interchangeable. If they are not, the perturbation resulting from the absence of MyoD1 for example should throw light on how transcriptional activation of muscle structural genes may depend on critical threshold levels of the different myogenic factors.

This work was supported by grants from the Pasteur Institute, CNRS and the AFM. G.L. held an NIH/CNRS fellowship from the Fogarty International Center, and C.B. a studentship from the AFM. M.O. was a sabbatical fellow in the laboratory from April 1988-March 1989 and benefitted from a Fogarty Senior Fellowship.

References

Barnea, E., Zuk, D., Simentov, R., Nudel, U. and Yaffé, D. (1990). Specificity of expression of the muscle and brain dystrophin gene promoters in muscle and brain cells. *Neuron* **5**, 881-888.
Bellairs, R., Ede, D. A. and Lash, J. W. (eds) (1986). *Somites in developing embryos.* New York: Plenum Press, pp. 320.

Blau, H. M. and Baltimore, D. (1991). Differentiation requires continuous regulation. *J. Cell Biol.* **112**, 781-783.

Bober, E., Lyons, G., Braun, T., Cossu, G., Buckingham, M. and Arnold, H. (1991). The muscle regulatory gene, myf-6, has a biphasic pattern of expression during early mouse development. *J Cell Biol.* **113**, 1255-1265.

Buckingham, M. E. (1985). Actin and myosin multigene families: their expression during the formation of skeletal muscle. *Essays in Biochemistry* **20**, 77-109.

Buckingham, M., Biben, C., Catala, F., Lyons, G. and Ott, M. O. (1992). Myogenesis in the mouse: the expression of regulatory and structural genes. In *Neuromuscular Development and Disease* (ed. A. M. Kelly and H. M. Blau). New York, Raven Press Ltd.

Chamberlain, J., Pearlman, J., Muzny, D., Gibbs, R., Ranier, J., Reeves, A. and Caskey, T. (1988). Expression of the murine Duchenne muscular dystrophy gene in muscle and brain. *Science* **239**, 1416-1418.

Chevallier, A., Kieny, M. and Mauger, A. (1977). Limb-somite relationship: origin of the limb musculature. *J. Embryol. Exp. Morph.* **41**, 245-258.

Christ, B., Jacob, H. J. and Jacob, M. (1977). Experimental analysis of the origin of the wing musculature in avian embryos. *Anat. Embryol.* **150**, 171-186.

Cox, R. and Buckingham, M. E. (1991). Actin and myosin genes are transcriptionally regulated during mouse skeletal muscle development. *Dev. Biol.* **149**, 228-234.

Cusella-de Angelis, M. G., Lyons, G., Sonnino, C., De Angelis, L., Vivarelli, E., Farmer, K., Wright, W. E., Molinaro, M., Bouché, M., Buckingham, M. and Cossu, G. (1992). MyoD1, myogenin independent differentiation of primordial myoblasts in mouse somites. *J. Cell Biol.* **116**, 1243-1255.

Davis, R. L., Weintraub, H. and Lassar, A. B. (1987). Expression of a single transfected cDNA converts fibroblasts to myoblasts. *Cell* **51**, 987-1000.

Donoghue, M., Ernst, H., Wentworth, B., Nadal-Ginard, B. and Rosenthal, N. (1988). A muscle-specific enhancer is located at the 3′ end of the myosin light chain 1/3 gene locus. *Genes Dev.* **2**, 1779-1790.

Ede, D. A. and El-Gadi, A. O. A. (1986). Genetic modifications of developmental acts in chick and mouse somite development. *In Somites in developing embryos* (ed. R. Bellairs, D.A. Ede, J.W. Lash) pp. 209-224 New York: Plenum Press.

Ervasti, J. M. and Campbell, K. P. (1991). Membrane organization of the dystrophin-glycoprotein complex. *Cell* **66**, 1121-1131.

Hoffman, E. and Kunkel, L. (1989). Dystrophin abnormalities in Duchenne/Becker muscular dystrophy. *Neuron* **2**, 1019-1029.

Houzelstein, D., Lyons, G. E., Chamberlain, J. and Buckingham, M. E. (1992). Localisation of dystrophin gene transcripts during mouse embryogenesis. *J. Cell Biol.* (in press).

Jacob, M., Christ, B. and Jacob, H. J. (1978). On the migration of myogenic stem cells into the prospective wing region of chick embryos. A scanning and transmission electron microscopic study. *Anat. Embryol.* **153**, 179-193.

Kaehn, K., Jacob, H. J., Christ, B., Hinricksen, K. and Poelmann, R. E. (1988). The onset of myotome formation in the chick. *Anat. Embryol.* **177**, 191-201.

Kelly, A. M. (1987). In *Handbook of Physiology* (L. D. Peachey Ed.) Section 10, pp. 507-537.

Le Douarin, N. (1982). *The neural crest.* Cambridge: Cambridge University Press.

Lyons, G. E., Buckingham, M. E. and Mannherz, H. G. (1991). α-actin proteins and gene transcripts are colocalized in embryonic mouse muscle. *Development* **111**, 451-454.

Lyons, G. E., Mühlebach, S., Moser, A., Masood, R., Paterson, B. M., Buckingham, M. E. and Perriard, J. C. (1991). Developmental regulation of creatine kinase gene expression by myogenic factors in embryonic mouse and chick skeletal muscle. *Development* **113**, 1017-1029.

Lyons, G. E., Ontell, M., Cox, R., Sassoon, D. and Buckingham, M. (1990b). The expression of myosin genes in developing skeletal muscle in the mouse embryo. *J. Cell Biol.* **3**, 1465-1476.

Lyons, G. E., Schiaffino, S., Sassoon, D., Barton, P. and Buckingham, M. (1990a). Developmental regulation of myosin gene expression in mouse cardiac muscle. *J. Cell Biol.* **111**, 2427-2436.

Milaire, J. (1976). Contribution cellulaire des somites à la genèse des bourgeons de membres postérieurs chez la souris. *Arch. biol.* **86**, 177-221.

Montarras, D., Chelly, J., Bober, E., Arnold, H., Ott, M. O., Gros, F. and Pinset, C. (1991).

Developmental patterns in the expression of myf-5, MyoD, myogenin and MRF4 during myogenesis. *The New Biologist* **3**, 592-600.

Numberger, M., Dürr, I., Kues, W., Koenen, M. and Witzemann, V. (1991). Different mechanisms regulate muscle-specific AChR γ- and ε-subunit gene expression. *EMBO J.* **10**, 2957-2964.

Olson, E. N. (1990). MyoD family: a paradigm for development? *Genes Dev.* **4**, 1454-1461.

Ordahl, C. P. (1991). Developmental regulation of sarcomeric gene expression. In: *The cytoskeleton in cell motility and development* (E. Bearer ed.). In press

Ott, M. O., Bober, E., Lyons, G., Arnold, H. and Buckingham, M. (1991). Early expression of the myogenic regulatory gene, myf-5, in precursor cells of skeletal muscle in the mouse embryo. *Development* **111**, 1097-1107.

Ott, M. O., Robert, B. and Buckingham, M. (1990). Le muscle d'où vient-il? *Médecine/Science* **6**, 653-663.

Rhodes, S. J. and Konieczny, S. F. (1989). Identification of MFR4, a new member of the muscle regulatory factor gene family. *Genes Dev.* **3**, 2050-2061.

Rugh, R. (1990). *The Mouse: its reproduction and development.* Oxford University Press.

Sassoon, D., Garner, I. and Buckingham, M. (1988). Transcripts of α-cardiac and α-skeletal actins are early markers for myogenesis in the mouse embryo. *Development* **104**, 155-164.

Sassoon, D., Lyons, G., Wright, W., Lin, V., Lassar, A., Weintraub, H. and Buckingham, M. (1989). Expression of two myogenic regulatory factors, myogenin and MyoD1, during mouse embryogenesis. *Nature* **341**, 302-307.

Selleck, M. and Stern, C. (1991). Fate mapping and cell lineage analysis of Hensen's node in the chick embryo. *Development* **112**, 615-626.

Theiler, K. (1989). *The House Mouse: Atlas of embryonic development.* New York: Springer-Verlag.

Wachtler, F., Christ, B. and Jacob, H. J. (1982). Grafting experiments on determination and migratory behaviour of presomitic and splanchnopleural cells in avian embryos. *Anat. embryol.* **164**, 369.

Wachtler, F. and Jacob, M. (1986). Origin and development of the cranial skeletal muscles. *Bibl. anat.* **29**, 24-46.

Weintraub, H., Davis, R., Tapscott, S., Thayer, M., Krause, M., Benezra, R., Blackwell, K., Turner, D., Rupp, R., Hollenberg, S., Zhuang, Y. and Lassar, A. (1991). The MyoD gene family: Nodal point during specification of the muscle cell lineage. *Science* **251**, 761-766.

Wright, W. E., Sassoon, D. A. and Lin, V. K. (1989). Myogenin, a factor regulating myogenesis has a domain homologous to MyoD. *Cell* **56**, 607-617.

Yutzey, K. E., Rhodes, J. J. and Konieczny, S. F. (1990). Differential transactivation associated with the muscle regulatory factors MyoD1, myogenin, and MRF4. *Mol. Cel. Biol.* **10**, 3934-3944.

Printed in Great Britain © *Society for Experimental Biology 1992*   219

MOLECULAR ANALYSIS OF PROTEIN SORTING DURING BIOGENESIS OF MUSCLE CYTOARCHITECTURE

JEAN-CLAUDE PERRIARD, PIERRE VON ARX, STEFAN BANTLE, HANS M. EPPENBERGER, MONIKA EPPENBERGER-EBERHARDT, MARIUS MESSERLI and THIERRY SOLDATI

Institute for Cell Biology, Swiss Federal Institute of Technology, CH 8093 Zürich, Switzerland

Summary

Isolated, rod-shaped adult rat cardiomyocytes (ARC) were kept in long-term cell cultures and the changes of the cardiomyocyte structure were investigated by confocal microscopy. The cells round up and make contact with the substrate by very flat, foot-like structures. After prolonged culture the amorphous cells regenerate a cardiomyocyte-like cytoarchitecture and myofibrils reemerge. In the perinuclear region myofibrils form continuously while in other cells discontinous myofibrillogenesis was observed, where short sarcomeric segments occur all over the cytoplasmic space. During the regeneration of myofibrils certain proteins like a smooth muscle actin sort to non sarcomeric region, while myomesin or heart C-protein localize on myofibrils with high specificity. This culture system combined with method of epitope-tagging of contractile proteins are ideally suited to monitor the intracellular localization sites of exogenously introduced constructs to different cytoskeletal, since ARC exhibit at the same time stress fiber-like filaments (SFLF) and nascent myofibrils.

The molecular properties of the different members of the myosin light chain isoprotein family were investigated by transfection experiments using epitope-tagged myosin light chain (MLC) cDNA. The sorting of the different types of MLC was shown to be isoprotein specific and with chimeric constructs it was shown that the isoprotein-specific incorporation into myofibrils was dependent on the presence of the middle domain of MLC-1f/3f. These MLC isoproteins can be arranged into a sequence of increasing affinity to myofibrils. A hierarchical order of myofibrillar assembly is postulated based on the association affinity.

Similar experiments with constructs containing α-cardiac, α-smooth muscle and γ-cytoplasmic actins have shown that expression of epitope-tagged actins in ARC result in different epitope staining patterns. While the α-cardiac actin showed a marked preference for sarcomeres, the α-smooth muscle isoproteins had an intermediate specificity and could either be preferentially incorporated into stress fiber-like filaments (SFLF) and in some cells to a lesser extent into myofibrils as well. Most striking results were obtained with γ-cytoplasmic actin carrying a 5 or 11-mer epitope. This actin gave rise to large cells, induced the formation of

Key words: myofibrillogenesis, cytoarchitecture, cardiac muscle, protein sorting, actin, myosin light chain.

filopodia filled with the transfected actin and depletion of the transfected actin
from the perinuclear myofibrillar region.

Introduction

During cytodifferentiation many cellular parameters change e.g. in myogenic
cells shift from an embryonic state to become typical for muscle. While the
proliferation competent embryonic precursor cells express non-muscle forms of
contractile proteins and have a cytoarchitecture like many mesodermal cells with a
well developed cytoskeleton which does not exhibit cell-type specific features, the
differentiated muscle cells are postmitotic, express a set of contractile isoproteins
typical for muscle and their cytoarchitecture changes dramatically as the highly
ordered myofibrils are assembled from the newly expressed muscle specific
proteins. During muscle maturation these organelles can even be reorganized and
instead of embryonic or neonatal contractile isoforms the definitive contractile
proteins are incorporated into the myofibrils.

Recent research has concentrated on the question concerning determination of
the muscle phenotype by myogenic factors and regulation of muscle specific genes.
Special regard was given to the molecular genetics of the many isoprotein
transitions observed in muscle as differentiation takes place. However, little
attention has been devoted to questions of how these newly programmed protein-
products interact with the cell's organization to yield such highly organized
structures as myofibrils. During muscle differentiation myofibrillogenesis is a most
prominent feature of the cellular organization and lends itself to the study of the
biogenesis of cytoarchitecture. The well characterized myofibril is also a model
system to investigate the structure-function relationship of the various muscle and
non-muscle contractile isoproteins at the molecular level (Epstein and Fischman,
1991).

In this contribution the cultured heart cardiomyocyte is exploited to yield
information of how myofibrillogenesis takes place and the turnover of these
structures is effected. Furthermore these cells are also used as living test tubes to
investigate the molecular interactions with cytoskeletal structures of contractile
proteins and their mutants generated *in vitro*. In this contribution special attention
is given to the muscle and non-muscle members of the myosin light chain (Barton
and Buckingham, 1985) and actin (Vandekerckhove and Weber, 1984) isoprotein
families.

Results

Degeneration and regeneration of myofibrils in ARC

The adult myocardium consists of a seemingly homogeneous population of
cardiomyocytes, which in rat are binucleate and adhere to each other by the
formation of intercalated disks, which link cells to each other in a tension resistant
manner (Eppenberger, Bächi, Vollenweider, Volk and Eppenberger, 1988a).

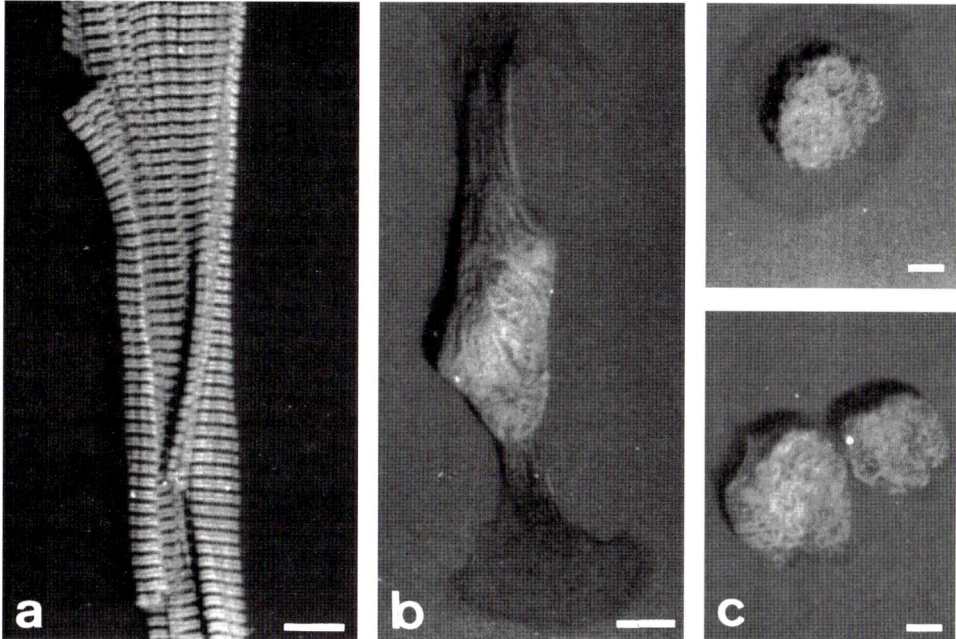

Fig. 1. Early stages of the degeneration of adult rat cardiomyocytes. Freshly isolated (Eppenberger et al., 1988b). ARC cells were immediately fixed and stained for the myofibrillar protein heart C-protein staining the A-band (a) or cultured for 2 days (b), (c) and stained for F-actin with RITC-phalloidin. The flat footlike protrusions are visible as darkly stained areas at the distal tips of the cell in (b), or around cells displaying a round morphology (c). In confocal sections of such cells there is no defined myofibrillar organization visible. The fluorescence pictures were recorded with BioRad MRC 600 confocal scanning device attached to a Zeiss Axioplan microscope. The pictures shown here are three-dimensional reconstructions (SFP, Van der Voort, Brackenhof and Baarslag, 1989) of the the stacks of optical section processed with the image processing software 'Imaris' developed by M. Messerli. Bar is 10 μm.

These cells are almost completely filled with myofibrils, the contractile organelles of cross-striated muscle. The highly organized myofibrils are made up of definitive myofibrillar proteins and their highly ordered array can best be demonstrated by the immunofluorescence stain of an individual building block like the cardiac C-protein, a component of thick filaments, or the protein myomesin occurring in the M-band of sarcomeres. In figure 1a, a freshly isolated adult rat cardiomyocyte (ARC) is shown as a three dimensional reconstruction of heart C-protein immunofluorescence produced by image processing of a stack of optical sections taken with a confocal microscope (Eppenberger-Eberhardt, Riesinger, Messerli, Schwarb, Müller, Eppenberger and Wallimann, 1991). These cells are bipolar and immediately after isolation show a highly organized cytoskeleton probably very similar to the one of the cells in the heart, although, as shown in this cell occasionally some of the myofibrils are detached from the membrane and are bent.

Many of these freshly isolated cells can be kept in long term culture for several weeks, although they do not proliferate. Some attach to the substrate by a foot like extrusion as seen in figure 1b, and upon continuous culture the cytoarchitecture of these cells becomes disorganized (Eppenberger, Hauser, Baechi, Schaub, Brunner, Dechenne and Eppenberger, 1988b). The myofibrils seem to detach from membranes, get wavy and with longer culture also loose their sarcomeric organization (figure 1b). At the same time many cells become rounded while others fail to attach and die. Changing the culture substrate or the medium appears also to affect the morphology of this inital phase of cardiomyocyte cultures. It is not clear on which criteria surviving cells are selected. After 2-4 days in culture the cardiomyocytic phenotype of the majority of cells is desintegrated, the myofibrillar organisation is almost completely abolished and the shape of the cells changes dramatically. It appears that the whole of the cytoplasmic organisation breaks down, all the existing myofibrils are desintegrated (figure 1c) and the cells cease to contract.

As will be shown in the next section these cells are capable of regenerating myofibrils and start to contract again spontaneously. The process of regeneration of the myofibrils is ideally suited to study myofibrillogenesis, since these cells are very large, flat, and display a variety of morphologically distinct types as shown in the next series of pictures. Furthermore, during the regeneration process in the cytoplasm the non-muscle cytoarchitecture coexists with the nascent myofibrils and thus allows to follow the transition from the non-muscle state to the myofibrillar organization within the same cell. It is not yet clear if the original myofibrils are breaking down to protein subunits and if these subunits are potentially reutilized. Alternatively, even the subunit peptides may be degraded and no reutilization would then occur at the protein level.

In the flattened cells the myofibrillar organization reappears

As these cells become reorganized newly formed myofibrils reappear in the cytoplasm and regenerate a heart type like phenotype (Eppenberger et al. 1988 a, b, Eppenberger-Eberhardt, Flamme, Kurer and Eppenberger, 1990) and on day 10-12 in culture they represent cells as shown in figure 2. Most of the genetic programme for the heart phenotype appears to be reexpressed and typical heart proteins are synthesized again. Simultaneously, some of the myofibrillar proteins are expressed as isoforms which normally are typical for fetal or embryonic cells like α-smooth muscle actin, a shift from α-MHC to β-MHC expression was observed and the M-CK was no longer present in the M-bands of the myofibrils (Eppenberger et al., 1988b). The pattern of incorporation into myofibrils of such newly occurring contractile proteins is variable but nevertheless indicates the sites of nascent myofibrillogenesis or non-muscle cytoskeletal structures.

Discontinuous assembly of myofibrils takes place

The exceedingly flat cells are ideally suited for the observation with immunofluorescence and the nascent organelles can be observed to occur in at least two different modes. In many cells the regeneration of the myofibrils takes place in the perinuclear region and yield finally a disc like region consisting of myofibrils (figure 2b, 2d). Thus the myofibrils appear to be localized precisely within a cellular domain and many of the myofibrillar marker molecules like heart C-protein (figure 2d) or myomesin (figure 2b), can be shown to localize in this region. This distribution may indicate the nucleation of myofibrillogenesis in the perinuclear zone and myofibrils could then continuously grow on both ends. The myofibrillar disc in such cells if often surrounded by a crown-like structure of fiber bundles, very likely SFLF, without a sarcomeric organization. In some cells α-smooth muscle actin is reexpressed and the actin accumulates preferentially in these filaments (Eppenberger-Eberhardt, et al., 1990) as well as α-actinin (figure 2a) which may occur in both the sarcomers and to an even higher extent in the stress-fiber like filaments of the crown. The myofibrils in most cases extend into the stress fiber-like filaments without an interruption and therefore a continous mode of myofibrillogenesis can be assumed. On the other hand α-smooth muscle actin is not always so clearly excluded from the participation in the myofibrils, because cells can be found which look similar to the cells transfected with α-smooth muscle actin as shown in figure 4b.

In many other cells, however, another mode of myofibrillar assembly appears to take place. Upon stainig with a sarcomeric marker antibody like the one directed against heart C-protein or myomesin, the myofibrillar assemblages appear scattered all over the cell (figure 2 e,f (Eppenberger *et al.*, 1988b)). Unlike in the cells shown previously sarcomeric assemblages as short as one sarcomere (Soldati and Perriard, 1991) occur in the cytoplasm surrounded by cytoskeletal structures with characterististics of stress fiber-like filaments. Again the resulting intermittently stained filaments appear to be continual with the nascent cross striated myofibrillar elements, which are scattered in between the SFLF. It is reasonable to assume that the SFLF may have a template function for myofibrillogenesis. Some of the myofibrillar proteins sort specifically to the sites of myofibrillar assembly, while other contractile proteins codistribute to both types of regions within these filaments.

Functional significance of protein diversity in myofibrillogenesis

As cells differentiate many new proteins are expressed which eventually participate in the formation of the new cytoarchitecture culminating in the elaboration of myofibrils. In the course of differentiative processes some of these contractile proteins occur also in non bound form, but are extracted with non ionic detergents which are commonly used in immuno fluorescent techniques. Several patterns occur. On one hand there are proteins like C-protein (figure 2d), myomesin (figure 2b) or muscle specific members of contractile isoprotein families e.g. α-skeletal actin, MLC-1f/3f or fast MHC etc. which are exclusively localized in sarcomeres. Such components may be crucial for the formation of stable myofibrils, therefore the knowledge of their sequence and structure is of great importance. In order to achieve this goal for myomesin (Grove, Kurer, Lehner,

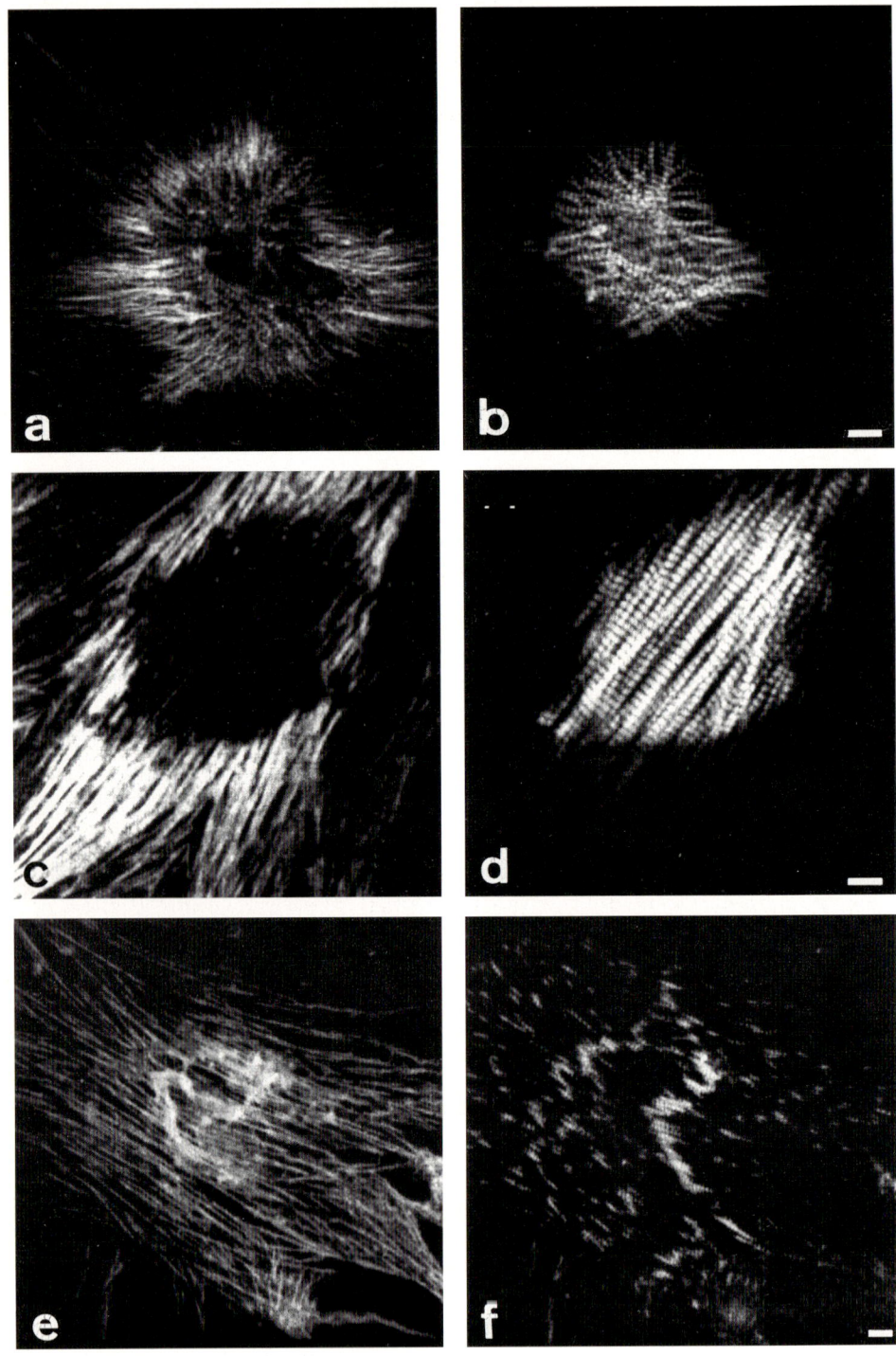

Fig. 2. Single confocal sections of ARC showing different modes of myofibrillar regeneration. of cells cultured for 12 d. Sections in a, b display cells doubly stained with antibodies against α-actinin showing Z lines and cytoskeletal structures (a) and polyclonal antibody against myomesin and M-protein staining exclusively the M-band of the myofibrils. Section c was stained with monoclonal antibody against α-smooth muscle actin and the corresponding section in d is the result of the heart C-protein stain and displays again the strict localization in the perinuclear myofibrils. A different mode of myofibrillar assembly is shown in e (antibody against α-smooth muscle actin, Skalli, Ropraz, Trzeciak, Benzonana, Gillessen and Gabbiani, 1986) and f (heart C-protein, Bähler, Moser, Eppenberger and Wallimann, 1985) the nascent myofibrillar assemblages are dispersed over the whole cell and occur on the cytoskeletal SFLF stained in e. Note that heart C-protein staining appears to exclude simultaneous staining with α-smooth muscle actin as is especially evident in c but also the staining of the strands in e are weaker in regions of abundant heart C-protein. Bar is 10 μm.

Doetschman, Perriard and Eppenberger, 1984) the molecular cloning of its cDNA is in progress (not shown).

On the other hand there are proteins that display a more ambiguous distribution like the α-smooth muscle actin. The antibody stain directed against this contractile protein is excluded from the perinuclear, myofibrillar region and concentrates in the crown-like structure surrounding the myofibrillar area (figure 2c) while staining for all actin isoforms with a reagent like RITC-phalloidin clearly demonstrate that actins are present in the myofibrillar region (Eppenberger, et al., 1990). Similarly the staining with an antibody detecting all α-actinin isoforms shows that these proteins are present in the Z-disk as well as in the other cytoskeletal structures (figure 2a). The existence of isoprotein families with muscle and non muscle members is notorious for many contractile proteins (e.g. myosins, actins, myosin light chains to name only a few) and they are usually expressed in the course of differentiation in a developmentally regulated sequence. The amino acid sequences of the members of these protein families are very similar as e.g. those of the actin isoprotein family and little is known about the functional significance of isoform diversity. On one hand, the different genes coding for these isoproteins may have evolved to allow differential gene control to operate during differentiation and in the various types of adult tissues. On the other hand, the multiple forms may provide the cells with functionally different polypeptides. In the next sections the question of functional significance for myofibrillar assembly of isoproteins diversity will be investigated.

Protein tagging allows to study the behaviour of any protein in any cellular background

As discussed above, in the regenerating rat cardiomyocyte several endogenous proteins sort to specialized structures, some more stringently than others. Since both stress fibers and nascent, regenerating myofibrils coexist in ARC these cells are not only a great asset for the investigations concerning biogenesis of cytoarchitecture, but are also ideally suited to serve for tests of specificity of protein sorting within the myocyte. A crucial technique for the understanding of regulation of expression and localization of the specific gene products within the cells's structures is the study of the behavior of exogenous proteins of interest reintroduced into living cells. The main difficulties of such an approach is the discrimination of the endogenous from the introduced protein. There are not too many reliable isoprotein and species-specific antibodies available against contractile isoproteins, and hence the detectability can not generally be guaranteed. Furthermore, it is almost impossible to study these molecules if they have been reintroduced into the homologous cellular background. We have therefore introduced the technique of protein epitope tagging (Albers and Fuchs, 1987; Munro and Pelham, 1984) to the study of myofibrillogenesis. Into the cDNA of any protein a short epitope of a few amino acids can be introduced which codes for a foreign peptide, in our case 5-11 amino acids of the C-terminus of the VSV G Protein. It has been shown that this engineered epitope can be recognized not only by polyclonal antibody from rabbit but also by a well characterized monoclonal antibody (Kreis, 1986) offering many possibilities for the recognition of the foreign protein and mutants thereof. So far the epitope has been introduced in orthotopical position at the C-terminal end of various myosin light chain and actin isoproteins. Here, we want to summarize some of the data on the isoproteins of the myosin light chain family, further work on mutants and some experiments with the actin isoproteins. It will be clear from these data that there is indeed specific interaction of the various contractile isoproteins with the myofibrils or the stress fiber-like filaments, which depends on the protein sequence.

LC1f/3f isoproteins sort preferentially to sarcomeric sites

In a first series of experiments we have examined the question of isoprotein sorting within the cytoplasmic compartment of various members of the myosin light chain isoprotein family. For the first experiments cDNA for chicken MLC-1f (Billeter, Quitschke and Paterson, 1988) has been derivatized with oligonucleotides including C-terminal tags from 5 to 11 amino acids of the VSV-G-Protein. For the cellular studies the epitope bearing derivatives were subcloned into the constitutive expression vector pSCT which is under the control of a strong CMV promoter. After transfection into a variety of cells the proteins could be recognized by their epitope-tag by both types of polyclonal and monoclonal antibodies. In fibroblasts the proteins localized on the stress fiber- cytoskeleton consisting of actomyosin microfilaments. The expression of MLC-1f-11mer TAG in primary cardiomyocytes indicated the exact localization of the tagged protein in the sarcomeres of myofibrils. In addition to the avian cardiomyocytes the same constructs were expressed in ARC, cells with a structural organization ideally suited for the study of questions of the biogenesis of muscle cytoarchitecture and more specifically the intracompartmental sorting of isoproteins (Soldati and Perriard, 1991).

In these cells it was shown that MLC-sorting is isoprotein-specific. The MLC-1f and MLC-3f (see figure 3a) bind almost exclusively to the A-band of myofibrils with rather high affinities. This conclusion was derived from experiments in cells with different levels of expression. In cells expressing moderate levels, the foreign protein bound strictly to the A-segment of the myofibrils leaving the H-zone undecorated as expected if such protein would interact only with the head portion of the myosin. On the other hand the MLC-3nm interacts with stress-fibers and sarcomeres equally well but the interaction with the myofibrils resulted in a fuzzy pattern and the H-zone was less well defined (Soldati and Perriard, 1991).

Replacement of the middle MLC domain specifies loss of function or gain of function for myofibrillar assembly

In order to resolve the question which of the different sequence features is responsible for the myofibril-specific interaction, domain mapping experiments with the various MLC isoproteins were carried out. As is evident from sequence comparison of the MLC isoprotein family the sequences can be divided into 3 segments of variability. First, the N-terminal segment where MLC 1f and 3f differ in length and sequence, the middle segment with many amino acid variations between the different isoproteins, and finally the C-terminus which is highly conserved among the MLC isoprotein family (Barton and Buckingham, 1985). The experiments with MLC-1f and MLC-3f gave the same results indicating that the sequence differences in this case in the N-terminus between the two MLC are not important for interaction with the myosin head portion. Initial experiments deleting the middle domain indicated that the binding specificity of MLC-1f was lost but there was still binding observed to most if not all sites containing myosin. However, if the C-terminal region of the MLC-1f molecule was deleted, the truncated protein was diffusely distributed within the cell indicating general importance of this portion of the molecule for interaction with myosin containing structures. As a working hypothesis, the middle domain was assumed to be responsible for the isoprotein specific interaction and hence exchange of this domain among the isoforms should result in alteration of sorting specificity (Soldati and Perriard, 1991).

In figure 3 the results of such domain exchange experiments are shown. As control in figure 3a an immunostained ARC injected with the MLC-3f-T11 construct is shown. The protein clearly sorts to myofibrillar sites and as shown before localizes in the A-segment of the sarcomere, while the isoprotein typical for nonmuscle cells, MLC-3nm-T11, as shown in figure 3b interacts equally well with myofibrils and with the cytoskeleton. Replacing the middle domain of the muscle specific MLC-1f-T11 with the corresponding domain of the MLC-3nm, the sorting specificity for the myofibrils is lost and the resulting distribution as shown (figure 3d) cannot be distinguished from the behavior of MLC-3nm (figure 3b), although the major part of the molecule is of the MLC-1f origin. Thus this replacement leads to a loss of function. If the opposite replacement was constructed by insertion of the middle domain from the muscle specific MLC-1f into the MLC-3nm, the resurrection of the myofibrillar specificity was observed indicating the gain of the muscle specific sorting function. It remains to be seen which of the 23 amino acid differences (Billeter, et al., 1988; Hailstones and Gunning, 1990) in this region are responsible for this specificity of sorting.

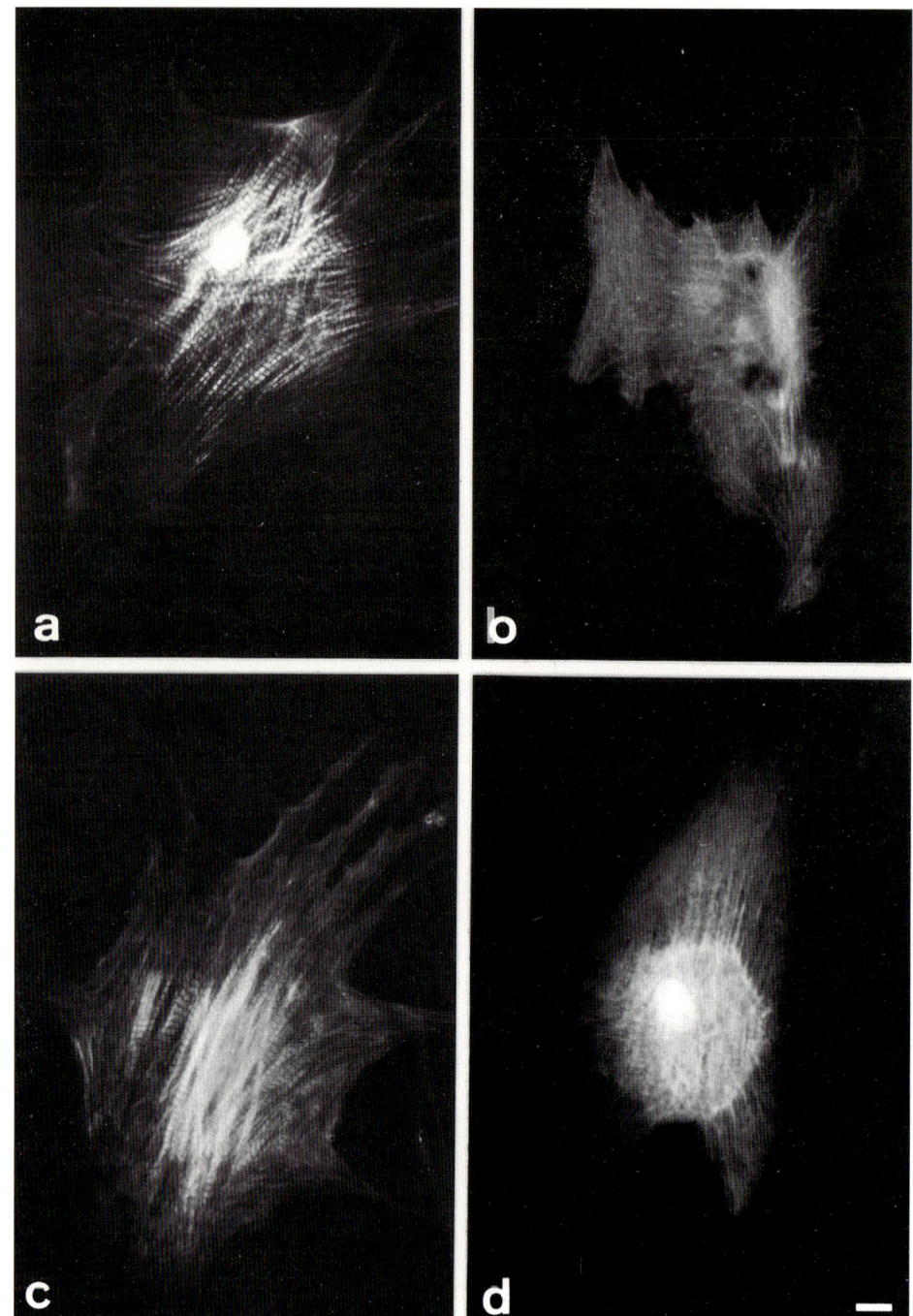

Fig. 3. Gain of function or loss of function of MLC chimeric constructs depends on the type of the middle segment of the MLC isoprotein. cDNA constructs all containing the 11-mer VSV-G protein epitope were cloned into the pSCT vector (Soldati and Perriard, 1991) and transfected into 8 d ARC cells by microinjection into the nucleus. After 20 h of further culture the cells were stained for the VSV-G protein with a polyclonal anti peptide antibody and also with monoclonal antibody B4 against myomesin indicating the M-band of the myofibrils (not shown). In (a) a control demonstrates the strict myofibrillar localization of the fast muscle MLC-3f while its nonmuscle counterpart MLC-3nm localizes equally well to the general cytoskeleton and myofibrils. In c the construct contained the non muscle MLC 3nm carrying an insertion of the middle segment of the MLC-1f, resulting in the gain of sorting specificity to myofibrils. Even the undecorated H-zone can be seen at higher magnifications (not shown here). The reversed construction was generated by insertion of the corresponding middle segment from the MLC-3nm into the MLC-3f the resulting protein has lost its sorting specificitiy and behaves like the MLC-3nm as shown in d. Bar is 10 μm.

The MLC isoproteins can be ordered into range of molecules with increasing affinity for myofibrillar association

As shown above the MLC isoproteins have distinct properties regarding their potential to sort specifically to myofibrillar sites. The two extreme types MLC-3nm and MLC-1f or MLC-3f have been compared with other MLC isoproteins. Although the differences found may be subtle, there is a tendency of the different molecules to compete for the binding on the myofibrillar or cytoskeletal myosin with disparate affinities. The results of such experiments in ARC cells were interpreted and represented in table 1.

The table summarizes the qualitative appraisal as observed by regular light microscopy as the analysis by confocal microscopy is still in progress. By the latter method we hope to be able to strengthen the statements by a more detailed analysis, but in general these statements made are well supported. As is true for all these experiments, the level of expression of the introduced proteins can have a certain influence on the localization of the competing exogenous protein (Soldati and Perriard, 1991), therefore the listings in the table show only the localization in the cells observed at all levels of expression, it may be, however, that cells expressing high levels will show substantial spillover into other locations. Since the material compiled in table 1 would take up too much space this simplified representation of the results was chosen rather than including exemplary micrographs for each of the MLC constructs, although some have been included in figure 3. For the sake of clarity these results have also been incorporated into this table. It is evident from this summary that the behavior of the epitope-tagged light chains can be interpreted as a continuous spectrum of affinities for the localization of the protein probes on the myofibrillar part of the cytoskeleton. The most extreme cases are the MLC-1f/3f with their tendency to associate with the myofibrils, while the MLC-3nm has the tendency to bind to any myosin containing structure. Interesting is the behavior of the two slow light chains, MLC-1sb, normally expressed in the adult heart and MLC-1sa, a slow light chain of skeletal muscle, which do not interact so well as compared with the skeletal light chains.

Table 1. *Protein sorting of MLC isoproteins in ARC*

MLC construct	Myo	Sarc	H	SFLF	M-P	Remarks
TS25 (MLC 1f-T11)	+++	+++	+++	− − −	+++	if overexpr. also nonsarcomeric localization
TS41 (MLC-3f-T11)	+++	+++	+++	− − −	+++	if overexpr. also nonsarcomeric localization
TS49 (MLC-1sb-T11)	+++	+	−?	+/−	++	slightly fuzzy sarcomeric stain, faint SFLF staining
TS48 (MLC-1sa-T11)	+++	−	−	+++	−	sarcomeric often fuzzy but better than TS47
TS47 (MLC-3nm-T11)	++	−	−	+++	−	sarcomeric often fuzzy, SFLF well defined
TS46 (MLC-1f/1sb-T11)	+++	++	++	+	++	sarcomeric often fuzzy but better than TS47
TS45 (MLC-1f/1sa-T11)	++	+	+	++	−	sarcomeric often fuzzy but better than TS47
TS44 (MLC-+f/3nm-T11)	++	+/−	−?	++	−	sarcomeric fuzzy, SFLF well defined
PA 2 (MLC-3nm/1f-T11)	+++	+++	+++	+/−	+++	gain of function, very faint staining of SFLF

The MLC constructs have been described in the text. The symbols are a qualitative assessments of the localization properties of the MLC constructs. The definitions of the columns are:

Myo=myofibrillar localization; Sarc=defined sarcomeric staining; H=H-zone in the A-band not stained; SFLF=stress fiber like filaments are stained; M-P=myofibrillar preference is evident over the association with SFLF cytoskeleton. +++=strong, ++=moderate, +=weak interaction; +/−=very weak, but variable; −=not visible; ?=not clear.

Although MLC-1sb is of the same type as the definitive endogenous protein, it does not compete so well for the binding sites as the exogenous MLC-1f/3f isoproteins and the association of the skeletal fast light chains may result in an even more stable structure as compared to the endogenous combination. In skeletal muscle, the developmental program includes the final expression of the skeletal muscle MLC isoproteins while this is not observed in the cardiac muscle cells and the cardiac MLC are the definitive isoproteins. It is likely that in cardiomyocytes the myosin accumulated is well adapted to the needs of cardiac physiology but the combination with the skeletal muscle isoproteins results in a more stable molecule not accumulated in the heart tissue.

Expression of different actins in ARC cells produce isoprotein-specific effects

As has been shown for the members of the MLC family there are different affinities for sites of assembly, which of course represent already sorted myosin molecules (Fallon and Nachmias, 1980), to which MLC can bind. It was therefore of interest to also test primary filament forming proteins, like the actin isoproteins. The question regarding the function of the diverse isoactins has not been answered. Similar experiments as above, were carried out with three types of actin. Thus, some of the actin isoproteins were also derivatized with the same epitope-tag as before and the resulting proteins were then analyzed in the same way as the tagged MLC proteins. The cDNA of the α-cardiac actin from human, α-smooth muscle from rat and the γ-cytoplasmic actin cDNA from human were constructed as fusions containing epitopes at their C-terminal ends with 5 or 11 amino acids. All of these proteins reacted in immunoprecipitation assays as predicted from the experience with the MLC constructs, and in transfected fibroblasts the actins visualized by tag-immunofluorescence bound to the cytoskeleton (data not shown). The results of transfection of these constructs into ARC are presented in figure 4. As expected the cardiac α-actin has a tendency to prefer myofibrillar binding sites, although the other cytoskeletal structures are interacting with this actin as well (figure 4a). In the area of the cell organized in sarcomeres the resulting pattern reveals a crisp pattern of sarcomeric staining. The pattern consists of narrow fluorescing bands and there is no broad A-band staining with an unstained H-zone as shown for the MLC-1f/3f (figure 3a). The overall distribution is like the one in figure 3a for MLC-3f although the myofibrillar preference is not as great and the cardiac actin appears to bind also, but with lower affinity to the non sarcomerically organized cytoskeleton. The α-smooth muscle actin with the tag interacts with both the myofibrils as well as with nonmuscle cytoskeleton as shown in figure 4b. In the perinuclear region, however, the staining is less dense than in the case of protein with myofibrillar preference as seen in figure 3a and 4a. The densest staining occurs in the region surrounding the myofibrillar area, also dubbed the crown area, which is reminiscent of the endogenous α-smooth localization in regenerating ARC as shown in figure 2b. In other cells injected with the epitope-labeled actin the labeling in the myofibrillar region is even much less evident and therefore it appears that the α-smooth muscle

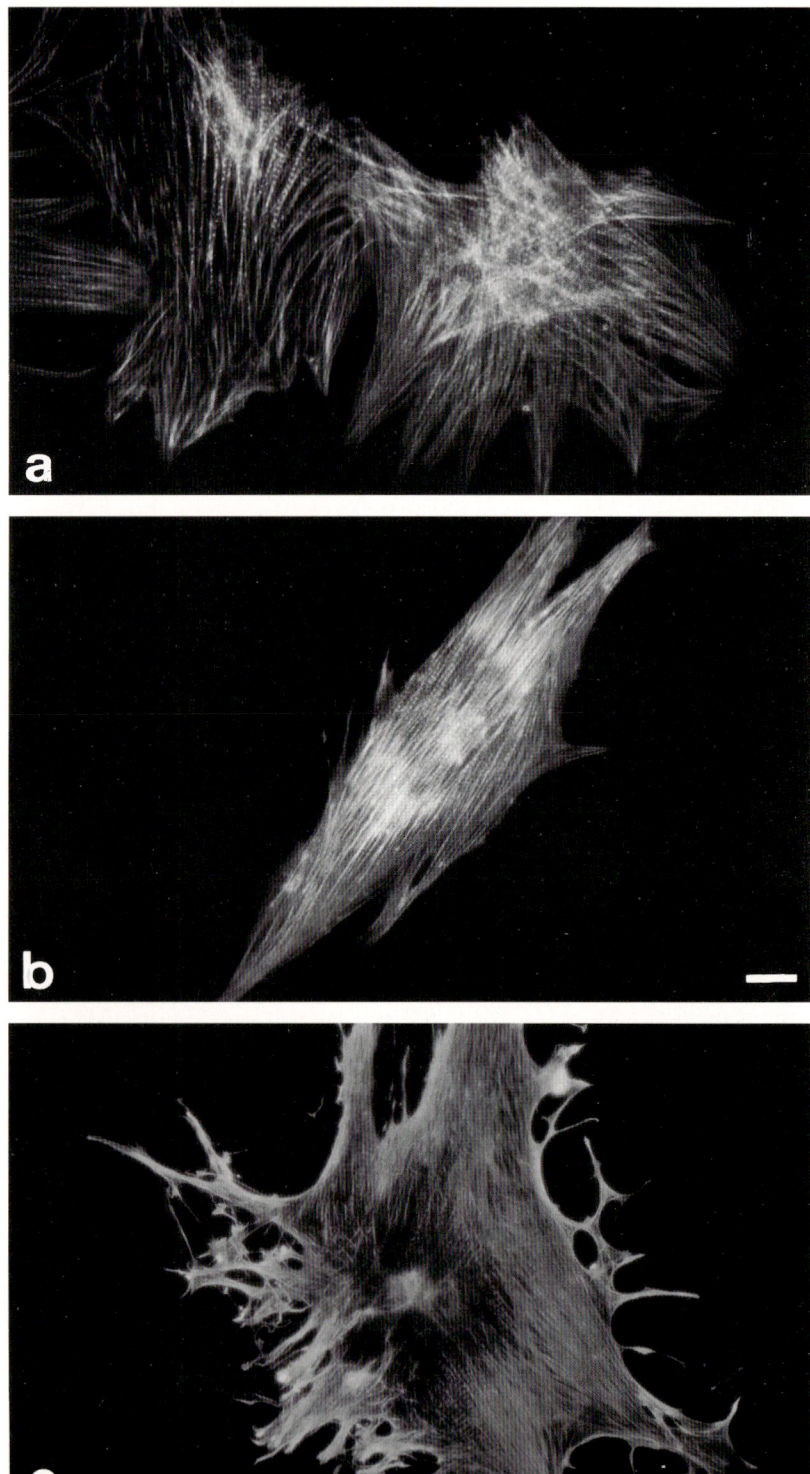

Fig. 4. Differences of localization of actin isoproteins in ARC. The cDNAs of three major actin isoproteins were derivatized with the VSV-G protein epitope containing 5 or 11 amino acids. The inclusion of the 5 or the 11-mer epitope in the constructions did not change the results. In (a) cells were microinjected with the construct carrying the rat α-cardiac actin cDNA and the cells was subsequently stained with the antibody recognizing the epitope and with anti myomesin (not shown) to localize endogenous myomesin in M-bands. The α-cardiac actin localizes preferentially to myofibrils especially at low levels of expression (cell in the middle) but interact also with the cytoskeletal structures not organized in sarcomeres. In (c) the staining pattern for the epitope is shown after the injection of a construct including the cDNA of rat α-smooth muscle actin including the 5 mer epitope. The distribution of the α-smooth muscle distributes to all actin containing sites and concentrates in the SFLF abutting the perinuclear myofibrillar regions, in crown-like fashion. These myofibrils appear to contain the protein in a lower abundance. In other cells no decoration of the myofibrils is observed corresponding to the distribution of the endogenous α-smooth muscle actin (see fig.2). In c the cells were transfected with the γ-cytoplasmic actin cDNA-5 mer epitope construct. This moderate phenotype is nevertheless an extremely flat cell showing the striking filopodia induced by the presence of the protein which is mostly contained in membrane near positions, while the central portion of the cells, where most of the myofibrils reside, is almost devoid of the exogenous actin. Again this actin interacts with both cytoplasmic and myofibrillar structures. Bar=10 μm.

isoprotein appears to be capable to participate in the formation of myofibrils, SFLF or structures of the general cytoskeleton.

The most striking effect, however, was produced by the transfection of γ-cytoplasmic actin tagged with either 5 or 11 amino acids into ARC cells. The major part of the tagged actin is found near the cellular edges. In many of these cells, the circumference of the ARC react with an extremely ununsual formation of filopodia, all of which are filled with the exogenous protein, while the center of the cells, where most of the myofibrils are found, are emptied from the transfected actin, although there are some myofibrils stained for this actin (figure 4c). This effect is not dependent on the length of the epitope, it was observed with both 5 mer as well as 11-mer tags and appears to be isoprotein specific for the γ-cytoplasmic actin species. It cannot be ruled at present if the striking effect may be produced by the combination of a γ-cytoplasmic actin property in combination with the the epitope. However, the effect is giving rise to a new striking phenotype which cannot be produced with any of the other two actins tested so far, thus there are also actin isoprotein specific effects on the biogenesis of cytoarchtitecture.

Discussion

In ARC cells the newly generated myofibrils can be studied from the newly expressed myofibrillar proteins to the appearance of the highly organized sarcomeres. During the early phases of cardiomyocyte cultures the original myofibrils from the fully diferentiated cells are degenerated and it remains to be investigated if the preexisting structures are broken down completely or if the liberated subunits are reutilized in the newly appearing structures. These studies however have given additional evidence that myofibrils arise in close association with stress fiber-like filaments and in some cells myofibrillogenesis takes place in dispersed segments along mixed filaments consisting of sarcomeric stretches and SFLF, which makes the SFLF good candidates for myofibrillar scaffolds. Further studies will have to be carried out to monitor proteins like titin, nebulin, muscle specific α-actinin etc. for determination of the relationship between the various types of cytoskeletal elements.

Some of the contractile proteins accumulate specifically at myofibrillar sites like the myomesin, C-protein and the muscle specific members of contractile isoprotein families (e.g. α-skeletal actin, MLC-1f/3f, MHC-f etc) some of which could not be investigated because of the lack of specific antibodies.

Other proteins show a more ambiguous localization like the α-smooth muscle actin. In many cells α-smooth muscle actin is found preferentially within non sarcomeric structures while in other cells it is found in both myofibrils and cytoskeleton. It may serve as a protein component which can fulfil multiple tasks and is not as highly constricted to its function as the previous group of proteins.

In the cases of the isoprotein families of the myosin light chains and some members of the actins an isoprotein specificity for assembly into myofibrils has been demonstrated and soon molecular parameters which are necessary for the specificity of assembly will be defined precisely (e.g. the sequence elements of these proteins). The observation that defined sequence elements are responsible for proper targeting of isoproteins leads to the conclusion that isoprotein differences are functionally significant and that there is intracompartmental protein sorting within the muscle cytoplasm (Soldati and Perriard, 1991).

It appears that the myosin light chains as well as some of the the actins, investigated so far, can be grouped into ranges of proteins that display an increase in affinity for the myofibrillar structures. The order in which the proteins of these ranges appear follows rather closely the order of their expression during development. The members yielding the less stable integration products beeing expressed early, while the isoproteins forming more stable myofibrillar assemblages, like the α-cardiac actin or MLC1f/3f appear later in muscle development. It is attractive to speculate that proteins forming complexes with increasing stability will result in more stable organelles and thus the newly synthesized proteins would also ensure the more efficient remodelling of the existing myofibrils by the expected increase in the exchange rates for the components with higher affinities replacing the ones with lower affinities.

A special thank goes to S. Keller for expert technical assistance and excellent photographic work. We are grateful to Drs. R. Billeter, T. Kreis, J. Lessard, S. Rusconi, P. Gunning, J. Léger and F. Stockdale for the precious gift of antibodies, cDNAs or vector DNA. We are also grateful to Drs. B. Schäfer, M. Schaub, T. Wallimann for discussions. The work was supported by grants n° 3.497-0.86 and 31,27756.89 from the Swiss National Science Foundation, a grant from Roche Research Foundation RRF to T. S. and a grant to JCP from the Muscle Dystrophy Association of America INC.

References

Albers, K. and Fuchs, E. (1987). The expression of mutant epidermal keratin cDNAs transfected in simple epithelial and squamous cell carcinoma lines. *J. Cell Biol.* **105**, 791-806.

Bähler, M., Moser, H., Eppenberger, H. M. and Wallimann, T. (1985). Heart C-protein is transiently expressed during skeletal muscle development in the embryo, but persists in cultured myogenic cells. *Dev. Biol.* **112**, 345-352.

Barton, J. R. and Buckingham, M. E. (1985). The myosin alkali light chain proteins and their genes. *Biochem. J.,* **231**, 249-261.

Billeter, R., Quitschke, W. and Paterson, B. M. (1988). Approximatively 1 kilobase of sequence 5' to the two myosin light-chain 1f/3f gene cap sites is sufficient for differentiation-dependent expression. *Mol. Cell. Biol.,* **8**, 1361-1365.

Eppenberger, H. M., Messerli, M., Müller, M., Schwarb, P. and Eppenberger-Eberhardt, M. E. (1990). Cultured adult rat cardiomyocytes as a model for differentiation. In D. Pette (Eds.), *The dynamic state of muscle fibers.* (pp. 193-204). Berlin: W.de Gruyter.

Eppenberger, M. E., Bächi, T., Vollenweider, I., Volk, T. and Eppenberger, H. M. (1988a). Myofibril regeneration and cell-cell interaction in cultures of adult rat cardiomyocytes. XVII Europ. Conference Muscle and Motility, 29.

Eppenberger, M. E., Hauser, I., Baechi, T., Schaub, M. C., Brunner, U. T., Dechenne, C. A. and Eppenberger, H. M. (1988b). Immunocytochemical analysis of the regeneration of myofibrils in long-term cultures of adult cardiomyocytes of the rat. *Dev. Biol.,* **130**, 1-15.

Eppenberger-Eberhardt, M., Flamme, I., Kurer, V. and Eppenberger, H. M. (1990). Reexpression of a-smooth muscle actin isoform in cultured adult rat cardiomyocytes. *Dev. Biol.,* **139**, 269-278.

Eppenberger-Eberhardt, M., Riesinger, I., Messerli, M., Schwarb, P., Müller, M., Eppenberger, H. M. and Wallimann, T. (1991). Adult Rat Cardiomyocytes Cultured in Creatine-deficient Medium Display Large Mitochondria with Paracristalline Inclusions, Enriched for Creatine Kinase. *J. Cell Biol.,* **113**, 289-302.

Epstein, H. F. and Fischman, D. F. (1991). Molecular Analysis of Protein Assembly in Muscle Development. *Science,* **251**, 1049-1044.

Fallon, J. R. and Nachmias, V. T. (1980). Localization of cytoplasmic and skeletal myosins in developing muscle cells by double-label immunofluorescence. *J. Cell Biol.,* **87**, 237-247.

Grove, B. K., Kurer, V., Lehner, C., Doetschman, T. C., Perriard, J.-C. and Eppenberger, H. M. (1984). A new 185.000-daltons skeletal muscle protein detected by monoclonal antibodies. *J. Cell Biol.,* **98**, 518-524.

Hailstones, D. L. and Gunning, P. W. (1990). Characterization of human myosin light chains 1sa and 3nm: implications for isoform evolution and function. *Mol. Cell. Biol.,* **10**, 1095-1104.

Kreis, T. E. (1986). Microinjected antibodies against the cytoplasmic domain of vesicular stomatitis virus glycoprotein block its transport to the cell surface. *EMBO J.,* **5**, 931-941.

Munro, S. and Pelham, H. R. B. (1984). Use of peptide tagging to detect proteins expressed from cloned genes: deletion mapping functional domains of Drosophila hsp 70. *EMBO J.,* **3**, 3087-3093.

Skalli, O., Ropraz, P., Trzeciak, A., Benzonana, G., Gillessen, D. and Gabbiani, G. (1986). A monoclonal antibody against α-smooth muscle actin: A new probe for smooth muscle differentiation. *J. Cell Biol.,* **103**, 2787-2796.

Soldati, T. and Perriard, J. C. (1991). Intracompartmental Sorting of Essential Myosin Light Chains: Molecular Dissection and In Vivo Monitoring by Epitope Tagging. *Cell,* **66**, 277-289.

Vandekerckhove, J. and Weber, K. (1984). Chordate muscle actins differ distinctly from invertebrate muscle actins. *J. Mol. Biol.,* **179**, 391-413.

Van der Voort, H. T. M., Brackenhof, G. I. and Baarslag, M. W. (1989). Three-dimensional visualisation methods for confocal microscopy. *J. Microsc.,* **153**, 123-132.

Printed in Great Britain © Society for Experimental Biology 1992   237

CARDIAC TROPONIN T GENE EXPRESSION IN MUSCLE

JANET H. MAR*, ROCCO C. IANNELLO†
and CHARLES P. ORDAHL

Department of Anatomy and Cardiovascular Research Institute, University of California,
San Francisco, San Francisco, CA 94143-0452

Summary

We have been analyzing the regulatory regions of the cardiac troponin T gene promoter as a mean toward understanding the mechanisms that govern the transcription of genes which are cross-expressed in cardiac and skeletal muscles during development. By analyzing the activities of mutant cardiac troponin T gene promoter by transient transfection of primary embryonic muscle cells, we showed that both common and distinct elements are required for activity of the cardiac troponin T promoter in these embryonic muscle cells.

In skeletal muscle the minimal promoter sufficient to direct activity of the cardiac troponin T promoter is only 99 nucleotides upstream from the transcription initiation site. Within the distal half of this promoter are two tandem copies of a conserved hexanucleotide sequence (5'-CATTCCT-3') we termed the 'M-CAT motif'. Since mutation of either one of the M-CAT motifs abolishes promoter activity, we concluded that both M-CAT motifs are essential for activity of the promoter.

The above minimal promoter is insufficient to confer promoter activity in embryonic cardiocytes. In these cells an additional 48-nucleotide region approximately 100 nucleotides upstream of the minimal promoter is needed for efficient promoter activity. We have named this region the 'cardiac element'. This element contains a conserved sequence motif found in other muscle gene promoter. The cardiac element can also act irrespective of orientation and is relatively independent of position, characteristics that are like transcriptional enhancers. This element alone, however, is insufficient to direct cardiac promoter activity. Activity of the 48-nucleotide cardiac element is dependent on either direct or indirect interaction with the downstream M-CAT motifs because mutation of either M-CAT motif also abolishes promoter activity in cardiac cells.

The third regulatory region (nucleotide position −550 to −268) is not essential for promoter activity but can enhance activity of the cTNT promoter or a heterologous promoter in both cardiac and skeletal muscle cells by three to five

*Corresponding author: Present address: The University of Texas Medical School, Departments of Internal Medicine and Biochemistry and Molecular Biology, Houston, Texas 77030.
†Present address: The Murdoch Institute, Royal Children's Hospital, Melbourne, Victoria, Australia 3052.

Key words: muscle, promoter, transcription.

folds. Within this region are sequences which show similarity to motifs found in other muscle gene enhancers.

In vitro DNA-protein binding studies showed that the above three regulatory regions interact with nuclear factors. A direct correlation exists between promoter activity and sequence specific binding of a nuclear factor we termed the 'M-CAT binding factor' to the M-CAT motifs. Similar interaction of nuclear regulatory molecules with sequences within the cardiac element and the upstream enhancer region is likely to be the mechanisms which control the action of these regulatory regions.

Our analysis of the regulation of the cardiac troponin T gene promoter has shown that similar as well as different elements govern the activity of this gene promoter in cardiac and skeletal muscles. Our role now is to determine how these sequences interact with the appropriate transcription regulatory factors to bring above the tissue and developmental specific expression of the cardiac troponin T gene.

Introduction

Of the three major types of muscle in vertebrates, cardiac and skeletal muscle can be differentiated from smooth muscle by their cross striations. The strong similarity in the structure, function and formation of the sarcomeric apparatus of these two striated muscle types suggests a common ancestry and common molecular mechanisms governing their muscle development. Despite this similarity, the origins of cardiac and skeletal muscles have been shown to derived from different precursor cells of the developing embryo and the proteins which make up the components of the contractile unit of the two muscle types are encoded by different genes. Why do these two related striated muscles evolve two different gene sets to produce seemingly similar proteins? One possibility is that cardiac muscle genes encode proteins which differ from their skeletal counterparts in some functionally important ways to fulfill distinct physiological needs of this particular muscle type. However, recent experiments with mutant mice expressing significantly lower levels of cardiac actin suggest that cardiac actin may be functionally replaced with skeletal actin during cardiac development (Alonso, Garner, Vanderkerckhove and Buckingham, 1990). Thus, the strict conservation and lineage specific expression of muscle genes may be less of a functional significance than of a regulatory advantage gained by compartmentalization of different gene sets for cardiac and skeletal muscles.

Attempts to differentiate cardiac and skeletal myogenesis is also further complicated by the observation that many members of these two gene sets are actually cross-expressed in both skeletal and cardiac muscles, especially during embryonic development. The cross-expression of cardiac and skeletal muscle genes during these early developmental stages may represent a vestigial regulatory mechanism due to common cell ancestry or a specific developmental requirement met by the cross-expressed gene products. In any event, the cross-expression of

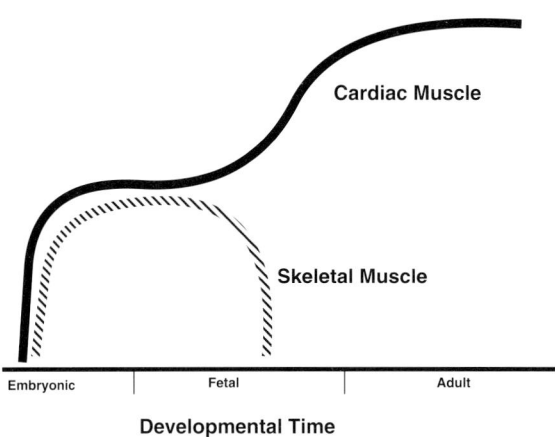

Fig. 1. Diagrammatic representation of the expression of the cardiac troponin T gene during development.

cardiac and skeletal contractile genes offers a unique opportunity to both compare the regulation and the role of the same gene in both cell types. Such analyses may ultimately shed light on questions regarding the intrinsic similarities and differences between cardiac and skeletal muscle cells.

Here we report on our recent efforts toward understanding the transcriptional mechanisms which govern expression of a single gene in both embryonic cardiac and skeletal muscle cells. The cardiac troponin T (cTNT) gene is expressed during embryonic and early fetal development of all striated muscles (Cooper and Ordahl, 1984; 1985; Ordahl, Kioussis, Ovitt and Fornwald, 1980; Toyota and Shimada, 1981). During early embryonic development, the cTNT gene is activated and transcribed at relatively low levels in both cardiac and skeletal muscle cells (Fig. 1). Expression of the cTNT gene continues in both cell types until mid-fetal development when transcription of this gene becomes divergently regulated. In cardiac cells, transcription of the cTNT gene is strongly upregulated; however, in skeletal cells it is repressed (Cooper and Ordahl, 1984; Long and Ordahl, 1989). The mechanisms which govern these different programs of cTNT gene expression during mid-fetal development have been suggested to involve hormonal or cellular interactions (Toyota and Shimada, 1983).

To begin to understand how the cTNT gene is regulated in embryonic cardiac and skeletal muscle cells, we focused our efforts towards defining the promoter and upstream regions of the cTNT gene which are required for its expression in these embryonic myocytes. Cardiac and skeletal myocytes cultured from early embryos were transiently transfected with a series of gene constructs consisting of varying lengths of cTNT promoter/upstream region functionally linked to a reporter gene (Mar, Antin, Cooper and Ordahl, 1988). The results of these experiments showed that the cTNT promoter/upstream region contains sufficient

Table 1. *cTNT promoter activity in muscle cells*

	Relative activity in:	
Promoters*	Cardiac	Skeletal
cTNT-129	12	100
cTNT-201	17	58
cTNT-268	100	65
cTNT-550	280	385

The cardiac troponin T gene promoter was used to direct expression of the bacterial gene encoding the enzyme chloramphenicol acetyltransferase (CAT; Gorman, Moffat and Howard, 1982). Each construct was transfected into primary embryonic cardiac and skeletal muscle and the resulting CAT enzyme activity was determined and normalized to the minimal promoter-CAT construction showing efficient activity in each respective cell type (cTNT-129 for skeletal muscle cells and cTNT-268 for cardiac cells).

 * number after cTNT indicate number of nucleotides from the transcription initiation.

information to direct muscle specific transcription. In skeletal muscle cells, specific expression of the reporter gene requires a cTNT gene fragment containing only 129 nucleotides upstream of the transcription initiation site (Table 1, cTNT-129). This fragment, however, is not sufficient for expression in cardiac cells. Expression of cardiac cells requires a cTNT gene fragment containing 268 upstream nucleotides (cTNT-268, Table 1). The different upstream regions required for cTNT promoter activity in cardiac and skeletal muscle cells indicate that expression of the same cTNT gene in each cell type requires different regulatory DNA sequences and transcription factors.

In addition to defining the approximate minimal cTNT promoter sequences required for activity in cardiac and skeletal muscle cells, these early deletion experiments also indicated the presence of an upstream region which can positively affect the overall activity of the minimal promoter fragments (Table 1, cTNT-550 and Mar et al., 1988). Initial experiments place this region between nucleotides -550 and -268 ($-550/-268$ region; numbering from the transcription initiation site at $+1$). Additional experiments showed that cTNT fragment deleted to -360 still retain this potentiating effect (Fig. 2c). This region is not essential for activity of the cTNT promoter, but its inclusion with the cTNT minimal promoter or with a heterologous promoter results in a three-fold increase in promoter activity in both cardiac and skeletal muscle cells but not in non-muscle cells (Mar et al, 1988 and unpublished data). Since this upstream region exerts a muscle-specific potentiating effect independent of position and orientation, it has characteristics that are typical of classically defined enhancers.

In the present paper, we present more detailed analysis of the regulatory sequences governing cTNT promoter activity in cardiac and skeletal muscle cells as well as evidence that the regulation of the cTNT promoter in these muscle cell types requires both common and divergent regulatory sequences and nuclear factors.

Results and discussion
The muscle-specific enhancer of the cTNT promoter

In addition to the upstream muscle-specific enhancer region we have identified for the cTNT promoter, several other muscle genes are known to contain muscle-specific enhancers (Donaghue, Ernest, Wentworth, Nadal-Ginard and Rosenthal, 1988; Horlick and Benfield, 1989; Jaynes, Johnson, Buskin, Gartside and Hauschka, 1988; Piette, Bessereau, Huchet and Changeux, 1990; Sternberg, Spizz, Perry and Olson, 1989; Wang, Xu, Wang, Ballivet and Schmidt, 1988; Yutzey, Kline and Konieczny, 1989). Of these, the most well characterized is that of the muscle creatine kinase gene enhancer located upstream of the promoter. Based on analysis of this enhancer and others, at least two different conserved sequence elements have been shown to be essential for enhancer activity. Almost all enhancers contain multiple MEF-1 motifs (or E-boxes) which have as their core sequence CANNTG. This motif has been shown to interact with the family of myogenic determination factors such as myoD, myogenin, myf-5 and others (see Olson, 1990 for review). Binding experiments demonstrated that interaction of the MEF-1 motifs with these muscle determination factors and perhaps other, as yet unidentified, factors is an essential step for activity of the enhancers. In addition to the MEF-1 motifs, an AT-rich segment located downstream of the MEF-1 motifs is also an essential component of the muscle creatine kinase and myosin light chain enhancers. This AT-rich region binds a factor which is different from the myogenic determination factors mentioned above (Gossett, Kelvin, Sternberg and Olson, 1989).

Sequence analysis of the cTNT enhancer and immediate region revealed the presence of sequence elements with homologies to the MEF-1 and AT-rich motifs. The locations of these motifs are similar to those within the muscle creatine kinase and myosin light chain gene enhancers (Fig. 2). To determine how these enhancers might be functionally related, we replaced the cTNT enhancer region with that of the muscle creatine kinase enhancer. Fig. 2d, e and f showed that the muscle creatine kinase enhancer can indeed increase the activity of the minimal cTNT promoter (cTNT-129, Fig. 2a) in skeletal muscle cells, albeit at a slightly lower level than that of the endogenous cTNT enhancer region (compare Fig. 2c with 2d,e,and f). The precise elements within the cTNT enhancer region responsible for its activity is not yet determined. However, it is interesting to note that preliminary DNase I footprint experiments show protection of the MEF-1 motifs within the cTNT enhancer region (J. Mar and C. Ordahl, unpublished observations) suggesting that this enhancer may function in a manner homologous to that of the muscle creatine kinase and myosin light chain gene enhancers in skeletal muscle cells.

The cTNT promoter containing the enhancer region also increases activity of the minimum cTNT promoter active in cardiac cells (Fig. 2c). Since the cTNT enhancer and the region specifying cardiac activity when taken together resembles the organization characteristic of other muscle enhancers (Fig. 2), we asked whether these other enhancers can replace this cTNT region in activating and

cTNT Promoter and Upstream Regulatory Elements

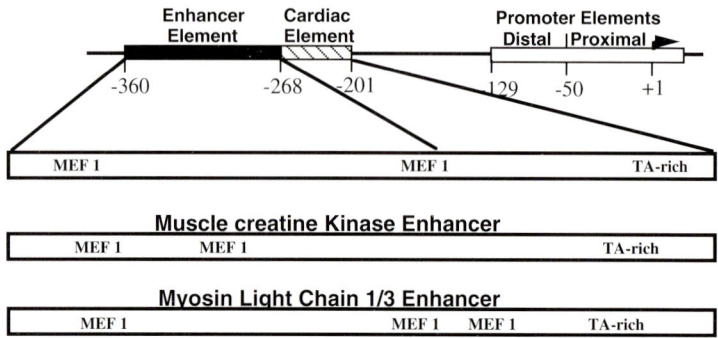

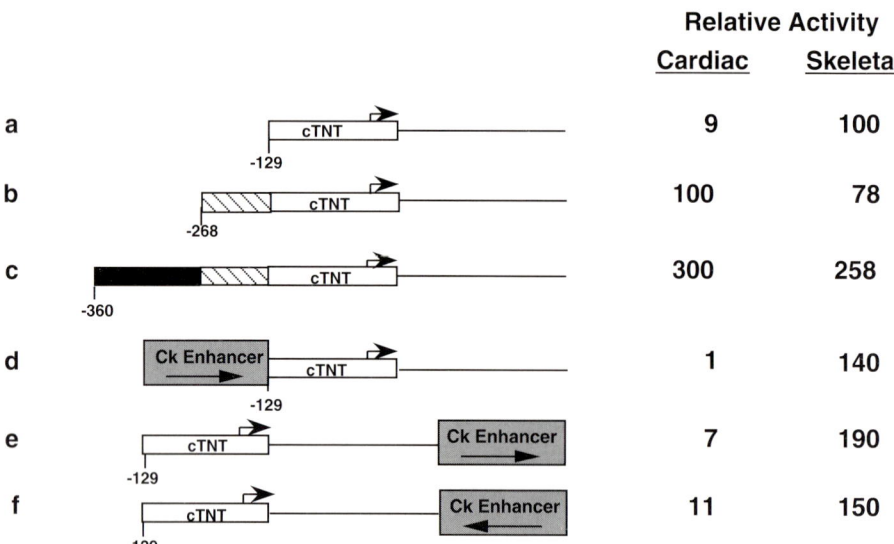

Fig. 2. Upper, boxed region: Schematic comparison of the structure of the enhancers of the cardiac troponin T, muscle creatine kinase and myosin light chain 1/3 genes. Lower Region: Activity of constructs containing either the endogenous enhancer region (construct 'c') or the mouse creatine kinase gene enhancer region located either directly in front of the promoter (construct 'd') or immediately after the bacterial CAT gene, (constructs 'e' and 'f'). Activities of promoter constructions in either embryonic cardiac or skeletal muscle cells have been normalized to the respective minimal promoter active in each cell type (construct 'a' in skeletal muscle and 'b' in cardiac cells).

specifying cardiac promoter activity. To address this question, the muscle creatine kinase enhancer was ligated either 5′ or 3′ of the cTNT-129 promoter which by itself is inactive in cardiac cells (Fig. 2a) and analyzed by transfection into embryonic cardiocytes. The results of these experiments show that the creatine kinase enhancer can not activate or enhance the activity of the cTNT-129 promoter in cardiac cells (Fig. 2d,e and f). Therefore, the cTNT-269 promoter (Fig. 2b) contains regulatory element(s) conferring cardiac specificity that is absent in the creatine kinase enhancer. In addition, these results suggest that muscle specific enhancers may exert their effect by interaction with different sets of regulatory elements or factors in cardiac and skeletal muscle cells.

The regulatory elements of the cTNT promoter

Skeletal muscle determinants

Previous experiments have demonstrated that activity of the cTNT promoter in skeletal muscle cells required only 129 nucleotides upstream of the transcription initiation site (cTNT- 129, Table 1). We have also used chimeric promoter experiments to show that the distal portion of this promoter contains muscle specific regulatory elements (Fig. 3 and Mar and Ordahl, 1988; 1989). Within this region are sequences resembling the CArG and MEF-1 motifs which have been shown to be important transcriptional regulatory elements of the actin and creatine kinase genes (Bergsma, Grichnik, Gosset and Schwartz, 1986, Minty and Kedes, 1986). However, activity of the cTNT promoter in skeletal muscle cells does not appear to required either the CArG or the MEF-1 motif. Deletion of both these elements from the cTNT promoter does not significantly affect its activity (cTNT-99, Fig. 3 and Mar and Ordahl, 1990).

Activity of the cTNT promoter in skeletal muscle cells, however, is absolutely dependent on the presence of two copies of the heptanucleotide 5'-CATTCCT-3' which is a conserved sequence element found in the regulatory region of many muscle genes (Nikovits, Kuncio and Ordahl, 1986). We have termed this element the M-CAT motif (Fig. 3). Mutation of either one or both of the M-CAT motifs, but not sites outside of these motifs, abolishes cTNT promoter activity in skeletal muscle cells (Fig. 3, cTNT-M1 and cTNT-M2 and Mar and Ordahl, 1990). Mutation of the M-CAT motifs does not result in activation of the cTNT promoter in non-muscle cells. Thus, these deletion and mutation experiments implicate the M-CAT motifs as essential positive-acting elements in regulating muscle transcription of the cTNT promoter.

Protein-DNA binding analyses suggest that activity of the M-CAT motifs involves binding interaction with nuclear factors. Fig. 3 shows the DNase I protection pattern observed for the distal portion of the cTNT promoter. In the presence of embryonic skeletal muscle nuclear extracts, three distinct regions of the distal portion of the cTNT-129 promoter are protected from nuclease digestion. The distal-most protected region centers around the CArG-like motif (5'-CCAAATAGC-3'). Since the CArG-like motif of the cTNT promoter is dispensable for activity in our transfection system, we can not yet ascribe a role or a significance to the binding interaction at this motif. The remaining two proximal DNase I protected sites are centered around the M-CAT motifs (Fig. 3).

Two series of experiments were performed to determine whether the M-CAT

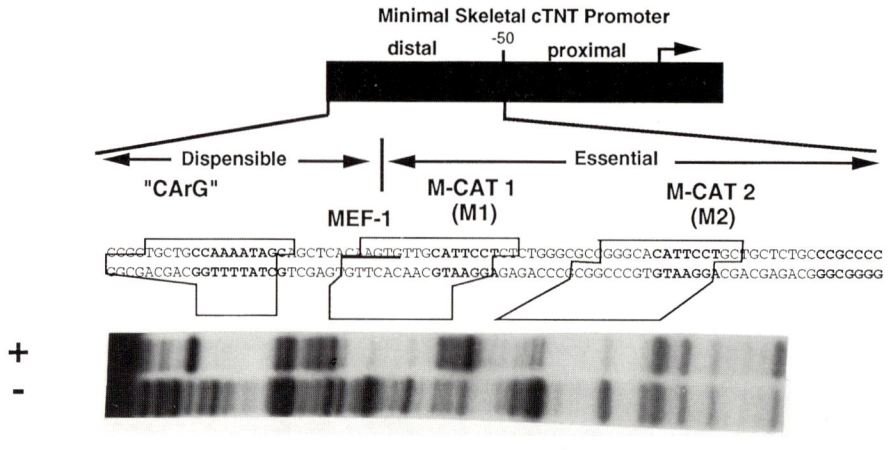

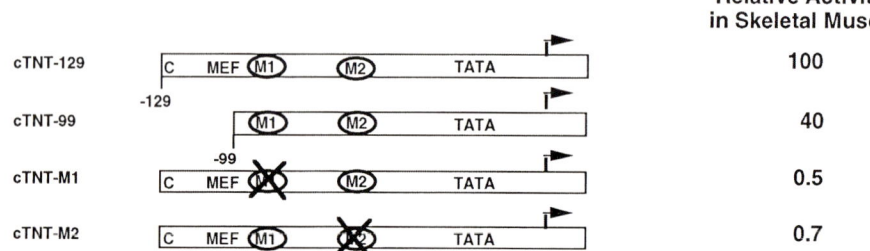

Fig. 3. DNase I footprint of the essential M-CAT motifs. Upper Portion: Autoradiograph shows the DNase I digestion pattern of the distal region of the cTNT promoter with (+) or without (−) muscle nuclear extracts. Footprint patterns of the lower DNA strand is shown. Nucleotide sequence of the distal region of the cTNT promoter is shown and protected nucleotides are blocked out and sequences of the CArG, MEF-1 and M-CAT motifs are shown in boldtype. Lower Portion: Relative activity of cTNT promoters containing mutated M-CAT motifs or deleted of MEF-1 and CArG motifs.

sequence are required for binding of the nuclear factors that results in protection of the regions at and around the M-CAT motifs (data not shown, see Mar and Ordahl, 1990). First, DNA fragments or oligonucleotides containing wildtype or mutant M-CAT motifs were used as binding competitors in the footprint experiments. Protection at the M-CAT motifs was abolished only when the unlabeled competitor contains a wildtype M-CAT motif. Second, footprint experiments with cTNT promoter fragments containing wildtype or mutant M-CAT motifs were performed. DNase I protection at a M-CAT region was observed only when that region contains an intact M-CAT motif but protection was not detected when the region contains a mutant M-CAT motif. Based on the results of these experiments, we conclude that the M-CAT motifs interact in a sequence specific manner with a nuclear factor which we have termed the M-CAT binding factor (or MCBF). Since cTNT promoters containing mutation in either or both

M-CAT motifs are transcriptionally inactive and do not show DNase I protection at the mutated M-CAT sites, we also conclude that MCBF-M-CAT interaction is essential for activity of the cTNT promoter in embryonic skeletal muscle cells.

At present we do not know how the two M-CAT motifs within the cTNT promoter exert their positive effect on transcription from this promoter. It is clear, however, that activity of the cTNT promoter requires two intact copies of the M-CAT motif and that both these motifs bind MCBF. Binding of MCBF to each M-CAT motif is independent and is unaffected by the spatial organization of the M-CAT motifs. However, activity of the cTNT promoter require strict spatial arrangement of the two M-CAT motifs (see Mar and Ordahl, 1990). These observations suggest that binding of MCBF at the M-CAT motifs is not itself sufficient for optimal activity of the cTNT promoter but that efficient promoter activity requires precise location of the two M-CAT motifs. This latter requirement strongly suggests that transcriptional activity of the cTNT promoter requires short-range interaction either between the MCBFs bound to the two M-CAT sites or these factors with a tertiary factor.

Cardiac muscle determinants

The cTNT promoter active in embryonic skeletal muscle (cTNT-129) is not sufficient for transcriptional activity in cardiac cells. What elements govern the activity of the cTNT promoter in embryonic cardiac myocytes? Since a cTNT promoter fragment containing 268 nucleotides upstream of the transcription initiation site (Fig. 4a) is active but one containing 201 nucleotides is inactive in cardiac cells (Table 1 and Iannello, Mar and Ordahl, 1991; Mar et al., 1988), we conclude that the region between −268 and −201 contain regulatory sequences that can confer cardiac activity to the cTNT promoter. To determine whether the −268/−201 fragment alone or portions of this fragment was sufficient for this activity, we cloned various sub-segments of this fragment immediately upstream of cTNT-129 and assayed the activity of the resulting promoter by transfection into cardiac cells. The final result of those experiments shows that cardiac activity can be circumscribed to the region between −247 and −201 (Fig. 4b,c,d,e and Iannello et al., 1991). Subdivision of this −247/−201 segment at position −215/−214 results in loss of myocardial activity (Fig. 4f, g) suggesting that two or more regulatory elements (see below) are located within this cTNT segment. Interestingly, this segment functions independent of orientation and position characteristics that are typical of transcriptional enhancer. We have termed this 47-nucleotide segment the 'cardiac element' of the cTNT promoter.

Although the cardiac element is required, it alone is not sufficient to direct promoter activity in embryonic cardiac cells. The cardiac element requires sequence elements within the distal portion of the cTNT promoter for activity. Within this distal region are the conserved CArG, MEF-1 and M-CAT motifs described above. We performed deletion and mutation experiments to determine whether any of these conserved motifs are required for activity of the cardiac element. Activity of the cardiac element is not significantly affected by deletion of

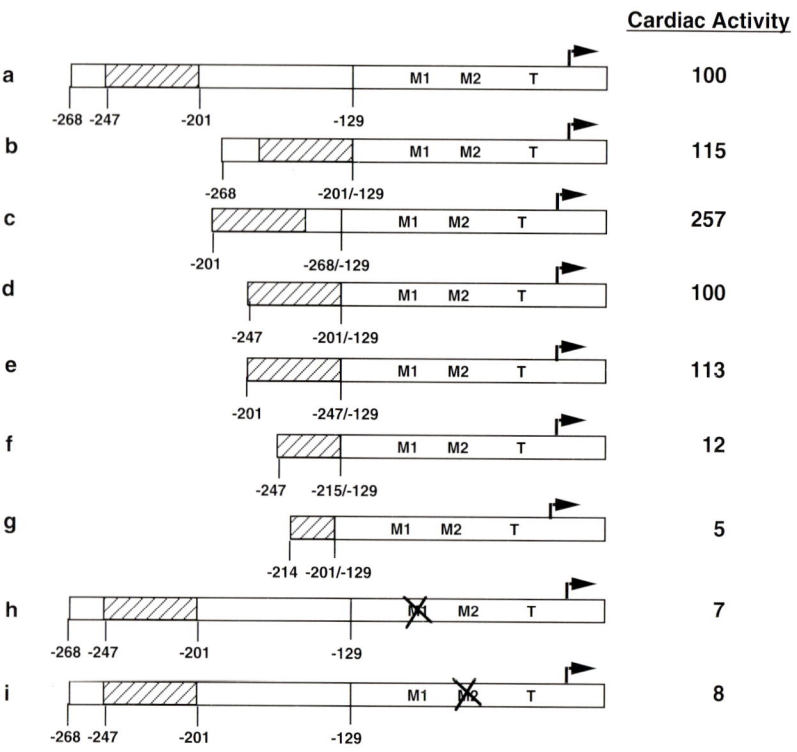

Fig. 4. Activity of cTNT promoter constructs in embryonic cardiac muscle cells. The CAT gene was placed under the transcriptional control of the cTNT promoter constructs as shown in figure. Plasmids were transfected and the CAT activity determined. Activity of each was normalized to cTNT-268 (construct 'a'). Hatched box represents the cardiac element as described in text. ×mark over M1 or M2 indicates mutation of the respective M-CAT motif. M1, M2 denote M-CAT-1 and M-CAT-2, T represents TATA box.

both the CArG and MEF-1 motifs (not shown, see Iannello et al., 1991). However, when the cardiac element is placed in front of the cTNT-129 promoter containing mutation of either one of the M-CAT motifs, the activity of the resulting promoter is severely diminished (Fig. 4h,i). These results indicate that activity of the cTNT promoter in cardiac cells must involve cooperative interaction of the cardiac element with the M-CAT motifs.

The requirement for both upstream (the cardiac element, located between −247 to −201 from the transcription initiation site) and downstream (the M-CAT motifs, located approximately 100 nucleotides downstream of the cardiac element) sequence elements for cTNT promoter activity in cardiac cells strongly suggests that cTNT promoter activity requires both DNA-protein and protein-protein interactions. To understand how these interaction may occur, we have begun to analyze the interaction of the cardiac element with nuclear factors. Gel shift assays with the cardiac element (not shown) demonstrated that nuclear factors from heart muscle bind to the cardiac element in a sequence specific manner. Our preliminary DNase I footprint experiments show that at least two sequence regions of the cardiac element are specifically protected from nuclease digestion presumably by binding of nuclear proteins to these regions. This footprint result corroborates the earlier deletion data and indicate that both sequences upstream and downstream of the $-214/-215$ junction are required for the activity of the cardiac element. The nuclear factor(s) binding to the cardiac elements is a prime candidate for being the transcription regulatory factor which governs the cardiac specific activity of the cTNT gene promoter. Isolation of the cardiac element binding factor (CEBF) and the M-CAT bind factor (MCBF) should enable us to perform detailed characterization of the protein-DNA and protein-protein interactions required for activity of the cTNT promoter in cardiac and skeletal muscles.

Conclusions

Although cardiac and skeletal muscle cells are closely related they arise from different embryonic lineages and express distinctive gene sets when terminally differentiated. However, during embryonic development both muscle types cross-express a number of genes. These cross-expressed genes offer us an opportunity to analyze the common or divergent mechanisms which govern gene expression in these two cell types during development. Our analysis of the sequence requirement for activity of the cTNT gene promoter outlined above clearly indicates that both common (the M-CAT motifs and the upstream enhancer) and distinct (the cardiac element) elements are required for expression of the cTNT gene in embryonic cardiac and skeletal muscle cells (Fig. 5). The interactions at these domains must then somehow be transmitted through to the transcription initiation complex.

Why should embryonic myocardial cells require an additional sequence element that is not required by skeletal muscle cells? The simplest explanation is that the cardiac cells lack the MCBF which is present in skeletal muscle cells. However, preliminary experiments indicate that MCBF and CEBF are present in both cardiac and skeletal muscle cells. We must assume, therefore, that interaction with cofactors not yet identified are responsible for this difference. Indeed, as recent results from other experiment systems have shown, such an interaction may be the norm for differential gene transcription.

Our analysis of the regulation of the cTNT gene promoter has allowed us to form a tentative picture of how embryonic cardiac and skeletal muscle cells may regulate the transcription of the cTNT gene during development. However, the functional significance of the cross-expression of the cTNT gene (or any other cross-expressed contractile genes) during embryonic development remains to be determined. In the case of the cTNT gene, the protein isoform expressed in embryonic cardiac and skeletal muscle cells is unique to this developmental stage.

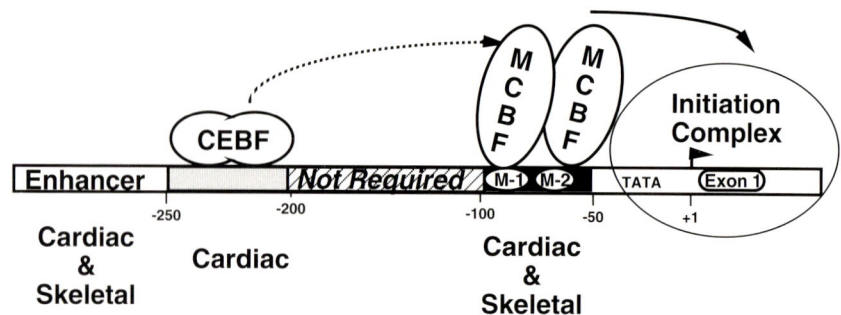

Fig. 5. Regulatory regions of the cTNT promoter. Diagrammatic represenation of the cTNT promoter region from nucleotide -550 to $+38$. Sequence elements are indicated. Regions essential in cardiac (stippled box) and/or skeletal muscle cells (black filled box) are noted. Enhancer region is shown as open box. Cross-hatched region indicates sequence which is not required for promoter activity. Nuclear factors MCBF and CEBF and the initiation complex are represented as ovals. Arrows denote potential interactions between the MCBF and the transcription initiation complex (solid arrow) and between MCBF and CEBF (dashed).

Therefore, it has been suggested that this cTNT isoform has a specialized role in embryonic cardiac and skeletal muscle cells. At present it is not clear what this unique function may be. What is clear, however, is that a complete understanding of the regulation of the cTNT gene and other contractile genes requires analysis of both the mechanism that govern their transcription and the function of these proteins during myogenesis.

Supported by NIH grants to CPO, and by postdoctoral fellowship awards from NIH and the Muscular Dystrophy Association to JHM and RCI, respectively.

References

Alonso, S., Garner, I., Vanderkerckhove, J. and Buckingham, M. (1990). Genetic analysis of the interaction between cardiac and skeletal actin gene expression in striated muscle of the mouse. *J. Molec. Biol.* **211**, 727-738.

Bergsma, D., Grichnik, J., Gosset, L. and Schwartz, R. (1986). Delimitization and characterization of cis-acting DNA sequences required for the regulated expression and transcriptional control of the chicken skeletal α-actin gene. *Molec. Cellul. Biol.* **6**, 2462-2475.

Buskin, J. N. and Hauschka, S. D. (1989). Identification of a myocyte nuclear factor that binds to the muscle-specific enhancer of the mouse muscle creatine kinase gene. *Molec. Cellul. Biol.* **9**, 2627-2640.

Cooper, T. A. and Ordahl, C. P. (1984). A single troponin T gene regulated by different programs in cardiac and skeletal muscle development. *Science* **226**, 979-982.

Cooper, T. A. and Ordahl, C. P. (1985). A single cardiac troponin T gene generates embryonic and adult isoforms via developmentally regulated alternate splicing. *J. Biol. Chem.* **260**, 11140-11148.

Donoghue, M., Ernest, H., Wentworth, B., Nadel-Ginard, B. and Rosenthal, N. (1988). A muscle specific enhancer is located at the 3′ end of the myosin light-chain 1/3 gene locus. *Genes and Develop.* **2**, 1779-1790.

Gorman, C. M., Moffat, L. F. and Howard, B. H. (1982). Recombinant genomes which express chloramphenicol acetyltransferase in mammalian cells. *Mol. Cellul. Biol.* **2**, 1044-1051.

Gossett, L. A., Kelvin, D. J., Sternberg, E. A. and Olson, E. N. (1989). A new myocyte-specific enhancer binding factor that recognizes a conserved element associated with multiple muscle-specific genes. *Molec. Cellul. Biol.* **9**, 5022-5033.

Horlick, R. and Benfield, P. (1989). The upstream muscle-specific enhancer of the rat muscle creatine kinase gene is composed of multiple elements. *Molec. Cellul. Biol.* **9**, 2394-2413.

Iannello, R. C., Mar, J. H. and Ordahl, C. P. (1991). Characterization of a Cardiac Transcriptional Control Element and its Interaction with Trans-Acting Factors. *J. Biol. Chem.* **266**, 3309-3316.

Jaynes, J. B., Johnson, J. E., Buskin, J. N., Gartside, C. L. and Haushka, S. D. (1988). The mouse creatine kinase gene is regulated by multiple upstream elements, including a muscle-specific enhancer. *Molec. Cellul. Biol.* **8**, 62-70.

Long, C. S. and Ordahl, C. P. (1989). Transcriptional repression of an embryo-specific muscle gene. *Dev. Biol.* **127**, 228-234.

Mar, J. H. and Ordahl, C. P. (1988). A conserved CATTCCT motif is required for skeletal muscle-specific activity of the cardiac troponin T gene promoter. *Proc. Natl. Acad. Sci. USA.* **85**, 6404-6408.

Mar, J. H. and Ordahl, C. P. (1988). A conserved CATTCCT motif is required for skeletal muscle-specific activity of the cardiac troponin T gene promoter. *Proc. Natl. Acad. Sci. USA.* **85**, 6404-6408.

Mar, J. H. and Ordahl, C. P. (1990). M-CAT Binding Factor, A Novel Trans-Acting Factor Governing Muscle-Specific Transcription. *Molec. Cellul. Biol.* **10**, 4271-4283.

Mar, J. H., Antin, P. B., Cooper, T. A. and Ordahl, C. P. (1988). Analysis of the upstream regions governing expression of the chicken cardiac troponin T gene in embryonic cardiac and skeletal muscle cells. *J. Cell Biol.* **107**, 573-585.

Minty, A. and Kedes, L. (1986). Upstream regions of the human cardiac actin gene that modulate its transcription in muscle cells: presence of an evolutionarily conserved repeated motif. *Molec. Cellul. Biol.* **6**, 2125-2136.

Nikovits, W., Jr, Kuncio, G. and Ordahl, C. P. (1986). The chicken fast skeletal troponin I gene: exon organization and sequence. *Nucl. Acids Res.* **14**, 3377-3390.

Olson, E. N. (1990). MyoD family: a paradigm for development? *Genes and Develop.* **4**, 1454-1461.

Ordahl, C. P., Kioussis, D., Tilghman, S. M., Ovitt, C. and Fornwald, J. (1980). Molecular Cloning of Developmentally Regulated, Low Abundance mRNA Sequences from Embryonic Muscle. *Proc. Natl. Acad Sci. USA* **77**, 4519-4523.

Piette, J., Bessereau, J.-L., Huchet, M. and Changeux, J-P. (1990). Two adjacent MyoD1-binding sites regulate expression of the acetylcholine receptor a-subunit gene. *Nature* **345**, 353-355.

Sternberg, E., Spizz, G., Perry, M. and Olson, E. (1989). A ras-dependent pathway abolishes activity of a muscle-specific enhancer upstream from the muscle creatine kinase gene. *Molec. Cellul. Biol.* **9**, 594-601.

Toyota, N. and Shimada, Y. (1981). Differentiation of troponin in cardiac and skeletal muscle in chicken embryos as studied by immunofluorescence microscopy. *J. Cell Biol.* **91**, 497-504.

Toyota, N. and Shimada, Y. (1983). Isoform variants of troponin in skeletal and cardiac muscle cells cultured with and without nerves. *Cell* **33**, 297-304.

Wang, Y., Xu, H. P., Wang, X. M., Ballivet, M. and Schmidt, J. (1988). A cell type specific enhancer drives expression of the chick acetylcholine receptor α-subunit gene. *Neuron* **1**, 527-534.

Yutzey, K., Kline, R. and Konieczny, S. (1989). An internal regulatory element control troponin I gene expression. *Molec. Cellul. Biol.* **9**, 1397-1405.

Printed in Great Britain © Society for Experimental Biology 1992   251

GENE EXPRESSION DURING CARDIAC DEVELOPMENT

PAUL J. R. BARTON, PANKAJ K. BHAVSAR, NIGEL J. BRAND,
PENNY S. CHAN-THOMAS, NINA DABHADE, HEND FARZA,
PHILIP J. TOWNSEND and MAGDI H. YACOUB

Department of Cardiothoracic Surgery, National Heart and Lung Institute, Dovehouse
Street, London SW3 6LY

Summary

The vertebrate heart forms as two concentric epithelial cylinders of myocardium and endocardium separated by an extended basement membrane matrix commonly referred to as cardiac jelly. Subsequent maturation involves a complex series of events including asymmetric changes in cell shape and division which contribute to bending and the formation of the bulboventricular loop, the formation of specialised tissues including endocardial cushion tissue of the atrioventricular (AV) and outflow tract regions, the development of conductive tissue and myocyte maturation leading to the overall pattern of expression characteristic of mature heart muscle. These processes depend on a precise spatial and temporal control of gene expression both of genes encoding regulatory molecules and those encoding structural components of the heart. In this chapter we address three aspects of cardiac development, namely, the determination of cell fate during formation of endocardial cushion tissue in the embryonic heart, transitions in troponin gene expression during fetal myocyte maturation, and the use of cloning techniques based on the polymerase chain reaction for identifying transcription factors present in the heart.

Early cardiac development

One of the important cellular events in early cardiogenesis is the formation of endocardial cushion tissue which, in the atrioventricular (AV) region marks the onset of AV (mitral and tricuspid) valve formation. In the outflow tract cushion tissue formation is involved in the septation of major vessels from a common trunk. Of particular interest in these events are genes which exhibit non-uniform expression in the heart and where the distribution is suggestive of a functional role in the induction of cushion tissue formation. Such genes include those encoding important regulators including growth factors, retinoic acid receptors and the recently described homeobox genes *Hox-7* and *Hox-8*. Here we describe

Author for correspondence: Paul J.R. Barton, National Heart and Lung Institute, Dovehouse Street, London SW3 6LY.

Key words: troponin, cardiac development, Homeobox gene, transcription factor, Polymerase chain reaction.

experiments aimed at defining the pattern of expression of *Hox-7* and *Hox-8* in the developing chick heart.

The formation of endocardial cushion tissue in the atrioventricular region occurs by the delamination of cells from the endocardial cell layer into the cardiac jelly separating endothelium and myocardium (Manasek, Icardo, Nakamura and Sweeney, 1986). These cells multiply to form the mesenchymal population of the developing cushion. Delamination occurs at a precise time during development and in a precise region of the heart. Experiments using explanted endothelium (Krug, Runyan and Markwald, 1985) have demonstrated that the process is induced by activating agent(s) derived from AV myocardium which traverse the extended basement membrane and induce the endothelial cells to delaminate and divide. Moreover it has been demonstrated that it is only endothelial cells from the AV region which are capable of responding to the inducer and only myocardial cells from the AV region which are capable of this inducing effect (Mjaatvedt, Lepera and Markwald, 1987; Krug, Mjaatvedt and Markwald, 1987). Hence, both the myocardial and endothelial cells in the AV region display specialised functions confined to this region. The molecular basis of this regional specialisation in the early heart is unknown.

Homeobox genes are known to be involved in regulating embryonic cell fate by acting as transcriptional regulators (reviewed by Gaunt, 1991). Surprisingly, none of the homeobox genes which constitute the four principal *Antennapedia*-like gene clusters in vertebrates (Graham, Papalopulu and Krumlauf, 1989) have been shown to be expressed in the heart, although mutation of *Hox-1.5* has been shown to result in cardiac malformations (Chisaka and Capecchi, 1991). Two recently identified genes (*Hox-7* and *Hox-8*) which do not form part of the *Antennapedia* gene clusters are expressed in early heart development. *Hox-7* was identified as a vertebrate homologue of the *Drosophila* gene *msh* and, in mouse, its pattern of expression was found to include the developing limb, the branchial arches and endocardial cushion tissue of the heart (Robert, Sassoon, Jacq, Gehring and Buckingham, 1989). A second, related, homeobox gene (*Hox-8*) has recently been described (Davidson, Crawley, Hill and Tickle, 1991; Robert, Lyons, Krabbenhoft, Simandl, Kuroiwa, Buckingham and Fallon, 1992) and is also expressed in the developing heart (Robert, personal communication). We therefore chose to examine the expression of these genes in relation to endocardial cushion formation.

Expression of *Hox-7* and *Hox-8* was examined by *in situ* hybridization of ^{35}S-labelled RNA probes on sections of paraffin embedded chick embryos essentially as described (Barton, Vallins, Lyons, Brand and Yacoub, 1991). Probes were produced by transcription from cloned cDNA fragments of mouse and chick *Hox-7* and chick *Hox-8* using standard techniques. Following hybridization and washing, slides were dipped in Kodak K2 emulsion and exposed for 7 - 10 days. Sections were then examined using both lightfield and darkfield optics. Figs 1A and 1B show a sagittal section through a stage 21 chick heart hybridized with the mouse *Hox-7* probe. Hybridization is clearly seen both in the cells of the

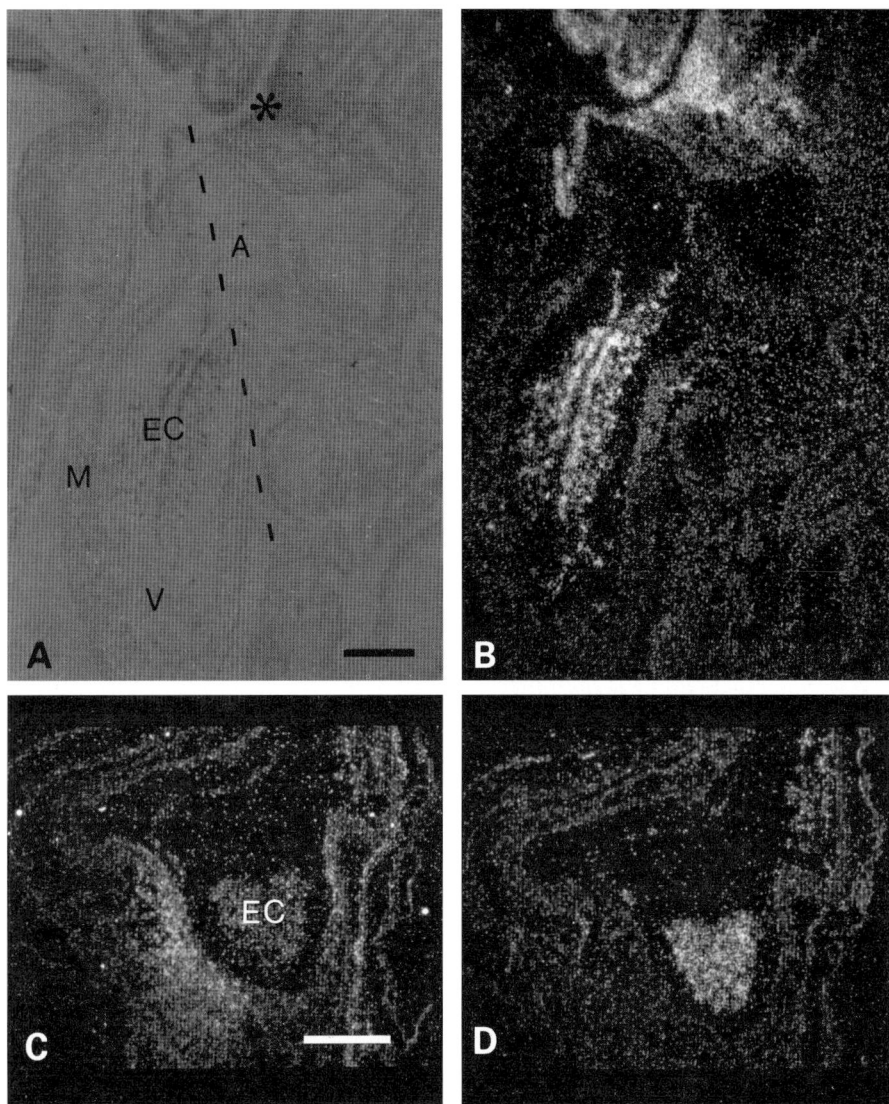

Fig. 1. Detection of *Hox-7* and *Hox-8* gene expression in the developing chick heart by *in situ* hybridization. A: Sagittal section through a stage 21 chick heart showing atrial (A) and ventricular (V) regions, myocardium (M) and atrioventricular endocardial cushion tissue (EC). Dotted line indicates the approximate plane of sectioning of embryon shown in in C and D. Bar=100 μm. B: Same section viewed under dark field optics following hybridization with mouse *Hox-7* probe. Hybridization is clearly seen in the endocardial cushion tissue. * indicates branchial arch tissue. C and D: Dark field images of serial frontal sections through inferior endocardial cushion tissue of a stage 20 chick embryo hybridized with chick *Hox-8* specific probe (C), and *Hox-7* specific probe (D). Bar=80 μm.

endothelial cell layer in the AV region and in those cells which have delaminated and are forming the mesenchyme population of the developing cushions. Hybridization is not evident in the myocardium or in the endothelial cell layer in the atrial region which is not participating in this process. Figs 1C and 1D show serial sections frontal through the cushion region hybridized with chick Hox-7 or Hox-8 specific probes. Again, *Hox-7* expression is limited to the cells of the developing cushion tissue. In contrast, *Hox-8* expression is seen in the myocardial cells subjacent to the cushion. Hox-8 expression is not apparent in the cushion tissue or in other regions of the myocardium.

The pattern of expression of these homeobox genes is intriguing in relation to what is known of the induction of cushion formation by AV myocardium. Experiments using explanted AV endothelial cells grown in culture on the surface of collagen blocks have shown that they can be induced to undergo an epithelial to mesenchyme transition (mimicking *in vivo* delamination by invading the collagen block) when stimulated by the presence of AV myocardial cells in co-culture, or by the addition to the culture medium of extracts derived from AV myocardium (Krug et al., 1985, 1987, Mjaatvedt et al., 1987). Given their distribution of expression, it is possible that *Hox-7* and *Hox-8* are directly involved in this induction process. Other regulatory genes also show intriguing patterns of expression at this stage. In the developing mouse heart several genes have been shown to be expressed in atrioventricular myocardium or cushion tissue including those of the transforming growth factors TGFβ-1, TGFβ-2 and TGFβ-3 (Akhurst, Lehnert, Faissner and Duffie, 1990; Millan, Denhez, Kondaiah and Akhurst, 1991; Schmid, Cox, Bilbe, Maier and McMaster, 1991; Pelton, Dickinson, Moses and Hogan, 1990), the bone morphogenic protein genes BMP2A (Lyons, Pelton and Hogan, 1990) and BMP4 (Jones, Lyons and Hogan, 1991), the retinoic acid receptor-γ (Ruberte, Dollé, Krust, Zelent, Morriss-Kay, and Chambon, 1990) and the retinol binding protein CRBP (Dollé, Ruberte, Leroy, Morriss-Kay and Chambon, 1990). In the case of TGFβ-1 direct involvement in the morphogenic induction of cushion formation has been implied through the use of endothelial cultures (Potts and Runyan, 1989) and it will be of great interest to determine how each of these regulatory factors relate to this process.

In summary *Hox-7* and *Hox-8* show a pattern of expression suggestive of a role in the induction of endocardial cushion tissue in the embryonic heart. These and other regulatory genes probably form part of a complex network of cell signalling events involved in this process. The use of cell culture techniques and genetic manipulation should allow the functional role of these genes to be examined in detail.

Troponin gene expression in the developing heart

The three troponin proteins troponin I (TnI), troponin T (TnT) and troponin C (TnC), form a protein complex on the thin filament of striated muscle which is involved in regulating contraction in response to intracellular calcium (reviewed by Winegrad, 1984). When calcium binds to troponin C there is a conformational change in the complex which allows interaction of the myosin head with actin thereby initiating muscle contraction. In the absence of calcium the troponin complex inhibits the interaction and muscle fibres become relaxed. Troponin I probably acts as a molecular switch during this process by altering its position on the thin filament (e.g. Tao, Gong and Leavis, 1990). Multiple isoforms of each of these proteins have been identified which are expressed in a tissue-specific and developmentally regulated manner. Cumulative data, principally from studies using troponin antibodies on rat and chick (see below), indicate that the three troponins show contrasting patterns of regulation during cardiac development. In the case of troponin C a single isoform is expressed at all stages of development and in the adult heart (see Parmacek and Leiden, 1991). For troponin I there is a transition from expression of a fetal isoform in the developing heart to expression of the adult isoform (TnIc) (Ausoni, de Nardi, Moretti, Gorza and Schiaffino, 1991; Murphy, Jones, Sims, Strauss, 1990; Bhavsar, Dhoot, Cumming, Butler-Browne, Yacoub and Barton, 1991). These isoforms are encoded by separate genes suggesting a developmental switch in gene transcription. For cardiac troponin T, two isoforms have been identified which are the result of developmentally regulated alternative splicing of a single cardiac TnT gene (Cooper and Ordahl, 1985; Jin and Lin, 1989). The troponin genes therefore provide a suitable model for studying developmental regulation of gene expression in the heart. Little information is currently available on gene expression during human cardiac development and we have therefore undertaken to examine troponin expression in the human heart.

In the case of troponin I, three principal isoforms have been described in striated muscle on the basis of protein sequencing (Leszyk, Dumaswala, Potter and Collins, 1988; Wilkinson and Grand, 1978), antibody studies (Cummins and Perry, 1978; Dhoot and Perry, 1979; Saggin, Auson, Gorza, Sartore and Schiaffino, 1988; Sabry and Dhoot, 1989a; Dhoot, Gell and Perry, 1978; Dhoot and Perry, 1979) and molecular cloning (e.g. Ausoni et al., 1991; Baldwin, Kittler and Emerson, 1985; Koppe, Hallauer, Karpati and Hastings, 1989; Murphy et al., 1990; Wade, Eddy, Shows and Kedes, 1990, and Vallins, Brand, Dabhade, Butler-Browne, Yacoub and Barton, 1990). These isoforms are associated with fast skeletal muscle, slow skeletal muscle and cardiac muscle and are referred to here as TnIf, TnIs and TnIc, respectively. We have examined the situation in man using antibodies to determine the distribution of isoforms in striated muscle. As illustrated in figures 2 and 3, adult cardiac muscle contains a single troponin I protein (TnIc) which is not detected in skeletal muscle. Extracts from fetal heart contain, in addition to TnIc, a developmental isoform which is indistinguishable, on the basis of size and immunoreactivity, from that found in slow skeletal muscle (i.e. TnIs). In early fetal samples the developmental isoform is the predominant troponin I detected, whereas later samples contain increasing amounts of the adult cardiac isoform. The developmental isoform remains abundant up to 15 day post-natal life but is not detectable by 9 months after birth (figure 3). These results

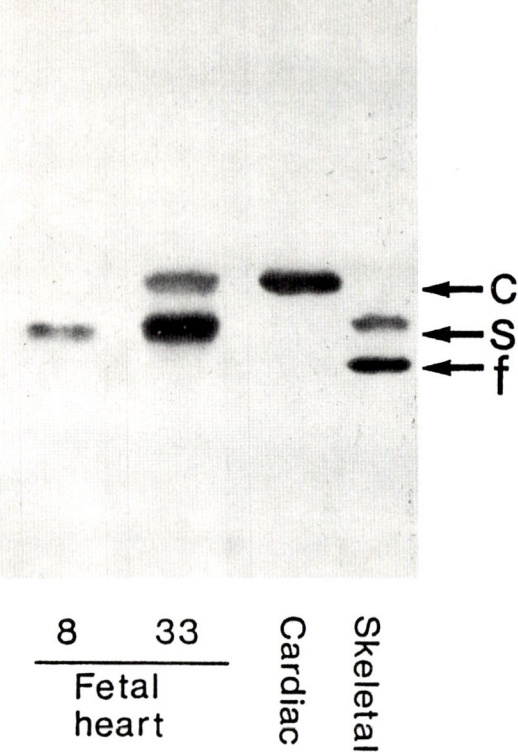

Fig. 2. Detection of troponin I isoforms in total protein extracts from human tissues. Fetal heart and skeletal muscle samples together with an adult heart (atrial) muscle sample were separated by SDS polyacrylamide gel electrophoesis, transferred to nitrocellulose and hybridized with a troponin I monoclonal antibody. C, S and F indicate the position of the cardiac muscle isoform, slow skeletal muscle isoform, and fast skeletal muscle isoforms respectively.

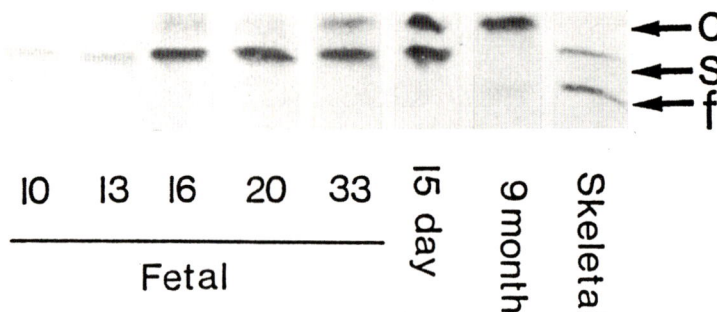

Fig. 3. Expression of troponin I isoforms in the developing human heart. Total protein extracts were analysed by immunoblotting as in figure 2. Note the transition from expression of TnIs to TnIc between 20 weeks fetal development and 9 months post-natal life.

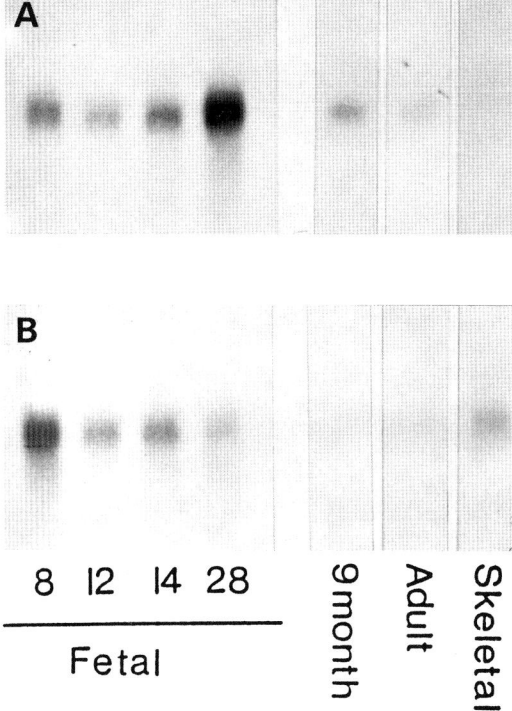

Fig. 4. Northern blot hybridization of RNA extracts from fetal heart, 9 month post-natal heart, adult heart (atrial) and fetal skeletal muscle. A: hybridization with cardiac (TnIc) cDNA probe. B: hybridization with slow skeletal muscle (TnIs) cDNA probe.

demonstrate a transition in troponin isoforms during human cardiac development that occurs between 20 weeks fetal development and 9 months post-natal life.

In order to examine TnI mRNA accumulation, we made use of the polymerase chain reaction (PCR) to clone fragments of the TnIc and TnIs mRNAs. The amino acid sequences of cardiac and skeletal isoforms from other species was compared and regions identified which are 100% conserved between all known isoforms. Degenerate sets of oligonucleotide primers corresponding to these regions were used in PCR amplifications from a single stranded cDNA template derived from microgram quantities of human ventricular RNA (Vallins et al., 1990). Resulting fragments were cloned and confirmed as encoding the human TnIc by sequencing. Cloned fragments were subsequently used to isolate full length cDNA clones from a human ventricular cDNA library as previously described (Vallins et al., 1990). For TnIs, a partial cDNA fragment was generated by PCR using genomic DNA and specific primers corresponding to the published mRNA sequence data (Wade et al., 1990). Troponin I mRNA accumulation was examined on Northern blots of total RNA extracts from adult and fetal heart (figure 4). Both TnIc and TnIs mRNA are detectable throughout fetal development whereas in the adult and 9 month post-natal samples only TnIc mRNA is detected indicating that the TnIs gene becomes transcriptionally quiescent after birth. The increased accumulation of TnIc mRNA after 20 weeks gestation correlates with the onset of TnIc protein accumulation at this time. However there is a significant level of TnIc mRNA in early samples, where the level of protein accumulation is extremely low, suggesting that post-transcriptional events also play a significant role in the accumulation of troponin I proteins.

In the case of troponin T, a single gene has been described in both rat and chick (Cooper and Ordahl, 1985; Jin and Lin, 1989). This gene undergoes developmental regulation through the use of alternative splicing which, in the adult, excludes a 30bp sequence derived from exon 5. The switch occurs close to the time of hatching in the chick, and studies in rat using TnT antibodies suggest that an equivalent switch occurs close to birth (Saggin et al., 1988; Sabry and Dhoot, 1989b). Preliminary data using a human adult cardiac troponin T cDNA indicates that this is also true for man (Barton and Wade, unpublished data): PCR analysis was carried out using primers spanning the region of the human cDNA sequence corresponding to the predicted boundary between exons 4 and 6 in the adult mRNA. Using these primers two transcript sizes are detected in human fetal heart of approximately 30bp difference in size, of which only the shorter persists in the adult heart. Troponin T expression in the developing human heart therefore appears to undergo developmentally regulated splicing similar to that seen in rat and chick.

The functional significance of alterations in troponin isoforms during cardiac development is not clear. The cardiac troponin I isoform differs from those of skeletal muscle in that it carries an extended N-terminal sequence including a pair of serine residues which are phosphorylated in response to β-adrenergic stimulation (Solaro, Moir and Perry, 1976). Phosphorylation at this site alters the calcium binding characteristics of the troponin complex such that it displays decreased affinity for calcium as part of the phosphorylation-mediated regulation of cardiac contractility. Expression of TnIs in the fetal heart would be expected to confer reduced responsiveness to adrenergic stimulation as it lacks the extended N-terminal sequence. The availability of cloned cDNAs encoding these isoforms should allow detailed functional analysis to be carried out by the use of suitable expression systems to produce modified troponin I protein *in vitro*, and through the use of transgenic techniques to examine function *in vivo*.

Use of PCR techniques to identify potential cardiac transcription factors

A major level for the control of gene expression is that of transcription and this is achieved through the interaction of specific DNA-binding transcription factors with DNA sequences located in the promoters of target genes (e.g. Wasylyk, 1988). Some transcription factors, such as the TATA-box binding factor TFIID, are expressed ubiquitously and are essential for accurate initiation of transcription by RNA polymerase II. In contrast, other factors are expressed only in cells of a certain lineage or at a particular stage of development, allowing precise and coordinated activation (or repression) of gene expression in a tissue-specific or stage-specific manner. This is exemplified by the restricted expression of myogenic regulatory factors such as MyoD1 (Tapscott, Davis, Thayer, Cheng, Weintraub and Lassar, 1988), myogenin (Wright, Sassoon and Lin, 1989; Edmonson and Olson, 1989) and *myf*-5 (Braun, Buschhausen-Denker, Bober, Tannich and Arnold, 1989) to differentiating skeletal myocytes (e.g. Sassoon et al., 1989).

Several families of transcription factors have been identified and these are classified according to whether they possess one or more of a limited number of protein domains (Harrison, 1991) which ascribe particular functions including DNA-binding and protein dimerisation. DNA-binding motifs include the zinc finger present in members of the steroid hormone receptor gene superfamily (Evans, 1988), the highly conserved helix-turn-helix motif present in homeoproteins (Gaunt, 1991) and the so-called POU-domain (Rosenfeld, 1991). Domains involved in protein dimerisation include the helix-loop-helix (HLH), present in many factors including MyoD1 and the related myogenic regulatory factors (Murre, McCaw and Baltimore, 1989), and the leucine zipper (Abel and Maniatis, 1989), present in a variety of oncogenes and the CREB (cAMP-responsive element binding protein) superfamily (Hai, Liu, Coukos and Green, 1989). The high degree of amino acid sequence homology in these domains has allowed many other transcription factors to be cloned. For example, cDNA clones for the human retinoic acid receptor α (RAR-α), a member of the steroid hormone receptor family, were isolated by synthesising oligonucleotide sets complementary in sequence to the most highly conserved part of the zinc finger domain and directly screening a human cDNA library (Petkovich, Brand, Krust and Chambon, 1987). Recently, the development of cloning strategies based on PCR has increased the sensitivity of this approach, allowing the cloning of cDNAs from extremely small quantities of tissue. Using this technique, He and co-workers isolated several POU-factor cDNAs expressed in a tissue-specific manner during rat development (He, Treacy, Simmons, Ingraham, Swanson and Rosenfeld, 1989).

One approach to understanding how cardiac gene expression is regulated is to identify potential transcription factors present in the heart by molecular cloning and to characterise their function. Using degenerate oligonucleotide primers complementary to conserved domains present in the MyoD-family of myogenic factors and to the recently defined domain present in the sex determination gene *SRY* (Sinclair, Berta, Palmer, Hawkins, Griffiths, Smith, Foster, Frischauf, Lovell-Badge and Goodfellow, 1990; Gubbay, Collignon, Koopman, Capel, Economou, Münsterberg, Vivian, Goodfellow and Lovell-Badge, 1990) we have carried out PCR-based cloning experiments to identify related sequences present in the heart. To date, no MyoD-related sequences have been isolated from either adult or fetal heart suggesting that cardiac muscle gene regulation is not effected by this group of transcription factors. This is surprising in view of their pivotal role in regulating gene expression during skeletal muscle development and is particularly intriguing when considering that many genes which are regulated by MyoD and its related factors in skeletal muscle are also expressed in heart. It remains to be seen whether transcription factors with related function but which do not process conserved sequence homology to this family exist in the heart.

In contrast to experiments using MyoD related sequences, PCR amplification using oligonucleotide primers complementary to the putative DNA-binding domain of the sex determination gene *SRY* have been successful in identifying several related sequences present in the heart. The *SRY* gene, and its murine homologue *sry*, is located on the Y chromosome. It is expressed only in testis and is thought to play a pivotal role in the development of the male phenotype. In mouse, *sry* is expressed in the Sertoli cells of the developing gonadal ridge and is the earliest known marker for testis development (Koopman, Münsterberg, Capel, Vivian and Lovell-Badge, 1990). Moreover, transgenic mice have been obtained by introducing the *sry* which although genotypically female develop with the male phenotype (Koopman, Gubbay, Vivian, Goodfellow and Lovell-Badge, 1991). The amino acid sequence derived from the *SRY* gene contains a region (termed the SRY-box) which shows similarity to a DNA-binding domain present in the ribosomal RNA gene upstream binding factor UBF (Jantzen, Admon, Bell and Tjian, 1990), the yeast mating factor Mc (Kelly, Burke, Smith, Klar and Beach, 1988) and a human transcription factor TCF-1 which is expressed specifically in T-lymphocytes (Van de Wetering, Oosterwegel, Dooijes and Clevers, 1991).

Using oligonucleotides complementary to this domain we have isolated six distinct cDNAs from human cardiac RNA which show significant homology to the

```
TCF-1        MKEMRAKVIAECTLKESAAINQILGRRWHALSREEQAKYYELARKERQLHMQLY
ubf-hmg3     SEEKRRQLQEERPELSESELTRLLAEMWNDLSEKKKAKYKAREAALKAQSERKP


SRY          SRDQRRKMALENPRMRNSEISKQLGYQWKMLTEAEKWPFFQEAQKLQAMHREKY
SOX2         --G------Q---K-H------R--AE--L-S-T--R--ID--KR-R---MK-H
SOX4         -QIE---IMEQS-D-H-A----R--KR----KDSD-I--I---ER-RLK-MAD-
SOX5         AK-RH--ILQAF-D-H--N---I--SR--AM-NL--Q-YYE-QAR-SKQ-L---
SOX6         AK-E---ILQAF-D-H--N---I--SR--SM-NQ--Q-YYE-QAR-SKI-L---
SOX8         AK-E-KRL-QQ--DLH-AVL--M--KA--E-NA---R--VE--E--RVQ-LRDH


consensus    A+d +++ΦΦqe P ΦHNA ΦS+ ΦG  Φ+ ΦS  D+ PF dE  KΦ+ ΦHΦ KY
             S e               S   T         T    E  Y e   R s Q   dh


                               **  **      *  *
             ----helix 1----   -turn-    ----helix 2----
```

Fig. 5. Comparison of sequences of the putative DNA-binding domain of SRY (the 'SRY-box' or SOX) with sequences derived from five SRY-related cDNA clones isolated from human heart. Identical residues are denoted by a hyphen. Also shown are sequences of equivalent regions from the human upstream binding protein UBF (UBF-HMG3), human T lymphocyte-specific transcription factor TCF-1 and a consensus sequence for the domain. Six positions in the consensus sequence are indicated by (*): these may allow the region to fold into a helix-turn-helix conformation reminiscent of DNA-binding factors such as homeoproteins or bacterial repressor proteins.

SRY-box. We have termed these sequences 'SOX' for <u>S</u>ry-<u>box</u> (Denny, Swift, Brand, Dabhade, Barton and Ashworth, 1992). Fig. 5 shows the sequence of 5 clones isolated in this way, aligned with human *SRY*, TCF-1 and UBF. It is clear that there is considerable homology between the SOX sequences and the prototype *SRY*. Structural considerations would suggest that these protein domains may adopt an extended helix conformation similar to that seen in other transcription factor families. In particular, several residues marked by asterisks in figure 5 are typically conserved residues at key positions in the helix-turn-helix motif in the homeodomain and bacterial repressor proteins.

One of the complications of PCR cloning is that sequence mutations may be introduced during the amplification procedure. Several observations argue that the SOX clones presented here are not the result of PCR-derived sequence mutation. Firstly each of the sequences presented has been observed in several independently isolated clones. Secondly, the sequence differences between the SOX clones are complex and are not easily explained by point mutation or rearrangement. Thirdly, homologues of several of these human clones have been identified in mouse (Gubbay et al., 1990; Denny et al., 1992).

In summary, SOX clones, related to *SRY*, form part of a novel multigene family that has been conserved throughout evolution, members of which are expressed in heart and may function as transcription factors.

We are grateful to R.P. Thompson for invaluable assistance with chick embryology and *in situ* hybridization to Tej Dhoot and Debbie Cumming for providing troponin antibodies and assistance with immunoblotting. We thank Benoit Robert for inspiration and for providing suitable *Hox7* clones and thank Atsushi Kuroiwa for providing the chick *Hox-8* clone. This work was supported by the British Heart Foundation.

References

Abel, T. and Maniatis, T. (1989). Action of leucine zippers. *Nature (London)* **341**, 24-25.

Akhurst, R. J., Lehnert, S. A., Faissner, A. and Duffie, E. (1990). TGF beta in murine morphogenetic processes: the early embryo and cardiogenesis. *Development* **108**, 645-656.

Ausoni, S., de Nardi, C., Moretti, P., Gorza, L. and Schiaffino, S. (1991). Developmental expression of rat cardiac troponin I mRNA. *Development* **112**, 1041-1051.

Baldwin, A. S., Jr, Kittler, E. L. W. and Emerson, C. P., Jr (1985). Structure, evolution, and regulation of a fast skeletal muscle troponin I gene. *Proc. Natl. Acad. Sci. USA.* **82**, 8080-8084.

Barton, P. J. R., Vallins, W. J., Lyons, G. E., Brand, N. J. and Yacoub, M. (1991). RNA detection methods in basic cardiac research. In *Genetic manipulation techniques and applications*, (Eds. J.M. Grange, A. Fox, and N.L. Morgan) pp. 265-277. Oxford: Blackwell Scientific Press.

Bhavsar, P. K., Dhoot, G. K., Cumming, D. V. E., Butler-Browne, G. S., Yacoub, M. H. and Barton, P. J. R. (1991). Developmental expression of troponin I isoforms in fetal human heart. *FEBS. Lett.* **292**, 5-8.

Braun, T., Buschhausen-Denker, G., Bober, E., Tannich, E. and Arnold, H. H. (1989). A novel human muscle factor related to but distinct from MyoD1 induces myogenic conversion in 10T1/2 fibroblasts. *EMBO J.* **8 (no. 3)**, 701-709.

Chisaka, O. and Capecchi, M. R. (1991). Regionally restricted developmental defects resulting from targeted disruption of the mouse homeobox gene *hox-1.5*. *Nature (London)* **350**, 473-479.

Cooper, T. A. and Ordahl, C. P. (1985). A single cardiac troponin T gene generates embryonic and adult isoforms via developmentally regulated alternate splicing. *J. Biol. Chem.* **260**, 11140-11148.

Cummins, P. and Perry, S. V. (1978). Troponin I from human skeletal and cardiac muscles. *Biochem. J.* **171** 251-259.

Davidson, D. R., Crawley, A., Hill, R. E. and Tickle, C. (1991). Position dependent expression of two related homeobox genes in developing vertebrate limbs. *Nature (London)* **352**, 429-431.

Denny, P., Swift, S., Brand, N., Dabhade, N., Barton, P. and Ashworth, A. (1992). A conserved family of genes related to the testis determining gene, SRY. *Nucl. Acids Res.* **20**, 2887.

Dhoot, G. K., Gell, P. G. H. and Perry, S. V. (1978). The localization of the different forms of troponin I in skeletal and cardiac muscle cells. *Exp. Cell Res.* **117**, 357-370.

Dhoot, G. K. and Perry, S. V. (1979). Distribution of polymorphic forms of troponin components and tropomyosin in skeletal muscle. *Nature (London)* **278**, 714-718.

Dollé, P., Ruberte, E., Leroy, P., Morriss-Kay, G. and Chambon, P. (1990). Retinoic acid receptors and cellular retinoid binding proteins. *Development* **110**, 1133-1151.

Edmondson, D. G. and Olson, E. N. (1989). A gene with homology to the *myc* similarity region of MyoD1 is expressed during myogenesis and is sufficient to activate the muscle differentiation program. *Genes and Dev.* **3**, 628-640.

Evans, R. M. (1988). The steroid and thyroid hormone receptor superfamily. *Science* **240**, 889-895.

Gaunt, S. J. (1991). Expression patterns of mouse hox genes: clues to an understanding of developmental and evolutionary strategies. *BioEssays* **13**, 505-513.

Graham, A., Papalopulu, N. and Krumlauf, R. (1989). The murine and drosophila homeobox gene complexes have common features of organization and expression. *Cell* **57**, 367-378.

Gubbay, J., Collignon, J., Koopman, P., Capel, B., Economou, A., Münsterberg, A., Vivian, N., Goodfellow, P. and Lovell-Badge, R. (1990). A gene mapping to the sex-determining region of the mouse Y chromosome is a member of a novel family of embryonically expressed genes. *Nature (London)* **346**, 245-250.

Hai, T., Liu, F., Coukos, W. J. and Green, M. R. (1989). Transcription factor ATF cDNA clones: an extensive family of leucine zipper proteins able to selectively form DNA-binding heterodimers. *Genes and Dev.* **3**, 2083-2090.

Harrison, S. C. (1991). A structural taxonomy of DNA-binding domains. *Nature (London)* **353**, 715-719.

He, X., Treacy, M. N., Simmons, D. M., Ingraham, H. A., Swanson, L. W. and Rosenfeld, M. G. (1989). Expression of a large family of POU-domain regulatory genes in mammalian brain development. *Nature (London)* **340**, 35-42.

Jantzen, H-M., Admon, A., Bell, S. P. and Tjian, R. (1990). Nucleolar transcription factor hUBF contains a DNA-binding motif with homology to HMG proteins. *Nature (London)* **344**, 830-836.

Jin, J-P. and Lin, J. J-C. (1989). Isolation and characterization of cDNA clones encoding embryonic and adult isoforms of rat cardiac troponin T. *J. Biol. Chem.* **264**, 14471-14477.

Jones, C. M., Lyons, K. M. and Hogan, B. L. M. (1991). Involvement of *bone morphogenetic protein-4 (BMP-4)* and *Vgr-1* in morphogenesis and neurogenesis in the mouse. *Development* **111**, 531-542.

Kelly, M., Burke, J., Smith, M., Klar, A. and Beach, T. (1988). Four mating-type genes control sexual differentiation in fission yeast. *EMBO J.* **7**, 1537-1547.

Koopman, P., Gubbay, J., Vivian, N., Goodfellow, P. and Lovell-Badge, R. (1991). Male development of chromosomally female mice transgenic for Sry. *Nature (London)* **351**, 117-121.

Koopman, P., Münsterberg, A., Capel, B., Vivian, N. and Lovell-Badge, R. (1990). Expression of a candidate sex- determining gene during mouse testis differentiation. *Nature (London)* **348**, 450-452.

Koppe, R. I., Hallauer, P. L., Karpati, G. and Hastings, K. E. M. (1989). cDNA clone and expression analysis of rodent fast and slow skeletal muscle troponin I nRNAs. *J. Biol. Chem.* **264**, 14327-14333.

Krug, E. L., Mjaatvedt, C. H. and Markwald, R. R. (1987). Extracellular matrix from embryonic myocardium elicits an early morphogenetic event in cardiac endothelial differentiation. *Dev. Biol.* **120**, 348-355.

Krug, E. L., Runyan, R. B. and Markwald, R. R. (1985). Protein extracts from early embryonic hearts initiate cardiac endothelial cytodifferentiation. *Dev. Biol.* **112**, 414-426.

Leszyk, J., Dumaswala, R., Potter, J. D. and Collins, J. H. (1988). Amino acid sequence of bovine cardiac troponin I. *Biochemistry.* **27**, 2821-2827.

Lyons, K. M., Pelton, R. W. and Hogan, B. L. M. (1990). Organogenesis and pattern formation in the mouse: RNA distribution patterns suggest a role for *bone morphogenetic protein-2A (BMP-2A)*. *Development* **109**, 833-844.

Manasek, F. J., Icardo, J., Nakamura, A. and Sweeney, L. (1986). Cardiogenesis: developmental mechanisms and embryology. In *The Heart and Cardiovascular System* (eds. H.A. Fozzard et al.), pp. 965-985. Raven Press, New York.

Millan, F. A., Denhez, F., Kondaiah, P. and Akhurst, R. J. (1991). Embryonic gene expression patterns of TGF β1, β2 and β3 suggest different developmental functions *in vivo*. *Development* **111**, 131-144.

Mjaatvedt, C. H., Lepera, R. C. and Markwald, R. R. (1987). Myocardial specificity for initiating endothelial-mesenchymal cell transition in embryonic chick heart correlates with a particulate distribution of fibronectin. *Dev. Biol.* **119**, 59-67.

Murphy, A. M., Jones II, L., Sims, H. F. and Strauss, A. W. (1990). Molecular cloning of rat cardiac troponin I and analysis of troponin I isoform expression in developing rat heart. *Biochemistry* **30**, 707-712.

Murre, C., McCaw, P. S. and Baltimore, D. (1989). A new DNA binding and dimerization motif in immunoglubulin enhancer binding, *daughterless*, *MyoD*, and *myc* proteins. *Cell* **56**, 777-783.

Parmacek, M. S. and Leiden, J. M. (1991). Structure, function, and regulation of troponin C. *Circulation* **84**, 991-1003.

Pelton, R. W., Dickinson, M. E., Moses, H. L. and Hogan, B. L. M. (1990). *In situ* hybridization analysis of TGFβ3 RNA expression during mouse development: comparative studies with TGFβ1 and β2. *Development* **110**, 609-620.

Petkovich, M., Brand, N. J., Krust, A. and Chambon, P. (1987). A human retinoic acid receptor which belongs to the family of nuclear receptors. *Nature (London)* **330**, 444-450.

Potts, J. D. and Runyan, R. B. (1989). Epithelial-mesenchymal cell transformation in the embryonic heart can be mediated, in part, by transforming growth factor β. *Dev. Biol.* **134**, 392-401.

Robert, B., Lyons, G., Krabbenhoft, K., Simandl, B. K., Kuroiwa, A., Buckingham, M. and Fallon, J. (1992). The apical ectodermal ridge regulates Hox-7 and Hox-8 gene expression in developing limb buds. (In Press).

Robert, B., Sassoon, D., Jacq, B., Gehring, W. and Buckingham, M. (1989). Hox-7, a mouse homeobox gene with a novel pattern of expression during embryogenesis. *EMBO J.* **8**, 91-100.

Rosenfeld, M. G. (1991). POU-domain transcription factors: pou-er-ful developmental regulators. *Genes and Dev.* **5**, 897-907.

Ruberte, E., Dollé, P., Krust, A., Zelent, A., Morriss-Kay, G. and Chambon, P. (1990). Specific spatial and temporal distribution of retinoic acid receptor gamma transcripts during mouse embryogenesis. *Development* **108**, 213-222.

Sabry, M. A. and Dhoot, G. K. (1989a). Identification and pattern of expression of a developmental isoform of troponin I in chicken and rat cardiac muscle. *J. Muscle Res. Cell Motil.* **10**, 85-91.

Sabry, M. A. and Dhoot, G. K. (1989b). Identification of and changes in the expression of troponin T isoforms in developing avian and mammalian heart. *J. Mol. Cell Cardiol.* **21**, 85-91.

Saggin, L., Auson, S., Gorza, L., Sartore, S. and Schiaffino, S. (1988). Troponin T switching in the developing rat heart. *J. Biol. Chem.* **263**, 18488-18492.

Sassoon, D., Lyons, G., Wright, W. E., Lin, V., Lassar, A., Weintraub, H. and Buckingham, M.

(1989). Expression of two myogenic regulatory factors myogenin and MyoD1 during mouse embryogenesis. *Nature (London)* **341**, 303-307.

Schmid, P., Cox, D., Bilbe, G., Maier, R. and McMaster, G. K. (1991). Differential expression of TGF β1, β2 and β3 genes during mouse embryogenesis. *Development* **111**, 117-130.

Sinclair, A. H., Berta, P., Palmer, M. S., Hawkins, J. R., Griffiths, B. L., Smith, M. J., Foster, J. W., Frischauf, A.-M., Lovell-Badge, R. and Goodfellow, P. N. (1990). A gene from the human sex-determining region encodes a protein with homology to a conserved DNA-binding motif. *Nature (London)* **346**, 240-244.

Solaro, R. J., Moir, A. J. G. and Perry, S. V. (1976). Phosphorylation of troponin I and inotropic effect of adrenaline in the perfused rabbit heart. *Nature (London)* **262**, 615-616.

Tao, T., Gong, B-J. and Leavis, P. C. (1990). Calcium-induced movement of troponin-I relative to actin in skeletal muscle thin filaments. *Science* **247**, 1339-1341.

Tapscott, S. J., Davis, R. L., Thayer, M. J., Cheng, P-F., Weintraub, H. and Lassar, A. B. (1988). MyoD1: A nuclear phosphoprotein requiring a Myc homology region to convert fibroglasts to myoblasts. *Science* **242**, 405-411.

Vallins, W. J., Brand, N. J., Dabhade, N., Butler-Browne, G., Yacoub, M. H. and Barton, P. J. R. (1990). Molecular cloning of human cardiac troponin I using polymerase chain reaction. *FEBS Lett.* **270**, 57-61.

Van de Wetering, M., Oosterwegel, M., Dooijes, D. and Clevers, H. (1991). Identification and cloning of TCF-1, a T lymphocyte-specific transcription factor containing a sequence-specific HMG box. *EMBO J.* **10**, 123-132.

Wade, R., Eddy, R., Shows, T. B. and Kedes, L. (1990). cDNA sequence, tissue-specific expression, and chromosomal mapping of the human slow-twitch skeletal muscle isoform of troponin I. *Genomics* **7**, 346-357.

Wasylyk, B. (1988). Transcription elements and factors of RNA polymerase B promoters of higher eukaryotes. *CRC Critical Reviews in Biochemistry* **23**, 77-120.

Winegrad, S. (1984). Regulation of cardiac contractile proteins. *Circ. Res.* **55**, 565-574.

Wright, W. E., Sassoon, D. A. and Lin, V. K. (1989). Myogenin, a factor regulating myogenesis, has a domain homologous to MyoD. *Cell* **56**, 607-617.

REGULATION OF MYOSIN HEAVY CHAIN AND ACTIN ISOGENES DURING CARDIAC GROWTH AND HYPERTROPHY

KETTY SCHWARTZ, LUCIE CARRIER, CATHERINE CHASSAGNE, CLAUDINE WISNEWSKY and KENNETH R. BOHELER

INSERM Unité 127, Hôpital Lariboisière, 41 Boulevard de la Chapelle, 75010 PARIS, FRANCE

Summary

Expression of myosin heavy chain (MHC) and actin multigene families changes in mammals during cardiac growth and hypertrophy, but whether or not there is a common regulatory pathway is unclear. To address this question, we have looked at the α- and β-MHC, and at the α-skeletal and α-cardiac actin (α-skel act and α-card act) isomRNA transitions during development and senescence, both in rat and human hearts. Since the precise amounts of each isoactin mRNA were not precisely known in the above situations, we first analyzed the time- course of accumulations of the two sarcomeric transcripts by primer extension assays, which allow an unambiguous quantification of the ratios of the two actin transcripts. In rats, both isogenes are expressed *in-utero*. α-skel act represents 40% of the total one week after birth, remains constant for 3 weeks, decreases to less than 5% at two months and does not re-accumulate thereafter. In humans, in contrast, α-skel act represents <20% *in-utero* and in neonates, increases to 48% during the first decade after birth and becomes the predominant isoform of adult hearts. In rats β-MHC mRNAs accumulate at birth, become undetectable at 3 weeks and reaccumulate to as much as 80% during senescence, and in humans β-MHC mRNAs predominate throughout all developmental stages. These data show that in both species, the multigene families encoding the major contractile proteins are not coordinately regulated during development and aging. During hypertrophy following a hemodynamic overload, the situation is also quite dissimilar because we previously showed that β-MHC and α-skel act mRNAs accumulate with different kinetics and with different cellular distributions. To differentiate between the possible levels of regulation, we have isolated myocyte nuclei from intact rat hearts and developed an *in-vitro* system which can distinguish between transcriptional and post-transcriptional events. Neonatal nuclei transcribe α- and β-MHC and α-card and α-skel act suggesting that these gene expression are primarily regulated at the level of transcription at birth but thereafter other factors may play important roles in maintaining their accumulations.

Key words: α-skeletal and α-cardiac actin, α- and β-myosin heavy chain, run-on assays, uncoordinated regulation.

Introduction

Our laboratory has been involved for a number of years in the pathophysiology of cardiac hypertrophy, with the aim of identifying the cellular and molecular mechanisms responsible for heart failure. Cardiac hypertrophy is common to almost all cardiovascular diseases (for example, hypertension or valvular diseases) that produce a sustained hemodynamic overload on the heart and which ultimately leads to congestive heart failure. It was hypothesized in the early thirties by Louis Katz that the hypertrophied heart was not simply an enlarged version of the normal heart. Indeed, Alpert and Gordon (1962) demonstrated that myosin ATPase activity is depressed in failing human hearts, and since that time, a growing number of molecular changes have been recognized in chronically overloaded cardiac myocytes. We proposed the term 'mechanogenic transduction' to describe these phenomena (Schwartz, 1990), and it is generally admitted that hypertrophy results in an enlarged organ better adapted to the new functional demand. The changes in gene expression involve proteins responsible for three primary functions of the heart, contraction, relaxation and endocrine function, and this paper will focus on contraction. In the hypertrophied heart, the β-myosin heavy chain (β-MHC) gene is activated and the α-myosin heavy chain (α-MHC) gene is deinduced (review in Lompré, Mercadier and Schwartz, 1990). Because β-MHC is predominant in rat fetal ventricles, the idea that reactivation of a fetal program occurred with hemodynamic overloading was developed. The transition from α-MHC to β-MHC (or isomyosin V1 to V3) results in a slower rate of ATP cycling by myosin, which accounts for the slower velocity of contraction in the hypertrophied fiber. The result is an improved economy of force development that is usually considered adaptive. Less was known concerning the isoactins, except that α-skeletal actin (α-skel act), which is expressed at birth in the rodent ventricle, is down-regulated in adults and that α-skel actin mRNAs reaccumulate with the onset of pressure overload induced hypertrophy (Schwartz, de la Bastie, Bouveret, Oliviero, Alonso and Buckingham, 1986; Izumo, Nadal-Ginard and Mahdavi, 1988). This was the second example of the reactivation of a fetal program by hemodynamic overload.

For the past several years our aim has been to delineate those events within the myocyte that contribute to the above isogene switches and determine if they are analogous to those mechanisms operating during normal cardiac development and aging. The purpose of this paper is therefore 1) to present our recent results concerning the pattern of expression for the two sarcomeric actins, α-skeletal and α-cardiac during cardiac growth, in rat and human hearts, and 2) to describe an efficient run-on assay which can be used to analyze the transcription of the isomyosin and the isoactin genes in nuclei from neonatal and young rat hearts.

Skeletal actin mRNA decreases in the rat heart and increases in the human heart during ontogeny

The respective mRNA levels of α-skel act and α-card act in developing rat and

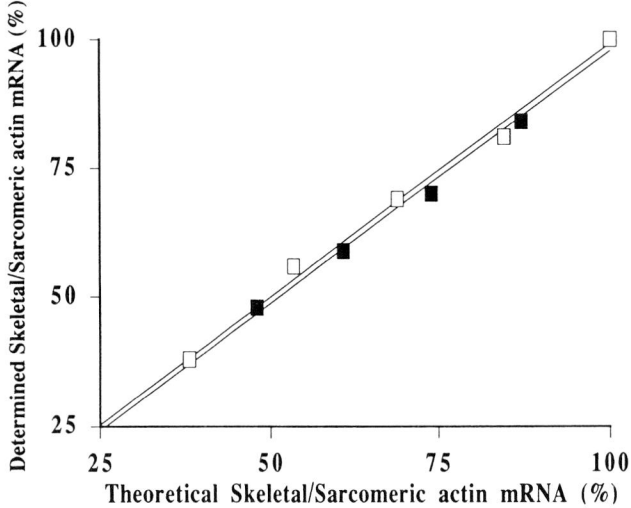

Fig. 1. Actin isomRNA analysis of human cardiac and skeletal muscles by primer extension and dot blot hybridization. Variable amounts of cardiac RNA were mixed with skeletal muscle RNA, and the determined ratios were compared with the theoretical ratios calculated from the value found in the 100% cardiac muscle RNA sample. Reproduced from the *Journal of Clinical Investigation*, 1991, 88, 323-330 by copyright permission of the American Society of Clinical Investigation.

human ventricles were analyzed by primer extension assays. Two 18 nucleotide primers were used, one in the rat and one in the human, each of which corresponded to a sequence common to both isogenes and coding for amino acids 31-37. The lengths of the extended fragments differed between the two species: in the rat, α-skel act and α-card act signals were 186 and 195 bases long, respectively, and in man the sizes were 222 and 174. All this corresponded well with the lengths predicted from the nucleotide sequences. Since this method had not been used before for human RNA and since RNA degradation is often a problem when dealing with human cardiac samples, the method was assessed by comparing the determined α-skel act/total sarcomeric RNA with the theoretical ratio using skeletal and cardiac RNA artificial mixes. In addition, the results were compared with those obtained, on the same RNA mixes, by dot blot analysis with human skeletal and cardiac specific cDNAs kindly given to us by P. Gunning and E. Hardeman. As can be seen in Fig. 1, both techniques show a linear accumulation of α-skel act transcripts with increasing amounts of skeletal muscle RNA and subsequent analyses were performed using only the primer extension assay.

In the rat, it was known from results published by the teams of D. Yaffe (Mayer, Czosnek, Zeelon, Yaffe and Nudel, 1984) and M. Buckingham (Minty, Alonso, Caravatti and Buckingham, 1982), that α-skel act is present in neonates, and Bishopric, Simpson and Ordhal (1987) estimated in one rat heart that it amounted to 50% of the total. We agree with this value and observed that neonatal rat ventricles contain 30 to 40% α-skel act mRNA (Boheler, Carrier, Chassagne, de la

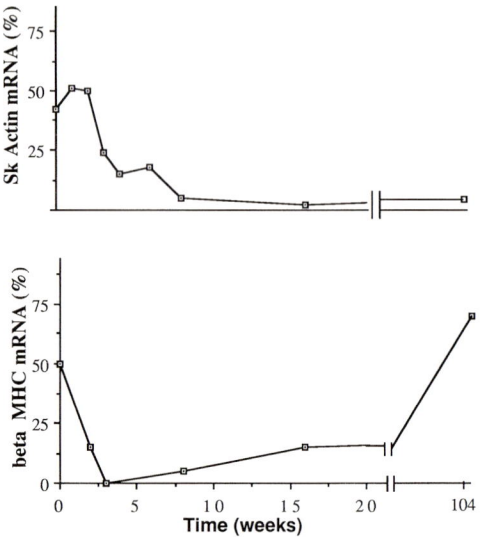

Fig. 2. Time course of β-MHC and α-skel act mRNA accumulations with rat development and aging.

Bastie, Mercadier, and Schwartz, 1991a) (Fig. 2). This relatively high level is maintained much longer than previously thought, decreasing slowly after 3 weeks of age, and only at 2 months of age are the transcripts for cardiac actin almost uniquely present. This figure does not change thereafter, even in 2 year old rats. Comparison of the time course of α-skel act and β-MHC mRNA accumulations during the lifespan of the rat shows an apparent dissociation (Fig. 2). Both β-MHC and α-skel act mRNAs accumulate to high levels at birth, but at 3 weeks, while α-skel act is still high, β-MHC has almost completely disappeared. β-MHC mRNAs slowly reaccumulate in adults and with senescence, and attain levels of as high as 80% of total sarcomeric MHC in two-year old animals (O'Neill, Holbrook and Lakatta, 1991). In contrast, we found that α-skel act is not reinduced during senescence. Thus in the rat ventricle, expression of the multigene families encoding two of the main contractile proteins is not coordinated during normal cardiac growth.

Very few studies had been conducted in the human heart. It was known from the analysis of only two adult hearts analyzed at the mRNA level (Gunning, Ponte, Blau and Kedes, 1983; Bennetts, Burnett and dos Remedios, 1986) and from two others at the protein level (Vandekerckhove, Bugaisky and Buckingham, 1986) that both sarcomeric actin are coexpressed. We have recently completed an analysis of cardiac tissue from 21 control patients (4 fetal, 5 juvenile, 12 adult) and observed that the isoactin pattern changes during cardiac development (Boheler, Carrier, de la Bastie, Allen, Komajda, Mercadier and Schwartz, 1991b). *In utero* (13 to 29 weeks) and in neonatal hearts, α-skel act mRNA is a minor component, representing <20% of total sarcomeric actin, it increases to around 50% during the first decade after birth and becomes the major isoform of adults, representing

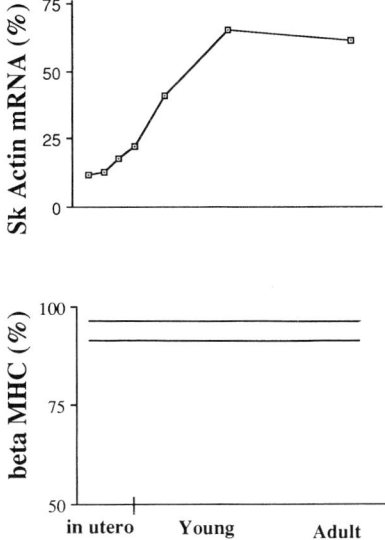

Fig. 3. Time course of β-MHC and α-skel act mRNA accumulations with development and aging, in humans.

approximately 60% of the total (Fig. 3). This pattern is opposite to that seen in the rat heart (see above), and is also different from what is known in chick and mouse. Man is the only species so far studied in which α-skel act mRNA increases in the heart after birth and accumulates in the adult to levels higher than those for cardiac actin mRNAs. Comparisons of the expression of MHC and actin multigene families in man show, as in rat, a complete dissociation (Fig. 3). Skeletal actin is upregulated during development, whereas βMHC more or less roughly remains the same, around 80 to 90%.

Analysis of the transcriptional step of myosin and actin isogenes

The expression of both MHC and actin multigene families is thus species specific and in each species, the two contractile proteins are independently regulated. If one wants to determine what factors control these independent expressions and ultimately control cardiac phenotype, the level at which each of these families is regulated, transcriptionally, post-transcriptionally or translationally, must be determined. With the team of V. Mahdavi, we showed several years ago during rat development and after a pressure overload, that the regulation of βMHC expression is pretranslational (Izumo, Lompré, Matsuoka, Koren, Schwartz, Nadal-Ginard and Mahdavi, 1987). Due to the inherent difficulties of working with cardiac tissue, only one study has been reported from control rats, in which total cardiac transcriptional activity was determined (MacCully and Liew, 1988). Since then we have begun analyzing the step of transcriptional elongation for each of the actin and myosin heavy chain isogenes by run-on assays.

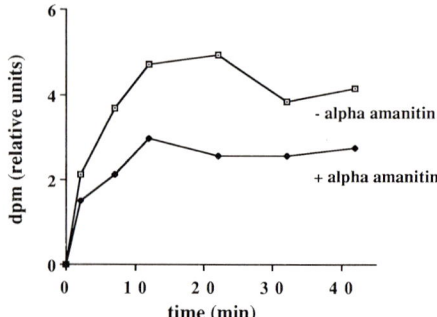

Fig. 4. Kinetics of UTP ^{32}P incorporation by nuclei isolated from 23 day old rat hearts during an *in vitro* transcription assay in the presence or absence of 12 $\mu g/ml$ α-amanitin. Relative units are dpm's (10^5) from 1.0×10^6 nuclei.

We have prepared myocyte nuclei from neonatal rats and from 23-day-old rats, with a modification of the techniques described in Liew, Jackowski, Ma and Sole (1983), Chassagne, Boheler and Schwartz (1991) and Long, Ordahl, and Simpson (1989). The run-on assays were performed with 500 μCi UTP^{32}P essentially as in Konieczny and Emerson (1985), in the presence or absence of α-amanitin. The rate of UTP^{32}P incorporation, as shown in Fig. 4, exhibited a characteristic two-stage response, with an initial phase of rapid incorporation during the first 10-20 minutes and a plateau phase thereafter. Incorporation was diminished by 20 to 30% in the presence of 12 $\mu g/ml$ α-amanitin, which suggested that the isolated nuclei had retained their capacity to elongate nascent mRNAs. Fig. 5 shows the patterns of hybridization of the labeled nascent transcripts to single stranded M13 derived DNA probes specific for each isogene. In neonatal rats both α- and β-MHC and α-card and α-skel actin isomRNAs are transcribed; but in 23 day old rats, β-MHC transcription is largely absent, whereas the α-skel act transcription is only diminished. Both α-MHC and α-card act transcription persist at important levels in 23 day old rats. These results suggest that at least for the β-MHC and α-skel actin genes, independent regulatory factors or differences in the threshold for this transcriptional inactivations, are responsible for their transcriptional activity in cardiac myocytes.

Concluding remarks

One would have liked to hypothesize that the two main contractile proteins of the cardiac sarcomere are highly and mutually coordinated at the transcriptional and/or translational levels during ontogeny and that this coordination would in turn lead to the most efficient state of contraction. The above studies do not support this hypothesis and show in contrast an independent pattern of expression, both in rat and man. The same dissociation occurs during hypertrophy due to increased workloads on the heart. It is now possible to explore the entire sequence of activation or deactivation of contractile proteins isogenes, which should provide

Nuclear Run on Assay

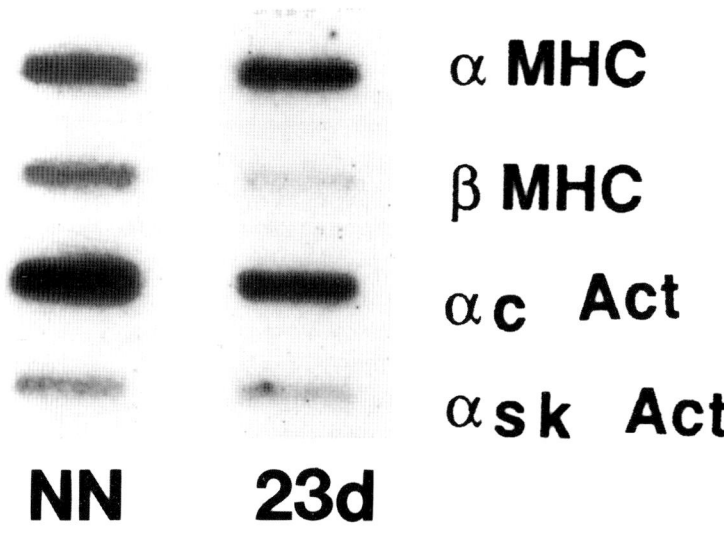

α **MHC**

β **MHC**

α**c Act**

α**sk Act**

NN   23d

Fig. 5. Autoradiogram from nuclear transcription run-on assays using neonatal (NN) or 23 day old (23d) rat myocyte nuclei. The isolated ^{32}P-labeled nascent RNAs from myocyte nuclei were hybridized to nylon membranes containing 20 μg of the indicated single stranded probes derived from M13. α MHC, α myosin heavy chain; β MHC, β myosin heavy chain; $α_c$ Act, cardiac α-actin; $α_s$ Act, skeletal α-actin.

further insights into the search for the regulatory factors that control these isogenes and that finally leads to a given cardiac phenotype.

Supported by the Institut National de la Santé et de la Recherche Médicale and the Association Française de Lutte contre les Myopathies. Drs. K. R. Boheler and L. Carrier are recipients of fellowships from l'Association Française contre les Myopathies.

References

Alpert, N. R. and Gordon, M. S. (1982). Myofibrillar adenosine triphosphatase activity in congestive heart failure. *Amer. J. Physiol.* **202**, 940-946.

Bennetts, B. H., Burnett, L. and Dos Remedios, C. G. (1986). Differential co-expression of α-actin genes within the human heart. *J. Mol. Cell. Cardiol.* **18**, 993-996.

Bishopric, N. H., Simpson, P. C. and Ordhal, C. P. (1987). Induction of the skeletal α-actin gene in α1-adrenoreceptor-mediated hypertrophy of rat cardiac myocytes. *J. Clin. Invest.* **80**, 1194-1199.

Boheler, K. R., Carrier, L., Chassagne, C., De la Bastie, D., Mercadier, J. J. and Schwartz, K. (1991a). Regulation of myosin heavy chain and actin isogenes expression during cardiac growth. *Molec. Cell. Biochem.* **104**, 101-107.

Boheler, K. R., Carrier, L., De la Bastie, D., Allen, P. D., Komajda, M., Mercadier, J. J. and Schwartz, K. (1991). Skeletal actin mRNA increases in the human heart during ontogenic development and is the major isoform of control and failing human hearts. *J Clin Invest.* **88**, 323-330.

Chassagne, C., Boheler, K. R. and Schwartz, K. (1991). Description of an *in vitro* transcription assay in nuclei isolated from control and hemodynamically overloaded rat cardiac myocytes. *Comptes-Rendus Acad. Sci. Paris.* **312**, III, 12.

Gunning, P., Ponte, P., Blau, H. and Kedes, L. (1983). α-skeletal and α-cardiac actin genes are coexpressed in adult human skeletal muscle and heart. *Mol. Cell. Biol.* **3**, 1985-1995.

Izumo, S., Lompré, A. M., Matsuoka, R., Koren, G., Schwartz, K., Nadal-Ginard, B. and Mahdavi, V. (1987). Myosin heavy chain messenger RNA and protein isoform transitions during cardiac hypertrophy. Interaction between hemodynamic and thyroid hormone - induced signals. *J. Clin. Invest.* **79**, 970-977.

Izumo, S., Nadal-Ginard, B. and Mahdavi, V. (1988). Protooncogene induction and reprogramming of cardiac gene expression produced by pressure overload. *Proc. Natl. Acad. Sci. USA.* **85**, 339-343.

Konieczny, S. F. and Emerson, C. P. (1985). Differentiation, not determination, regulates muscle gene activation: transfection of troponin I genes into multipotential and muscle lineages of 10T1/2 cells. *Mol Cell Biol.* **5**, 2423-2432.

Liew, C. C., Jackowski, G., Ma, T. and Sole, M. J. (1983). Nonenzymatic separation of myocardial cell nuclei from whole heart tissue. *Amer J Physiol.* **244**, C3-C10.

Lompré, A. M., Mercadier, J. J. and Schwartz, K. (1990). Changes in gene expression during cardiac growth. *Intern. Rev. Cytol.* **124**, 137-186.

Long, C. S., Ordhal, C. P. and Simpson, P. C. (1989). α1-adrenergic receptor stimulation of sarcomeric actin isogene transcription in hypertrophy of cultured rat heart muscle cells. *J. Clin. Invest.* **83**, 1078-1082.

MacCully, J. D. and Liew, C. C. (1988). RNA transcription in myocardial-cell nuclei during post-natal development. *Biochem. J.* **256**, 441-445.

Mayer, Y., Czosnek, H., Zeelon, P. E., Yaffe, D. and Nudel, U. (1984). Expression of the genes coding for the skeletal muscle and cardiac actins in the heart. *Nucleic Acids Res.* **12**, 1087-1100.

Minty, A. J., Alonso, S., Caravatti, M. and Buckingham, M. (1982). A fetal skeletal muscle actin mRNA in the mouse and its identity with cardiac actin mRNA. *Cell* **30**, 185-192.

O'Neill, L., Holbrook, N. and Lakatta, E. G. (1991). Progressive changes from young adult age to senescence in mRNA for rat cardiac myosin heavy chain genes. *Cardioscience* **2**, 1-5.

Schwartz, K. (1990). Phenoconversion and mechanogenic transduction of the mammalian heart. *Medecine/Science* **6**, 664-673.

Schwartz, K., De la Bastie, D., Bouveret, P., Oliviero, P., Alonso, S. and Buckingham, M. (1986). α-skeletal muscle actin mRNA's accumulate in hypertrophied adult rat hearts. *Circ. Res.* **59**, 551-555.

Vandekerckhove, J., Bugaisky, G. and Buckingham, M. (1986). Simultaneous expression of skeletal muscle and heart actin proteins in various striated muscle tissues and cells. *J. Biol. Chem.* **61**, 1838-1843.

Printed in Great Britain © Society for Experimental Biology 1992   273

CARBONIC ANHYDRASE 3 (CA3), A MESODERMAL MARKER

YVONNE H. EDWARDS[1], SUSAN TWEEDIE[1], NICK LOWE[2] and GARY LYONS[3]

[1]MRC Human Biochemical Genetics Unit, The Galton Laboratory, University College London, Wolfson House, 4 Stephenson Way, LONDON NW1 2HE
[2]Biochemistry Department, University College London, Gower Street, LONDON
[3]Department of Molecular Biology, URA CNRS 1148, Pasteur Institute, 28 Rue du Docteur Roux, F-757244, Paris Cedex 15, FRANCE

Abstract

Carbonic anhydrase 3 (CA3) is an abundant muscle protein characteristic of adult type 1, slow-twitch, fibres. The protein plays an important role in facilitated CO_2 diffusion and diverse processes involving H^+ and HCO^-_3 transport. Nucleotide sequence comparisons have identified putative promoter and enhancer regions in the 5′ flanking sequences of the CA3 gene. Functional assays show that 2.8kb of 5′ flanking sequence efficiently promotes transcription of a reporter gene in a muscle specific manner. Removal of sequences 5′ to −722bp leads to a major loss of activity and this result implies that the proximal promoter region which includes a GArG box and four potential MyoD1 binding sites is not adequate for maximal transcription. The longest CA3 promoter construct is also active in 10T1/2 cells, which are precursor mesodermal cells and do not normally express CA3. In situ hybridization to mRNA in developing mouse embryos reveals a pattern of expression in myotomes and pre-muscle masses of the limb buds which is consistent with the regulation of CA3 by myogenic determination factors. These studies also showed that CA3 expression is not confined to cells of the muscle lineage since it is expressed in primitive mesoderm prior to the onset of myogenesis. Later in embryogenesis CA3 defines a subset of mesodermal cell types which includes not only skeletal muscle but also notochord and adipocytes.

Introduction

The carbonic anhydrases (CA) are a family of related zinc metallo-enzymes that catalyze the interconversion of CO_2 and HCO_3^-, $[CO_2 + H_2O = HCO_3^- + H^+]$. This activity is very ancient and may be essential to life since CA is found not only in all animal species but also in plants, blue-green algae and bacteria (Tashian, 1989).

At least seven mammalian CA genes have been identified whose products show characteristic kinetic properties and tissue distribution [for review see Tashian, 1989; Fernley, 1988; Edwards, 1989]. The carbonic anhydrase isozymes, (CA1-

Key words: carbonic anhydrase, myogenesis, notochord, adipocytes, 10T1/2 cells, in situ hybridization.

CA7, alternative in previous publications CAI-CAVII) are not only important in the transport of carbon dioxide but also in diverse ion transport and secretory functions where bicarbonate ions and protons are essential (for recent review see Dodgson, Tashian, Gros and Carter, 1991).

Carbonic anhydrase 3 (CA3) is a prominent sarcoplasmic protein, accounting for around 15% and 50% of soluble protein in mature human soleus (Jeffery, Edwards and Carter, 1980) and rat soleus respectively (Shiels, Jeffery, Wilson and Carter, 1984). This muscle enzyme has been implicated in facilitated CO_2 transport (Gros, Ganhoff, Schied, Siffert, Teske and Kruger, 1984), maintenance of acid/base homeostasis by the production of bicarbonate anions and protons, and may play a role in stabilising optimal conditions for oxidative phosphorylation in type 1 fibres where glycolytic capacity is low (Fremont, Lazure, Tremblay, Chretien and Rogers, 1987). Trace amounts of CA3 have been reported in red cells, lungs and at several other sites (Tashian, 1989) and exceptionally amongst mammals CA3 is also abundant in male rodent perivenous hepatocytes (Jeffery, Wilson, Mode, Gustafsson and Carter (1989)).

While CA3 is abundant in skeletal muscle it is present only at low levels or is absent from smooth and cardiac muscle (Jeffery, Edwards and Carter, 1980; Lyons, Buckingham, Tweedie and Edwards, 1991). CA3 is especially abundant in type 1, slow twitch fibres but present at only low levels in type 2A and 2B fibres (Brownson, Isenberg, Brown, Salmons and Edwards, 1988). Studies of CA3 protein and mRNA in human fetal muscle (Lloyd, MacMillan, Hopkinson and Edwards, 1986) and in a rodent myogenic cell lines (Tweedie, Morrison, Charlton and Edwards, 1991) indicate that CA3 expression occurs unusually early in myogenesis, prior to the differentiation of adult muscle fibre types. It appears that this early expression of CA3 in myogenesis undergoes later modulation at the time of adult fibre type differentiation so that expression is confined to type 1 fibres (Lyons et al., 1991). Indeed CA3 has proved a useful marker of fibre type switching in studies involving either electrostimulation (Brownson et al., 1988) or denervation (Carter, Wistrand, Isenberg, Askmark, Jeffery, Hopkinson and Edwards, 1988) of rabbit and rat muscle.

In this paper we describe how DNA sequence comparisons, promoter function studies using cultured cells, and in vivo expression studies with in situ hybridization to RNA, begin to throw some light on the mechanisms and factors which regulate the expression of the CA3 gene.

The CA3 gene promoter

The CA3 gene lies in a cluster together with the CA1 and CA2 genes on the long arm of human chromosome 8 (8q22) These genes have been mapped using pulse field gel electrophoresis (Lowe, Edwards, Edwards and Butterworth, 1991) and lie in the order CA2, CA3 and CA1 with CA2 and CA3 transcribed in the same direction and away from CA1. CA3 and CA2 are relatively close together with a gap of about 20kb between the 3′ end of CA3 and the 5′ end of CA2. CA1 lies further away with the 5′ end about 80kb from the 5′ end of CA3.

The 5′ proximal promoter of the human CA3 gene has been structurally characterised (Edwards, Charlton and Brownson, 1988; Edwards, 1991). It exhibits a number of sequence motifs considered to be characteristic of "housekeeping" genes and shares some of these in common with the CA2 gene, which unlike CA3 is ubiquitously expressed. There is a 41bp sequence spanning the TATA box which shows 90% homology between CA2 and CA3 and fairly good alignment to a consensus found in this region in a number of housekeeping genes. At −173bp in the CA3 gene, an Sp1 site is flanked by the repeated element CACCTC, a motif configuration which is conserved in the CA2 gene at −110bp (Shapiro, Venta and Tashian, 1987). The CA3 proximal promoter, like that of CA2, is embedded in a methylation free CpG cluster (HTF island, Bird, 1986). These CpG rich sequences occur at the 5′ ends of constitutively expressed genes and are not commonly associated with genes such as CA3, which show a limited tissue expression. It seems unlikely that the CpG rich island plays any role in muscle specific expression and one possibility is that tissue specificity is achieved through a separate mechanism in which negative control has a major role (Colantuoni, Pirozzi, Blance and Cortese, 1987).

We have searched the 2.8kb sequence immediately flanking the 5′ end of the human CA3 gene for potential regulatory sequences homologous to those identified as important for tissue specific regulation of other muscle genes, Table 1. In addition to the basal regulatory elements there is a 200bp sequence (−1520bp to −1320bp relative to the cap site) showing homology to the SV40 early enhancer (Zenke, Grundstrom, Mattes, Wintzerith, Schatz, Wildeman and Chambon, 1986). Motifs from this enhancer have been found in the immediate area of tissue specific enhancers in vertebrate genes, in particular within the MCK enhancer (Jaynes, Johnson, Buskin, Gartside and Hauschke, 1988). There is also a single copy of a CArG box [CC(AT rich)$_6$GG] (Miwa and Kedes, 1987) at −57bp; this position is fairly typical for these elements and although they often occur more than once, a single CArG box at −91bp has been implicated in human dystrophin expression (Klamut, Gangopadhyay, Worton and Ray, 1990). The CA3 CArG box is immediately 5′ and adjacent to a 14bp sequence which shows good homology with the consensus sequence (CANNTG) proposed for muscle specific binding of myogenic determination factors from the MyoD1, myf6, myf5, myogenin family (Weintraub et al., 1991). The combined CArG box and MyoD1 motifs in CA3 are close to an Sp1 binding site at −77bp and this particular configuration is reminiscent of that described for the cardiac α actin promoter where each of the three elements has been shown to be functionally important in muscle expression (Sartorelli, Webster and Kedes, 1990). There are four other copies of the MyoD1 binding site one of them lying within the upstream SV40 enhancer-like region of the CA3 gene. These muscle specific sequences are not found in the promoter of the ubiquitously expressed CA2 gene or in the erythroid or colon promoters of the CA1 gene.

Table 1. *Putative transcription regulatory elements in the 5' region of the human CA3 gene*

Element	Position	Sequence
TATA box	- 22bp	CATAAA
CAAT box	- 92bp	CCAAT
SP1 binding sites	- 74bp	CCGCCC
	- 168bp	CCGCCC
SV40 enhancer	-1518bp	several motifs within 200bp
CArG[CC(A+T)$_6$GG] box	- 57bp	CCTAATAAGG
MyoD binding sites*	- 46bp	CA T GCAAGTGT G CG
	- 155bp	T CGACAGCTGTCCC
	- 317bp	GAAGCAGGTG AGG G
	- 490bp	CG C ACCAAGTGTCC A
	-1325bp	ATTC CACATGCCT T

*Underlined lettering indicates the core sequence CANNTG (Murre et al., 1989) and boxes delineate homology with the 14bp consensus sequence proposed for muscle specific binding of factors encoded by the MyoD1, myf5, myf6, myogenin family (Weintraub et al., 1991).

In order to begin to define cis-acting control elements required for the expression of the CA3 by a functional approach, various segments of the 5' flanking promoter of the human CA3 gene were subcloned into the pGCAT-A vector upstream of the CAT coding sequence and the modified plasmids transfected into CA3-expressing cells (Tweedie, Morrison, Charlton and Edwards, 1991). The longest construct contained 2810bp of human CA3 promoter, two others were shorter at the 5' end, containing only 715bp and 254bp of flanking sequence, Table 2. The same pattern of promoter function was observed in three myogenic cell lines (G8, C2C12, 23A2). The longest CA3 promoter, which includes the enhancer-like region, directs CAT expression moderately well at an efficiency of about 17-25% of that shown by the strong constitutive β-actin promoter. The efficiency of two shorter CA3/CAT constructs was very similar to each other but considerably less, on average about 10%, of that shown by the longest construct, Table 2. Neither the longest promoter construct nor the shorter constructs were functional in the non-CA3 expressing cell types, 1RE3 or HeLa.

Table 2. *CAT activity in cells C2C12, 23A2 and 10T/12 transfected with CA3 promoter/CAT plasmids*

| | CA3/CAT plasmid | % acetylated products | | |
		−2817	−722	−261
C2C12		17.0	1.0	1.0
23A2		25.0	1.5	1.5
10T1/2		25.0	2.5	4.0

The proportion of unacetylated [C14] chloramphenical converted to acetylated products is expressed as percent relative to the pβAcCAT control set at 100%. Data taken from Tweedie et al. (1991). The 5′ extent of CA3 5′ flanking sequence included in each plasmid construct is indicated (bp) relative to the normal initiation site. All terminated at the 3′end at −7bp.

These results appear to be relatively straightforward and support the simple proposal that the longest CA3 construct contains all the sequences that are necessary to direct muscle specific transcription. Furthermore if the presence of the CpG island across the CA3 promoter requires the binding of a negative acting element in cells not expressing CA3, then these must also be present within the 2.8kb sequence flanking the 5′ end of the CA3 gene. The relatively very low efficiency of the two shorter constructs leads to the conclusion that the proximal sequences which include the CArG box and a number of MyoD1 binding sites are not able to function at maximal levels without sequences upstream of −722bp.

Results from transfection studies using 10T1/2 fibroblasts suggest that the regulation of CA3 may be more complex than proposed in this simple model. 10T1/2 cells do not express CA3 as judged by both protein and mRNA analyses (Tweedie, Morrison, Charlton and Edwards, 1991) yet the pattern of transcriptional activity observed after transfection with the various CA3 promoter/CAT constructs is very similar to that observed after transfection of myogenic cells, Table 2. This observation was unexpected and implies that the suppression of CA3 expression in 10T1/2 cells is mediated by factors which differ from those which operate in HeLa and 1RE3 cells and that the sequences necessary for binding such factors are absent from the CA3/CAT constructs. 10T1/2 cells are thought to be mesodermal stem cells blocked in their differentiation; furthermore they can be induced to become myoblasts, which express CA3 (10T1/2 cells were the precursors of the myogenic line 23A2, Konieczny and Emerson, 1984) by low concentration 5-azacytidine (Taylor and Jones, 1979). This tempts the speculation that the transcriptional activity of the artificial CA3/CAT plasmid in 10T1/2 cells might depend on the presence of a developmental-stage-specific factor, and that this is related to the predisposition of 10T1/2 cells to become muscle cells (and to express CA3).

These results are preliminary and a more detailed functional analysis of the 5′ flanking sequences and other regions of the CA3 gene, is required before firm conclusions can be drawn. It may be relevant that similar results have emerged from a study of the erythroid promoter function of the CA1 gene. In these studies a 5' CA1 promoter/reporter gene plasmid was found to be transcriptionally active, not only in the erythroid cell lines HEL and MEL, which express CAI, but also in K562 cells which do not normally express CA1 and are judged to be akin to fetal erythroid cells on the basis of their haemoglobin phenotype. The action of a developmental-stage-specific factor has been invoked to explain these results (Sowden, 1991 and Brady, Sowden, Edwards, Lowe and Butterworth, 1989).

CA3 expression in early development

CA3 in muscle

In contrast to many other muscle genes, whose transcripts and protein products accumulate only after the formation of multinucleate myotubes, CA3 is expressed at moderate levels in cultured proliferating myoblasts prior to fusion. Some rodent myogenic cell lines show expression of CA3 which seems inappropriately high, reaching levels greater than found in 9 week human fetal muscle. CA3 mRNA is also present in vivo, in unfused myoblasts of mouse developing skeletal muscle. However, in situ hybridization to mRNA (Lyons et al., 1991) does not detect CA3 transcripts in vivo prior to the expression of other myogenic markers such as α-actin and myogenin. CA3 mRNAs are detected in the myotomes of the somites after myf5 and myogenin transcripts and approximately at the same time as MyoD1 mRNAs. Similarly, in developing limb buds CA3 transcripts are detected at 11.5 days a little after myf5 (Ott, Bober, Lyons, Arnold and Buckingham, 1991), and simultaneously with myogenin and MyoD1 mRNAs. This sequence of expression of the myogenic differentiation factors, viz: before or at the same time as CA3, (see summary Table 3), suggests that these factors may regulate the expression of CA3 in early myogenesis.

Table 3. *Transcript accumulation* of CA3, myogenic regulatory sequences and other muscle genes during myogenesis in the mouse*

Days	Myf-5	Myogenin	MyoDI	Cardiac α actin	MHC	CA3
			Rostral somites			
8.5	+	+ +	−	+	−	−
9	+ +	+ + +	−	+	−	−
9.5	+ +	+ + +	−	+ +	+	+
10.5	+ +	+ + +	+ +	+ +	+ +	+ +
11.5	+ +	+ + +	+ + +	+ + +	+ + +	+ + +
			Limb buds			
9	−	−	−	−		−
9.5	−	−	−	−		−
10.5	+	−	−	−		−
11.5	+	+ +	+ +	+ +		+ +

*data taken from Lyons et al. (1991), Sassoon, Garner and Buckingham (1988), Sassoon et al. (1989) and Ott et al. (1991).

CA3 is expressed in all primary muscle fibres in the mouse embryo but after 15 days p.c. expression becomes delimited to those fibres which will become slow twitch fibres in the adult mouse. The timing of this delimitation of expression is very similar to that shown by other markers of slow twitch fibres such as ventricular myosin light chain 1 and myosin heavy chain B. The molecular signals which initiate and maintain fibre type gene regulation are not yet understood but the CA3 promoter is clearly responsive to such signals. CA3 is also expressed in the muscle fibres encircling the fetal ureter. This muscle is often regarded as smooth muscle, and thus is unique in representing the only smooth muscle site for CA3 expression It seems most likely that CA3 has some role in the specialized function of this muscle, since there is no evidence that bicarbonate transport or H^+ secretion occurs in the ureter.

CA3 in non-muscle tissues

In situ hybridization studies of CA3 in the mouse revealed a feature of its expression which may be important with regard to its regulation. CA3 is not confined to cells committed to the muscle lineage. This gene is expressed in primitive mesoderm prior to the onset of myogenesis, and at later stages of embryogenesis its expression defines a subset of mesodermal cell types which includes not only skeletal muscle but also notochord and adipocytes.

During late gastrulation in the mouse CA3 is expressed in lateral and central precursor mesodermal cells destined to give rise to somites and notochord. It is continuously expressed in the developing and definitive notochord but is apparently down-regulated in somitic mesoderm and not subsequently expressed in the myotomes until 10 days p.c. These observations suggest that CA3 undergoes at least two phases of expression involving one set of factors in precursor mesoderm and another in the somites. There may also a be distinct mechanism operating in the notochord.

CA3 mRNA is more abundant in the notochord than in muscle at all stages of fetal development, Fig. 1, but its physiological function in this site is unknown. An attractive possibility is that CA3 plays a role in ion transport and fluid secretion. The importance of one or more of the carbonic anhydrase isoforms in ion movement and fluid secretion for example, in the secretion of aqueous humour and cerebrospinal fluid, is well documented (see for review, Maren, 1988). Various lines of evidence (for references see Lyons et al., 1991) show that in early development the notochord is critical for the organisation of the mesoderm to form somites and that this is probably mediated by a diffusable growth factor secreted by the notochordal cells (McCaig, 1986); During early tailbud stages of *Xenopus* the internal hydrostatic pressure of the notochord increases two to three fold by osmotic activity (Adams, Keller and Koehl, 1990); later in development, during the formation of the nucleus pulposus the notochord cells produce large quantities of a semi-fluid mucoid matrix (Crelin, 1981). It seems very likely that CA3 could have a role in these events, ranging from the buffering of the cellular enviroment to the transport of ions in the secretory/osmotic processes.

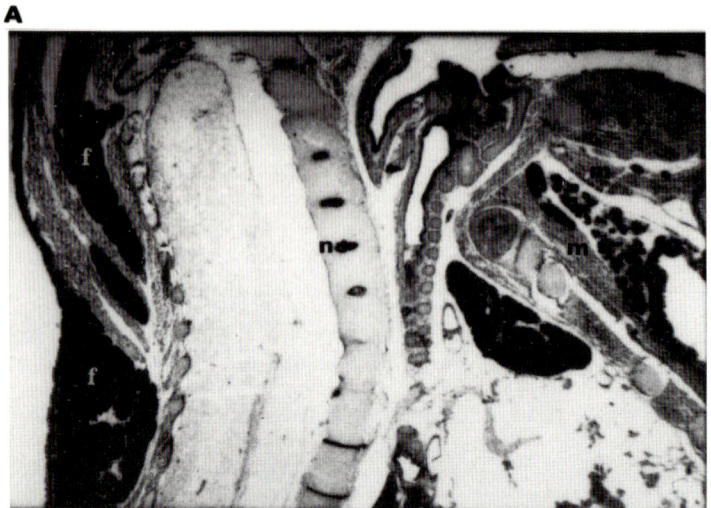

Fig. 1. CA3 mRNA expression in muscle, adipocytes and notochord in the mouse embryo detected using a radio-labelled antisense RNA probe. A. Phase-contrast and B dark-field, micrographs of a sagittal section of a 17.5 day embryo in the cervical region. Strong signals can be detected in muscle (m), cervical brown fat (f), and nucleus pulposus (notochord, n).

In addition to muscle and notochord, CA3 is also expressed in a limited number of other tissues of mesodermal origin (Lyons et al., 1991). CA3 is particularly abundant in brown fat both in the embryo and infant mouse, Fig. 1. A possible role for CA3 in this tissue, is the augmentation of fatty acid metabolism by facilitating the synthesis of oxaloacetate and citrate and their incorporation into fatty acids. At present there is no evidence to indicate how tissue specific regulation is achieved in either notochord or adipocytes. It is known that the myogenic determination factors are confined to skeletal muscle cells and and thus cannot be implicated in CA3 regulation in non-muscle cells.

Two points are worthy of attention in this context. Firstly, the pattern of CA3 expression in prenotochordal mesoderm and notochord is similar but not identical to that described for the T (Brachury) gene (Wilkinson, Bhatt and Hermann, 1990). It has been proposed that the T gene product is required to maintain the notochord and its precursors, and that it has a critical role upstream of a signalling pathway which specifies other mesodermal derivatives (Rashbass, Croke, Hermann and Beddington, 1991). Evidence is accumulating that the T protein acts as a transcription factor. Whether CA3 expression is regulated directly or indirectly by T is a question for early investigation.

Secondly, it is intriguing to recall that when 10T1/2 cells are treated with 5-azacytidine the two principal expressed mesenchymal phenotypes are myoblasts and adipocytes (Taylor and Jones, 1979). This implies that 10T1/2 cells are committed further towards these two particular mesodermal derivatives than towards others and leads us to speculate whether such predisposition involves a common determination factor(s). If this is so, could it be related to the pattern of expression of CA3 in embryogenesis and the ectopic use of the exogenous human CA3 promoter construct in 10T1/2 cells?

References

Adams, D., Keller, R. and Koehl, M. (1991). The mechanics of notochord elongation, staightening and stiffening in the embryo of Xenopus laevis. *Development*, **110**, 115-130.

Bird, A. P. (1986). CpG-rich islands and the function of DNA methylation. *Nature* **321**, 209-213.

Brady, H. J. M., Sowden, J. C., Edwards, M., Lowe, N. and Butterworth, P. H. W. (1989). Multiple GF-1 binding sites flank the erythroid specific transcription unit of the human carbonic anhydrase I gene. *FEBS Letters* **257**, 451-456.

Brownson, C., Isenberg, H., Brown, W., Salmons, S. and Edwards, Y. (1988). Changes in skeletal muscle gene transcription induced by chronic stimulation. *Muscle Nerve* **11**, 1183-1189.

Carter, N. D., Wistrand, P. J., Isenberg, H., Askmark, H., Jeffery, S., Hopkinson, D. and Edwards, Y. (1988). Induction of carbonic anhydrase III mRNA and protein by denervation of rat muscle. *Biochem. J.* **256**, 147-152.

Colantuoni, V., Pinnozz, A., Blance, C. and Cortese, R. (1987). Negative control of liver specific gene expression: cloned retinol binding-protein gene is repressed in HeLa cells. *EMBO J.* **6**, 631-636.

Crelin, E. (1981). Development of the musculoskeletal system. *CIBA Clinical Symposia* **33**, 2-36.

Dodgson, S. J., Tashian, R. E., Gros, G. and Carter, N. D. (eds) (1991). The Carbonic Anhydrases. Plenum Publishers.

Edwards, Y. (1989). Structure and expression of mammalian carbonic anhydrases. *Biochem. Soc. Trans.* **18**, 171-175.

Edwards, Y. (1991). In the *Carbonic Anhydrases: Cellular physiology and Mol. Genet. Edition* Structure and expression of the carbonic anhydrase III gene. 215-224.

Edwards, Y., Charlton, J. and Brownson, C. (1988). A non-methylated CpG-rich island associated with the human muscle-specific carbonic anhydrase III gene. *Gene* **71**, 473-481.

Fernley, R. (1988). Non-cytoplasmic carbonic anhydrases. *TIBS* **13**, 356-359.

Freemont, P., Lazure, C., Tremblay, R. R., Chretien, M. and Rogers, P. A. (1987). Regulation of carbonic anhydrase III by thyroid hormone: opposite modulation in slow- and fast-twitch skeletal muscle. *Biochem. Cell Biol.* **65**, 790-797.

Gros, G., Ganhoff, F., Scheid, P., Siffert, W., Teske, W. and Kruger, D. (1984). Concentration, properties and functional significance of skeletal muscle carbonic anhydrase III. *Pflügers Arch.* **400**, R58.

Jaynes, J., Johnson, J., Buskin, J., Gartside, C. and Hauschka, S. (1988). The muscle creatine kinase gene is regulated by multiple upstream elements, including a muscle-specific enhancer. *Mol. Cell Biol.* **8**, 62-70.

Jeffery, S., Edwards, Y. and Carter, N. (1980). Distribution of CAIII in fetal and adult human tissue. *Biochem. Genet.* **18**, 843-849.

Jeffery, S., Wilson, C., Mode, A., Gustafsson, J. and Carter, N. (1986). Effects of hypophysectomy and growth hormone infusion on rat hepatic carbonic anhydrases. *J. Endocrinol.* **110**, 123-126.

Klamut, H. J., Gangopadhyay, S. B., Worton, R. J. and Ray, P. N. (1990). Molecular and functional analysisi of the muscle specific promoter region of the Duchenne Muscular dystrophy gene. *Mol. Cell. Biol.* **10**, 193-205.

Konieczny, S. and Emerson, C. (1984). 5′ azacytidine induction of stable mesodermal cell lineages from 10T1/2 cells; evidence for regulatory genes controlling determination. *Cell*, **38**, 791-800.

Lloyd, J., MacMillan, S., Hopkinson, D. and Edwards, Y. (1986). Nucleotide sequence and derived amino acid sequence of a cDNA encoding human muscle carbonic anhydrase. *Gene* **41**, 233-239.

Lowe, N., Edwards, Y. H., Edwards, M. and Butterworth, P. H. W. (1991). Physical mapping of the human carbonic anhydrase gene cluster on chromosome 8. *Genomics* **10**, 882-888.

Lyons, G., Buckingham, E., Tweedie, S. and Edwards, Y. (1991). Carbonic anydrase III, an early mesodermal marker, is expressed in embryonic mouse skeletal muscle and notochord. *Development.* **111**, 233-244.

Maren, T. (1988). The kinetics of HCO_3^- synthesis related to fluid secretion, pH control and CO_2 elimination. *A. Rev. Physiol.* **50**, 695-717.

McCaig, C. (1986). Myoblasts and notochord influence the orientation of somitic myoblasts from *Xenopus laevis. J. Embryol. exp. Morph.* **93**, 121-131.

Miwa, T. and Kedes, L. (1987). *Mol. Cell. Biol.* **7**, Duplicated CArG box domains have positive and mutually dependent regulatory roles in expression of the human cardiac actin gene. 2803-2813.

Ott, M., Bober, E., Lyons, G., Arnold, H. and Buckingham, M. (1991). Early expression of the myogenic regulatory gene, *myf*-5, in precursor cells of skeletal muscle in the mouse embryo. *Development III* **4**, 1097-1107.

Rashbass, P., Cooke, L. A., Herrmann, B. G. and Beddington, R. S. P. (1991). A cell autonomous function of *Brachyury* in T/T embryonic stem cell chimaeras. *Nature.* **353**, 348-350.

Sartorelli, V., Webster, K. A. and Kedes, L. (1990). Muscle-specific expression fo the cardiac α-actin gene requires MyoD1, CArG-box binding factor, and Sp1. *Genes and Develop.* **4**, 1811-1822.

Sassoon, D., Garner, I. and Buckingham, M. (1988). Transcripts of α-cardiac and α-skeletal actins are early markers for myogenesis in the mouse embryo. *Development* **104**, 155-164.

Sassoon, D., Lyons, G., Wright, W. E., Lin, V., Lassar, A., Weintraub, H. and Buckingham, M. (1989). Expression of two myogenic regulatory factors myogenin and MyoD1 during mouse embryogenesis. *Nature* **341**, 303-307.

Shapiro, L. H., Venta, P. J. and Tashian, R. E. (1987). Molecular analysis of G+C-rich upstream sequences regulating transcription of the human carbonic anhydrase II gene. *Mol Cell Biol.* **7**, 4589-4593.

Shiels, A., Jeffery, S., Wilson, C. and Carter, N. (1984). Radioimmunoassay of carbonic anhydrase III in rat tissues. *Biochem J.* **218**, 281-284.

Sowden, J. (1991). The regulation of expression of the carbonic anhydrase I gene. *University of London Thesis.*

Tashian, R. (1989). The carbonic anhydrases: widening perspectives on their evolution, expression and function. *BioEssays* **10**, 186-192.

Taylor, S. M. and Jones, P. A. (1979). Multiple new phenotypes induced in 10T1/2 and 3T3 cells treated with 5-azacytidine. *Cell* **17**, 771-779.

Tweedie, S., Morrison, K., Charlton, J. and Edwards, Y. H. (1991). CAIII a marker for early myogenesis: analysis of expression in cultured myogenic cells. *Somatic Cell and Mol. Genet.* **17**, 215-228.

Weintraub, H., Davis, R., Tapscott, S., Thayer, M., Krause, M., Benezra, R., Blackwell, K., Turner, D., Rupp, R., Hollenberg, S., Zhuang, Y. and Lassar, A. (1991). The *myo*D Gene Family: Nodal point during specification of the muscle cell lineage. *Science* **251**, 761-766.

Wilkinson, D., Bhatt, S. and Herrmann, B. (1990). Expression pattern of the mouse T gene and its role in mesoderm formation. *Nature.* **343**, 657-658.

Zenke, M., Grundstrom, T., Mattes, H., Wintzerith, M., Schatz, C., Wildeman, A. and Chambon, P. (1986). Multiple sequence motifs are involved in SV40 enhancer function. *EMBO J.* **5**, 387-397.

Printed in Great Britain © *Society for Experimental Biology 1992*

A MOLECULAR APPROACH TOWARDS THE UNDERSTANDING OF EARLY HEART DEVELOPMENT: AN EMERGING SYNTHESIS

ANTOON F. M. MOORMAN and WOUTER H. LAMERS

Department of Anatomy and Embryology, AMC, University of Amsterdam,
Meibergdreef 15, 1105 AZ, Amsterdam (The Netherlands)

Summary

In the past decade we have made an inventory of the changing three-dimensional patterns of expression of a number of key proteins involved in contraction, energy metabolism and conduction in developing and adult chicken, rat, bovine and human hearts. These integrated morphological and immunohistochemical studies were complemented with electrophysiological studies in developing chicken hearts and have resulted in a preliminary model of heart development, that explains how the embryonic heart can function without valves and without an atrioventricular conduction system that is indispensable for the adult heart.

Cardiomyocyte-specific proteins are first expressed in the cardiogenic plate when 6 somites have developed, while electrical activity becomes detectable only slightly later. Development proceeds as follows:

1. Upon its formation 'primary' myocardium is characterised by anteroposterior gradients in gene expression. Therefore cardiogenesis resembles many other developmental processes in the embryo. It serves as source for endocardial cells and cells specialized in mechanical contraction and in impulse generation/conduction supporting the view that a single population of cells (the 'primary' myocardium) serves as a precursor for these distinct cell types.

2. 'Primary' myocardium is characterized by the expression of α and β myosin, acetylcholinesterase and the absence of fast sodium channels and of connexin 43. It has a peristaltoid contraction form due to a relatively slow propagation of the impulse.

3. In the looping stage, two cardiac segments appear due to the development of atrial and ventricular working myocardium, that is characterized by the expression of either α or β myosin, connexin 43, fast sodium channels, the dissappearance of acetylcholinesterase and by a relatively fast conduction. The atrial and ventricular segments remain initially flanked by 'primary' myocardium with slow conduction, allowing the embryonic heart to function without valves.

Most parts of the 'primary' myocardium, such as the outflow tract will disappear, but remnants are found in the sinuatrial and atrioventricular nodes.

Key words: heart development, myosin, acetylcholinesterase, connexin 43, conduction.

Introduction

Molecular analysis of cardiac development is currently advancing along two major fronts. On the first front, as pointed out in detail by P. Barton in this volume, the regulatory genes that are expected to be involved in the onset and maintenance of the cardiac myogenic program, are being identified and characterized. The second major approach, to which this review is limited, is largely focused on the identification of phenotypically homogeneous cellular compartments within the embryonic heart, that may predict testable models of cardiac function in this very early stage of development. This approach has been initiated by our laboratory during the last decade by the three-dimensional mapping of a number of enzymes that can be directly related to contraction, energy metabolism and conduction. In addition, some electrophysiological features of the developing heart that parallel the developmental changes in the molecular phenotype, have been mapped. The combined results have led to the emergence of a general model of the development of form and function of the heart, that appears to apply both to avian and mammalian species including man, and that will be presented in this brief survey. Due to its limited size many important contributions have been omitted from the references, but we hope that we have provided a balanced selection from which further references may be obtained. A more detailed review is given in (Lamers et al. 1991).

Basic concepts and problems

The adult vertebrate heart is a chamber pump with contractile myocardial walls. In the two most advanced vertebrate classes, birds and mammals a perfect double-circuited heart has evolved. It consists of four chambers: two atria and two ventricles that are electrically insulated from the atria. Rhythmical contractions of the walls move the blood out of the chambers. One-way valves that are present at the atrioventricular junction, and in the aorta and pulmonary trunc, are essential to prevent backflow of blood. Contraction begins in the sinus node or pacemaker of the heart that is positioned at the entrance of the superior vena cava in the right atrium. The impulse, which is rapidly spread over the atrium, is transmitted by the atrioventricular node with a delay to the ventricles. Within the ventricle the ventricular conduction system, consisting of the atrioventricular bundle and bundle branches, ensures simultaneous contraction of both ventricles.

The four-chambered heart develops from a single tubular heart. The embryonic heart is initially a peristaltic pump that moves blood ahead as a result of a unidirectional wave of myocardial contractions along the heart tube (Patten, 1949). Subsequently, it develops via a single-tubed chamber pump, without valves into the adult dual-circuited heart with valves. Similar to the adult condition, activation of the embryonic heart is initiated in the inflow region of the heart. Furthermore, the atrial and ventricular chambers contract sequentially (Patten, 1949; van Mierop, 1967), indicating a coordination of the contraction. This, in turn, requires a coordinated regulation of the propagation of the depolarisation wave over the heart tube, as is reflected in the developing electrocardiogram (van Mierop, 1967), the presence of which may have erroneously suggested the existence of an adult type of conduction system in the early embryonic heart. As a result research has focused primarily on the early recognition of adult structures such as the precursors of the conduction system, that were reported to be present as rings of 'specialized' tissue at the boundaries of segments in the tubular heart (Wenink, 1976; Kim and Yasuda, 1980; Virágh and Chalice, 1983; Williams, Warwick, Dyson and Bannister, 1989). Not surprisingly therefore, current ideas on cardiac morphogenesis represent barely disguised examples of the preformation theory.

Identification of the spatio-temporal changes in the molecular phenotype during early cardiogenesis is an important element to understand how the early embryonic heart functions, without valves and without a morphologically recognizable conduction system. In this paper we shall attempt to summarize our present concept.

Differentiation and molecular phenotype of the 'primary' myocardium

The development of the heart as an organ is a very early event. In chicken, in the 4-somite stage (H/H stage 8) (Hamburger and Hamilton, 1951) acetylcholinesterase (Lamers, Geerts and Moorman 1990) and contractile proteins such a titin (Tokayasu and Maher, 1987) and myosins (de Jong, Geerts, Lamers, Los and Moorman, 1990) are already expressed in the cardiogenic plates before they are transformed into the cardiac tube. The formation of this 'primary' myocardium proceeds according to cranio-caudal and latero-medial gradients, as might be expected from the progression of gastrulation along similar gradients. Thus, myosin expression can first be detected in the cranial and lateral parts of the cardiogenic plates (de Jong et al. 1990). The atrial and ventricular myosin isoforms are first expressed in opposite gradients within these cardiogenic plates. The ventricular isoform is highest in concentration in the cranial and medial part of the developing heart, the atrial isoform is highest in concentration in the caudal and lateral parts. Later in development both isomyosins become almost homogeneously expressed within the heart tube (de Jong et al. 1990; Sweeney, Nag, Eisenberg, Manasek and Zak, 1985; Zhang, Shafiq and Bader, 1986). Coexpression of the atrial and ventricular myosin isoforms in the early tubular heart seems to be a general feature in avian and mammalian species, including man (de Jong et al. 1987; de Groot, Lamers and Moorman, 1989; Wessels, Vermeulen, Viragh, Kalman, Lamers and Moorman, 1991b). Other phenotypic properties that are expressed in gradients are acetylcholinesterase that is predominantly present in the cranial part of the embryonic chicken and rat heart (Lamers, Te Kortschot, Moorman and Los 1987; Lamers, Geerts and Moorman, 1990) and the frequency of the intrinsic cellular beat rate that that decreases going from the inflow to the outflow tract (Kamino, Hirota and Fujii, 1981; Satin, Fujii and de Haan, 1988). In contrast to the myosin isoform expression pattern, the gradients in the expression of acetylcholinesterase and in the intrinsic beat rate persist during subsequent development.

The molecular and electrophysiological phenotype of the myocytes comprising the 'primary' myocardium of the tubular heart seems to be determined by local positional cues, the endoderm of the developing foregut being frequently implied (Orts Llorca and Ruano Gil, 1965; Manasek, 1976). There is good experimental evidence that distinct cell lineages have not yet formed within this population of cardiomyocytes. Transplantation experiments have clearly demonstrated a change in the expression of myosin isoforms and in the beat rate as a function of the position of the explant in the cranio-caudal gradient (Orts Llorca and Jiminez Collado, 1967; Satin, Bader and de Haan, 1987; Satin et al. 1988), indicating that the cardiomyocytes have not yet obtained a distinct positional value, by which they have become committed. In line whith Patten's view on cardiac differentiation (Patten, 1956), the 'primary' myocardium, therefore, still constitutes a homogeneous population of cells in the same state of differentiation. This population serves as precursor for both the cardiomyocytes specialized in impulse generation and conduction, and those cells particularly involved in mechanical contraction (see Fig. 1). These two distinct cell types have been proposed by others to originate from distinct cell lineages (de Haan, 1961).

Interestingly, the 'primary' myocardium also gives rise to at least part of the endocardium, by cell detachment from the cardiogenic plates (Virágh, Szabo and Challice, 1989). This interpretation is strongly supported by the myosin expression in the morphologically recognizable endocardial cells (Tokayasu and Maher, 1987; de Jong et al. 1990). The endocardium may, therefore, be considered as the first differentiation product of the 'primary' myocardium.

It should be realized that formation of the cardiac tube from the cardiogenic plates is realized within a very short period (in chicken only approx. 12 h), but that the formation of 'primary' myocardium continues at the inflow and outflow extremities of the cardiac tube, even though an atrial and ventricular working myocardium are developing in the 'older', established parts of this tissue. Hence, both processes are not entirely separated in time and functional characteristics are to some extent superimposed upon another.

Development of cardiac segments and molecular phenotype of working myocardium

The development of the working myocardium (Fig. 2), that is the second differentiation product of the 'primary' myocardium (Fig. 1), is well illustrated by the spatio- temporal changes in isomyosin expression both in chicken, rat and human tubular hearts (de Jong, Geerts, Lamers, Los and Moorman, 1987; de Groot et al. 1989; Wessels et al. 1991b). Within the 'primary' myocardium of the heart tube that coexpresses the atrial (α) and ventricular (β) myosin heavy chain (MHC) isoforms, atrial and ventricular segments develop that are characterized by the mono-expression of either the atrial or the ventricular MHC isoform,

Development of primary myocardium

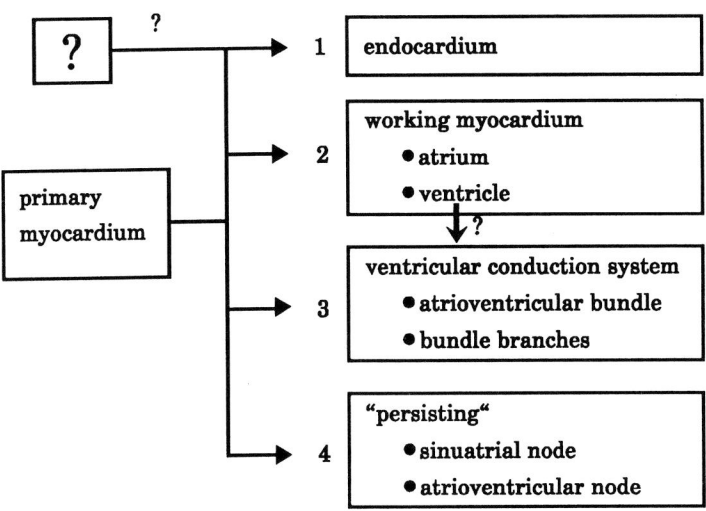

Fig. 1. Development of 'primary' myocardium. See text for additional explanation. The question marks deserve some additional comment. First, it is not yet unambiguously established that the 'primary' myocardium is the only source of endocardium. Second, the precursor cells of the ventricular conduction system can first be recognized in the ventricular segment of the heart tube. The ventricle can be defined as such, because the ventricular myosin isoform is no longer expressed in the atrial segement of the heart (Wessels et al. 1991a). These precursor cells are present in myocardium that surrounds the communication between the embryonic right and left ventricle and initially resemble the ventricular trabeculae in their expression pattern (Wessels et al. 1991b; Wessels et al. 1991a). Concerning the fourth differentiation product, the term 'persisting' 'primary' myocardium does not imply that all characteristics of the 'primary' myocardium are necessarily conserved in this myocardium; it just means that a number of characteristic features of the 'primary' myocardium are still present, such as a low abundance of gap junctions, slow conduction, coexpression of myosin isoforms and a typical morphology of the cells.

respectively (Fig. 3, panel A). These developing segments are flanked by atrial and ventricular MHC coexpressing 'primary' myocardium. Hence, consecutive segments can be recognized, alternatingly consisting of 'primary' and working myocardium, that we dub, following the direction of the blood flow: inflow tract (sinuatrium), *embryonic* atrium, atrioventricular junction, *embryonic* ventricle and outflow tract (Fig. 2). The developmental changes in chicken (de Jong et al. 1987), rat (de Groot et al. 1989) and human (Wessels et al. 1991b) are essentially the same, although minor temporal differences are observed.

The first contractions occur in embryos possessing approx. 10 pairs of somites (Sabin, 1920) in myocardium that still coexpresses α and β MHC (de Jong et al. 1990), but that shortly thereafter ceases to express acetylcholinesterase and only shortly later α MHC (Lamers et al. 1987; Lamers et al. 1990; Fig. 3, panels A and B). This myocardium will give rise to the ventricular segment of the heart. The

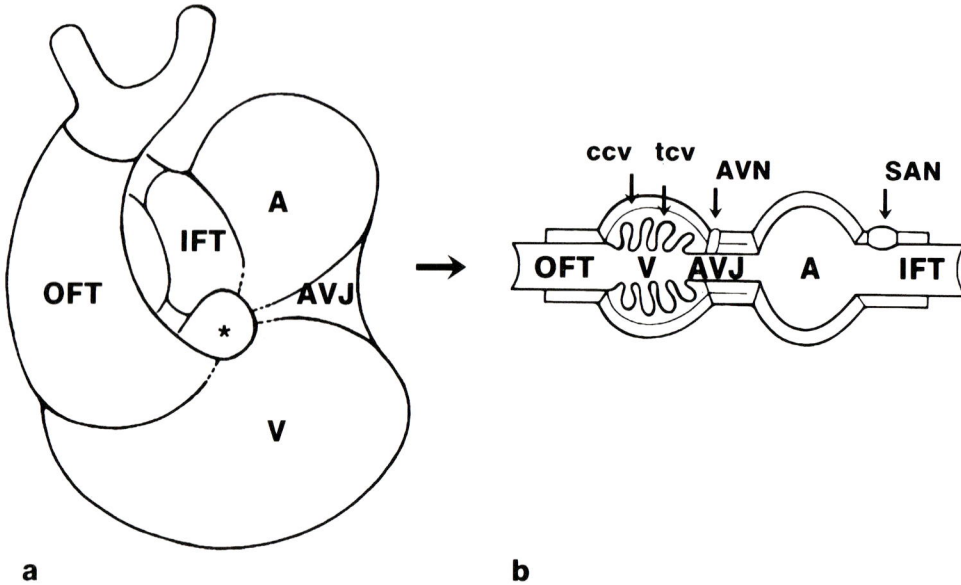

Fig. 2. a. Scheme of a looped heart of approx. Carnegie stage 14 in which segments
have developed. b. A cross section of such a heart after straightening of the loop, as
presented in Fig. 3. IFT, inflow tract; SAN, sinuatrial node; A, atrium, AVJ,
atrioventricular junction; AVN, atrioventricular node; v, ventricle; ccv, compact
myocardial component of the ventricle; tcv, trabecular component of the ventricle;
OFT, outflow tract; *, lesser curvature.

difference in timing of the changes in expression of acetylcholinesterase and MHC
isoforms may be accounted for by the longer half-life of the myofibrillar
complexes. Acetylcholinesterase and α MHC remain expressed in the adjacent
parts of the cardiac tube, i.e. the atrioventricular junction and the outflow tract.
When the ventricular trabeculae develop, they are characterized by an abundant
expression of both α and β MHC and of acetylcholinesterase (vide infra).
Although the ventricular segment can already be recognized when the atrial
segment is still being formed as the caudal part of the heart, it is clear from Fig. 3
that the atrial compartment already acquires its adult phenotype during its
formation. Towards the end of the embryonic period the ventricular architecture
will be remodelled extensively and valves have to develop before it obtains the
adult pattern of gene expression.

Recently, the distribution of connexin-43 in developing and adult rat hearts has
been established (Navaratnam, Kaufman, Shepper, Barton and Guttridge, 1986;
van Kempen, Fromaget, Gros, Moorman and Lamers, 1990; Fig. 3, panel C).
Since connexin-43 is the major component of cardiac gap junctions in the rat
(Yancey, John, Lal, Austin and Ravel, 1989; El Aoumari, Fromaget, Dupont,
Reggio, Durbac, Briand, Boller, Kreitman and Gros, 1990), this distribution
reflects the presence of gap junctions. Gap junctions are thought to be responsible

for the electrical coupling of the cardiomyocytes (Spray and Burt, 1990). Hence, their distribution is of interest, as it probably reflects the pattern of conduction. The changes in the pattern of expression of connexin-43 parallel precisely the developmental changes leading to the formation of cardiac segments as determined by the changing pattern of expression of the myosin isoforms. Thus connexin-43 cannot be detected immunohistochemically in the 'primary' myocardium of the early tubular heart and of the junctional myocardial regions in later stages, reflecting a low abundance of gap junctions in these areas (see Fig. 3, panel C). Subsequently it can be detected in the atrial working myocardium and in the trabecular component of the ventricle, while later in development connexin-43 becomes predominant in the compact ventricular myocardium as well. This distribution potentially allows faster conduction velocities in the atrial and ventricular segments relative to the junctional myocardium (see next section) and would account for the electrocardiogram that is present in these early stages of heart development (van Mierop, 1967).

In the adult situation both the atrio-ventricular node and the sinuatrial node reveal low velocities of the propagation of the impulse. Interestingly, both structures maintain a low abundance of gap junctions, a characteristic of the 'primary' myocardium (see Fig. 1).

The inner layer or trabecular component of the ventricle may have to be considered as a separate cardiac compartment, with a distinct pattern of gene expression (see Fig. 3 and Table 1). Thus, at Carnegie stage 16, they not only express α MHC, β MHC and acetylcholinesterase, but also connexin-43. In addition, relatively high levels of creatine kinase M are observed in the ventricular trabeculae (Wessels, Vermeulen, Viragh, Kalman, Morris, Nguen Thi Man, Lamer and Moorman, 1990; Hasselbaink, Labruyere, Moorman and Lamers, 1990; Lamers, Geerts, Moorman and Dottin, 1989). Together these observations suggest an important functional role for the ventricular trabeculae in the embryonic ventricle, which is in line with the notion that most of the mechanical force is developed by these strucures that traverse the ventricular lumen (Challice and Virágh, 1974).

Table 1. *Molecular phenotype of the ventricular segment**

Protein	Ventricular trabeculae	Compact (outer) layer
β myosin heavy chain[1]	+	+
α myosin heavy chain[1]	+	−
acetylcholinesterase[2]	+	−
creatine kinase M[3]	+	−
connexin-43[4]	+	−

*, Carnegie stage 16; [1]), Wessels et al. 1991b; de Groot et al. 1989; de Jong et al. 1987; [2]), Lamers et al. 1990; Lamers et al. 1987; [3]), Wessels et al. 1990; Hasselbaink et al. 1990; Lamers et al. 1989; [4]), van Kempen et al. 1990.

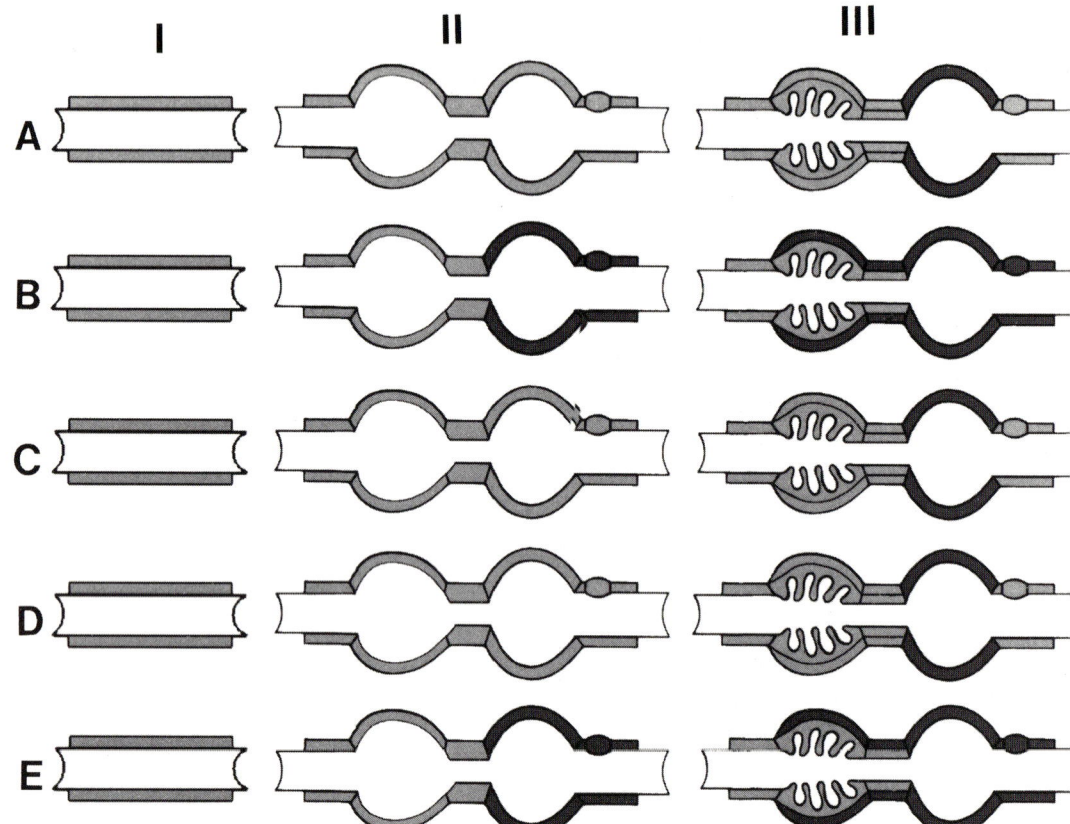

Fig. 3. Development of cardiac segments. Stages I-V indicate successive 'prototypical' developmental stages to illustrate the phenotypic changes in the different parts of the developing heart tube. They correspond approximately to the following staging:

I. *Carnegie stage 9.* 'The straight heart tube'. (human, 20 embryonic days (ED) (1-3 pairs of somites); rat, 10 ED (0-4 pairs of somites); chicken, stage 10 (H/H) (8-10 pairs of somites)).

II. *Carnegie stage 11.* 'Tubular heart in which segments have just started to develop'. (human, 24 ED (13-20 pairs of somites); rat, 10.5 ED (4-13 pairs of somites); chicken, stage 13 (H/H) (12-14 pairs of somites)).

III. *Carnegie stage 14.* 'Tubular heart in which ventricular trabeculae are being formed'. (human, 32 ED; rat, 12 ED; chicken stage 23 (H/H)).

IV. *Carnegie stage 16.* 'Typical embryonic heart' (human, 38 ED; rat, 13 ED; chicken, stage 26 (H/H)).

V. *Carnegie stage 23.* 'Early fetal heart' (human, 56 ED; rat, 15 ED; chicken, stage 30 (H/H).

Light-shaded areas indicate the molecular phenotype of the 'primary' myocardium; Dark-shaded areas indicate the working myocardium. If no shading is applied, myocardium is no longer present as is the case for the outflow tract after stage 23, or fibrous tissue has been interposed as is the case for the atrioventricular insulation in stage 23. A: Developmental changes in the pattern of expression of α and β MHC isoforms, as based on (Wessels et al. 1991b; de Groot et al. 1989; de Jong et al. 1990; de Jong et al. 1987; Sanders, Moorman and Los, 1984; de Groot et al. 1988). In the 'primary' myocardium both isoforms are expressed, whereas single expression is first observed in the atrial part (α MHC), then in the compact ventricular myocardium (β MHC) and then in nearly the entire heart tube except the periphery of the sinuatrial node. B: Developmental changes

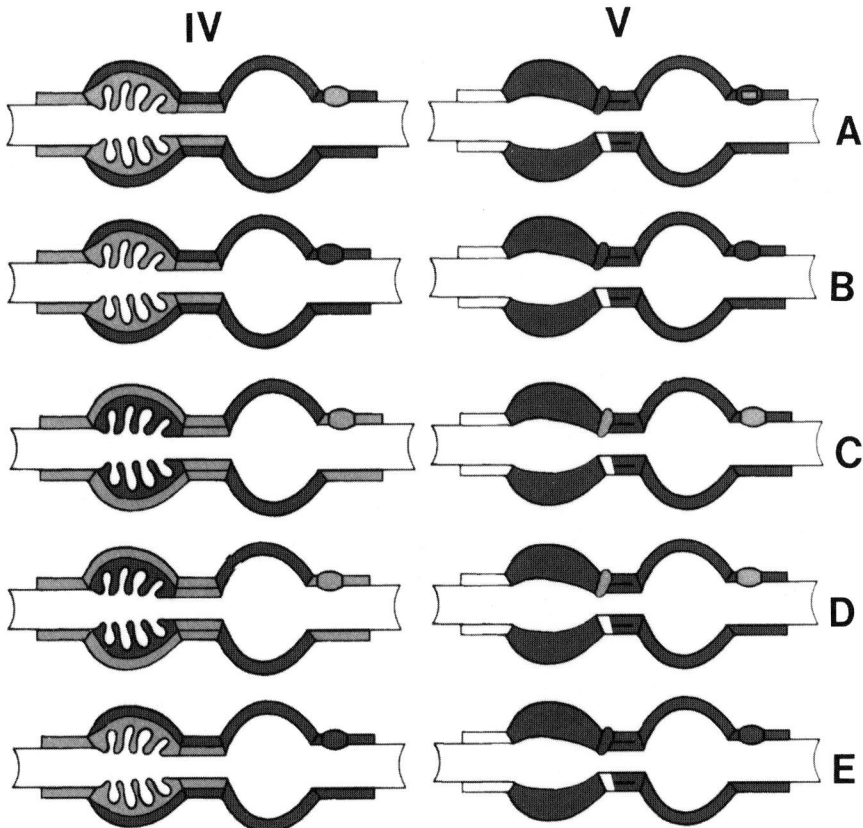

in the pattern of expression of acetylcholinesterase based on (Lamers et al. 1987; Lamers et al. 1990). Initially, the entire 'primary' myocardium expresses acetylcholinesterase. Its expression becomes rapidly confined to the downstream, anterior part of the heart tube, viz. the inner layer of the atrioventricular junction, the trabecular component of the ventricle and the outflow tract. In fully developed myocardium acetylcholinesterase is no longer expressed. C: Developmental changes in the pattern of expression of Connexin-43 based on (van Kempen et al. 1990). In the 'primary' myocardium no Connexin-43 can be detected. Subsequently it becomes detectable in the atrium and the trabecular component of the ventricle and much later in development it becomes expressed in all parts of the myocardium except the sinuatrial node and the atrioventricular node both in rat (van Kempen et al. 1990), human (Oosthoek, van Kempen, Wessels, Lamers and Moorman, 1990) and cow (P. W. Oosthoek et al., manuscript in preparation). D. Developmental changes in the pattern of conduction based on (de Jong et al. 1988) and De Jong et al., manuscript submitted for publication. The 'primary' myocardium reveals a slow propagation of the impulse, whereas the developing atrial working myocardium and the trabecular component of the ventricle reveals a relatively high conduction velocity. In the adult low conduction velocities persist in the sinuatrial node and in the atrioventricular node. E. Developmental changes in the duration of contraction (see also Fig. 4). The upstream, posterior part of the heart (inflow tract and atrium) shows brief contractions whereas the downstream, anterior part of the heart (ventricle and outflow tract) shows contractions with a long duration. The inner layer of the ventricle, i.e., the trabecular component of the ventricle is held responsible for the long duration of the contraction as the inner layer has been reported to be primarily responsible for the force production (Challice and Virágh, 1974), in line with the presence of high concentrations of creatine kinase M (Lamers et al. 1989; Hasselbaink et al. 1990; Wessels et al. 1990).

Development of the pattern of embryonic conduction

Rhythmical activation of the developing myocardium is initiated in the inflow tract of the heart (van Mierop, 1967; Kamino, Komuro, Sakai and Hirota, 1988). Rhythmical action potentials can be registered well before the first contractions are observed at stage 10 (H/H) (Fujii, Hirota and Kamino, 1981), showing that excitation and contraction are poorly coupled initially. In this stage the cardiomyocytes are characterized by slow voltage-gated calcium ion channels (de Haan, 1980; Sperelakis, 1984). Conduction velocities are very low (approx. 1 cm/s) (de Jong, Opthot, Wilde, Janse, Charles, Lamers and Moorman, 1992) and similar along the cardiac tube. As a consequence, the contraction pattern is peristaltic as described by Patten (Patten, 1949).

Subsequently, atrial and ventricular segments develop (Fig. 3, panel D), in which a higher conduction velocity evolves (Lieberman and Paes de Carvalho, 1965a; Lieberman and Paes de Carvalho, 1965b; Arguello, Alanis, Pantoja and Valenzuela, 1986; de Jong, Sanders, de Groot, Wessels, Lamers and Moorman, 1988; de Jong et al., 1992). The action potentials in these areas have a fast-rising phase and a high amplitude, characteristic of fast voltage-gated sodium ion channels (Galper and Catterall, 1978; de Haan, 1980; Sperelakis, 1984). The flanking regions of 'primary' myocardium still have low conduction velocities (van Mierop, 1967; Lieberman and Paes de Carvalho, 1965a; Lieberman and Paes de Carvalho, 1965b; Arguello et al. 1986; de Jong et al., 1992) and are characterized by action potentials with a slow-rising phase, low amplitude and long duration (Galper and Catterall, 1978; de Haan, 1980; Sperelakis, 1984). The development of a multiphasic electrocardiogram (van Mierop, 1967) reflects the segmental differences in conduction velocity in the myocardium and parallels the development of a relatively high conduction velocity in the atrium (Arguello et al. 1986), rather than representing the development of an adult type of conduction system.

Taken together, the heart has developed into a series of segments of alternatingly low and relatively high conduction velocities. As a consequence, sequential contractions of the atrial and ventricular compartments of the embryonic heart do occur. Initially, connexin-43 cannot be detected in the compact layer of the ventricular myocardium. This indicates a low abundance of gap junctions and suggests that the impulse is slowly conducted. In fact we have observed that the compact layer of the ventricular myocardium is activated by the more rapidly conducting trabeculae, thus making the entire ventricular segment rapidly conducting and contracting. Later in development the compact ventricular myocardium becomes fast-conducting too (see Fig. 3, panel D).

Development of the pattern of embryonic contraction

Sequential, rapid contractions of atrium and ventricle, imposed by the alternating differences in conduction velocities, are not sufficient for a proper function of the embryonic heart, as it will not guarantee a unidirectional propagation of the blood. Contraction of the downstream ventricular segment should not occur prior to the termination of the contraction of the upstream atrial segment. On the other hand, relaxation of an atrial or ventricular segment prior to contraction of a downstream junctional segment is not permitted either, if it has to act as a substitute of valve function to prevent backflow of blood.

During development the pattern of contraction changes from peristaltic to synchronous. In the peristaltic heart, a contraction wave of the myocardial muscle, that is wrapped around the cardiac jelly, moves along the cardiac tube, thereby reducing the lumen of the endocardial tube and pushing the blood ahead toward the arterial pole of the heart. Such a contraction form obviates the need of one-way valves to ensure a unidirectional flow of blood. Although the low conduction velocities that are essential for this type of contraction persist in the flanking segments, a peristaltic contraction form remains only identifiable in the outflow tract. This contraction pattern is realized due to the relatively great length of the outflow tract in combination with low conduction velocities and long-lasting contractions. These properties assure that the contraction wave in the outflow tract is not yet completed when the next cardiac cycle is already initiated. In this way backflow of blood from the arteries is prevented. One could thus phrase that in the embryonic heart the outflow tract substitutes for the semilunar valve function as present in the adult condition (de Jong et al. 1988; de Jong et al., 1992). Competence of cardiac function will not be achieved when the heart cycles do no longer overlap, as can be observed when cardiac rhythm slows down at lower ambient temperatures.

The function of the atrioventricular junction can only fully be appreciated if not only its molecular phenotype but also its geometry is taken into account. As illustrated in Fig. 2, the segmented heart is looped. As a consequence the segments are wedge-shaped and the impulse will pass the consecutive segments preferentially along the shortest route in the lesser curvature. This implies that the downstream fast-conducting ventricular segment is already activated, when activation still continues in the atrioventricular junction. In this junctional segment activation spreads from the lesser to the greater curvature. Hence, the myocardium of the atrioventricular junction in the greater curvature remains contracted during ventricular contraction and actually has a sphincter-like valve function.

Analysis of the changing contraction form during development (Fig. 3, panel E and Fig. 4) shows that the atrial segment acquires a pattern of rapid and short-lasting contractions relatively early in development. As a consequence, the atrium is relaxed during most of the heart cycle and functions as a sink for afferent blood. On the other hand, the ventricular part of the heart changes gradually from a peristaltic to a synchronous pattern of long-lasting contractions. Interestingly, the downstream part of the heart is characterized by the expression of acetylcholin-esterase. This enzyme might be part of a muscarinic receptor-mediated regulatory system that has recently been demonstrated in the embryonic chicken heart (Oettling, Schmidt and Drews, 1989; Schmidt, Oettling and Drews, 1988a;

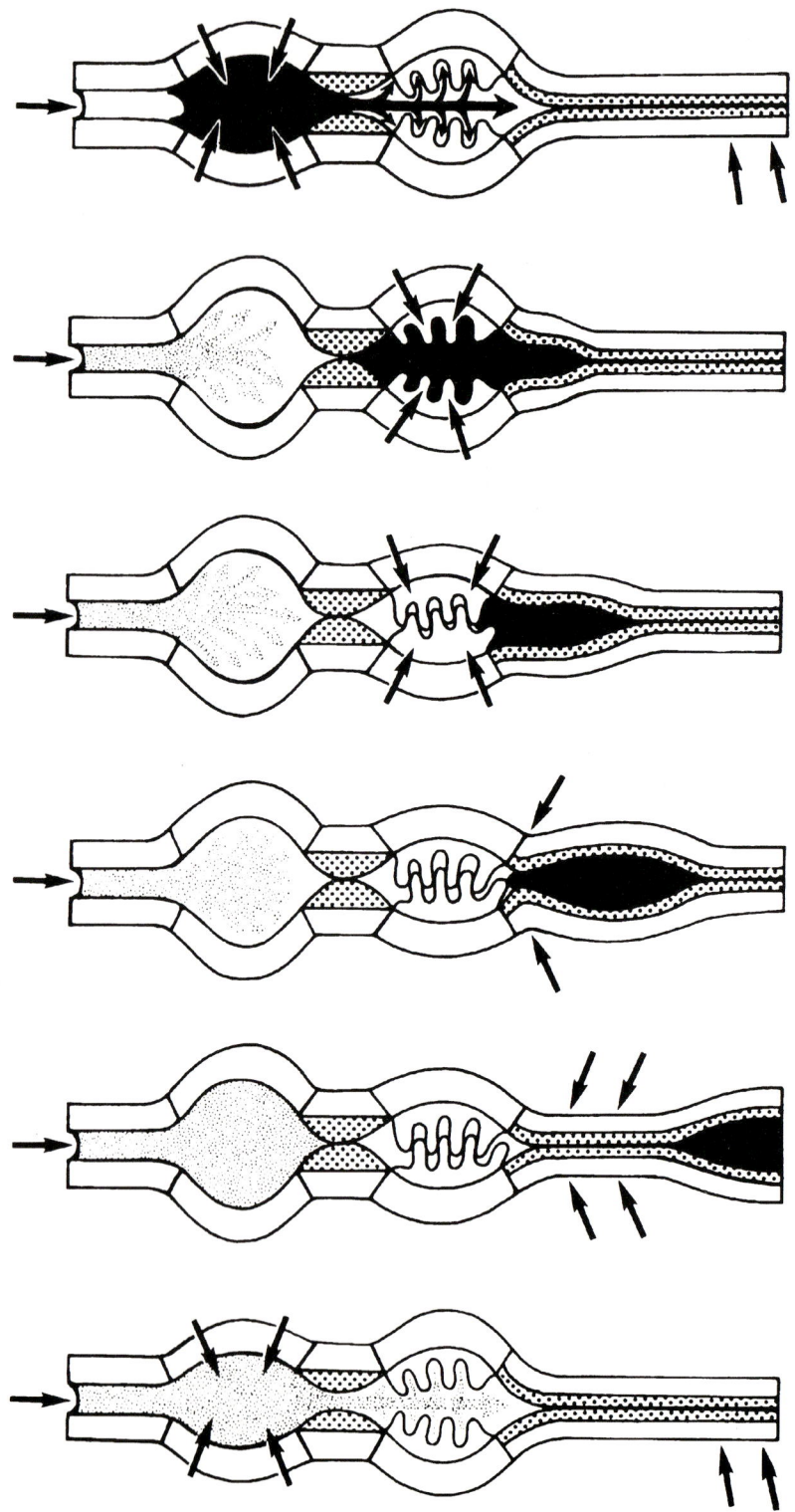

Fig. 4. Cartoon, illustrating the pattern of contraction of the embryonic heart. The proximal part of the heart (inflow tract and atrium) contracts rapidly and with a short duration. As a consequence the atrium is filled with blood most of the cardiac cycle. The distal part of the heart shows contractions with a long duration and is, hence, contracted most of the time. The ventricle contracts rapidly, whereas the outflow tract has a peristaltic contraction form.

Schmidt, Oettling and Drews, 1988b). This system is coupled to phospholipase C, recruits intracellular Ca^{2+} and might as such be responsible for the long-lasting contractions observed in the downstream part of the heart. It would explain why this part of the heart remains contracted during most of the heart cycle. Obviously identification of other components of such a system will be essential to verify its possible role in the contraction of the embryonic heart.

CODA

We now return to the questions that motivated our studies. How functions the embryonic heart, in terms of the underlying molecular mechanism, without valves and without a morphologically recognizable conduction system? Although our 'molecular' reconstruction is far from complete, the molecular and electrophysiological data show that the peristaltically-contracting primary heart tube develops via a number of ordered steps into a chamber pump. The development of rapidly-conducting chambers of working myocardium that remain flanked by slowly-conducting junctional segments of 'primary' myocardium is likely to be essential for the pumping function of the embryonic heart. This basic functional architecture of the early embryonic heart appears to follow a phylogenetically old pattern as it is also observed in the hearts of adult amphibians (Alanis, Benitez, Lopez and Martinez-Palomo, 1973; Kanno, 1963; Irisawa et al. 1965).

Paradoxically, in more advanced vertebrates the junctional segments were supposed to consist of so-called 'cardiac specialized tissue' that would give rise to the definitive adult conduction system (Wenink, 1976; Kim and Yasuda, 1980; Virágh and Chalice, 1983). However, most parts of the 'primary' myocardium, such as the outflow tract will disappear, and remnants of the 'primary' myocardium are only found in the atrioventricular and in the sinuatrial node (Lamers et al. 1991). The development of the more rapidly-conducting ventricular conduction system (see Fig. 1) to ensure a simultaneous contraction of both ventricles is intimately associated with ventricular septation. It is both in terms of evolution (Lamers and Moorman, 1990) and of ontogenesis (Vassall-Adams, 1978; Wessels, Vermeulen, Verbeek, Viragh, Kalman, Lamers and Moorman, 1992) a relatively late event.

Finally, one might imagine that molecular biological techniques will provide the probes that will allow to establish with high spatial resolution the pattern of distribution of ion channels, receptors and modulatory factors that are essential parameters to understand the functioning of the embryonic heart in more detail.

We are indebted to our colleagues for experimental results, advice and stimulating discussions. This work was supported in part by the Netherlands Heart Foundation.

References

Alanis, J., Benitez, D., Lopez, E. and Martinez-Palomo, A. (1973). Impulse propagation through the cardiac junctional regions of the axolotl and the turtle. *Jap. J. Physiol.* **23**, 149-164.

Arguello, C., Alanis, J., Pantoja, O. and Valenzuela, B. (1986). Electrophysiological and ultrastructural study of the atrioventricular canal during the development of the chick embryo. *J. Mol. Cell. Cardiol.* **18**, 499-510.

Challice, C. and Viragh, S. Z. (1974). The phylogenetic and ontogenetic development of the mammalian heart: some theoretical considerations. *Acta Biochim. Biophys. Acad. Sci.* **9**, 131-140.

De Groot, I. J. M., Lamers, W. H. and Moorman, A. F. M. (1989). Isomyosin expression pattern during rat heart morphogenesis: an immunohistochemical study. *Anat. Rec.* **224**, 365-373.

De Groot, I. J. M., Wessels, A., Viragh, S. Z., Lamers, W. H. and Moorman, A. F. M. (1988). The relation between isomyosin heavy chain expression pattern and the architecture of sinoatrial nodes in chicken, rat and human embryos. In 'Sarcomeric and non-sarcomeric muscles: basic and applied research prospects for the 90s' (ed. Carraro U.), pp. 305-310. Unipress, Padova.

De Haan, R. L. (1961). Differentiation of the atrioventricular conducting system of the heart. *Circulation* **24**, 458-470.

De Haan, R. L. (1980). Development of rhythmic activity in cardiac cells. In 'Physiology of atrial pacemakers and conductive tissues' (ed. Little R. C.), pp. 21-53. Futura Publ. Co., Mount Kisco, NY.

De Jong, F., Geerts, W. J. C., Lamers, W. H., Los, J. A. and Moorman, A. F. M. (1987). Isomyosin expression patterns in tubular stages of chicken heart development: a 3-D immunohistochemical analysis. *Anat. Embryol.* **177**, 81-90.

De Jong, F., Geerts, W. J. C., Lamers, W. H., Los, J. A. and Moorman, A. F. M. (1990). Isomyosin expression pattern during formation of the tubular chicken heart: a 3D immunohistochemical analysis. *Anat. Rec.* **226**, 213-227.

De Jong, F., Opthot, T., Wilde, A. A. M., Janse, M. J., Charles, R., Lamers, W. H. and Moorman, A. F. M. (1992). Persisting zones of slow impulse conduction in developing chicken hearts. *Circ. Res.* **71**, 240-250.

De Jong, F., Sanders, E., De Groot, I. J. M., Wessels, A., Lamers, W. H. and Moorman, A. F. M. (1988). Changing patterns of gene expression during heart development and their functional implications. In 'Fetal and neonatal development' (ed. Jones C. T.), pp. 158-161. Perinatology Press.

El Aoumari, A., Fromaget, C., Dupont, E., Reggio, H., Durbac, P., Briand, J. P., Boller, K., Kreitman, G. and Gros, D. (1990). Conservation of a cytoplasmic carboxy-terminal domain of connexin 43, a gap junctional protein in mammalian heart and brain. *J. Membrane Biol.* **115**, 229-240.

Fujii, S., Hirota, A. and Kamino, K. (1981). Optical indications of pacemaker potential and rhythm generation in early embryonic chick heart. *J. Physiol.* **312**, 253-263.

Galper, J. B. and Catterall, W. A. (1978). Developmental changes in the sensitivity of embryonic heart cells to tetrodotoxin and D600. *Dev. Biol.* **65**, 216-227.

Hamburger, V. and Hamilton, J. L. (1951). A series of normal stages in the development of the chick embryo. *J. Morphol.* **88**, 49-92.

Hasselbaink, H. D. J., Labruyere, W. T., Moorman, A. F. M. and Lamers, W. H. (1990). Creatine kinase isozyme expression in prenatal rat heart. *Anat. Embryol.* **182**, 195-203.

Irisawa, H., Hama, K. and Irisawa, A. (1965). Mechanism of slow conduction at the bulboventricular junction. *Circ. Res.* **17**, 1-10.

Kamino, K., Hirota, A. and Fujii, S. (1981). Localization of pacemaking activity in early embryonic heart monitored using voltage-sensitive dye. *Nature* **290**, 595-597.

Kamino, K., Komuro, H., Sakai, T. and Hirota, A. (1988). Functional pacemaking area in the early embryonic chick heart assessed by simultaneous multiple-site optical recording of spontaneous action potentials. *J. Gen. Physiol.* **91**, 573-591.

Kanno, T. (1963). Electrical activity of the atrioventricular conducting tissue of the toad, studied by a minute section electrode. *Jap. J. Physiol.* **13**, 97-111.

Kim, Y. and Yasuda, M. (1980). Development of the cardiac conducting system in the chick embryo. *Zbl. Vet. Med. C. Anat. Histol. Embryol.* **9**, 7-20.

Lamers, W. H., De Groot, I. J. M., De Jong, F. and Moorman, A. F. M. (1991). The development of the avian conduction system. *Eur. J. Morph.* **29**, 233-253.

Lamers, W. H., Geerts, W. J. C. and Moorman, A. F. M. (1990). Distribution pattern of acetylcholinesterase in early embryonic chicken hearts. *Anat. Rec.* **228**, 297-305.

Lamers, W. H., Geerts, W. J. C., Moorman, A. F. M. and Dottin, R. P. (1989). Creatine kinase isozyme expression in embryonic chicken heart. *Anat. Embryol.* **179**, 387-393.

Lamers, W. H. and Moorman, A. F. M. (1990). Neural tissue antigen identifies the evolutionary origin of the left and right ventricle. In 'Muscle and motility' (eds. Marechal G. and Carraro U.), Vol. 2nd, pp. 91-98. Intercept, Andover, Hampshire.

Lamers, W. H., Te Kortschot, A., Moorman, A. F. M. and Los, J. A. (1987). Acetylcholinesterase in prenatal rat heart: a marker for the early development of the cardiac conductive tissue? *Anat. Rec.* **217**, 361-370.

Lieberman, M. and Paes de Carvalho, A. (1965a). The electrophysiological organization of the embryonic chick heart. *J. Gen. Physiol.* **49**, 351-363.

Lieberman, M. and Paes de Carvalho, A. (1965b). The spread of excitation in the embryonic chick heart. *J. Gen. Physiol.* **49**, 365-379.

Manasek, F. J. (1976). Glycoprotein synthesis and tissue interactions during establishment of the functional embryonic chick heart. *J. Mol. Cell. Cardiol.* **8**, 389-402.

Navaratnam, V., Kaufman, M. H., Shepper, J. N., Barton, S. and Guttridge, K. M. (1986). Differentiation of the myocardial rudiment of mouse embryos: an ultrastructural study including freeze-fracture replication. *J. Anat.* **146**, 65-85.

Oettling, G., Schmidt, E. and Drews, U. (1989). An embryonic Ca++ mobilizing muscarinic system in the chick embryo heart. *J. Dev. Physiol.* **12**, 85-94.

Oosthoek, P. W., Van Kempen, M. J. A., Wessels, A., Lamers, W. H. and Moorman, A. F. M. (1990). Distribution of the cardiac gap junction protein, connexin-43 in the neonatal and adult human heart. In 'Muscle and motility' (eds. Marechal G. and Carraro U.), Vol. 2nd, pp. 85-90. Intercept, Andover, Hampshire.

Ortsllorca, F. and Jiminez Collado, J. (1967). Determination of heart polarity (arterio-venous axis) in the chicken embryo. *Wilhelm Roux' Arch. Entwicklungsmech.* **158**, 147-163.

Ortsllorca, F. and Ruano Gil, D. (1965). Influence of the endoderm on heart differentiation. *Wilhelm Roux' Arch. Entwicklungsmech.* **156**, 368-370.

Patten, B. M. (1949). Initiation and early changes in the character of the heartbeat in vertebrate embryos. *Physiol. Rev.* **29**, 31-47.

Patten, B. M. (1956). The development of the sinoventricular conduction system. *Univ. Mich. Med. Bull.* **22**, 1-21.

Sabin, F. R. (1920). Studies on the origin of blood vessels and of red blood corpuscles as seen in the living blastoderm of chicks during the second day of incubation. *Contrib. Embryol.* **9**, 213-262.

Sanders, E., Moorman, A. F. M. and Los, J. A. (1984). The local expression of adult chicken heart myosins during development. I. The three days embryonic chicken heart. *Anat. Embryol.* **169**, 185-191.

Satin, J., Bader, D. and De Haan, R. L. (1987). Local cues influence atrial and ventricular differentiation of precardiac mesoderm. *J. Mol. Cell. Cardiol.* **19**, S16a. (Abstract)

Satin, J., Fujii, S. and De Haan, R. L. (1988). Development of cardiac heartbeat in early chick embryos is regulated by regional cues. *Dev. Biol.* **129**, 103-113.

Schmidt, H., Oettling, G. and Drews, U. (1988a). An embryonic Ca++ mobilizing muscarinic system in the chick embryo heart. *Roux's Arch. Dev. Biol.* **197**, 37-39.

Schmidt, H., Oettling, G. and Drews, U. (1988b). Muscarinic receptor-mediated intracellular Ca++ mobilization in embryonic chick heart cells. *FEBS Letters* **230**, 35-37.

Sperelakis, N. (1984). Developmental changes in membrane electrical properties of the heart. In 'Physiology and pathophysiology of the heart' (ed. Sperelakis N.), pp. 543-573. Martinus Nijhoff Publ., Den Haag.

Spray, D. C. and Burt, J. M. (1990). Structure-activity relations of the cardiac gap junction channel. *Am. J. Physiol.* **258**, 195-205.

Sweeney, L. J., Nag, A. C., Eisenberg, B., Manasek, F. J. and Zak, R. (1985). Developmental aspects of cardiac contractile proteins. *Basic Res. Cardiol.* **80 Suppl. 2**, 123-137.

Tokayasu, K. T. and Maher, P. A. (1987). Immunocytochemical studies of cardiac myofibrillogenesis in early chick embryos. I. Presence of immunofluorescent titin spots in premyofibril stages. *J. Cell Biol.* **105**, 2781-2793.

Van Kempen, M. J. A., Fromaget, C., Gros, D., Moorman, A. F. M. and Lamers, W. H. (1990). Spatial distribution of the cardiac gap-junction protein, connexin-43 in the developing and adult rat heart. *Circ. Res.* **68**, 1638-1651.

Van Mierop, L. H. S. (1967). Localization of pacemaker in chick embryo heart at the time of initiation of heartbeat. *Am. J. Physiol.* **212**, 407-415.

Vassall-Adams, P. R. (1978). The development of the atrioventricular bundle and its branches in the avian heart. Thesis, London.

Viragh, S. Z. and Chalice, C. E. (1983). The development of the early atrioventricular conduction system in the embryonic heart. *J. Pharm. Pharmac.* **61**, 775-792.

Viragh, S. Z., Szabo, E. and Challice, C. E. (1989). Formation of primitive myo- and endocardial tubes in the chicken embryo. *J. Mol. Cell. Cardiol.* **21**, 123-137.

Wenink, A. C. G. (1976). Development of the human cardiac conduction system. *J. Anat.* **121**, 617-631.

Wessels, A., Vermeulen, J. L. M., Verbeek, F. J., Viragh, S. Z., Kalman, F., Lamers, W. H. and Moorman, A. F. M. (1991a). Spatial distribution of 'tissue-specific' antigens in the developing human heart and skeletal muscle: III. An immunohistochemical analysis of the distribution of the neural tissue antigen GLN in the embryonic heart; implications for the development of the atrioventricular conduction system. *Anat. Rec.* (In Press)

Wessels, A., Vermeulen, J. L. M., Viragh, S. Z., Kalman, F., Lamers, W. H. and Moorman, A. F. M. (1992). Spatial distribution of 'tissue specific' antigens in the developing human heart and skeletal muscle: III. An immunohistochemical analysis of distribution of the neural tissue antigen GIN in the embryonic heart; implications for the development of atrioventricular conduction system. *Anat. Rec.* **232**, 97-111.

Wessels, A., Vermeulen, J. L. M., Viragh, S. Z., Kalman, F., Morris, G. E., Nguen thi man lamer, W. H. and Moorman, A. F. M. (1990). Spatial distribution of 'tissue-specific' antigens in the developing human heart and skeletal muscle. I. An immunohistochemical analysis of creatine kinase isoenzyme expression patterns. *Anat. Rec.* **228**, 163-176.

Williams, P. L., Warwick, R., Dyson, M. and Bannister, L. H. (1989). Gray's anatomy. Churchill Livingstone, New York.

Yancey, S. B., John, S. A., Lal, R., Austin, B. J. and Ravel, J. P. (1989). The 43-kD polypeptide of heart gap junctions: immunolocalization, topology and functional domains. *J. Cell Biol.* **108**, 2241-2254.

Zhang, Y., Shafiq, S. A. and Bader, D. (1986). Detection of a ventricular-specific myosin heavy chain in adult and developing chicken heart. *J. Cell Biol.* **102**, 1480-1484.

Printed in Great Britain © Society for Experimental Biology 1992

RECIPROCAL CHANGES IN MYOSIN ISOFORM EXPRESSION IN RABBIT FAST SKELETAL MUSCLE RESULTING FROM THE APPLICATION AND REMOVAL OF CHRONIC ELECTRICAL STIMULATION

CAROL BROWNSON, PAULINE LITTLE, CAROLINE MAYNE*, JONATHAN C. JARVIS* and STANLEY SALMONS*

School of Life Sciences, The University of North London, Holloway Road, London N7 8DB and *Department of Human Anatomy and Cell Biology and The Muscle Research Centre, University of Liverpool, P.O. Box 147, Liverpool L69 3BX, U.K.

Summary

Chronic indirect electrical stimulation of adult mammalian skeletal muscle brings about a transformation from the fast-twitch to the slow-twitch type. Underlying this transformation there is a sequence of profound changes in the expression of proteins involved in all the major molecular systems of the muscle. These include qualitative changes in the expression of myosin light and heavy chain isoforms. The time course of these changes has been studied in some detail at the protein level, both during chronic stimulation and during the recovery process that follows the cessation of stimulation. Here we report on the use of cDNA probes to study corresponding changes in myosin heavy chain (MHC) and light chain (MLC) mRNAs in rabbit fast-twitch muscles during continuous electrical stimulation at 10 Hz and during the first 12 days of recovery after cessation of 6 weeks of stimulation. At an early stage of the response to stimulation, fast MHC mRNA is replaced by slow MHC mRNA. During recovery this process occurs in reverse but takes longer. Broadly similar changes are seen for MLC mRNAs, although the time course is somewhat different. These experiments contribute to a growing body of evidence that many of the protein changes induced by chronic stimulation are the result of regulatory events that take place at a pre-translational level.

Introduction

The ability of adult mammalian fast skeletal muscle to undergo a transformation of type was discovered through changes in physiological properties that were found to take place in response to cross-reinnervation (Buller, Eccles and Eccles, 1960) and chronic low-frequency stimulation (Salmons and Vrbová, 1969). In the early 70's, histochemical and biochemical evidence began to emerge for the phenotypic changes in stimulated fast muscle that were responsible for the

Key words: muscle, stimulation, recovery, myosin, protein isoforms, mRNA, molecular biology.

observed development of a slower contractile speed and more fatigue-resistant behaviour. We now know that the transformation consists of a sequence of profound changes in the expression of proteins involved in all the major molecular systems of the muscle: contraction, calcium transport and storage, and the generation of ATP (reviewed by Salmons and Henriksson, 1981; Pette and Vrbová, 1985). The most unexpected aspect of the transformation was that it involved qualitative changes in protein profile, for which the first evidence was the expression in fast muscle of myosin light chain isoforms normally found only in slow muscle (Sréter et al., 1973).

The time course of these changes in myosin light and heavy chains has since been studied in some detail, both during chronic stimulation (Brown, Salmons and Whalen, 1983) and during the recovery process that follows the cessation of stimulation (Brown, Salmons and Whalen, 1985; Brown, Henriksson and Salmons, 1989; Salmons, 1990). The starting point for the latter type of experiment is normally a stage of stimulation - such as 6 weeks - at which most of the changes in the physiological, ultrastructural and biochemical properties associated with long-term stimulation are already complete. On their recovery from such stimulation, rabbit muscles show a complete reversion to fast-muscle character-istics, with a time course more prolonged than that of the original transformation. In general, the associated changes in type-specific proteins occur in a reverse sequence to that seen during fast-to-slow transformation. In particular, calcium-activated myosin ATPase activity regains control fast levels in about 6-8 weeks (Salmons, 1990). Histochemical demonstration of myofibrillar ATPase activity indicates that the proportion of Type 1 fibres declines to that of the unstimulated contralateral muscle by about 4 weeks, and the proportions of Type 2B and Type 2A fibres reach control levels by 12 weeks (Brown, Henriksson and Salmons, 1989).

Recently we have been using probes to study the corresponding mRNAs and we have been able to demonstrate changes in the levels of mRNAs encoding fast (Type 2B) and slow (Type 1) myosin heavy chain (MHC) and fast light chain (MLC) isoforms (Brownson et al., 1988, 1989). We will report here some further observations on the transitions that take place in the MLC mRNAs. We also show that when stimulation is suddenly removed from a muscle that has acquired the slow phenotype as a result of continuous stimulation at 10 Hz for 6 weeks, the mRNAs revert to their original levels, albeit with a slower time course.

Methods

Electrical stimulation and recovery of skeletal muscle

Fast muscles of adult New Zealand White rabbits were chronically stimulated via the common peroneal nerve (Brown, Salmons and Whalen, 1983). An internally powered miniature stimulator (Salmons and Jarvis, 1991) was implanted subcutaneously on the left flank; stimulation consisted of a continuous train of supramaximal pulses at a frequency of 10 Hz. For recovery experiments, muscles were stimulated for 6 weeks before the stimulator was turned off via a remote optical link. After periods of stimulation up to 6 weeks and periods of recovery from 4 to 12 days, the extensor digitorum longus (EDL) and tibialis anterior (TA) muscles were removed, frozen rapidly in liquid N_2 and stored at $-70°C$.

Probes

For the heavy chain studies we used rabbit-specific cDNA probes to fast and slow muscle MHC mRNA. Although MHCs are derived from a multigene family whose members are highly homologous, the 3′-untranslated regions show no significant homology within a species (Saez and Leinwand, 1986). A probe for fast (Type 2B) MHC mRNA (pMHC450-F; Maeda, Sczakiel and Wittinghofer, 1987) was already available to us. We used pMHC450-F and another probe from the same laboratory, pMHC600-F, to screen a cDNA library constructed from rabbit soleus muscle. Preliminary sequencing indicated that a cDNA fragment, pMHC1450-S (1450 bp), from a clone that hybridised strongly to pMHC600-F, contained the poly(A)$^+$ tail, and hence coding and non-coding sequences at the 3′-untranslated region of a MHC gene. To increase the specificity for slow muscle MHC mRNA, we used a 450 bp fragment (pMHC450-S), derived from pMHC1450-S by *Hinf I* digestion, as the probe in our studies; its isolation, sequence and specificity are reported elsewhere (Brownson et al., 1992).

For the light chain studies we used the MLC probes pA29, a 540 bp cDNA clone of mouse slow MLC1s (Barton et al., 1985) and p450/4, a 470 bp genomic DNA clone derived from a pseudogene corresponding to mouse fast myosin alkali light chain MLC1f/MLC3f (Robert et al., 1988).

Preparation and analysis of RNA

Total RNA was extracted from rabbit skeletal muscles in 8M urea and 4M LiCl. The crude RNA recovered by centrifugation was extracted with phenol and precipitated with ethanol. Poly(A)$^+$ RNA was isolated by passage through an oligo-dT Sepharose column.

Total or poly(A)$^+$ RNA isolated from muscles was analysed by Northern and slot-blotting techniques as described previously (Brownson et al., 1988) using the two different myosin cDNA probes, pMHC450-S and pMHC450-F. The densities of the bands on the slot-blot autoradiographs were measured with a Joyce Loebl Scanning Densitometer and related to μg of total or poly(A)$^+$ RNA. Integrated peak areas/μg of RNA isolated from stimulated or recovery muscles were expressed as a percentage of the value obtained for control muscles. The integrity of all RNA preparations was checked on urea gels. Neither these gels, nor Northern blots, showed any evidence of RNA degradation after stimulation, with or without recovery, of TA and EDL muscles.

For RNase protection assays, pMHC450-S and pMHC450-F were subcloned into pGEM 3Z and anti-sense RNA probes were transcribed using T7 and SP6 DNA polymerase for pMHC450-S and pMHC450-F respectively in the presence of 20 μCi of $[\gamma\text{-}^{32}P]UTP$. RNase-resistant fragments were denatured in formamide, resolved on formaldehyde agarose gels and transferred to nylon membranes for autoradiography.

Results

Myosin heavy chain mRNA levels in rabbit fast skeletal muscles after stimulation and recovery

The results, shown in Fig. 1A, confirm the previous demonstration (Brownson et al., 1988) of a dramatic decline in the levels of fast MHC mRNA in EDL muscle in the first week of stimulation.

In earlier studies, with pMHC1450-S as the probe, we had found slow MHC mRNA to be 4 times more abundant in stimulated EDL muscle than in the contralateral control muscle (Brownson, Salmons and Edwards, 1989). We confirmed this finding with the more specific pMHC450-S probe (Fig. 1B). Expression of slow MHC mRNA was clearly evident after 9 days of stimulation and was further elevated after 42 days of stimulation. Similar trends were observed whether poly(A)$^+$ RNA (2.9-fold increase) or total RNA (5-fold increase) was used for the analysis. RNase protection assays with poly(A)$^+$ RNA confirmed that the mRNAs detected in the slot-blots contained sequences that were homologous to the two cDNA probes (Brownson et al., 1992).

Specific MHC mRNA levels were also analysed in total and poly(A)$^+$ RNA during recovery after 6 weeks of stimulation. Slot-blot analysis with pMHC450-F as probe showed that, whereas fast MHC mRNA disappeared rapidly following the onset of electrical stimulation, the reappearance of this mRNA during recovery was relatively slow. After 4 days of recovery, fast MHC mRNA in the experimental EDL muscle had reached only 2% of the control value and even after 12 days it had risen to only 42% of this value (Fig. 1A). RNase protection assays indicated that the actual levels may have been even lower, since analysis of total RNA on slot-blots could include a contribution due to cross-hybridization with any intermediate species that may be present during the transition. Both Northern analysis and RNase protection assays confirmed that there was a more rapid return to control levels of fast MHC in TA muscle than in EDL muscle.

The concurrent decline in slow MHC mRNA was monitored with pMHC450-S as a probe. The level appeared to be sustained up to 4 days in EDL muscle, after which it declined more rapidly to 30% of the initial value at 12 days (Fig. 1B). The results of slot- and Northern-blot analysis and RNase protection assays suggested that the decline in slow MHC mRNA of the TA muscle was similar but had a more rapid time course.

Myosin light chain mRNA levels in rabbit fast skeletal muscles after stimulation and recovery

We evaluated the specificity of the mouse MLC mRNA probes in the rabbit by slot-blot analysis. The distribution of MLC mRNAs in TA and soleus is seen in Table 1.

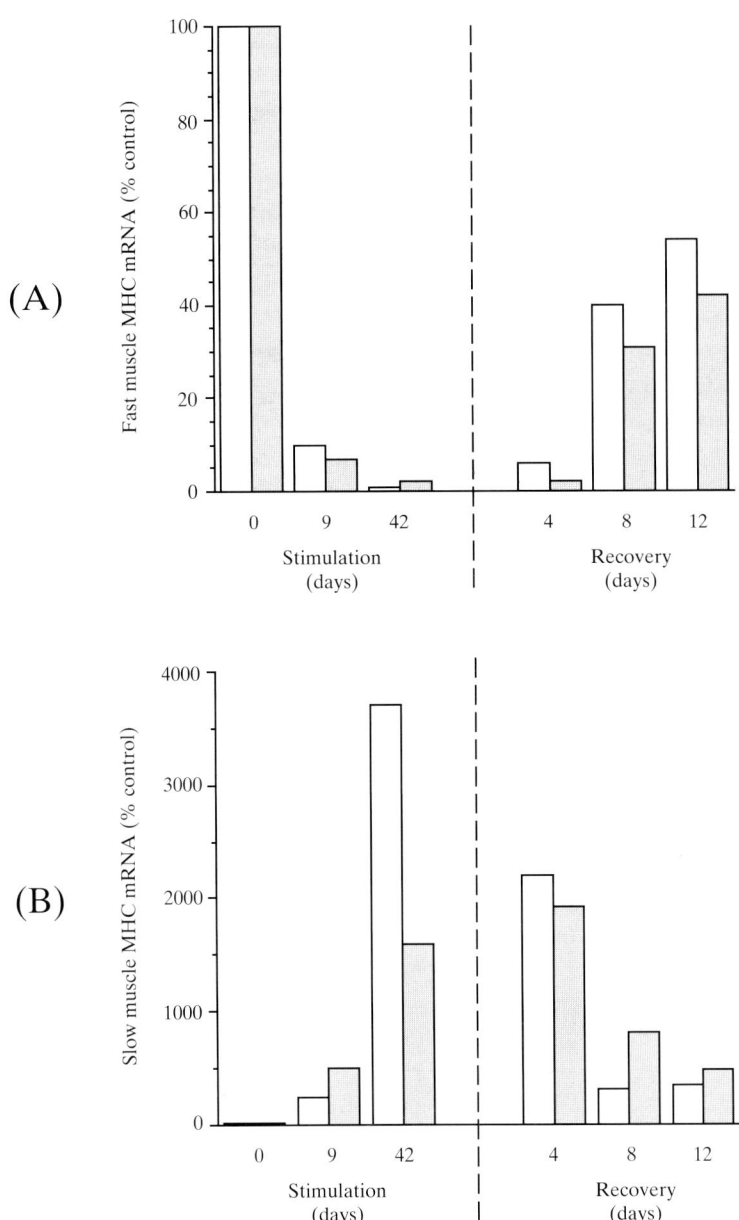

Fig. 1. MHC mRNA levels in TA (unshaded columns) and EDL (shaded columns) after stimulation and recovery. Figures are integrated peak areas/μg total RNA determined by slot blot analysis, expressed as a percentage of the value for the same-day control muscle. (A) Probed by pMHC450-F for fast muscle MHC mRNA. (B) Probed by pMHC450-S for slow muscle MHC mRNA.

Table 1. *Distribution of MLC mRNAs in TA and soleus muscles by slot-blot analysis*

Probe for:	MLC1/3f	MLC1s
TA	100	15
Soleus	18	100

Integrated peak areas per μg mRNA were determined and expressed as a percentage of the value in the muscle in which the particular mRNA is most abundant.

Table 2. *Effect of stimulation and post-stimulation recovery on MLC mRNA levels*

Probe	MLC1/3f	MLC1s
Stimulation (days)		
4	46±1	53±25
10	51±11	41±8
21	58±4	52±17
35	29	56
56	–	138
Recovery (days)		
4	41	128
6	41	119
8	80	72

The figures given are integrated peak areas per μg poly(A)$^+$ RNA (stimulated) or total RNA (recovery) determined from slot-blot analysis, expressed as a percentage of the value for the same-day control muscle.

Data are average of determinations on 3-4 animals; where no S.E.M. is specified, the determination was made on a single animal.

Slot-blot analysis was then used to examine the relative levels of fast (MLC1/3f) and slow (MLC1s) muscle MLC mRNA in TA muscles following stimulation and recovery. In the experiments on muscle recovery, all muscles were initially stimulated continuously at 10 Hz for 6 weeks, by which time type transformation is known to be substantially complete (Salmons, 1990). The results are given in Table 2.

Discussion

To analyse changes in specific myosin mRNA levels in skeletal muscle, one needs probes that can differentiate unambiguously between the members of this multigene family. In this study we used a complementary pair of probes: a fast muscle MHC probe that had been described previously (Maeda, Sczakiel and Wittinghofer, 1987) and a rabbit slow muscle MHC probe containing coding and non-coding sequences at the 3'-end of the gene, which we isolated from a rabbit soleus cDNA library (Brownson et al., 1992).

The study presented here was concerned with changes in the myosin mRNAs during indirect electrical stimulation and recovery of the rabbit fast skeletal muscles TA and EDL. In a previous paper (Brownson et al., 1988), we reported that continuous stimulation of these muscles at 10 Hz reduced their fast MHC mRNA content to 34% (TA) and 58% (EDL) of control levels after only 4 days of stimulation and to 6-8% after 10 days. Here we have measured these levels during the recovery process that follows an extended period of stimulation with the same pattern. The fast MHC mRNA reappeared with a time course that was slower than that for stimulation, so that even after 12 days the level had reached only 54% (TA) and 42% (EDL) of control levels. This gradual increase in fast myosin mRNA is consistent with the time course for the reappearance of the protein product in sections processed histochemically for the demonstration of myosin ATPase (Brown, Henriksson and Salmons, 1989).

We had shown earlier (Brownson, Salmons and Edwards, 1989) that stimulating EDL muscle for 10 days produced a 4-fold increase in slow-muscle MHC mRNA, as revealed by the 1450 bp probe. The more specific probe pMHC450-S used in the present work revealed increases in slow MHC mRNA of approximately 16-fold for TA and 37-fold for EDL after 6 weeks' stimulation. When stimulation was discontinued after 6 weeks, these elevated levels of slow MHC mRNA did not respond immediately, but they had begun to decline fairly rapidly after 4 days and at 12 days had reached 50% (EDL) and 10% (TA) of their former levels in the long-term stimulated muscles. Although the actual events were similar in TA and EDL muscles, the time course appeared to be different: the rate at which slow muscle mRNA disappeared and fast muscle mRNA reappeared during recovery seemed to be faster in TA than in EDL. At all events, the changes in both fast and slow MHC mRNA levels occur less rapidly during recovery than during the initial period of stimulation.

Under broadly similar conditions of stimulation, rat muscles behave somewhat differently to rabbit muscles, and show no appreciable induction of Type 1 myosin. There is, however, a well-marked transition from Type 2B to Type 2A myosin (Termin, Staron and Pette, 1989). Corresponding mRNA changes have been observed in the rat: Type 2A MHC mRNA (the predominant isoform in this species after stimulation) declines noticeably over 6 days of recovery, with a concomitant reappearance of Type 2B MHC mRNA (Kirschbaum et al., 1990). These transitions echo the time course of the more extensive changes (between Type 2B and Type 1) that are seen in the rabbit.

When we examined the changes in MLC mRNAs, we were initially puzzled by the low levels of MLC1s, which appeared to be only about half those in the control muscles over the first 21 days of stimulation. Brown et al. (1985) had shown that the protein isoform composition of the MLCs of TA muscle changes very little during this period. The result was, however, consistent, and we now believe that this apparent decrease reflects a more rapid increase in other species of mRNA, rather than a decline in the absolute amount of MLC1s mRNA. Furthermore the probe used in these experiments detects MLC1s mRNA, and at the protein level it is known that this isoform increases at a later stage than MLC2s (Brown, Salmons and Whalen, 1983, 1985). After 56 days of stimulation, MLC1s mRNA had increased to 138% of control levels. The results of Kirschbaum et al. (1989), obtained by *in vitro* translation, indicated that, in rabbit muscles stimulated at 10 Hz for 12 hours/day for 7-12 days, the level of MLC1s mRNA could increase up to two-fold. There was, however, considerable variation between animals - for example, some animals showed no change at all after 12 days - and even after 35 days of stimulation the level of this particular isoform was only 12% of the total MLC population. This could account for the slight differences between our observations.

During the recovery after 6 weeks of stimulation there was a rapid decrease in the level of MLC1s mRNA. This is consistent with the behaviour of the corresponding protein isoforms (Brown, Salmons and Whalen, 1985).

In the TA muscle there was an apparent decrease of about 50% in the level of MLC1/3f mRNA over the first 21 days of stimulation, confirming results reported previously for the stimulated EDL muscle (Brownson et al., 1989). The similarity between this and the early reduction in MLC1s mRNA referred to above tends to reinforce the interpretation given there, that it reflects changes in the types and amounts of total mRNA rather than an absolute decline in MLC mRNA levels. After 35 days' stimulation, however, there was a significant decrease in MLC1/3f mRNA. Again the probe that we used detects alkali light chain mRNAs, and the corresponding protein isoforms are known to persist for a longer period during stimulation than the phosphorylatable light chain MLC2f (Brown, Salmons and Whalen, 1983, 1985). In the study by Kirschbaum et al. (1989), the mRNAs for MLC1f and MLC3f could be measured separately and the changes in MLC3f appeared to be the earlier of the two; however, after 35 days of stimulation neither of these MLCf mRNAs had declined below about 50% of control values. This may be attributable to the intermittent pattern of stimulation employed in their experiments, or to differences arising from the *in vitro* translation technique used.

Our findings show little evidence of coordination between the switching of fast and slow MHC and MLC genes in the rabbit. It must be said that while it has frequently been assumed that such coordination exists, this appears to be based less on actual evidence than on a tacit faith that Nature would wisely opt to produce proteins in strictly stoichiometric amounts. It is possible that coordinated regulation of switching takes place at a different level, between one or more transient forms of MHC and the two predominant forms of MHC (Type 2B and Type 1). However, it is equally possible that there is only limited coordination at the level of myosin gene switching, and that control of subunit composition is exercised mainly at the level of myofibrillar assembly. Further work, with other probes, may shed light on these possibilities.

The results presented here show that an alteration in the pattern of activity sustained by fast muscles triggers events that lead to changes in gene products associated with the contractile apparatus. Our experiments, and related work from other groups, make a strong case for the proposition that many of the protein changes induced by chronic stimulation are the result of regulatory events taking place at a pre-translational level. The intracellular signals responsible for initiating these changes are unknown. However, our studies of events in the first few hours of stimulation suggest that the signalling pathway for transformation is independent of the generation of force and the utilization of metabolic energy, but may be related to events associated with the primary depolarisation of the muscle membrane (Mayne et al., 1991).

We thank the Wellcome Trust and the British Heart Foundation for grant support.

References

Barton, P., Cohen, A., Robert, B., Fiszman, M. Y., Bonhomme, F., Guenet, J-L., Leader, D. P. and Buckingham, M. (1985). The myosin alkali light chains of mouse ventricular and slow skeletal muscle are indistinguishable and are encoded by the same gene. *J. Biol. Chem.* **260**, 8578-8584.

Brown, J. M. C., Henriksson, J. and Salmons, S. (1989). Restoration of fast muscle characteristics following cessation of chronic stimulation: physiological, histochemical and metabolic changes during slow-to-fast transformation. *Proc. R. Soc. B* **235**, 321-346.

Brown, W., Salmons, S. and Whalen, R. (1983). The sequential replacement of myosin subunit isoforms during muscle type transformation induced by long term electrical stimulation. *J. Biol. Chem.* **258**, 14686-14692.

Brown, W., Salmons, S. and Whalen, R. (1985). Mechanisms underlying the asynchronous replacement of myosin light chain isoforms during stimulation-induced fibre-type transformation of skeletal muscle. *FEBS Lett.* **192**, 235-238.

Brownson, C., Isenberg, H., Brown, W., Salmons, S. and Edwards, Y. (1988). Changes in skeletal muscle gene transcription induced by chronic stimulation. *Muscle Nerve* **11**, 1183-1189.

Brownson, C., Little, P., Jarvis, J. C. and Salmons, S. (1992). Reciprocal changes in myosin isoform mRNAs of rabbit skeletal muscle in response to the initiation and cessation of chronic electrical stimulation. *Muscle Nerve* **15**, 694-700.

Brownson, C., Salmons, S. and Edwards, Y. (1989). Changes in the concentrations of selected mRNA transcripts in response to continuous electrical stimulation of skeletal muscle. In Carraro, U. (ed.): *Sarcomeric and Non-sarcomeric Muscles: Basic and Applied Research Prospects for the 90's*. Unipress, Padova, pp. 353-359.

Buller, A. J., Eccles, J. C. and Eccles, R. M. (1960). Interactions between motoneurons and muscles in respect of the characteristic speeds of their responses. *J. Physiol.* **150**, 417-439.

Kirschbaum, B. J., Heilig, A., Härtner, K-T. and Pette, D. (1989). Electrostimulation-induced fast-to-slow transitions of myosin light and heavy chains in rabbit fast-twitch muscle at the mRNA level. *FEBS Lett.* **243**, 123-126.

Kirschbaum, B. J., Schneider, S., Izumo, S., Mahdavi, V., Nadal-Ginard, B. and Pette, D. (1990). Rapid and reversible changes in myosin heavy chain expression in response to increased neuromuscular activity of rat fast twitch muscle. *FEBS Lett.* **268**, 75-78.

Maeda, K., Sczakiel, G. and Wittinghofer, A. (1987). Characterization of cDNA coding for the complete light meromyosin portion of a rabbit fast skeletal muscle myosin heavy chain. *Eur. J. Biochem.* **167**, 97-102.

Mayne, C. N., Jarvis, J. C. and Salmons, S. (1991). Dissociation between metabolite levels and force fatigue in the early stages of stimulation-induced transformation of mammalian skeletal muscle. *Basic Appl. Myol.* **1**, 63-70.

Pette, D. and Vrbová, G. (1985). Neural control of phenotypic expression in mammalian muscle fibres. *Muscle Nerve* **8**, 676-689.

Robert, B., Daubas, P., Akimenko, M-A., Cohen, A., Garner, I., Guenet, J-L. and Buckingham, M. (1984). Single locus in the mouse encodes both myosin light chains 1 and 3, a second locus corresponds to a related pseudogene. *Cell* **39**, 129-140.

Saez, L. and Leinwand, L. A. (1986). Characterization of diverse forms of myosin heavy chain expressed in adult human skeletal muscle. *Nucl. Acid Res.* **14**, 2951-2969.

Salmons, S. (1990). On the reversibility of stimulation-induced muscle transformation. In Pette, D. (ed.): *The Dynamic State of Muscle Fibres*. Walter De Gruyter, Berlin: NY, pp. 401-414.

Salmons, S. and Henriksson, J. (1981). The adaptive response of skeletal muscle to increased use. *Muscle Nerve* **4**, 94-105.

Salmons, S. and Jarvis, J. C. (1991). Simple optical switch for implantable devices. *Med. Biol. Engng Comput.* **29**, 554-556.

Salmons, S. and Vrbová, G. (1969). The influence of activity on some contractile characteristics of mammalian fast and slow muscles. *J. Physiol.* **201**, 535-549.

Sréter, F. A., Gergely, J., Salmons, S. and Romanul, F. (1973). Synthesis by fast muscle of myosin light chains characteristic of slow muscle in response to long-term stimulation. *Nature* **241**, 17-19.

Termin, A., Staron, R. S. and Pette, D. (1989). Changes in myosin heavy chain isoforms during chronic low-frequency stimulation of rat fast hindlimb muscles. *Eur. J. Biochem.* **186**, 749-754.

Printed in Great Britain © *Society for Experimental Biology 1992*

FAST-TO-SLOW TRANSITION IN MYOSIN HEAVY CHAIN EXPRESSION OF RABBIT MUSCLE FIBRES INDUCED BY CHRONIC LOW-FREQUENCY STIMULATION

SIGRID AIGNER and DIRK PETTE

Faculty of Biology, University of Konstanz, Konstanz, Germany

Summary

An *in situ*-hybridization assay using a digoxigenin-labeled cRNA probe specific for the slow myosin heavy chain (HCI) was established. Type I fibres of normal rabbit muscles were stained with this probe. The reaction product was confined to the perinuclear regions of the subsarcolemmal space and extended along the I-bands into the fibre core region. Myosin HCI mRNA was also detected in transforming fibres of low-frequency stimulated rabbit fast-twitch muscles. Its intracellular distribution resembled that of normal type I fibres, but higher amounts of the message were present in fibres undergoing a fast-to-slow transition. The number of HCI mRNA-positive fibres in stimulated muscles increased in a time-dependent manner and correlated with the amount of myosin HCI protein in these muscles. These findings support the notion that enhanced transcription of the slow myosin HC gene leads to an increased translation of HCI mRNA during the stimulation-induced fibre transformation. Finally, the progressive increase in fibres expressing myosin HCI mRNA indicates that the fast-to-slow fibre conversion occurs in a sequential manner. The pre-existing type IIA fibres appear to transform first, whereas fibre types IIB and IID have to first reach the IIA state.

Adult muscle fibers represent versatile entities and may be transformed in response to altered functional demands. Although the majority of normal muscle fibers express only a single myosin HC isoform (for review see Pette and Staron, 1990), the coexistence of two or more myosin HC isoforms has been shown in transforming adult muscle fibers (Staron, Gohlsch, Pette, 1987; Termin, Staron, Pette, 1989). The predominance of a newly expressed HC isoform over an existing isoform may thus ultimately result in a fibre type transition. In the present study, we have used chronic low-frequency stimulation of rabbit fast-twitch muscle to study the replacement of the fast myosin HCIIa by the slow myosin HCI at both the mRNA and protein level. Previous investigations have shown that the expression of this isoform represents the final step of the sequential fast-to-slow transition induced by chronic low-frequency stimulation. Thus, HCI mRNA was shown to be the last isoform to appear in chronically stimulated rabbit fast twitch muscle (Kirschbaum, Heilig, Härtner, Pette 1989). Similarly, the expression of HCI protein and significant increases in the percentage of type I fibres have been

Key words: chronic low-frequency stimulation, *in situ*-hybridization, muscle fibre transformation, rabbit fast-twitch muscle, slow myosin heavy chain mRNA.

observed only after prolonged stimulation periods (Staron et al. 1987; Maier, Gorza, Schiaffino, Pette 1988). The purpose of the present study was to follow these processes at the cellular level.

In situ-hybridization was used to detect HCI mRNA in paraformaldehyde-fixed and paraffin-embedded cross and longitudinal muscle sections. The specific HCI mRNA was detected using a digoxigenin-labeled cRNA probe. The template for transcription of the probe was derived from rabbit cardiac β-myosin cDNA clone pMHCA3/48 kindly supplied by Dr. R. Zak, University of Chicago. This clone is specific for the 3'end of the βHC gene. The 350 bp SacI fragment from pMHCA3/48 was ligated into the T3/T7 transcription vector pBluescribe pBS in such an orientation that T7 RNA polymerase transcribed the hybridising cRNA, and T3 RNA polymerase transcribed the nonhybridising mRNA (Aigner and Pette, 1990). The specificity of the digoxigenin-labeled probe was assessed by Northern blot analysis. The cRNA gave a strong signal with RNA preparations from cardiac ventricle and slow-twitch soleus muscle, but reacted only weakly with total RNA from fast-twitch tibialis anterior (TA) and extensor digitorum longus (EDL) muscles due to the low percentage of type I fibres. No specific signal was obtained with the mRNA probe (Aigner and Pette 1990).

Distribution of myosin HCI mRNA in normal soleus and EDL muscles

As documented by mATPase histochemistry, normal rabbit EDL and TA muscles contain only a minor percentage (3-5%) of slow type I fibres. In agreement with this, the majority of the fibres was unreactive in the *in situ*-hybridization for HCI mRNA and only a small percentage of fibres showed a moderately positive reaction in cross sections of these muscles (Fig. 1a). Conversely, the majority (approximately 95%) of the fibres in the soleus muscle

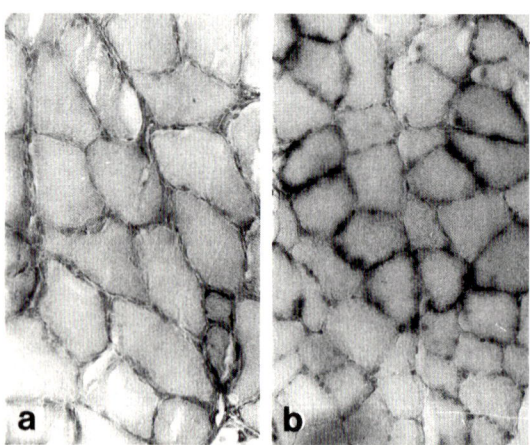

Fig. 1. *In situ*-hybridization of slow myosin heavy chain mRNA in 10 μm thick cross sections of normal (a) and 30-day low-frequency stimulated (b) rabbit tibialis anterior muscle. Magnification 170-fold.

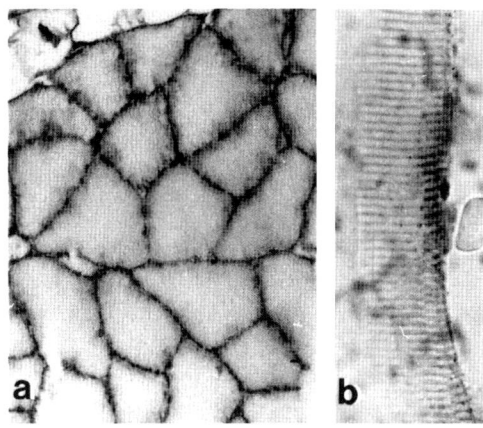

Fig. 2. *In situ*-hybridization of slow myosin heavy chain (HCI) mRNA in 10 μm thick cross (a) and longitudinal (b) sections of normal rabbit soleus muscle. Magnification is 170-fold in a) and 870-fold in b).

reacted positively for HCI mRNA (Fig. 2a). The positive reaction for HCI mRNA was abolished when sections were treated with RNase before the hybridization step.

The distribution of the specific mRNA was restricted to the subsarcolemmal space and perinuclear regions of the individual fibres (Fig. 2a). As seen in longitudinal sections of soleus muscle, the positive reaction was around the subsarcolemmal nuclei and extended in a cross-striational pattern with decreasing intensity into the fibre core region (Fig. 2b). The stained cross striations were identified as the isotropic zones (I-bands) by means of phase contrast and polarization microscopy (Aigner and Pette 1990). Interestingly, this distribution pattern corresponds to the localization of polysomes at the site of the I-bands previously established by electron microscopy (Eisenberg, Dix, Kennedy, 1988). However, using monoclonal antibodies against ribosomal subunits, Horne and Hesketh (1990) showed a preferential distribution of ribosomes in the A-bands of rat skeletal muscle.

Distribution of HCI mRNA in transforming muscle

A significant increase in the amount of the slow-twitch type I fibres was observed in low-frequency stimulated rabbit fast-twitch muscle after stimulation periods longer than three weeks (Maier et al. 1989). This agrees with the onset of increases in the amount of HCI mRNA. As shown by S1-nuclease mapping with a cDNA probe specific for HCI mRNA, low amounts of this mRNA isoform become detectable in RNA preparations from 21-day stimulated rabbit fast-twitch muscle and progressively increase with ongoing stimulation (Kirschbaum et al. 1989). Similar observations were made in the present study using *in situ*-hybridization. As illustrated in Fig. 3, the number of HCI mRNA-positive fibres increased in a time-

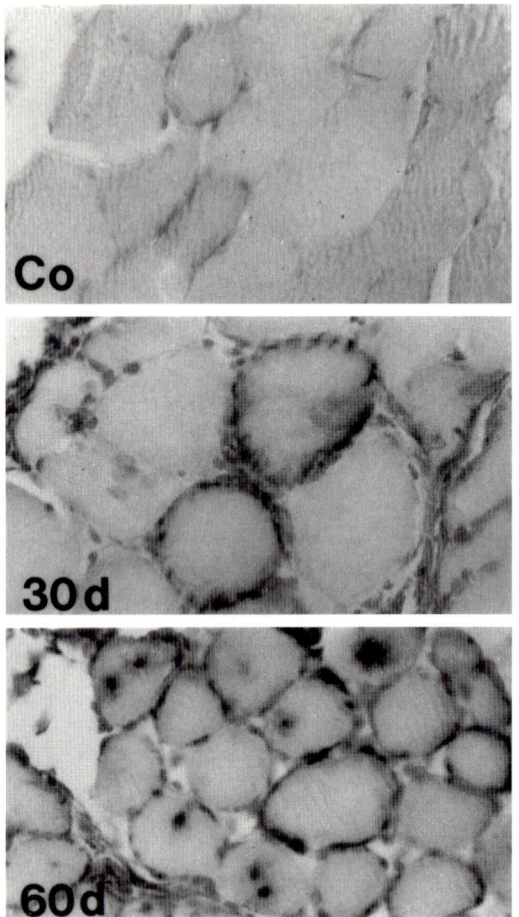

Fig. 3. Increase in HCI mRNA-positive fibres as visualized by *in situ*-hybridization performed on cross sections of unstimulated control (Co), 30-day-, and 60-day-stimulated rabbit extensor digitorum muscles. Note central nuclei in some fibres of the 60-day stimulated muscle. Magnification is 170-fold.

dependent manner in muscles that had been subjected to low-frequency stimulation during longer time periods.

As shown by *in situ*-hybridization studies on cross sections of 30-day stimulated TA muscle, the expression of HCI mRNA was not uniform, but was confined to a certain number of fibres (Fig. 1b, Fig. 3). Also, the positively reacting fibres varied by the intensity of their staining. The intracellular distribution of the HCI mRNA corresponded to that seen in normal type I fibres, although, as judged from the staining intensity, some fibres appeared to contain higher amounts of the specific reaction product. In addition to the staining of the perinuclear regions, these fibres displayed nuclei that contained considerable amounts of the reaction product. The distribution of the reaction product within the stained nuclei was dot-like. A

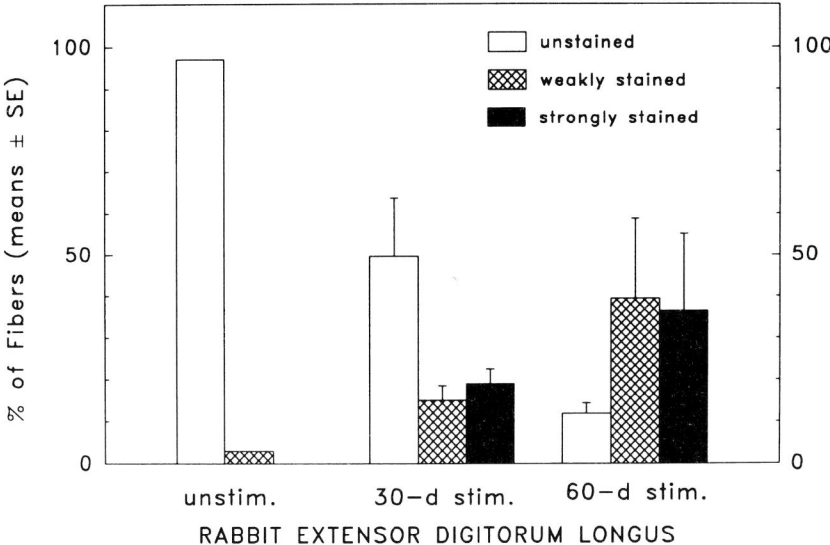

Fig. 4. Time-dependent increases in myosin HCI mRNA-positive fibres as induced by chronic low-frequency stimulation in rabbit extensor digitorum longus muscles.

strong intranuclear and perinuclear staining was also observed in regenerating fibres characterized by their centrally located nuclei (Fig. 3). Regenerating fibres were previously identified in low-frequency stimulated rabbit muscles (Maier, Gambke, Pette, 1986; Maier et al. 1988). Intranuclear staining which was not seen in normal type I fibres, is interpreted as indicating an enhanced expression of HCI mRNA.

The number of HCI mRNA-positive fibres increased with longer stimulation periods (Fig. 3), and amounted to approximately 80% of the total fibre population in 60-day stimulated EDL muscle (Fig. 4). Differences in the staining intensity of the HCI mRNA- positive fibres were also seen in 60-day stimulated muscles. It may be speculated that the strongly reactive fibres have been fully transformed by this time, whereas the less reactive fibres are still undergoing transformation. Therefore, the less reactive fibres most likely represent type C fibres which coexpress the fast HCIIa and the slow HCI (for review see Pette and Staron 1990).

The finding that HCI mRNA-positive fibres were seen only in long-term stimulated muscles and that their number increases with stimulation periods up to 60 days supports the notion that chronic stimulation-induced fibre type transitions occur in a graded and sequential manner (Maier et al. 1988; Termin et al. 1989; Pette 1990). As studied by mATPase histochemistry, chronic low-frequency stimulation of rabbit fast-twitch muscle initially induces a decrease in type IIB fibres concomitant with an increase in type IIA and type IIC/IC fibres. Prolonged stimulation ultimately leads to an increase in type I fibres at the expense of type IIA fibres and type C fibres. Sequential fast-to-slow fibre type conversions have also been deduced from single fibre analyses indicating sequential transitions in

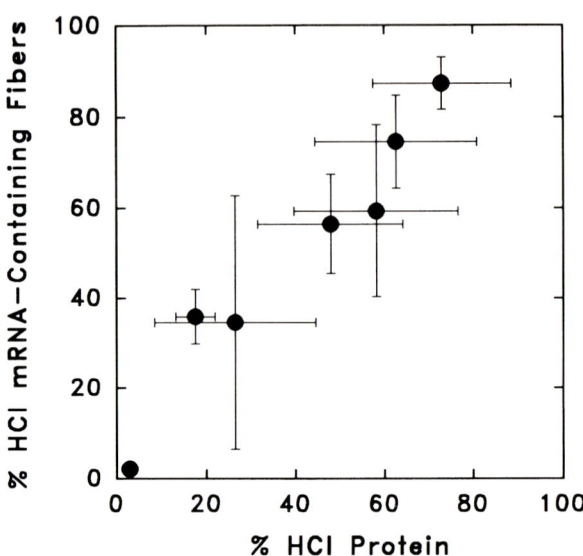

Fig. 5. Correlation between the percentage of myosin HCI mRNA- positive fibres and the percentage of myosin HCI protein in rabbit extensor digitorum longus and tibialis anterior muscles subjected to low-frequency stimulation for different time periods.

the expression of the various fast and slow myosin heavy chain isoforms (Staron et al. 1987; Termin et al. 1989).

In addition, it was of interest to correlate the appearance of HCI mRNA with changes in the protein amount of HCI in chronically stimulated muscles. For this reason, homogenates obtained from the same muscles used for *in situ*-hybridization were analysed electrophoretically for their myosin heavy chain composition. A linear relationship was found between the percentages of the HCI mRNA-positive fibres and the percentage of HCI protein in the same muscles (Fig. 5). These data indicate that the enhanced transcriptional activity of the myosin HCI gene is followed by an increased translation of the transcript.

In conclusion, the induced fast-to-slow transition of rabbit skeletal muscle fibres occurs in a time dependent manner as demonstrated by the progressive increase in the number of fibres expressing the mRNA of the slow myosin HCI. This final step in the transformation process relates to the fibre type IIA-to-type I conversion, i.e., the HCIIa-to-HCI transition. Because this process occurs in a graded manner only after long-term low-frequency stimulation, it appears that type IIA fibers do not transform at the same time. Taking into account that rabbit EDL and TA muscles contain a relatively small percentage of type IIA fibres, it is conceivable that only these native type IIA fibres appear to be capable of performing the final step in the transformation process. As other fast-type fibres (types IIB and IID) will have reached the type IIA state by sequential transitions, these newly formed IIA fibres may then also begin to express myosin HCI. The proposed model of sequential fibre type transitions (Fig. 6) implies that the genes coding for the

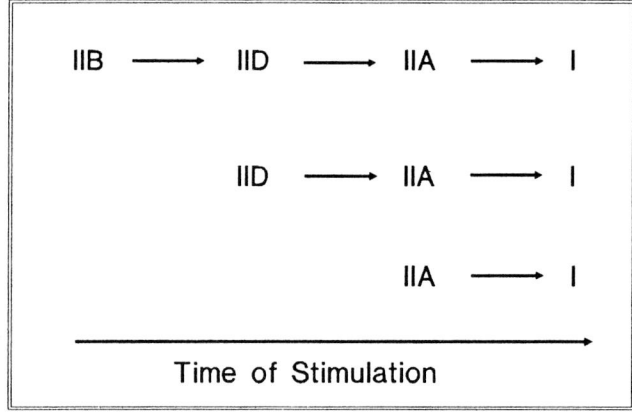

Fig. 6. Schematic representation of the proposed model of sequential fibre type transitions in low-frequency stimulated fast-twitch muscle.

different myosin heavy chain genes respond in a graded manner to low-frequency stimulation and are sequentially activated due to differences in threshold. Thus, the gene for myosin HCI appears to have the highest threshold as it is activated last in the fast-to-slow fibre conversion.

References

Aigner, S. and Pette, D. (1990). In situ hybridization of slow myosin heavy chain mRNA in normal and transforming rabbit muscles with the use of a nonradioactively labeled cRNA. *Histochemistry* **95**, 11-18.

Eisenberg, B. R., Dix, D. J. and Kennedy, J. M. (1988). Physiological factors influencing the growth of skeletal muscle. In *Plasticity of the Neuromuscular System, Ciba Foundation Symposium*, Vol. 138, pp. 3-13. Chichester, New York: J. Wiley.

Horne, Z. and Hesketh, J. (1990). Immunological localization of ribosomes in striated rat muscle. Evidence for myofibrillar association and ontological changes in the subsarcolemmal:myofibrillar distribution. *Biochem. J.* **268**, 231-236.

Kirschbaum, B. J., Heilig, A., Härtner, K.-T. and Pette, D. (1989). Electrostimulation-induced fast-to-slow transitions of myosin light and heavy chains in rabbit fast-twitch muscle at the mRNA level. *FEBS Lett.* **243**, 123-126.

Maier, A., Gambke, B. and Pette, D. (1986). Degeneration-regeneration as a mechanism contributing to the fast to slow conversion of chronically stimulated fast-twitch rabbit muscle. *Cell Tissue Res.* **244**, 635-643.

Maier, A., Gorza, L., Schiaffino, S. and Pette, D. (1988). A combined histochemical and immunohistochemical study on the dynamics of fast to slow fiber transformation in chronically stimulated rabbit muscle. *Cell Tissue Res.* **254**, 59-68.

Pette, D. (1990). Dynamics of stimulation-induced fast-to-slow transitions in protein isoforms of the thick and thin filament. In *The Dynamic State of Muscle Fibers*, (ed. D. Pette), pp. 415-428. Berlin New York: W. de Gruyter.

Pette, D. and Staron, R. S. (1990). Cellular and molecular diversities of mammalian skeletal muscle fibers. *Rev. Physiol. Biochem. Pharmacol.* **116**, 1-76.

Staron, R. S., Gohlsch, B. and Pette, D. (1987). Myosin polymorphism in single fibers of chronically stimulated rabbit fast-twitch muscle. *Pflügers Arch.* **408**, 444-450.

Termin, A., Staron, R. S. and Pette, D. (1989). Changes in myosin heavy chain isoforms during chronic low-frequency stimulation of rat fast hindlimb muscles - A single fiber study. *Eur. J. Biochem.* **186**, 749-754.

Printed in Great Britain © *Society for Experimental Biology 1992*   319

REGULATION OF INSULIN-LIKE GROWTH FACTOR 1 GENE EXPRESSION IN SKELETAL MUSCLE

PAUL T. LOUGHNA and PAUL MASON

Department of Veterinary Basic Science, Royal Veterinary College, Royal College St.,
London NW1 0TU

PETER C. BATES

AFRC Institute of Animal Physiology and Genetics, Babraham Hall, Babraham, Cambridge

Summary

Insulin-like growth factor 1 (IGF-1) is implicated in the growth processes of many tissues in the adult animal. This hormone can act in an endocrine manner or can be produced in the specific tissues in response to growth promoting stimuli to act in an autocrine/paracrine manner. We have examined, in the rat, changes in serum concentrations of IGF-1 and muscle IGF-1 mRNA levels in several studies in which muscle growth has been significantly altered. In the first study we examined the interactions of growth hormone (GH) and under-nutrition upon muscle growth. We observed that when GH was administered to hypophysectomised rats the anabolic effect of this hormone was independent of dietary intake. In a similar manner muscle IGF-1 mRNA levels were also elevated by GH but unaffected by food intake. In contrast serum IGF-1 levels were markedly reduced by under-nutrition. These data suggested that the anabolic action of GH on muscle could be mediated through the autocrine/paracrine action of the IGF-1 hormone. Similarly we observed in other studies that muscle hypertrophic stimuli of work-overload and passive stretch are associated with significantly increased muscle IGF-1 mRNA levels. In contrast insulin dramatically affected muscle protein synthesis rates but had no measurable effect upon muscle IGF-1 mRNA levels, which suggests that the anabolic action of this hormone is not mediated through the autocrine/paracrine action of IGF-1. These studies suggest that IGF-1 may mediate growth in muscle in response to variety of stimuli by autocrine/paracrine action or in response to certain stimuli possibly by endocrine action.

Introduction

Muscle, the major component of lean body tissue and the largest protein store in the body, is capable of undergoing rapid and extensive alterations in mass. Such alterations in the growth of this tissue can be produced in response to a variety of different stimuli, including altered nutritional and endocrine status as well as neural and mechanical influences. Over the past few years our laboratories have

Key words: insulin-like growth factor 1, muscle hypertrophy, protein synthesis, growth hormone.

examined the effects of many factors which modulate muscle growth studying changes in contractile protein gene expression in the rat. From these studies and from work carried out in other laboratories and in other species, it has been observed that quantitatively similar changes in mass, produced in response to different factors, can be associated with distinctly different patterns of contractile protein gene expression. This was first clearly shown in hypertrophy of the rat cardiac ventricle; the response to volume overload is associated with an induction of the β myosin heavy chain (β-MHC) and skeletal α-actin isoforms whereas a similar level of hypertrophy produced by elevated thyroxine concentration causes a reduced expression of the β-MHC gene with no change in skeletal α-actin expression (Izumo et al., 1988). We have subsequently observed similar distinct patterns of contractile protein gene expression in rat muscle undergoing protein accumulation in response to a variety of factors including passive stretch, work-overload, growth hormone (GH), insulin and the β adrenoceptor agonist clenbuterol. A preliminary analysis of this data suggests that completely different and independent intracellular metabolic pathways are involved to produce muscle anabolism in response to different growth inducing factors.

However a further analysis of the data, from our laboratories and those of other workers, reveals two other characteristics of contractile protein gene expression during non-stable growth states. Firstly the alterations in expression of genes coding for isoforms of different contractile proteins are discoordinate. Secondly that although changes in the expression of isoforms of various contractile proteins may alter during growth, the total amount of mRNA coding for all isoforms of a protein probably maintain stoichiometry with total mRNA for each of the other contractile proteins (Wade et al., 1990).

Thus the mechanisms by which different classes of anabolic agents stimulate growth could be different at the physiological level but may share common intracellular pathways. Such pathways could upregulate the total amounts of each protein and any switching in expression of isoforms may result from a 'fine tuning' process. Several molecular signals that might be involved in mediating the anabolic signal in response to a growth-inducing stimuli have been identified. Muscle cells are known to be targets for the action of several peptide growth factors, which may act in either an endocrine or an autocrine/paracrine manner to mediate the hypertrophic process. Present evidence suggests that one factor which may be particularly involved in muscle hypertrophy is insulin-like growth factor 1 (IGF-1). We have therefore tried to examine the changes in concentrations of IGF-1 and its mRNA which are associated with different anabolic states.

IGF-1 and its gene

IGF-1 is a single peptide chain, 70 amino acid long and similar to proinsulin, contains an amino-terminal B domain and an A domain that are connected by a short C region. Unlike proinsulin, IGF-1 also contains a D region extension peptide at the carboxy-terminus. Furthermore the nascent prepro-IGF-1 has an amino terminal E peptide which is presumably cleaved during processing as it is absent in the mature circulating form of IGF-1. Elucidation of the structure of the rat IGF-1 gene has shown that it consists of at least six exons distributed over more than 50 Kb (Shimatsu and Rotwein, 1987; Roberts et al., 1987). There are at least two leader exons 1 and 2 which contain different 5′-untranslated region sequences and sequences which probably encode different signal peptides. These exons are mutually exclusive and are alternatively spliced to exon 3 which encodes some signal peptide sequence and the majority of the B domain. The reminder of the mature IGF-1 peptide as well as part of the E domain is encoded by exon 4. Exon 6 encodes the remainder of the E domain and all of the 3′ untranslated region. Exon 5, in the rat, is a 'cassette' exon, whose sequence is retained in some IGF-1 mRNAs but not in others. A number of different mRNAs are produced from the IGF-1 gene by differential splicing of the primary transcript. Exons 1 and 2 are mutually exclusive and their inclusion produces class 1 and class 2 mRNAs respectively. Exon 5 determines the nature of the E peptide and it's absence or presence produces IGF-1 mRNAs termed Ea or Eb respectively. On Northern blots of mammalian RNA, a number of IGF-1 mRNA species of between 0.8-7.5 Kb are observed, all of which are large enough to encode the IGF-1 precursor. Multiple polyadenylation signals are present in rat exon 6 and the mRNA size heterogeneity has been shown to result primarily from the use of progressively downstream polyadenylation sites in exon 6, resulting in variable lengths of 3′-UTRs in IGF-1 mRNAs. These different mRNAs are expressed in a tissue-specific manner and may be subject to translational control or regulation at the level of transcript stability. The factors involved in determining the various transcripts remain to be elucidated. However while some transcripts may encode differing prohormones all of these differing mRNAs produce the same functional hormone.

Regulation of IGF-1 gene expression

IGF-1 is a ubiquitous protein with activity detectable in tissue extracts as well as in conditioned media of many cell types in culture. Although in the adult animal the liver is thought to be the main source of circulating IGF-1, which would be involved in growth regulation in an endocrine manner, other tissues are known to produce IGF-1 suggesting that this peptide can act as an autocrine/paracrine hormone. The relative importance of endocrine versus paracrine/autocrine action of IGF-1 in their influence upon muscle anabolism is still poorly understood. We have carried out a number of studies to investigate changes in serum levels and tissue specific gene expression of IGF-1 where marked alterations in muscle growth are induced in response to various stimuli.

(i) Growth hormone and nutrition

The principal regulators of IGF-1 postnatally are growth hormone and nutrition. These two factors are also highly influential in determining the rate of

Table 1. *Liver*

	Hx0	Hx7	GH	GH-PF
% Body weight	3.90	4.10	4.38	3.05*[#]
Protein (mg)	1344	1322	1453	1123*[#]
RNA/Protein	45.2	46.0	50.9*	51.0*

Liver mass, protein and RNA content of rats hypophysectomised rats subjected to GH treatment and altered food intake. Male rats (body wt. 191 ± 3g; mean ± S.E.M.). After hypophysectomy they were divided into four groups; the first group were killed after 14 days (Hx0; n=4) and the remaining 3 groups (n=5 per group) were treated for 7 days with saline (Hx7), hGH (GH; 60mU/day) and fed *ad libitum* or hGH but with food intake restricted to match that of Hx7 (GH-PF).

Values are means ± S.E.M. and statistical analysis was carried out using Student's t-test (* compared to Hx7, P<0.01; # comparison between GH and GH-PF, P<0.01).

skeletal muscle protein accretion. Daughaday et al. (1972) proposed that the anabolic actions of GH were mediated via circulating IGF-1, produced in the liver in response to GH and transported in blood to the site of its action as an endocrine hormone. However, it is also clear that IGF-1 production within skeletal muscle is altered by GH (Turner et al., 1988). Studies in human volunteers have provided insight into nutritional regulation of IGF-1 concentrations in serum. Fasting decreases serum IGF-1 concentrations and it would appear that IGF-1 acts as a critical link between nutritional status and post-natal growth. In rats fasting causes a reduction in binding of GH to its receptors in the liver and a fall in serum IGF-1 concentrations (Maes et al., 1983). We carried out studies to investigate further the interrelationship between GH and food intake in regulating muscle growth and to determine whether IGF-1, acting in either an endocrine or paracrine manner may mediate such regulation. Male rats were hypophysectomised (Hx) and maintained for 14 days to ensure no growth and completeness of Hx; they were then randomly allocated into four treatment groups. The first group (Hx0) were killed, the second group were treated for seven days with saline (Hx7) and the remaining two groups were administered human GH and were either fed ad libitum (GH) or pair-fed to match the reduced food intake levels of the Hx7 group (GH-PF). A fifth group of non-hypophysectomised rats of equivalent body weight were included as a control group. The liver, gastrocnemius muscle and serum were removed from each animal for further analysis. Liver weights were increased in hypophysectomised rats administered GH when compared to saline administered rats (Hx7) whereas restricted food intake prevented this increase in GH-PF (Table 1). In contrast muscle weights were increased by GH administration irrespective of food intake (Table 2). Serum levels of IGF-1 increased in GH but not with restricted food intake (GH-PF) (Table 3). Thus although the serum concentrations of IGF-1 were markedly different in GH administered animals dependant upon dietary level, muscle growth was unaffected; this suggests that muscle growth in response to GH administration was not under the control of circulating IGF-1 acting in an endocrine manner. Muscle IGF-1 mRNA concentrations fell

Table 2. *Gastrocnemius muscle*

	Hx0	Hx7	GH	GH-PF
% Body weight	0.49	0.48	0.48	0.53*[#]
Protein (mg)	142*	128	147*	148*
K_G (%/d)	–	−1.58	0.87*	0.85*
RNA/Protein	7.21*	6.63	8.14*	7.46*

Growth parameters of the gastrocnemius muscle from hypophysectomised rats fed *ad libitum*, with (GH) and without (hx7) administration and given GH on a restricted food intake. Statistical analysis as described above (Table 1).

dramatically from Hx0 to Hx7 but were maintained in GH irrespective of dietary intake (Fig. 1). Thus the levels of IGF-1 mRNA in the gastrocnemius closely paralleled the growth rates for this tissue. These data suggest that the most important mediator of GH influence upon skeletal muscle growth is the autocrine/paracrine action of IGF-1; the endocrine action is considerably less influential.

(ii) Insulin

The most thoroughly investigated role for insulin action upon muscle is its effects upon glucose utilisation and uptake; however, it has been evident for many years that this hormone is involved in regulating muscle growth. Manchester and Young (1960) reported that insulin stimulated amino acid uptake in isolated diaphragm muscle. Subsequently a number of *in vivo* and *in vitro* studies have indicated a highly significant influence of insulin on muscle protein synthesis rates. Studies on muscle cells in culture, have however generally needed supraphysiological levels of insulin to produce anabolic effects and thus it has been suggested that these effects could be mediated by the IGF-1 receptor.

We carried out a study to investigate the *in vivo* effects of insulin upon muscle protein anabolism and to examine a possible role for IGF-1 in mediating these effects. Rats were made diabetic by administration of streptozotocin and randomly divided into four groups, two of which were administered with insulin and two which were not, muscle protein synthesis rates and IGF-1 mRNA concentrations

Table 3.

	Hx0	Hx7	GH	GH-PF
Serum				
Insulin (pmol/l)	104	102	159*	43*[#]
IGF-1 (nmol/l)	3.0	3.1	12.4*	3.2[#]

Interactions of GH and undernutrition upon serum insulin and IGF-1 concentrations in hypophysectomised rats. Experimental design and statistical analysis as described above (Table 1). Insulin and IGF-1 concentrations measured by RIA.

Gastrocnemius muscle concentration of IGF-1 mRNA

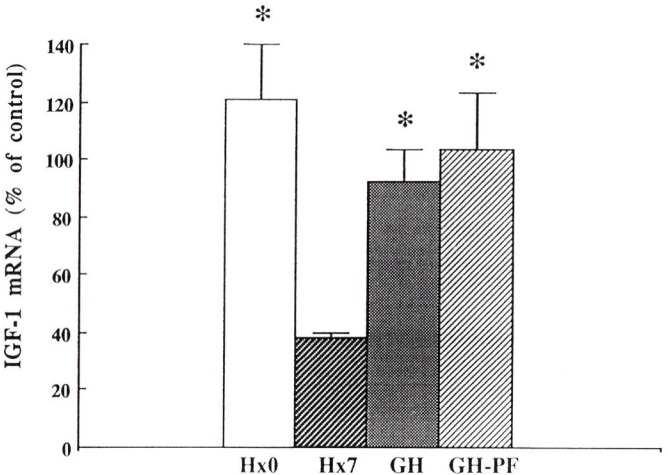

Fig. 1. Relative IGF-1 mRNA levels in muscles from hypophysectomised rats with and without growth hormone administration. Total RNA was extracted using a modification of the hot phenol procedure (Soeiro et al., 1966). IGF-1 mRNA concentrations were measured by standard slot blot analysis using a 5′ end-labelled exon 2 specific oligonucleotide as a probe. Values presented as mean ± SEM. Statistical significance from Hx7 analysed by Student's t-test (* P<0.01).

were measured after 1 and 3 days. At both time points insulin administration was observed to cause a significant elevation of muscle protein synthesis rates (Fig. 2). However despite the increase in protein synthesis and muscle growth produced by the administration of insulin to diabetic rats, no alteration in IGF-1 concentrations or IGF-1 mRNA levels could be detected (data not shown). This suggests that alterations in protein synthesis and growth produced in response to altered serum levels of insulin are not mediated by IGF-1 acting in an autocrine/paracrine manner. In contrast serum levels of IGF-1 were significantly higher at both time points in diabetic animals administered insulin (Table 4). Muscle protein synthesis rates could therefore be influenced by insulin and/or IGF-1 acting in an endocrine manner on muscle tissue. In contrast to the serum levels of insulin, however, the percentage increase in serum IGF-1 levels in response to insulin administration (134% after 1 day and 224% after 3 days) was very similar to the percentage elevation in protein synthesis rates in response to insulin administration at these time points (Table 4, Fig. 2). These data therefore suggest that the endocrine action of IGF-1 may be more important in elevating protein synthesis in muscle when insulin is administered to diabetic rats than the direct action of insulin itself on this tissue.

(iii) Work-overload and mechanical influences

Muscle mass is highly susceptible to altered patterns of activity. Muscle inactivity, provided the muscle is not subjected to passive stretch, will rapidly lead

Effects of Insulin on Muscle Protein Synthesis Rates

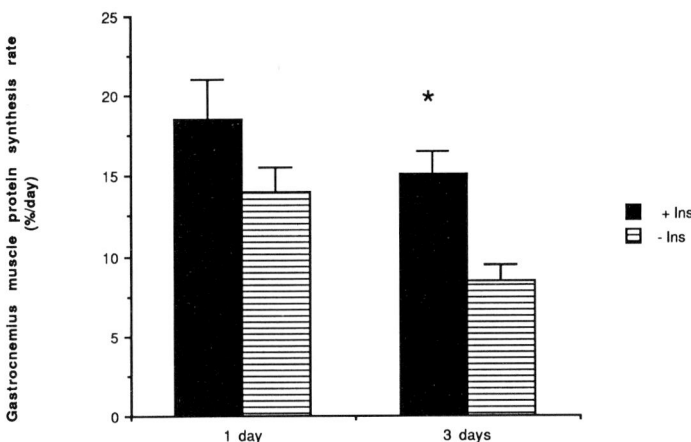

Fig. 2. Fractional rates of protein synthesis in muscle from diabetic rats, with and without administration of insulin. Protein synthesis was measured using a flooding dose of radioactively labelled phenylalanine, administered intra-peritonealy, as described previously (Bates and Holder, 1988). Values presented as mean ± SEM. Statistical significance of insulin administration measured by Student's t-test (* P<0.01).

Table 4. *Serum*

	+Ins 1d	−Ins 1d	+Ins 3d	−Ins 3d
Insulin conc (pmol/l)	196 + 16	92 + 15	351 + 64	15 + 7
IGF-1 (nmol/l)	32.0 + 2.1	23.8 + 2.1	34.1 + 0.6	15.2 + 2.4

Serum insulin and IGF-1 concentrations in diabetic rats after administration of insulin (+Ins) or saline (−Ins) over a 1 or 3 day time period.

to atrophy due to increased protein degradation rates and decreased synthesis rates. In contrast increased tension and certain patterns of increased muscle activity can produce a hypertrophic response. One animal model for inducing muscle hypertrophy that has undergone intensive investigation involves overload of the rat soleus and plantaris muscles by tenotomy of the synergistic gastrocnemius. In this model increased growth in both muscles is associated with increased protein synthesis rates and alterations in contractile protein isoforms at both the mRNA and protein level. Furthermore, DeVol et al. (1990) have recently shown that hypertrophy of the soleus and plantaris in this model is associated with elevated IGF-1 mRNA levels in these muscles.

We have carried out a similar study, in which we induced overload in the plantaris and soleus muscles of the rat by tenotomy of the gastrocnemius muscle

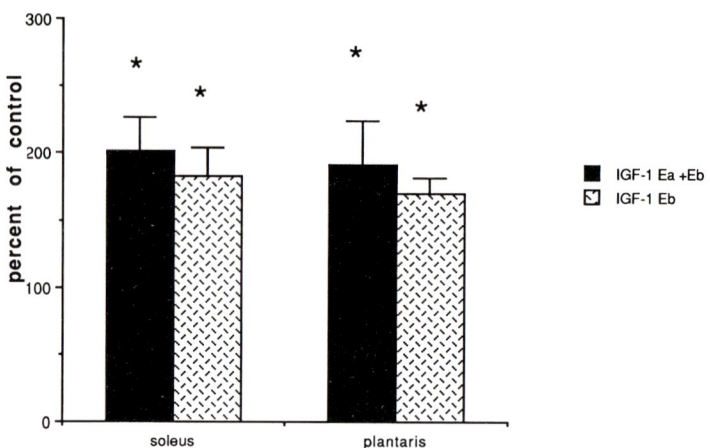

Fig. 3. Relative levels of IGF-1 Eb and total IGF-1 mRNA levels in work-overloaded soleus and plantaris muscles, produced by tenotomy of the synergistic gastrocnemius muscle, compared to muscles from sham operated control animals. RNA was extracted and total IGF-1 mRNA concentrations measured by slot blot analysis as described in Fig. 1. IGF-1 Eb concentrations were measured in the same manner using an exon 5 specific oligonucleotide as a probe. Values are presented as mean ± SEM. Statistical significance from control musclesanalysed by Student's t-test (* P<0.01).

and measured IGF-1 mRNA levels after 5 days. We observed similar increase in muscle IGF-1 mRNA levels to those observed by DeVol et al. (1990) (Fig. 3). These data suggest that IGF-1 may be acting in an autocrine/paracrine manner during work-overload induced muscle hypertrophy. A preliminary examination of the possibility that specific mRNA transcripts were preferentially induced in muscle in response to work-overload was also undertaken by measuring the relative levels of IGF-1 Eb using an exon 5 specific oligonucleotide as a probe. However, no significant difference in the percentage elevation of these transcripts was observed when compared to total IGF-1 mRNA (Fig. 3).

The 'trigger' for increased muscle growth in this model is not fully established but, while factors such as the activity of the nerve may play a contributory role, there is growing evidence that the mechanical influence of tension is the most crucial factor. We have previously shown that tension induced by passively stretching a muscle induces rapid muscle hypertrophy, increases protein synthesis and causes similar qualitative changes in myosin heavy chain gene expression to those observed in the work-overload model described above (Laurent et al., 1978; Loughna et al., 1986; Loughna et al., 1990). To examine the effects of passive stretch upon IGF-1 gene expression we immobilised rat hind-limbs in a dorsi-flexed position such that the slow postural soleus and the fast phasic plantaris muscles were both held inactive at greater than resting length for 2 days and 5

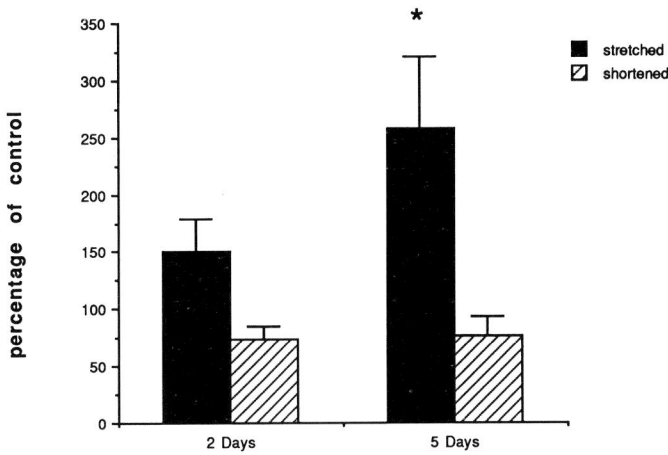

Fig. 4. Relative levels of IGF-1 mRNA/poly A RNA in soleus muscles from rat limbs immobilised in either the dorsi-flexed (stretched) or plantar-flexed (shortened) position compared to levels in the soleus of contralateral control limbs. IGF-1 mRNA concentration was measured as described in Fig. 1 and poly A levels were measured using oligo-dt as described previously (Morgan and Loughna, 1989). Values are given as mean ± SEM. Statistical significance from control (contralateral limb) muscles analysed by Student's t-test (* P<0.01).

days. To disassociate the effects of disuse from those of passive stretch in another group of animals one hind-limb was immobilised in the plantar-flexed position such that these muscles were held inactive at less than resting length. In each animal of both groups muscles from the non-immobilised contralateral limb were used as controls. In the soleus inactivity without stretch (immobilisation in the shortened position) produced no significant difference in levels of IGF-1 mRNA when compared to contralateral controls after either 2 days or 5 days (Fig. 4). In contrast at both time points passive stretch induced a significant increase in IGF-1 mRNA concentrations. Similarly in the plantaris stretch also caused a significant rise in IGF-1 mRNA levels; however, surprisingly, after 5 days of immobilisation in the shortened position there was also an increase of IGF-1 mRNA in this muscle (Fig. 5). Therefore, hypertrophy due to stretch causes an increase in IGF-1 mRNA concentration similar to that produced in the work-overload model. It is probable that in both situations the IGF-1 peptide could be involved in an autocrine/paracrine manner to induce an increase in muscle growth.

Discussion

Within each of the skeletal muscle anabolic situations described it is known that there are changes in the relative levels of the various fast and slow isoforms of the contractile proteins. However, in order to maintain the appropriate proportions of

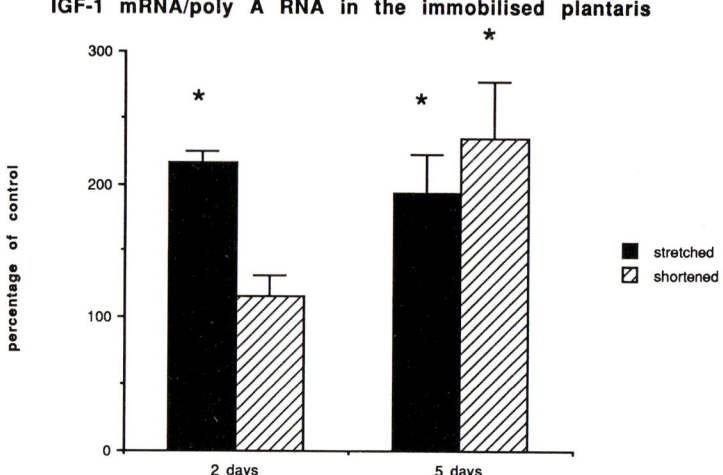

Fig. 5. Relative levels of IGF-1 mRNA/poly A in plantaris muscles immobilised in either a lengthened or a shortened position compared to muscles from the contralateral control limb. Values given as mean ± S.E.M. Statistical significance analysed as for Fig. 4.

the various contractile proteins there must be some form of coordination. Thus the regulation of growth may be seen as a pathway where various steps may be influenced to alter protein accumilation and that the changes in the relative levels of specific isoforms within a gene family is a fine tuning process and is independantly regulated. It is clear that the action of IGF-1 on its receptor can influence the anabolic pathway in skeletal muscle (Pell and Bates, 1992). With GH treatment and mechanical stress there is an increase in the levels of mRNA for IGF-1 within the tissue itself wheras with insulin there appears to be an increase in the endocrine action of IGF-1. Paracrine/autocrine regulation of muscle growth can obviously be regulated at the level of tissue production of this peptide. However selective response of individual cells within a tissue to either paracrine or endocrine action of IGF-1 can be regulated by receptor levels and also by the local production of IGF-binding proteins (IGFBPs). At present six IGFBPs have been identified in the rat but their biological actions wnen they have interacted with IGF-1 are unknown. The IGFBPs have been shown to have both stimulatory and inhibitory effects on IGF action *in vitro* (Elgin et al., 1987; Rutanen et al., 1988). It is likely that the enhancement or inhibition of IGF-1 action by IGFBPs will depend upon the type of IGFBPs, the target cell type and the microenvironment of interaction. The different IGFBPs can interact with IGFs at the endocrine, paracrine or autocrine level. While little is known about the biological actions of IGFBPs or about the regulation of IGFBP production by skeletal muscle it has been observed that serum levels of certain IGFBPs and their respective mRNA concentrations in the liver are subject to modulation by insulin (Unterman et al., 1989; Unterman et al., 1990; Ooi et al., 1990).

A great deal of further work will be required to determine exactly how IGF-1 influences accretion of muscle protein and whether different transcripts are produced in response to different stimuli thus channeling responses to production of the same IGF-1 protein. Other possible common mediators of muscle growth promoting stimuli may also be identified but at least in some cases IGF-1 would appear to play an important role.

References

Bates, P. C. and Holder, A. T. (1988). The anabolic actions of growth hormone and thyroxine on protein metabolism in Snell dwarf and normal mice. *J. Endocrinol.* **119**, 31-41.

Daughaday, W. H., Hall, K., Raben, M. S., Salmon, W. D., Van den Brande, J. L. and Van Wyk, J. J. (1972). Somatemedin: proposed designation for sulphation factor. *Nature* **235**, 107-108.

DeVol, D. L., Rotwein, P., Sadow, J. L., Novakofski, J. and Betchel, P. J. (1990). Activation of insulin-like growth factor gene expression during work-induced skeletal muscle growth. *Am. J. Physiol.* **259** (Endocrinol. Metab.22), E89-E95.

Elgin, R. C., Busby, W. H. and Clemmons, D. R. (1987). An insulin-like growth factor (IGF) binding protein enhances the biologic response to IGF-1. *Proc. Natl. Acad. Sci. USA* **84**, 3254-3258.

Izumo, S., Nadal-Ginard, B. and Mahadavi, V. (1988). Protooncogene induction and reprogramming of cardiac gene expression produced by pressure overload. *Proc. natn. Acad. Sci. U.S.A.* **85**, 339-343.

Laurent, G. J., Sparrow, M. P., Bates, P. C. and Millward, D. J. (1978). Turnover of muscle protein in the fowl: changes in rates of protein synthesis and breakdown during hypertrophy of the anterior and posterior latissimus dorsi muscles. *Biochem. J.* **176**, 407-417.

Loughna, P. T., Goldspink, G. and Goldspink, D. F. (1986). Effect of inactivity and passive stretch on protein turnover in phasic and postural rat muscles. *J. Appl. Physiol.* **61**(1), 173-179.

Loughna, P. T., Izumo, S., Goldspink, G. and Nadal-Ginard, B. (1990). Disuse and passive stretch cause rapid alterations in expression of developmental and adult contractile protein genes in skeletal muscle. *Development* **109**, 217-233.

Maes, M., Underwood, L. E. and Ketelslegers, J. M. (1983). Plasma somatomedin-C in fasted and refed rats: close relationship with changes in liver somatogenic but not lactogenic binding sites. *J. Endocrinol.* **97**, 243-252.

Manchester, K. I. and Young, F. G. (1960). The influence of the induction of alloxan-diabetes on the incorporation of amino acids into protein of rat diaphragm. *Biochem. J.* **77**, 386-399.

Morgan, M. J. and Loughna, P. T. (1989). Work overload induced changes in fast and slow skeletal muscle myosin heavy chain gene expression. *FEBS Letts.* **255**, 427-430.

Ooi, G. T., Orlowski, C. C., Brown, A. L., Becker, R. E., Unterman, T. G. and Rechler, M. M. (1990). Tissue distribution and hormonal regulation of mRNAs encoding rat insulin-like growth factor binding proteins rIGFBP-1 and rIGFBP-2. *Mol. Endocrinol.* **4**, 321-328.

Pell, J. M. and Bates, P. C. (1992). Differential actions of growth hormone and insulin-like growth factor on tissue protein metabolism in dwarf mice. *Endocrinology* **130** (4), In press.

Roberts Jr, C. T., Lasky, S. R., Lowe Jr, W. L., Seaman, W. T. and LeRoith, D. (1987). Molecularcloning of rat insulin-like growth factor I complementary deoxyribonucleic acids: Differential messenger ribonucleic acid processing and regulation by growth hormone in extrahepatic tissues. *Mol. Endocrinol.* **1**, 243-248.

Rutanen, E. M., Pekonen, F. and Makinen, T (1988). Soluble 34K binding protein inhibits the binding of insulin-like growth factor I to its cell receptors in human secretory phase endometrium: Evidence for autocrine/paracrine regulation of growth factor action. *J. Clin. Endocrinol. Metab.* **66**, 173-180.

Shimatsu, A. and Rotwein, P. (1987). Mosaic evolution of the insulin-like growth factors. *J. Biol. Chem.* **262**, 7894-7900.

Soeiro, R. H., Birnboim, C. and Darnell, J. E. (1966). Rapidly labelled HeLa cell nuclear RNA

II. Bast composition and cellular localization of a heterogenous RNA fraction. *J. molec. Biol.* **19**, 362-372.

Turner, J. D., Rotwein, P., Novkofski, J. and Betchel, P. (1988). Induction of mRNA for IGF-I and II during growth hormone-stimulated muscle hypertrophy. *Am. J. Physiol.* **255**, E513-E517.

Unterman, T. G., Oehler, D. T. and Becker, R. (1989). Identification of a type 1 insulin-like growth factor binding protein (IGF BP) in serum from rats with diabetes mellitus. *Biochem. Biophys. Res. Commun.* **163**, 882-887.

Unterman, T. G., Patel, K., Kumar Mahathre, V., Rajamohan, G., Oehler, D. T. and Becker, R. E. (1990). Regulation of low molecular weight insulinlike growth factor binding proteins (IGF BPs) in experimental diabetes mellitus. *Endocrinology* **126**, 2614-2624.

Wade, R. W., Sutherland, C., Gahlmann, R., Kedes, L., Hardeman, E. and Gunning, P. (1990). Regulation of contractile protein gene family mRNA pool sizes during myogenesis. *Develop. Biol.* **142**, 270-282.

Printed in Great Britain © *Society for Experimental Biology 1992*   331

ACTIVATION OF MUSCLE-SPECIFIC TRANSCRIPTION BY MYOGENIC HELIX-LOOP-HELIX PROTEINS

ERIC OLSON

Department of Biochemistry and Molecular Biology, The University of Texas
M. D. Anderson Cancer Center, 1515 Holcombe Blvd., Houston, Tx 77030

Abstract

Myogenin is a muscle-specific transcription factor that acts as a molecular switch to induce myogenesis. Myogenin shares homology with MyoD and other myogenic regulatory proteins within a basic region and helix-loop-helix (HLH) motif that mediate binding to a conserved DNA sequence (CANNTG) present in the regulatory regions of numerous muscle-specific genes. Binding of myogenin and other members of the MyoD family to DNA can be augmented upon heterodimerization with the widely expressed HLH protein E12. We have used the muscle creatine kinase (MCK) enhancer as a target to study the mechanism whereby myogenin activates muscle-specific transcription. Full activity of the MCK enhancer requires cooperative interactions between myogenin (or other myogenic HLH proteins that bind the same site) and a complex array of ubiquitous and cell type-specific nuclear factors. To define the domains of myogenin responsible for sequence-specific DNA binding, activation of muscle-specific transcription, and cooperativity with other transcription factors, we have generated an extensive series of mutants by site-directed mutagenesis and domain swapping. These mutants have revealed strong transcriptional activation domains in the N- and C-termini of myogenin that rely on a specific amino acid sequence within the DNA binding domain for activity. Myogenin's ability to induce muscle-specific transcription is subject to negative regulation by growth factor and oncogenic signals. Mechanisms through which growth signals may repress myogenin function are discussed.

Introduction

Activation of the muscle differentiation program involves the coordinate induction of an array of genetically unlinked muscle-specific genes that encode the proteins required for the specialized functions of the mature myofiber. The cloning of the muscle-specific regulatory factor MyoD (Davis, Weintraub and Lassar, 1987), and subsequent identification of the related myogenic factors myogenin (Wright, Sassoon and Lin, 1989; Edmondson and Olson 1989), myf5 (Braun, Buschhausen-Denker, Bober, Tannich and Arnold, 1989) and MRF4 (Rhodes and Konieczny 1989)/herculin (Miner and Wold 1990)/myf6 (Braun, Bober, Winter, Rosenthal and Arnold, 1990a) has led to rapid progress toward understanding the

Key words: differentiation, muscle creatine kinase, myogenesis, helix-loop-helix proteins.

molecular mechanisms that underlie the establishment of this complex cellular phenotype (for reviews, see Olson 1990; Tapscott and Weintraub 1991). Our group has focused on the role of myogenin in the activation of muscle-specific transcription. To approach this problem, we have used the muscle creatine kinase (MCK) enhancer as a prototypical muscle-specific regulatory element and have attempted to identify the cellular factors with which myogenin collaborates to activate transcription from this regulatory region. We have also created a series of myogenin mutants in order to define the protein domains responsible for its ability to induce the myogenic program. The results of these studies and their implications for understanding the molecular events associated with formation of differentiated skeletal muscle and perhaps of other specialized cell types will be addressed in this review.

The MCK enhancer is comprised of multiple regulatory elements that serve as binding sites for muscle-specific and ubiquitous nuclear factors

The MCK enhancer lies between -1050 and -1250 base pairs upstream of the MCK gene and is among the most well-characterized muscle-specific control regions (Jaynes, Johnson, Bushkin, Gartside and Hauschka, 1988; Sternberg, Spizz, Perry, Vizard, Weil and Olson, 1988; Horlick and Benfield 1989). The enhancer is silent in nonmuscle cell types and in myoblasts exposed to growth medium (GM) containing high concentrations of fetal bovine serum or type-beta transforming growth factor (TGF-beta). The MCK enhancer is also silenced by activated RAS gene products (Sternberg, Spizz, Perry and Olson, 1989), which extinguish expression of myogenin and MyoD (Lassar, Thayer, Overell and Weintraub, 1989a). Following withdrawal of serum from myoblasts, the enhancer is rapidly activated. Activation of the enhancer upon mitogen withdrawal is a cell type-specific response and does not occur in nonmyogenic cells under the same conditions.

The MCK enhancer, like other cellular and viral enhancers, is comprised of a complex array of elements that serve as binding sites for cell type-specific and ubiquitous transcription factors (Fig. 1). The cell type-specific pattern of transcription characteristic of this enhancer results from the combinatorial interactions among these heterologous factors. The MCK enhancer contains a central core surrounded by peripheral activating elements. These latter elements lack significant enhancer activity alone, but they increase activity of the core 5- to 10-fold. Within the enhancer core are two conserved sequence motifs known as E-boxes (CANNTG); they have been referred to as the left and right sites, and bind myogenic HLH proteins with low and high affinity, respectively (Lassar, Buskin, Lockshon, Davis, Apone, Hauschka and Weintraub, 1989b; Weintraub, Davis, Lockshon and Lassar, 1990; Brennan and Olson, 1990a; Chakraborty, Brennan and Olson, 1991a). A ubiquitous nuclear factor that we refer to as C-rich binding factor (CRBF) binds to a site between the left and right E-boxes, as well as to a second lower-affinity site $3'$ of the enhancer core (D. Kelvin and E. Olson,

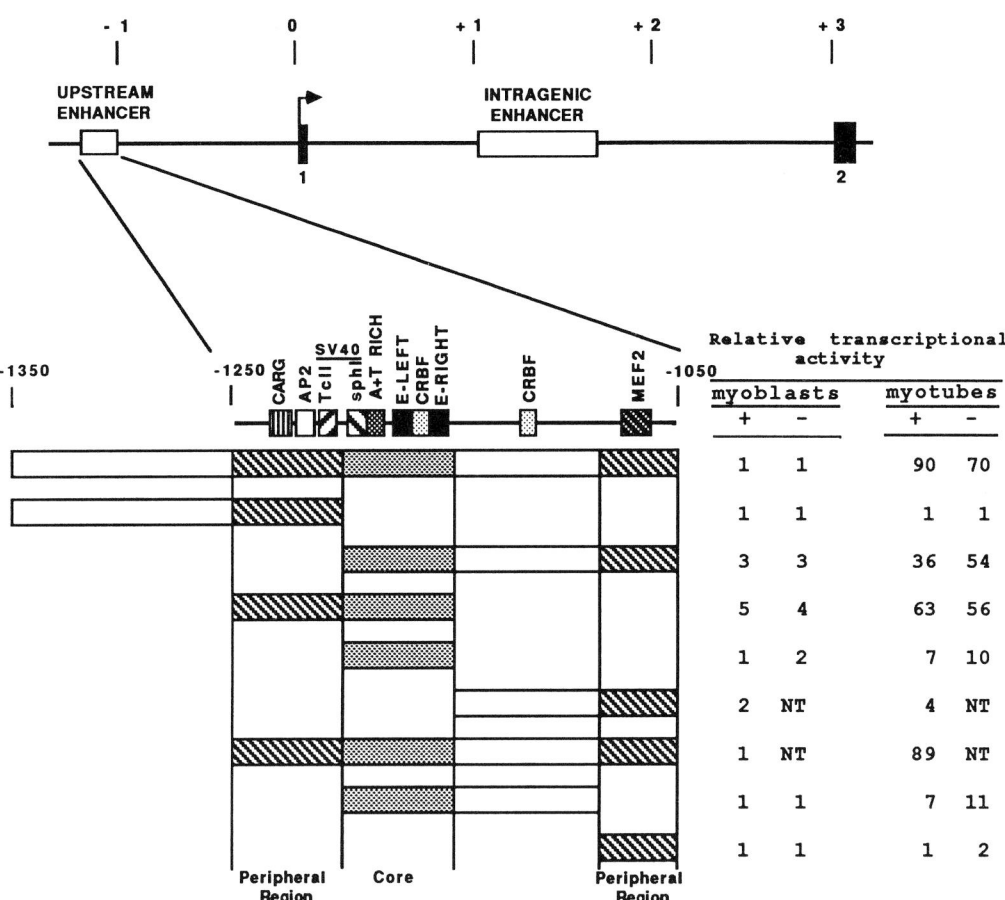

Fig. 1. Schematic representation of the MCK 5' enhancer. The 5' region of the mouse MCK gene is shown. The first and second exons are separated by a large intron that contains a weak muscle-specific enhancer, the exact boundaries and properties of which remain to be fully defined (Sternberg et al. 1988). The 5' enhancer lies between −1050 and −1250 bp upstream of the MCK gene. The enhancer contains a central core that possesses weak muscle-specific activity and two peripheral activating regions that augment activity of the core 5- to 10-fold. Each region of the enhancer shown was tested for activity by insertion into the BamHI site 3' of the chloramphenicol acetyltransferase gene (*CAT*) in the vector pCK246CAT which contains the MCK basal promoter immediately 5' of *CAT*. Each reporter plasmid was assayed following transient transfection into the C2 muscle cell line in growth medium (GM) or differentiation medium (DM). Values represent the average of at least three separate experiments and did not vary by more than 20% of the mean. NT, not tested.

unpublished results). Immediately 5' of the left E-box is an A+T-rich element and an element with homology to the SphI motif in the SV40 enhancer; both elements are essential for full enhancer activity. The more 5' peripheral activating region contains a CArG box, an AP-2 site, and an element with homology to the TcII motif in the SV40 enhancer. The more 3' peripheral activating region contains a binding site for the muscle-specific enhancer-binding factor MEF-2 (Gossett et al. 1989).

We have focused considerable attention on the MEF-2 site because MEF-2 is a muscle-specific DNA binding activity that is rapidly induced when myoblasts are triggered to differentiate upon exposure to growth factor-deficient medium (Gossett, Kelvin, Sternberg and Olson, 1989). Whereas a single MEF-2 site upstream of a basal promoter does not exhibit significant enhancer activity, multimers of the MEF-2 site can function as a strong enhancer in muscle cells. Intriguingly, myogenin can trans-activate a reporter gene linked to a basal promoter and a multimerized MEF-2 site in fibroblasts (P. Cserjesi and E. Olson), but myogenin does not interact directly with the MEF-2 site. The latter observation suggested that myogenin might indirectly activate transcription through the MEF-2 site, perhaps by inducing MEF-2 activity. Indeed, forced expression of myogenin (Cserjesi and Olson 1991) or MyoD (Lassar, Davis, Wright, Kadesch, Murre, Voronova, Baltimore and Weintraub, 1991) in nonmuscle cell types leads to induction of MEF-2 DNA binding activity. Induction of MEF-2 by myogenin occurs in transfected 10T1/2 cells that have been converted to muscle by myogenin, as well as in CV-1 kidney cells that do not activate the myogenic program in response to myogenin (Cserjesi and Olson 1991). Activation of MEF-2 expression by myogenin in both cell types is dependent on withdrawal of mitogens from the medium, indicating that a mitogen-dependent regulatory mechanism inhibits the ability of myogenin to induce MEF-2. The ability of myogenin to induce MEF-2 activity in cells that do not activate 'downstream' genes associated with terminal differentiation demonstrates that myogenin retains limited function within cell types that are nonpermissive for myogenesis and suggests that MEF-2 is regulated independently of other muscle-specific genes (see Weintraub, Tapscott, Davis, Thayer, Adam, Lassar and Miller, 1989).

Mutagenesis of myogenin reveals cooperative interactions between the basic region and transcriptional activation domains

A schematic representation of the myogenin protein and its functional domains is shown in Fig. 2. Deletion analysis of myogenin indicates that the basic-HLH region mediates DNA binding and dimerization, but does not efficiently convert fibroblasts to myoblasts or trans-activate exogenous muscle-specific target genes (Schwarz, Chakraborty, Martin, Zhou and Olson, 1992). The basic-HLH region of myogenin shares about 80% homology with the corresponding regions of MyoD, myf5 and MRF4. Similar, but more divergent basic-HLH regions are present in a rapidly growing number of transcription factors that regulate cell fate and proliferation. All HLH proteins that have been shown to bind DNA recognize the E-box consensus sequence (CANNTG). However, individual HLH proteins exhibit selectivity with respect to the E-boxes they recognize, the specificity being determined by the central nucleotides of the consensus, as well as by nucleotides flanking the consensus sequence (Blackwell and Weintraub 1990). As discussed

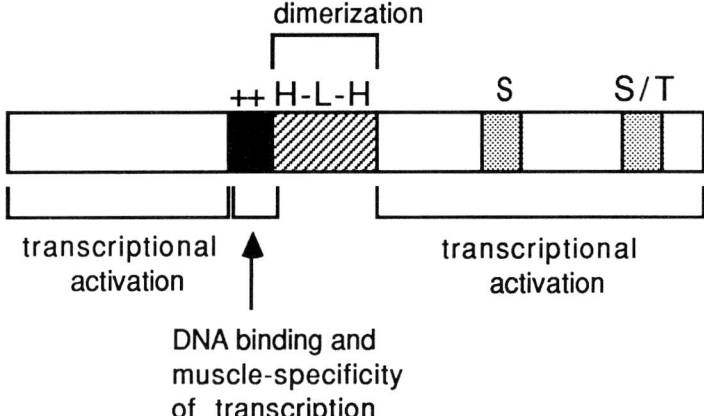

Fig. 2. Schematic representation of the myogenin protein and its functional domains. The basic region (++) and HLH motif mediate DNA binding and dimerization, respectively. The DNA binding domain also is essential for muscle-specificity of transcription. The N- and C-termini of myogenin contain transcriptional activation domains that can act independently of the basic-HLH region when fused to a heterologous DNA binding domain such as that of GAL4. The C-terminal activation domain contains serine (S)-, and serine-, threonine- (S/T)-rich subdomains.

below, HLH proteins also differ in their abilities to induce transcription from a given E-box even though they may show equivalent affinities for the same site. This specificity of DNA binding and transcriptional activity appears to be dictated, at least in part, by the nonconserved residues in the basic region and perhaps within the HLH motif, which could affect the conformation of the DNA binding domain upon dimerization.

Analysis of the DNA binding properties of myogenic HLH proteins indicates that alone, they show relatively weak affinity for the E-box sequence, whereas in the presence of the widely expressed HLH protein E12 encoded by the E2A gene (Murre, McCaw and Baltimore, 1989a), they form heterodimers and acquire high-affinity for DNA (Murre, McCaw, Vaessin, Caudy, Jan, Jan, Cabrera, Buskin, Hauschka, Lassar, Weintraub and Baltimore, 1989b; Davis, Cheng, Lassar and Weintraub, 1990; Brennan and Olson, 1990a; Chakraborty, Brennan, Edmondson and Olson, 1991b). Recent studies support the notion that E2A gene products are essential for activation of myogenesis by MyoD (Lassar et al. 1991).

Full transcriptional activity of myogenin requires the N- and C-termini, both of which contain transcriptional activation domains (TADs). These TADs can induce high levels of transcription when fused to the DNA binding domain of the yeast transcription factor GAL4 and assayed for trans- activation of a reporter gene linked to a multimerized GAL4 binding site. The precise residues that contribute to transcriptional activation remain to be determined; however, studies thus far indicate that the N-terminal TAD is contained within the first 70 residues of myogenin. This region of the protein contains a preponderance of acidic residues, suggesting that the N-terminal TAD may function as an acidic activator. Myf5 (Braun, Winter, Bober and Arnold, 1990b) and MyoD (Weintraub, Dwarki, Verma, Davis, Hollenberg, Snider, Lassar and Tapscott, 1991) also contain TADs near their amino termini and are relatively acidic throughout this region. The C-terminal TAD is contained in the last 60 residues of myogenin. Within the C-terminal TAD are two serine-, threonine-rich regions that may cooperate to activate transcription.

Studies with other transcription factors have revealed TADs that function in cell type-restricted fashions, as well as those that function independent of cell type. We have taken two approaches to investigate whether the myogenin TADs rely on cell type-restricted factors for activity. First, we have assayed for transcriptional activity of myogenin-GAL4 chimeras in cell backgrounds that are nonpermissive for myogenic conversion by myogenin and MyoD (e.g. HeLa and CV-1 cells). In both cell types, the TADs function as effectively as in cell types that are permissive for myogenin and MyoD function (e.g. 10T1/2 cells). We have also examined whether the myogenin TADs are functional within cells that are exposed to growth factors such as TGF-beta or high serum, which suppress the transcriptional functions of myogenin and MyoD. Again, the TADs are fully functional under these conditions, indicating that intracellular growth factor signals do not suppress the functions of myogenic HLH proteins by repressing activity of their TADs. As an independent test for cell type-specificity of the myogenin TADs, we have examined whether the strong acidic coactivator VP16 can substitute for the TADs of myogenin by creating a chimera of the myogenin basic-HLH region and VP16. This chimera is able to strongly activate myogenesis (Schwarz, Chakraborty, Martin, Zhou, and Olson, 1992), further indicating that the TADs of myogenin are unimportant for cell-type specificity of transcription. Similar observations have been made with MyoD (Weintraub et al. 1991).

In contrast to the apparent lack of muscle-specific activity of the myogenin TADs, the DNA binding domain of myogenin, like that of MyoD, is essential for muscle-specificity of transcription. Systematic mutagenesis of the basic region of myogenin and MyoD has defined a 12-amino acid subdomain that mediates recognition of the CANNTG consensus sequence and directs muscle-specific transcription (Davis et al. 1990; Brennan, Chakraborty and Olson, 1991a; Weintraub et al. 1991). By swapping nonconserved residues of myogenin or MyoD with residues at the corresponding positions in the basic region of E12, two adjacent amino acids, alanine-threonine, in the center of the basic region have been identified that are essential for efficient activation of muscle-specific genes. Substitution of these residues with those present at the corresponding position of other HLH proteins abolishes the ability of myogenin and MyoD to activate myogenesis, but does not affect their affinity for DNA. These results indicate that binding of these proteins to a muscle-specific regulatory region is not by itself sufficient to induce transcription and suggest that additional events are involved in activation of the myogenic program. Alanine-threonine are conserved at this position in the basic region of all known myogenic regulatory factors identified to date, in species ranging from *Drosophila* to man, suggesting that these residues constitute part of an ancient myogenic regulatory motif.

It is presently unclear how the basic region controls the activity of the TADs of myogenin when DNA binding is mediated by the CANNTG consensus sequence, but not when DNA binding is mediated by a heterologous DNA binding domain, such as that of GAL4, interacting with a different target sequence. One possibility is that a coregulator that recognizes the basic region of myogenin and MyoD is required for transcriptional activation through the CANNTG consensus and that the activation domains are released from their dependence on such a coactivator when DNA binding is mediated by a different sequence. There is precedent for the existence of coactivators that recognize the DNA binding domains of other transcription factors and cooperatively induce transcription. Alternatively, or in addition, DNA binding through the CANNTG consensus may be coupled to allosteric activation of the protein that unmasks the activation domains so that they can efficiently induce transcription. Perhaps myogenin and MyoD can only respond to their appropriate target sequence when they contain the correct combination of amino acids (i.e. alanine-threonine) in the basic region.

Target gene specificity for transcriptional activation by myogenic HLH proteins is mediated by nonconserved domains outside the bHLH motif

An important question that remains to be answered is whether different members of the MyoD family carry out distinct functions within the myogenic lineage or whether they are functionally redundant. A suggestion that MyoD and myogenin may be functionally distinct was provided by the observation that the BC3H1 cell line, which does not express MyoD and is fusion-defective, can be induced to form myotubes by forced expression of exogenous MyoD (Brennan, Edmondson and Olson, 1990b). Comparison of the abilities of different myogenic regulatory factors to trans-activate exogenous muscle-specific genes also suggests that MRF4 differs from myogenin, MyoD and myf5. Whereas the latter three factors can efficiently trans-activate the MCK and Troponin (Tn) I enhancers, MRF4 cannot activate these enhancers (Yutzey, Rhodes and Konieczny, 1990; Chakraborty et al. 1991b). All four factors, however, can bind to the MCK enhancer as heterodimers with E12, suggesting that differential trans-activation by these factors is separable from DNA binding. The inability of MRF4 to trans-activate the MCK and TnI enhancers could, in principle, be due to the nonconserved N- and C-termini or it could be attributable to subtle amino acid differences in its basic-HLH region. To distinguish between these possibilities, we have created tripartite chimeras of myogenin and MRF4 and have compared their relative abilities to trans-activate reporter genes linked to the complete MCK 5′ enhancer or to a multimerized E-box. The results showed that all chimeras possessed equivalent transcriptional activity when assayed against a multimerized E-box, but only myogenin and a chimera in which the N- and C-termini of MRF4 were replaced with those of myogenin were able to induce transcription from the

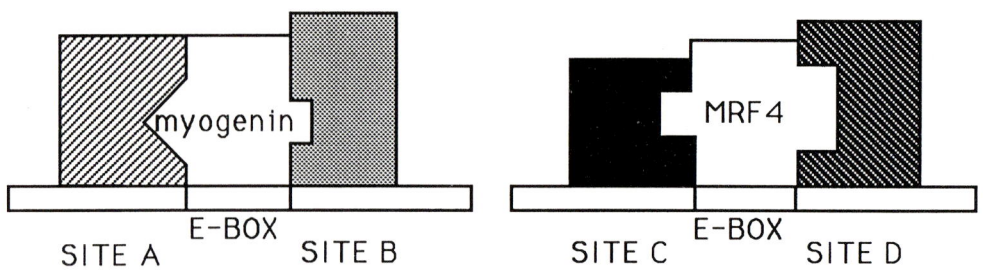

TARGET ENHANCERS

Fig. 3. Hypothetical model for the roles of the N- and C-termini of myogenic HLH proteins in differential trans-activation. Myogenin and MRF4 bind the same E-box consensus sequence, which may be surrounded by binding sites for different transcription factors in different muscle-specific regulatory regions. The abilities of myogenin and MRF4 to activate a target enhancer may be determined by their abilities to cooperate with heterologous transcription factors that bind adjacent sites. These cooperative interactions may depend on the divergent N- and C-termini. The transcriptional specificities of myogenin and MRF4 can be interchanged by swapping their N- and C-termini.

MCK enhancer (Chakraborty and Olson, 1991). These results suggest that the basic-HLH region of MRF4 is permissive for activation of the MCK enhancer, but the N- and C-termini are unable to cooperate with other transcription factors that bind this complex enhancer. A model to account for these observations is shown in Fig. 3. Although this model is greatly simplified, it is consistent with the data obtained thus far. According to this idea, the N- and C-termini of myogenin, or other HLH proteins, may mediate specific protein-protein interactions required for enhancer activation. Since each muscle-specific enhancer may contain distinct sets of target sequences for ubiquitous and cell type-specific transcription factors, the ability of a particular enhancer to respond to a given myogenic HLH protein could depend on the binding sites that surround the E-boxes (see also Sartorelli, Webster and Kedes, 1990). This type of mechanism could also be extended to other types of HLH proteins and might explain, at least in part, the observation that different HLH proteins that bind the same DNA sequence show differences in transcriptional activity.

Repression of myogenin function by growth signals involves multiple mechanisms

The ability of myogenin and other members of the MyoD family to induce muscle-specific transcription is subject to negative regulation by growth factor and oncogenic signals. Little is known of the mechanisms through which growth factor signals disrupt the mechanism for activation of muscle-specific transcription. One mechanism that has been implicated in growth factor-dependent repression involves the HLH protein Id, which lacks a functional basic region and attenuates the DNA binding activity of E12 and MyoD (Benezra et al. 1990). Id may mediate the negative effects on myogenesis of high serum and perhaps certain other growth factors, but it does not appear to explain the repression of myogenin and MyoD function by TGF-beta. In the presence of TGF-beta, myogenin is unable to activate muscle-specific genes; however, myogenin retains its ability to bind DNA (Brennan, Edmondson, Li and Olson, 1991b). The behavior of myogenin in cells exposed to TGF-beta is reminiscent of the behavior of basic domain mutants that can bind DNA but cannot activate muscle-specific transcription (Brennan et al. 1991a). Thus, we favor the notion that the lack of myogenic activity in these two situations may reflect a common underlying mechanism. One possible explanation for the lack of activity of myogenin in the presence of TGF-beta is that TGF-beta may repress the expression or activity of a coactivator required for muscle-specific transcription. Since, as discussed above, there is reason to believe that such a coactivator may interact with the basic region, it is conceivable that this is the target for repression by TGF-beta.

Summary

Rapid progress has been made toward understanding the mechanisms through which members of the MyoD family induce muscle-specific transcription. It is clear that these myogenic regulatory proteins do not act alone, but rather that they interact directly and indirectly with a broad array of cellular factors that positively and negatively modulate their activity. An important challenge for the future will be to identify and further characterize the factors that collaborate with and antagonize the actions of the myogenic HLH proteins. Elucidation of these additional 'players' in the myogenic regulatory pathway, will undoubtedly shed light not only on the molecular events responsible for the formation of skeletal muscle, but also other cellular phenotypes during development.

Work in the author's laboratory is supported by grants from the National Institutes of Health and American Cancer Society. ENO is an Established Investigator of the American Heart Association.

References

Benezra, R., Davis, R., Lockshon, D., Turner, D. and Weintraub, H. (1990). The protein Id: A negative regulator of helix-loop-helix DNA binding proteins. *Cell* **61**, 49-59.

Blackwell, K. and Weintraub, H. (1990). Differences and similarities in DNA-binding preferences of MyoD and E2A protein complexes revealed by binding site selection. *Science* **250**, 1104-1110.

Braun, T., Bober, E., Winter, B., Rosenthal, N. and Arnold, H. H. (1990a). *Myf-6*, a new member of the human gene family of myogenic determination factors: evidence for a gene cluster on chromosome 12. *EMBO J.* **9**, 821-831.

Braun, T., Buschhausen-Denker, G., Bober, G., Tannich, E. and Arnold, H. H. (1989). A novel human muscle factor related to but distinct from MyoD1 induces myogenic conversion in 10T1/2 fibroblasts. *EMBO J.* **8**, 701-709.

Braun, T., Winter, V., Bober, E. and Arnold, H. H. (1990b). Transcriptional activation domain of the muscle-specific gene regulatory protein myf5. *Nature* **346**, 663-665.

Brennan, T., Edmondson, D. G., Li, L. and Olson, E. N. (1991b). TGF-β represses the actions of myogenin through a mechanism independent of DNA binding. *Proc. Natl. Acad. Sci. U.S.A.* **88**, 3822-3826.

Brennan, T. J., Chakraborty, T. and Olson, E. N. (1991a). Mutagenesis of the myogenin basic region identifies an ancient protein motif critical for activation of myogenesis. *Proc. Natl. Acad. Sci. U.S.A.*, **88**, 5675-5679.

Brennan, T. J., Edmondson, D. G. and Olson, E. N. (1990b). Aberrant regulation of MyoD1 contributes to the partially defective myogenic phenotype of BC₃H1 cells. *J. Cell Biol.* **110**, 929-937.

Brennan, T. J. and Olson, E. N. (1990a). Myogenin resides in the nucleus and acquires high affinity for a conserved enhancer element on heterodimerization. *Genes Dev.* **4**, 582-595.

Chakraborty, T., Brennan, T. J., Li, L., Edmondson, D. G. and Olson, E. N. (1991b). Inefficient homooligomerization contributes to the dependence of myogenin on E2A products for efficient DNA binding. *Mol. Cell. Biol.* **11**, 3633-3641.

Chakraborty, T., Brennan, T. J. and Olson, E. N. (1991a). Differential trans-activation of a muscle-specific enhancer by myogenic helix-loop-helix proteins is separable from DNA binding. *J. Biol. Chem.* **266**, 2878-2882.

Chakraborty, T. and Olson, E. N. (1991). Domains outside of the DNA-binding domain impart target gene specificity to myogenin and MRF4. *Mol. Cell. Biol.* **11**, 6103-6108.

Cserjesi, P. and Olson, E. N. (1991). Myogenin induces the myocyte-specific regulatory factor muscle-specific independently of other muscle-specific gene products. *Mol. Cell. Biol.* **11**, 4854-4862.

Davis, R. L., Cheng, P-F., Lassar, A. B. and Weintraub, H. (1990). The MyoD DNA binding domain contains a recognition code for muscle-specific gene activation. *Cell* **60**, 733-746.

Davis, R. L., Weintraub, H. and Lassar, A. B. (1987). Expression of a single transfected cDNA converts fibroblasts to myoblasts. *Cell* **51**, 987-1000.

Edmondson, D. G. and Olson, E. N. (1989). A gene with homology to the *myc* similarity region of MyoD1 is expressed during myogenesis and is sufficient to activate the muscle differentiation program. *Genes Dev.* **3**, 628-640.

Gossett, L. A., Kelvin, D. J., Sternberg, E. A. and Olson, E. N. (1989). A new myocyte-specific enhancer-binding factor that recognizes a conserved element associated with multiple muscle-specific genes. *Mol. Cell. Biol.* **9**, 5022-5033.

Horlick, A. and Benfield, P. A. (1989). The upstream muscle-specific enhancer of the rat muscle creatine kinase gene is composed of multiple elements. *Mol. Cell. Biol.* **9**, 2396-2413.

Jaynes, J. B., Johnson, J. B., Bushkin, J. M., Gartside, C. L. and Hauschka, S. D. (1988). The muscle creatine kinase gene is regulated by multiple elements, including a muscle-specific enhancer. *Mol. Cell. Biol.* **6**, 2855-2864.

Lassar, A., Davis, R., Wright, W., Kadesch, T., Murre, C., Voronova, A., Baltimore, D. and Weintraub, H. (1991). Functional activity of myogenic helix-loop-helix proteins requires heterooligomerization with E12/E47-like proteins *in vivo. Cell* **66**, 305-315.

Lassar, A. B., Buskin, J. N., Lockshon, D., Davis, R. L., Apone, S., Hauschka, S. D. and Weintraub, H. (1989b). MyoD is a sequence-specific DNA binding protein requiring a region of *myc* homology to bind to the muscle creatine kinase enhancer. *Cell* **58**, 823-831.

Lassar, A. B., Thayer, M. J., Overell, R. W. and Weintraub, H. (1989a). Transformation by activated RAS or FOS prevents myogenesis by inhibiting expression of MyoD1. *Cell* **5**, 659-667.

Lin, H., Yutzey, K. and Konieczny, S. F. (1991). Muscle-specific expression of the troponin I gene requires interactions between helix-loop-helix muscle regulatory factors and ubiquitous transcription factors. *Mol. Cell. Biol.* **11**, 267-280.

Miner, J. H. and Wold, B. (1990). Herculin, a fourth member of the *MyoD* family of myogenic regulatory genes. *Proc. Natl. Acad. Sci. U.S.A.* **87**, 1089-1093.

Murre, C., McCaw, P. S. and Baltimore, D. (1989a). A new DNA binding and dimerization motif in immunoglobulin enhancer binding, *daughterless*, MyoD, and *myc* proteins. *Cell* **56**, 777-783.

Murre, C. P., McCaw, P. S., Vaessin, H., Caudy, M., Jan, L. Y., Jan, Y. N., Cabrera, C. V., Buskin, J. N., Hauschka, S. D., Lassar, A. B., Weintraub, H. and Baltimore, D. (1989b).

Interactions between heterologous helix-loop-helix proteins generate complexes that bind specifically to a common DNA sequence. *Cell* **58**, 537-544.

Olson, E. N. (1990). MyoD Family: A paradigm for development? *Genes Dev.* **4**, 1454-1461.

Rhodes, S. J. and Konieczny, S. F. (1989). Identification of MRF4: a new member of the muscle regulatory factor gene family. *Genes Dev.* **3**, 2050-2061.

Sartorelli, V., Webster, K. A. and Kedes, L. (1990). Muscle-specific expression of the cardiac-actin gene requires MyoD1, CArG-box binding factor, and Sp1. *Genes Dev.* **4**, 1811-1822.

Schwarz, J. J., Chakraborty, T., Martin, J. and Olson, E. N. (1992). The basic region of myogenin cooperates with two transcription activation domains to induce muscle-specific transcription. *Mol. Cell. Biol.* **12**, 266-275.

Sternberg, E., Spizz, G., Perry, M. E. and Olson, E. N. (1989). A *ras*-dependent pathway abolishes activity of a muscle-specific enhancer upstream from the muscle creatine kinase gene. *Mol. Cell Biol.* **9**, 594-601.

Sternberg, E. A., Spizz, G., Perry, W. M., Vizard, D., Weil, T. and Olson, E. N. (1988). Identification of upstream and intragenic regulatory elements that confer cell-type-restricted and differentiation-specific expression on the muscle creatine kinase gene. *Mol. Cell. Biol.* **8**, 2896-2909.

Tapscott, S. J. and Weintraub, H. (1991). MyoD and the regulation of myogenesis by helix-loop-helix proteins. *J. Clin. Invest.* **87**, 1133-1138.

Weintraub, H., Davis, R., Lockshon, D. and Lassar, A. (1990). MyoD binds cooperatively to two sites in a target enhancer sequence: Occupancy of two sites is required for activation. *Proc. Natl. Acad. Sci. U.S.A.* **87**, 5623-5627.

Weintraub, H., Dwarki, V., Verma, I., Davis, R., Hollenberg, S., Snider, L., Lassar, A. and Tapscott, S. (1991). Muscle-specific transcriptional activation by MyoD. *Genes Dev.* **5**, 1377-1386.

Weintraub, H., Tapscott, S. J., Davis, R. L., Thayer, M. J., Adam, M. A., Lassar, A. B. and Miller, A. D. (1989). Activation of muscle specific genes in pigment, nerve, fat, liver and fibroblast cell lines by forced expression of MyoD. *Proc. Natl. Acad. Sci. U.S.A.* **86**, 5434-5438.

Wright, W. E., Sassoon, D. A. and Lin, V. K. (1989). Myogenin, a factor regulating myogenesis, has a domain homologous to MyoD1. *Cell* **56**, 607-617.

Yutzey, K. E., Rhodes, S. and Konieczny, S. F. (1990). Differential trans-activation associated with the muscle regulatory factors MyoD1, myogenin and MRF4. *Mol. Cell. Biol.* **10**, 3934-3944.

Printed in Great Britain © Society for Experimental Biology 1992   343

POSITIVE AND NEGATIVE GENE REGULATION IN MUSCLE

YO-ICHI NABESHIMA, TAICHI UETSUKI, TOHRU KOMIYA, YOKO NABESHIMA, ATSUSHI ASAKURA, KEIJU KAMIJO, TAKAKO YAGAMI and ATSUKO FUJISAWA-SEHARA

Division of Molecular Genetics, National Institute of Neuroscience, National Center of Neurology and Psychiatry, Ogawahigashi-cho 4-1-1, Kodaira-shi, Tokyo 187, Japan

Summary

By the analysis of cis and trans-acting element involved in transcriptional regulation of chicken myosin alkali light chain genes, we have identified the MLC box, muscle specific enhancer element and negative regulatory element. The MLC box is an essential element for the expression of MLC genes located at approximately 100 bp upstream from mRNA start sites. The core sequence of MLC box is similar to the consensus of actin gene CArG box and SRE of c-fos oncogene. In vitro DNA-protein binding assay has revealed that the MLC box, CArG box and SRE might bind to a common or a similar protein complex. CMD1, cMyogenin, and cMRF4 transactivate the promorter with an intact MLC box, but not the promoter lacking MLC box, indicating that the MLC box itself is transactivated by the myogenic regulatory factors. This transactivation must have been due to the indirect effect of the myogenic regulatory factors, because chicken myogenic factors do not bind to the MLC box.

A cis element identified at about 150 bp upstream from the cap site of cardiac MLC gene suppresses the cardiac MLC gene expression in skeletal muscle cells but not in cardiac muscle cells. The protein(s) bound to NRE might be identical with one of proteins bound to SRE. NRE may block the function of MLC box and resultantly inhibits the expression of cardiac MLC1 gene in skeletal muscle cells.

Skeletal muscle enhancer at −2 kb of skeletal MLC1f gene is composed of two subelements P and D, cooperative action between them is required for sufficient enhancer activity. CMD1 and myogenin bind to the enhancer sequences of skeletal MLC1 gene and MCK gene and transactivate these genes preferentially in skeletal muscle cells. In addition to the CMD1 responsible enhancer, another cis-element is required for transactivation of the MLC1f gene by cMyogenin. An E-box adjacent to MLC box may co-work with the enhancer to increase the expression of MLC1f gene.

Muscle specific and developmentally regulated expression of MLC gene family is regulated by the combination of these cis and trans-acting elements.

Key words: myosin light chain gene family, transcriptional regulation, myogenic regulatory factors.

Introduction

Myoblast cells are developed from multi-potential mesodermal cells in the somite and differentiate to myotubes. Recently, MyoD and three related proteins were found to convert embryonic fibroblasts to myoblasts when their genes were expressed in the cells under the control of a viral promoter (Davis, Weintraub and Lassar 1987, Wright, Sassoon and Lin, 1989, Edmonson and Olson, 1989, Rhodes and Konieczny, 1989, Miner and Wold, 1990, Braun, Buschhausen-Denker, Bober, Tannich, and Arnord, 1989, Braun, Bober, Winter, Rosenthal, and Arnold, 1990). These factors belong to a family of regulatory proteins, designated as helix-loop-helix proteins, which can bind *in vitro* to the consensus sequence refered to as an E-box (Murre, McCaw, and Baltimore 1989, reviewed in ref. Weintraub, Davis, Tapscott, Thayer, Krause, Benezera, Blackwell, Turner, Rupp, Hollenberg, Zhuang, and Lassar 1991, Olson 1990). MyoD, myogenin, and Myf-5 were shown to activate muscle-specific genes including the muscle creatine kinase (MCK) gene and the myosin alkali light chain (MLC) gene presumably by acting directly on their regulatory regions containing E-boxes (Lassar, Buskin, Lockshon, Davis, Apone, Hauschka, and Weintraub 1989, Chakraborty, Brennan, and Olson 1991, Rosenthal, Berglund, Wentworth 1990).

We have previously cloned a series of chicken MLC genes including the adult type (LC1/LC3) (Nabeshima, Fujii-Kuriyama, Muramatsu and Ogata, 1984), the embryonic type (L23) (Nabeshima, Nabeshima, Kawashima, Nakamura, Nonomura, and Fujii-Kuriyama 1988), the smooth muscle-type (Nabeshima, Nabeshima, Nonomura, and Fujii-Kuriyam, 1987), and the cardiac muscle type genes (Nakamura, Nabeshima, Kobayashi, Nabeshima, Nonomura, and Fujii-Kuriyama 1988). Northern blot analysis revealed that these MLC genes are developmental stage and tissue-specifically regulated in chicken. To study how the transcription of skeletal muscle MLC gene is dramatically increased and continuously expressed in skeletal muscle myotubes, and how cardiac muscle MLC and embryonic MLC (L23) genes are transiently expressed during myogenesis, we have analysed the regulatory elements involved in the expression of above three genes and identified the MLC box, negative regulatory element (NRE) and muscle specific enhancer.

We report here the character and the biological importance of three cis-elements and the effect of the expression of three myogenic factors, CMD1 (Lin, Dechesne, Eldridge, and Paterson, 1989), c-Myogenin (Fujisawa-Sehara, Nabeshima, Hosoda, and Nabeshima, 1990), and cMRF4, on the transcription of the MLC genes.

Results and discussion

MLC box: A conserved common element in MLC genes

To analyze the cis-acting elements of L23 gene (Nabeshima et al 1988), we have constructed fusion genes containing a series of deletion mutants of the 3.7 kb 5′ upstream region of the L23 gene linked to bacterial chloramphenicol

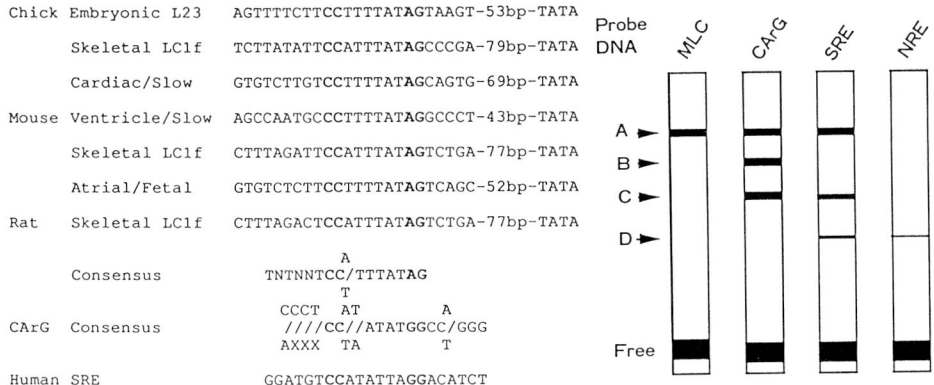

```
Chick  Embryonic L23    AGTTTTCTTCCTTTTATAGTAAGT-53bp-TATA
       Skeletal LC1f    TCTTATATTCCATTTATAGCCCGA-79bp-TATA
       Cardiac/Slow     GTGTCTTGTCCTTTTATAGCAGTG-69bp-TATA
Mouse  Ventricle/Slow   AGCCAATGCCCTTTTATAGGCCCT-43bp-TATA
       Skeletal LC1f    CTTTAGATTCCATTTATAGTCTGA-77bp-TATA
       Atrial/Fetal     GTGTCTCTTCCTTTTATAGTCAGC-52bp-TATA
Rat    Skeletal LC1f    CTTTAGACTCCATTTATAGTCTGA-77bp-TATA

                                    A
       Consensus        TNTNNTCC/TTTATAG
                                    T
                        CCCT   AT        A
CArG   Consensus        ////CC//ATATGGCC/GGG
                        AXXX   TA        T
Human  SRE              GGATGTCCATATTAGGACATCT
```

Fig. 1. Comparison of the nucleotide sequences of MLC boxes found in the promoter regions of chicken, mouse, and rat MLC genes. The results of *in vitro* DNA-protein binding assay are summarized.

acetyltransferase (CAT) gene and introduced them into the primary cultured cells prepared from the breast muscles of 11 day chick embryos. The rate of transcription gradually increased as the 5′ upstream region became shorter and transcriptional activity decreased to almost the background level when the sequence between −110bp and −83bp was deleted. Therefore, there may be two cis-acting regulatory elements, one of which is a negative regulatory element between −3.7 kb and −375bp and the other of which is a positive regulatory element between −110 bp and −80 bp. In the primary cultured fibroblast cells, the same series of fusion genes showed very weak or negligible transcriptional activities compared with those in the muscle cells. From these results, we concluded that the cis- acting region between −110 bp and −80 bp is an essential positive element for the expression of the L23 gene in muscle cells and used this region for the homology research analysis. Interestingly, the well conserved 16 bp element has been found in the 5′ promoter regions of chick, mouse, and rat MLC genes (Nabeshima et al 1984, Nakamura et al 1988, Daubas, Robert, Carner, and Buckingham, 1985, Strehler, Periasamy, Strehler-Page, and Nadal-Ginard, 1985) expressed in striated muscles and their consensus sequences are listed in Fig. 1. The analyzes of cardiac MLC1 and skeletal MLC1f genes revealed that homologous elements played an important role in the efficient expression of these MLC genes in skeletal muscle cells. Therefore, we designated this element as 'MLC box' (Uetsuki, Nabeshima, Fujisawa-Sehara, and Nabeshima, 1990).

To determine whether sequence alteration at various positions in the MLC box would interfere with its function, we analyzed the transcriptional efficiency, using synthesized double-stranded oligonucleotides which contained the mutated sequences of the MLC box. Almost alterations in the MLC box interfered with the transcriptional activity. Especially, in several mutants, only a single base exchange in the sequence of MLC box dramatically reduced the activity. Exceptionally, one mutant, the reverse oriented sequence of which is just same as CArG box sequence of actin gene, showed a property of up-mutation, because CArG box has been reported to be active in reverse orientation. The MLC box could function in both orientations but it was less efficient in reverse orientation. From these results, we concluded that the MLC box spanning 16 bp was a minimal cis-acting element which is essential for MLC gene expression.

In an attempt to detect the nuclear protein which specifically recognizes the MLC box sequence, we performed a gel mobility retardation assay using DNA fragment of MLC box which corresponds to the sequence between the -110bp and -75bp of L23 (probe MLC) and nuclear extracts of skeletal muscle cells (Fig. 1). From the result of competition assays, we concluded that the protein corresponding to the retardation of upper band (band A) was associated with the transcriptional activity and preferentially present in skeletal muscle cells rather than non-muscle cells (liver cells). As mentioned above, one base alteration in MLC box fulfilled the consensus of actin CArG box sequence and Boxer, Prywes, Roeder, and Kedes, 1989 have reported that the CArG box (Boxer et al 1989) and serum responsible element (SRE) of c-fos oncogene (Norman, Runswick, Pollock, and Treisman, 1988) could be functionally interchangeable. Therefore, to test the possibility that MLC box, CArG box and SRE could share a common protein as a trans-acting factor, the competition assay was performed using synthetic double-stranded oligonucleotides containing the sequences of the chicken skeletal α-actin CArG box, c-fos SRE and MLC box. Both oligonucleotides of CArG box and SRE inhibited the MLC box-protein formation (band A) to the same degree as probe MLC box. The gel mobility retardation assay with the oligonucleotide CArG as a probe revealed that one (band A) of three CArG-protein complexes (band A,B,C) had the same mobility as that of MLC box-protein complex. Furthermore this band showed the same competition patterns as the probe MLC box-protein complex. Similarly, a band (band A) of three SRE-protein complexes (band A,C,D) had the similar property with the band of MLC or CArG-protein complex. These results suggested that the MLC box binding protein could bind to both CArG box and c-fos SRE (Fig. 1). We assume that the protein(s) that recognizes the CArG and MLC boxes is involved in the coordinate expression of contractile proteins in muscle development. c-fos SRE has been reported to be involved in inducible transcription by stimulation with serum or growth factors, wherease the expression of MLC and actin genes occurs during terminal muscle cell differentiation, which is promoted by serum deprivation in vitro. How can this contradiction be explained? One possibility is that the CArG, MLC-box binding factor is different from SRE binding factor, although the DNA sequences that they specifically recognize are quite similar. Norman et al. (1988) reported the possibility that many SRF-like genes exist in the genome by cross-hybridization analysis at low stringent condition with SRF cDNA. Ryan, Franza, and Cilman (1989) also demonstrated that there exist at least two different SRE-binding factors: 67K and 62K proteins, both of which bind to SRE and function in the serum induction of c-fos gene. The second possibility is that SRE-binding factor and the CArG/MLC-box binding factors are the same but are modified in different manner or work co-operatively with the different factor(s) between muscle development and serum stimulation. Third possibility is that a single E-box adjacent to the MLC box plays a critical role for gene expression in muscle cells, because it may be possible that myogenic regulatory factors bind to E-box and interact to MLC box binding factors and/or to the protein of transcriptional initiation machinary to activate the MLC gene expression. To test the importance of MLC box and E box in the promoter region for muscle specific expression, we analysed the effect of myogenic factors on the MLC box and on E box by transfection of a pair of repoter genes in which either the MLC box or E box was disrupted. The expression of these reporter genes showed that the disruption of either cis-element reduced transactivation of L23 gene promorter by CMD1. The reporter plasmid which had no E box was strongly transactivated by CMD1, however the plasmid lacking MLC box was weakly transactivated. These results pinpoint the area of transactivation to the MLC box. Furthermore, the influence of the MLC box and the E box on the transactivation is synergistic rather than additive, suggesting the cooperative effect of these cis elements on transactivation of the L23 gene by myogenic regulatory factors (Fig. 2).

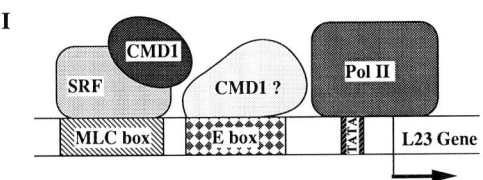

CMD1 is a component of MLC box binding protein complex

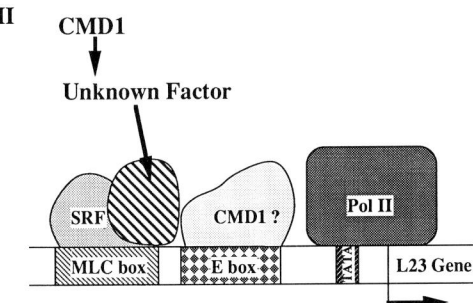

CMD1 activates unknown component(s) of MLC box binding complex

Fig. 2. Myogenic factors transactivate MLC box. There are two possible mechanisms, one of which is that the myogenic factor interacts with the MLC box binding protein to form a complex as an active transcriptional regulator, and another of which is that the MLC box binding factor itself is transcriptionally or post-transcriptionally activated by myogenic factors.

The negative regulatory element

A negative regulatory element (NRE) was discovered in the promoter region of cardiac myosin alkali light chain gene. To investigate the mechanism of suppression of cardiac LC1 gene in the skeletal muscle development, we analysed transcriptional activities of the constructs of a series of deletion mutant genes of the upstream region of cardiac LC1 gene in the skeletal muscle cells. The constructs containing the 5′ upstream region up to −150 are sufficiently expressed. However, by the addition of more upstream region, the constructs abolished their activities. These results suggested that the negative regulatory element lies in the sequence of 5′ upstream to −150 bp of cardiac LC1 gene promoter. Subsequently the suppressor activity was examined precisely and core sequence of NRE was identified between −200bp to −170bp of cardiac MLC1 gene promoter. The NRE inhibited the L23 gene expression, when the core sequence was placed at the upstream of MLC box sequence of L23 gene, but do not inhibit the promoter function of SV40 and tk genes. To clarify the effect of the NRE on the MLC box, we constructed a reporter gene in which MLC box was disrupted. NRE inhibited the expression of the reporter gene having the MLC box, but do not inhibit the reporter gene lacking the MLC box. Furthermore, NRE blocked the transactivation of L23 gene by MyoD through the activation of MLC box. NRE inhibited the enhancer activity of skeletal MLC1f gene (see below) when MLC box sequence was present in the promoter region of reporter plasmid, but do not inhibit the enhancer functon when MLC box was deleted from promoter region of reporter plasmid. The NRE binding protein(s) was detected in skeletal muscle nuclear extracts, but not in cardiac muscle nuclear extracts. As shown in Fig. 3,

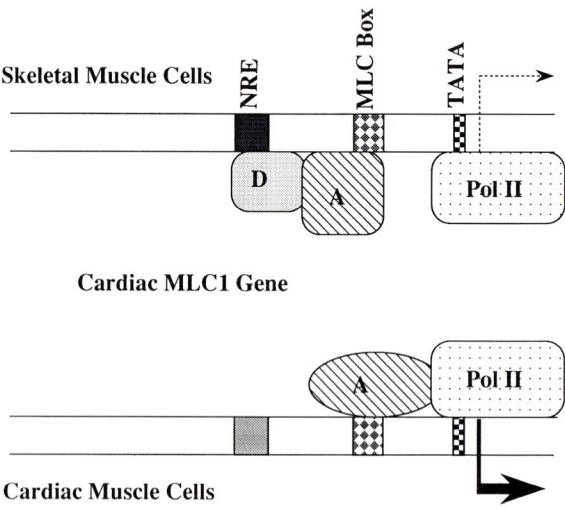

Fig. 3. Negative regulatory element inhibits the cardiac MLC1 gene expression in skeletal muscle cells, but not in cardiac muscle cells.

NRE binding protein(s) probably interacts with the MLC box binding factor and resultantly the function of MLC box is interrupted by NRE in skeletal muscle cells.

Enhancer element of skeletal muscle LC1f gene

We have previously shown that a cis-acting sequence was located in the 5' upstream region, between −2096 and −1936 of chick skeletal muscle LC1 gene (Shirakata, Nabeshima, Konishi, and Fujii-Kuriyama, 1988). This element is necessary for the inducible expression in response to muscle cell differentiation from myoblasts to myotubes. To characterize the upstream regulatory region, we inserted the region of interest, into the upstream or downstream of MLC1f promoter of the fusion gene, pLC299 (a construct containing DNA from −299 to +63 of LC1f promoter) in either direction and examined its activity. Transcriptional activities of four fusion genes were dramatically enhanced independently of its location and orientation. This result clearly showed that the upstream regulatory region works as an enhancer element in muscle cells. We named this region as ELC1 (Enhancer of LC1). Moreover, the expression of this fusion gene was two fold higher than that of pLC3381 which inserted all of the 5' flanking region up to −3381. This indicated that tissue and developmental regulation of LC1f gene expression is controlled by two regions from −2096 to −1936 and from −299 to +63.

To refine the enhancer region, we constructed fusion genes containing a series of external and internal deletion mutants of enhancer sequence and assayed their enhancing activities. The enhancer region is divided into two elements, one of which, designated as the P element works without the D element, although its enhancing activity is reduced to less than half of the native one and the other of which, designated as D element, can not function alone but works coordinately with the P element to increase the whole activity of ELC1. P region is composed of tandemly repeated two homologous sequences, right and left E-boxes and these two are essential for enhancer activity, because the disruption of one of these two core sequences strikingly reduces the enhancer activity.

To elucidate the importance of this sequence and to understand the coordinate expression of contractile proteins, we analyzed the sequence similarity with previously reported cis-acting elements and found a conserved tandemly repeated sequence among the P region of ELC1, enhancer sequence of muscle type creatine kinase (MCK gene) of mouse (Buskin and Hauschka, 1989) and enhancer region of rat skeletal muscle MLC1f/MLC3f, located downstream of the structural gene (Donoghue, Ernst, Wentworth, and Nadal-Ginard, 1988). Interestingly, Buskin and Hauschka reported that the myo D (Davis,Weintraub, and Lassar, 1987) probably bound to the tandemly repeated two core sequences present in the enhancer sequence of MCK gene, the binding sites of which were just homologous with E-boxes of ELC1. To examine the property of proteins which recognize the enhancer core sequence, *in vitro* DNA-protein binding assay was performed using DNA fragments of right E-box of ELC1 and enhancer core of MCK gene and nuclear extracts of skeletal muscle cells and non-muscle liver cells. The mobilities of protein(s) bound to both DNA fragments seemed to be similar. E-box DNA-protein complex formation was clearly competed with by the excess amounts of DNA of right and left E-boxes and MCK-E box, but not inactive mutants of right E-box and MCK-E box. Furthermore, MCK-E box-protein complex formation was also competed with by the addition of left and right E-box and MCK-E box, but not mutants of right E-box and MCK-E box. These results indicate that the protein components which recognize E-boxes of ELC1 and MCK enhancer are likely to be identical. Subsequently, we examined whether CMD1 (Lin et al 1989), the chick homologue of MyoD, chick myogenin (Fujisawa-Sehara et al 1990) and chick MRF4 could bind to E-box or not. CMD1, cMyogenin and cMRF4 synthesized in bacterial T7 system were used for *in vitro* DNA-protein binding assay. The assay was done in the presence or absence of nuclear extracts of non-muscle cells to detect the heterodimer of CMD1, cMyogenin or cMRF4 and ubiquitous factor(s) such as E12 included in the nuclear extracts. CMD1 and cMyogenin sufficiently bound to the DNA fragments of right E-box and MCK enhancer in the presence of nuclear extracts. Without the addition of non-muscle nuclear extracts, CMD1 and cMyogenin also recognized right E-box. In contrast, even in the absence of nuclear extracts of non-muscle cells, CMD1, cMyogenin and cMRF4 bound to MCK enhancer. Right E-box-protein complex detected in the presence of nuclear extracts seemed to be similar with the complex of right E-box and factor(s) present in nuclear extracts of skeletal muscle cells. However, without the addition of non-muscle nuclear extracts, the shifted band corresponding to this nature was not detected. Next important question is whether CMD1, cMyogenin and cMRF4 can activate the expression of MLC1f gene? To examine this question, CMD1 or cMyogenin expressing plasmid and the reporter plasmids were co-transfected into chicken primary cultured fibroblast cells and then the CAT activity was assayed. The result of the CAT assay clearly showed that CMD1 and cMyogenin, but not cMRF4, activated the construct containing the 3.4 Kb upstream region of skeletal muscle MLC1f gene. When the MLC1f promoter with a deleted enhancer was transfected as a reporter gene, neither CMD1 nor cMyogenin activated the reporter gene, implying that these myogenic regulatory factors required the enhancer to activate the transcription of the MLC1f gene. However, transfection of a reporter plasmid containing the enhancer ligated to a truncated 0.3 Kb MLC1f promorter revealed that CMD1, but not cMyogenin could activate the enhancer. Considered together with the strong affinity of cMyogenin to the enhancer, it is concievable that a cis-element(s) other than the enhancer is required for the transactivation of the MLC1f gene by cMyogenin. Alternatively, there may be a distinct cMyogenin resposible enhancer in the proximity of the CMD1 resposive enhancer (Fig. 4).

To test the importance of MLC box and E-box in the promoter region for enhancer activity, we analysed the effect of enhancer on the reporter genes in which either of the MLC-box or E-box was disrupted. The plasmid containing MLC box (E box disrupted) was weakly enhanced, however the plasmid having E box (MLC box disrupted) was strongly enhanced. From these results, we are

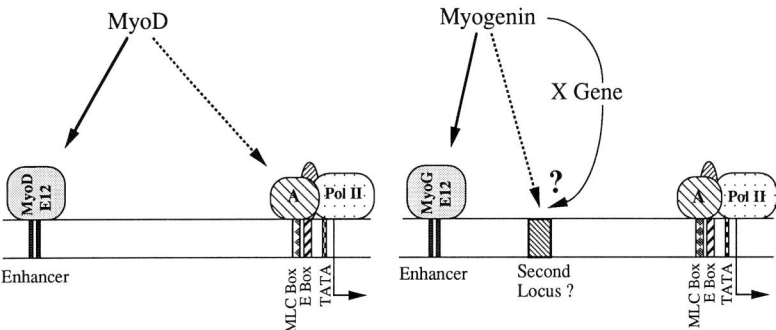

Fig. 4. CMD1 and cMyogenin activate the skeletal MLC1f gene expression. CMD1 and cMyogenin synthesized in skeletal muscle cells bind to the enhancer sequence of MLC1f gene and gives rise to preferential expression in matured skeletal muscle cells. In addition to CMD1 responsible enhancer, a second locus is required for transactivation of MLC1f gene by cMyogenin. The formation of heterodimer with a ubiquitous factor, such as E12/E47 is necessary for binding and transactivation.

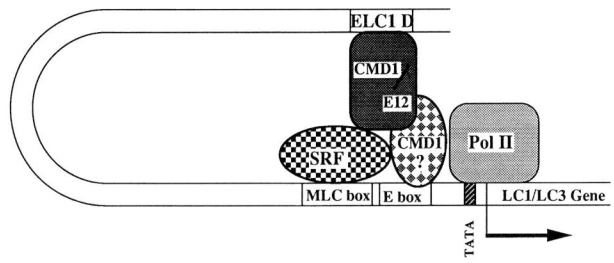

Fig. 5. E-box adjacent to the MLC box palys an important role for enhancement of gene expression.

speculating that the enhancer binding protein interacts with the protein bound to E box to form a higher order complex required for transcriptional enhancement (Fig. 5).

References

Boxer, L. M., Prywes, R., Roeder, R. G. and Kedes, L. (1989). The sarcomeric actin CArG-binding factor is indistinguishable from the c-fos serum response factor. *Mol. Cell. Biol.* **9**, 515-522.

Braun, T., Bober, E., Winter, B., Rosenthal, N. and Arnold, H. H. (1990). Myf-6, a new member of the human gene family of myogenic determination factors; evidence for a gene cluster on chromosome 12. *EMBO J.* **9**, 821-831.

Braun, T., Buschhausen-Denker, G., Bober, E., Tannich, E. and Arnold, H. H. (1989). A novel human muscle factor related to but distinct from MyoD 1 induces myogenic conversion in 10T1/2 fibroblasts. *EMBO J.* **8**, 701-709.

Buskin, J. N. and Hauschka, D. (1989). Identification of a myocyte nuclear factor that binds to the muscle-specific enhancer of the mouse muscle creatine kinase gene. *Mol. Cell. Biol*, **9**, 2627-2640.

Chakraborty, T., Brennan, T. and Olson, E. (1991). Differetial trans-activation of a muscle-specific enhancer by myogenic helix-loop-helix proteins is separable from DNA binding. *J. Biol. Chem.* **266**, 2878-2882.

Daubas, P., Robert, R., Carner, I. and Buckingham, M. (1985). A comparison between mammalian and avian fast skeletal muscle alkali myosin light chain genes: regulatory implications. *Nucleic Acids Res.* **13**, 4623-4643.

Davis, R. L., Weintraub, H. and Lassar, A. B. (1987). Expression of a single transfected cDNA converts fibroblasts to myoblasts. *Cell*, **51**, 987-1000.

Donoghue, M., Ernst, H., Wentworth, B., Nadal-Ginard, B. and Rosenthal, N. (1988). A muscle-specific enhancer is located at the 3' end of the myosin light-chain 1/3 gene locus. *Genes and Dev.* **2**, 1779-1790.

Edmondson, D. G. and Olson, E. N. (1989). A gene with homology to myc similarity region of MyoD1 is expressed during myogenesis and is sufficient to activate the muscle differentiation program. *Genes Dev.* **3**, 628-640.

Fujisawa-Sehara, A., Nabeshima, Y., Hosoda, Y. and Nabeshima, Y. (1990). Myogenin contains two domains conserved among myogenic factors. *J. Biol. Chem.* **265**, 15219-15223.

Lasser, A. B., Buskin, J. N., Lockshon, D., Davis, R. L., Apone, S., Hauschka, S. D. and Weintraub, H. (1989). MyoD is a sequence-specific DNA binding protein requiring a region of myc homology to bind to the muscle creatine kinase enhancer. *Cell* **58**, 823-831.

Lin, Z., Dechesne, C. A., Eldridge, J. and Paterson, B. M. (1989). An avian muscle factor related to MyoD1 activates muscle-specific promoters in non-muscle cells of different germ layer origin and BrdU-treated myoblasts. *Genes and Dev.* **3**, 986-996.

Miner, J. H. and Wold, B. (1990). Herculin, a fourth member of the MyoD family of myogenic regulatory genes. *Proc. Natl. Acad. Sci. USA* **87**, 1089-1093.

Murre, C., McCaw, P. S. and Baltimore, D. (1989). A new DNA binding and dimerization motif in immunoglobulin enhancer binding, daughterless, MyoD, and myc proteins. *Cell* **56**, 777-783.

Nabeshima, Y., Fujii-Kuriyama, Y., Muramatsu, M. and Ogata, K. (1984). Alternative transcription and two modes of splicing result in two myosin light chains from one gene. *Nature* **308**, 333-338.

Nabeshima, Y., Nabeshima, Y., Kawashima, M., Nakamura, S., Nonomura, Y. and Fujii-Kuriyama, Y. (1988). Isolation of the chick myosin light chain gene expressed in embryonic gizzard muscle and transitional expression of the light chain gene family *in vivo*. *J. Mol. Biol.* **204**, 497-505.

Nabeshima, Y., Nabeshima, Y., Nonomura, Y. and Fujii-Kuriyama, Y. (1987). Nonmuscle and smooth muscle myosin light chain mRNAs are generated from a single gene by the tissue-specific alternative RNA splicing. *J. Biol. Chem.* **262**, 10608-10612.

Nakamura, S., Nabeshima, Y., Kobayashi, H., Nabeshima, Y., Nonomura, Y. and Fujii-Kuriyama, Y. (1988). Single chicken cardiac myosin alkali light-chain gene generates two different mRNAs by alternative splicing of a complex exon. *J. Mol. Biol.* **203**, 895-904.

Norman, C., Runswick, M., Pollock, R. and Treisman, R. (1988). Isolation and properties of cDNA clones encoding SRF, a transcription factor that binds to the c-fos serum respose element. *Cell* **55**, 989-1003.

Olson, E. N. (1990). MyoD family: a paradigm for development? *Gene Dev.* **4**, 1454-1456.

Rhodes, J. H. and Konieczny, S. F. (1989). Identification of MRF4: A new member of the muscle regulatory factor gene family. *Gene and Dev.* **3**, 2050-2061.

Rosenthal, N., Berglund, E. B., Wentworth, B. M., et al (1990). A highly conserved enhancer downstream of the human MLC1/MLC3 locus is a target for multiple myogenic determination factors. *Nucleic Acids Res.* **18**, 6239-6246.

Ryan, W. A., Jr, Franza, B. R., Jr and Cilman, M. Z. (1989). Two distinct cellular phosphoproteins bind to the c-fos serum response element. *EMBO J.*, **8**, 1785-1792.

Shirakata, M., Nabeshima, Y., Konishi, K. and Fujii-Kuriyama, Y. (1988). Upstream regulatory region for inducible expression of the chicken skeletal myosin alkali light-chain gene. *Mol. Cell. Biol.*, **8**, 2581-2588.

Strehler, E. E., Periasamy, M., Strehler-Page, M. A. and Nadal-Ginard, B. (1985). Myosin light-chain 1 and 3 gene has two structurally distinct and differentially regulated promoters evolving at different rates. *Mol. Cell. Biol.* **5**, 3168-3182.

Uetsuki, T., Nabeshima, Y., Fujisawa-Sehara, A. and Nabeshima, Y. (1990). Regulation of the chicken embryonic light chain (L23) gene: Existence of a common regulatory element shared by myosin alkali light chain genes. *Mol. Cell. Biol.* **10**, 2562-2569.

Weintraub, H., Davis, R., Tapscott, S., Thayer, M., Krause, M., Benezere, R., Blackwell, K., Turner, D., Rupp, R., Hollenberg, S., Zhuang, Y. and Lassar, A. (1991). The myoD gene family: Nodal point during specification of the muscle cell lineage. *Science,* **251**, 761-766.

Wright, W. E., Sassoon, D. A. and Lin, V. K. (1989). Myogenin, a factor regulating myogenesis, has a domain homolgous to MyoD. *Cell* **56**, 607-617.

Printed in Great Britain © *Society for Experimental Biology 1992*   355

CIS REGULATING ELEMENTS WHICH CONTROL IN VIVO ALTERNATIVE SPLICING OF THE CHICKEN β TROPOMYOSIN PRIMARY TRANSCRIPT

MARC Y. FISZMAN, DOMENICO LIBRI and LAURENT BALVAY

Pasteur Institute, 28 rue du Dr Roux, 75724 Paris Cedex 15, France

Summary

The β tropomyosin gene of the chicken contains a pair of alternatively spliced mutually exclusive exons the use of which is developmentally regulated. Exon 6A is used by non muscle and undifferentiated muscle cells (myoblasts) while exon 6B is exclusively used in differentiated skeletal muscle cells. A complex array of cis acting sequence elements are involved in the regulation of this alternative splicing process. Transfection assays of quail muscle cells in culture were used to define these cis acting elements. We show that, in undifferentiated muscle cells, exon 6B is skipped as a result of a negative control on its selection while exon 6A is spliced as a default choice. We provide evidence that this negative control involves a secondary structure of the primary transcript around the 5' end of exon 6B as well as intronic sequence elements located between the branch point and the acceptor splice site of exon 6B. In differentiated muscles, both exons are accessible to the splicing machinery and the preferential use of exon 6B depends on the existence of a competition between the two exons for the selection of the flanking splice sites. In particular, we show that the donor splice site of exon 6A is a weak splice site while the branch point associated with exon 6B is a strong branch point.

Introduction

Eukaryotic genes require a precise process of assembly for their dispersed coding information which is known as the splicing mechanism. In most cases, the genetic information contained in all exons will constitute the mature transcript. However, in some cases, not all exons are used and splicing proceeds according to alternative patterns which may depend of the type of tissue or the stage of development at which the gene is expressed. This mechanism of alternative splicing provides a very efficient mechanism of post transcriptional control of gene expression and a growing body of evidence indicates that it is very widely used.

Most of our current understanding of splicing regulation rests upon the knowledge of the cis elements involved in the selection of splice junctions and the definition of splice site strength has been a starting point toward the establishment of the role played by splice site competition in alternative splicing.

The chicken β tropomyosin gene has provided us with a model system to study

Key words: alternative splicing, secondary structures, exon competition, β tropomyosin.

alternative splicing. This gene produces three transcripts by using alternative promoters, alternative signals for termination of transcription and alternative splicing of a pair of internal exons (exons 6A and 6B) (Libri, Lemonnier, Meinnel and Fiszman 1989; Libri, Mouly, Lemonnier and Fiszman 1990). Exon 6A is shared by RNAs expressed in smooth muscle and non muscle tissues (the two species differ by their different transcription start site) while exon 6B is present only in skeletal muscle specific transcripts. When muscle cells differentiate in culture, myoblasts show a preeminent use of exon 6A while myotubes use preferentially exon 6B and in no cases are the two exons spliced together. We have previously shown that a minigene containing the genomic fragment spanning from exons 5 to 7 of the gene contains all the necessary and sufficient cis information to faithfully reproduce splicing regulation (Libri, Marie, Brody and Fiszman 1989). The present report is a review of our current knowledge of the cis-acting elements which are involved in the regulation of splicing of exons 6A and 6B during muscle cells differentiation. We will show that in myoblasts the use of exon 6B is under a negative control which involves a secondary structure of the primary transcript around exon 6B as well as sequences located in the intron between exons 6A and 6B. In myotubes, on the other hand, exon 6A is out competed by exon 6B and we will show that suboptimal exon 6A associated splice sites as well as a stronger exon 6B associated branch point are involved in this competition.

Results and discussion

A competition between exons 6A and 6B is responsible for preferential use of exon 6B in myotubes but not of exon 6A in myoblasts.

The alternative use of exons 6A or 6B in the context of different state of differentiation could be due to the existence of a competition mechanism. To assess whether this was the case, we have constructed two different minigenes: MutΔ6A and MutK6B. MutΔ6A are a serie of mutants in which exon 6A has been deleted together with either one of the two introns which flank it, while in MutK6B both splice sites of exon 6B have been destroyed. In this latter case, we did not excise exon 6B to avoid the loss of potential structural informations. The various minigenes were transfected into cultured quail muscle cells and the transcripts isolated before (myoblasts) and after differentiation (myotubes) were analyzed by cDNA-PCR. Amplifications were performed between amplimers directed against exon 5 (sense) and SV40 sequences (antisense). The result of such an experiment is presented in figure 1. When exon 6A is deleted together with its preceding intron (MutΔ6A1), the major transcript which accumulates in myoblasts contains exon 5 spliced directly to exon 7 as evidenced by the presence of a product of 187 bp (MutΔ6A, lane Mb). In myotubes, however, the major transcript now contains exons 5, 6B and 7 (MutΔ6A, lane Mt). This result thus indicates that preferential splicing of exon 6A in myoblasts is not the result of a competition between the two exons but rather results from a direct inhibition of the use of exon 6B. Interestingly, the same result was obtained when a second mutant, MutΔ6A2, was

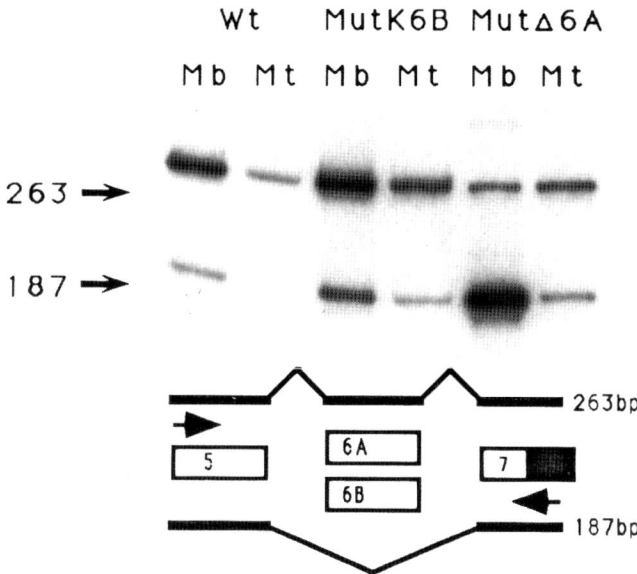

Fig. 1. Effect of mutations deleting or inactivating exons 6A or 6B. The transcripts derived from cells transfected with the various constructs were analyzed by cDNA-PCR and the products of amplification were separated by electrophoresis on a 6% polyacrylamide gel. The minigenes which are used and the cell types (myoblasts, Mb, or myotubes, Mt) are indicated for each lane. Amplification was performed between oligonucleotides complementary to the SV 40 sequence and to exon 5. A diagramatic representation indicates the size of the expected products from the amplification: a 187 bp product corresponds to the skipping of exon 6 while a 263 bp product indicates its inclusion.

used (data not shown). In this mutant, exon 6A was deleted together with the intron which separates exons 6A and 6B. This result indicates that inhibition of splicing of exon 6B can be produced even in the absence of intron sequences.

A different situation is obtained when MutK6B is used to transfect muscle cells. In this case, both in myoblasts and myotubes, the major transcript which is observed contains exons 5, 6A and 7 (MutK6B, lanes Mb and Mt). This result indicates that the down regulation observed in myotubes on splicing of exon 6A depends on the presence of an active exon 6B, and, as a consequence, that competition between the two exons is the most probable mechanism underlying the preferential use of exon 6B in this splicing environment.

Two regions are involved in the negative regulation of splicing of exon 6B in myoblasts: 1) role of a secondary structure of the transcript around exon 6B.

The fact that we could delete exon 6A and the intron located between exons 6A and 6B indicates that sequences of exon 6B itself are implicated in the regulation of splicing of exon 6B. A computer analysis of the sequence from exon 5 to exon 7 had prompted us to propose the possible existence of secondary structures which could play an important role in controlling the mechanism of splicing. In particular, it was striking to note that the sequence located around the 5′ end of

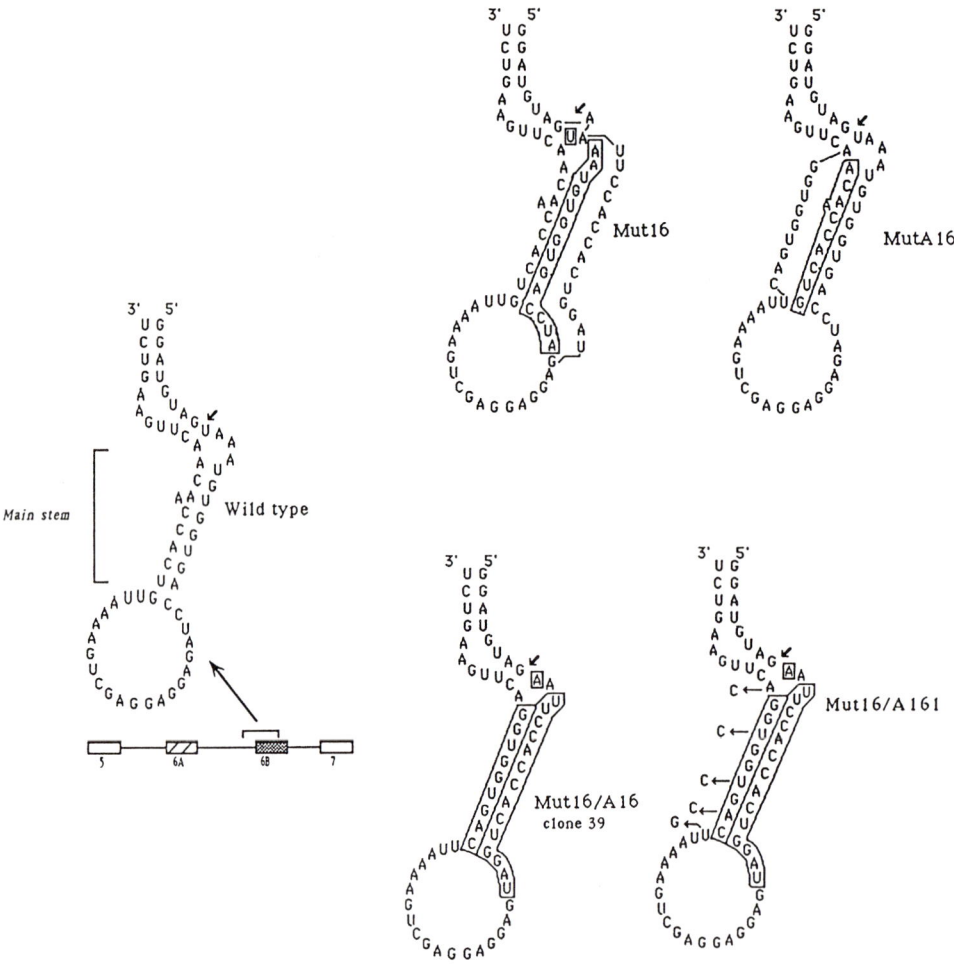

Fig. 2. Schematic representation of the secondary structure around exon 6B and localization of the sequence changes which were performed.

exon 6B could form a very stable stem and loop structure. We tested the existence of this small secondary structure by constructing mutant minigenes harboring changes in either one of the two strands and on both strands at the same time. As diagrammed on figure 2, mutation of either strand alone should induce splicing of exon 6B in myoblasts (Mut16 and MutA16). On the other hand, when the two strands are mutated, the result should depend on whether or not the two mutated regions can (Mut16/A16) or cannot (Mut16/A161) base pair with one another. This is precisely what has been obtained (Libri, Piseri and Fiszman 1991). Lastly, mutations in the loop region are not able to induce splicing of exon 6B in myoblasts (data not shown). We feel that all these data provide strong evidence for the existence and for a crucial role played by this secondary structure.

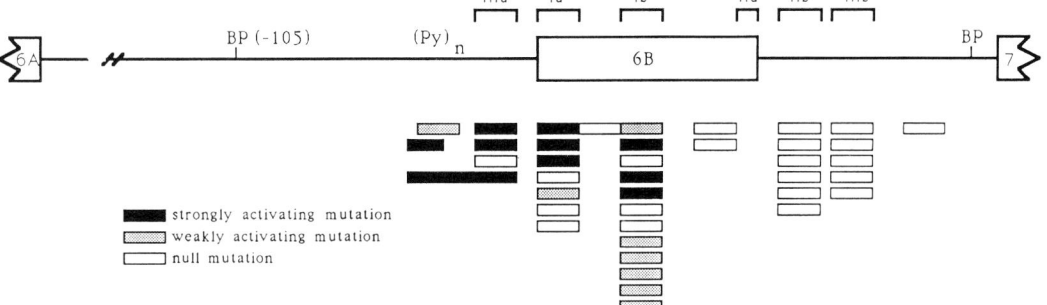

Fig. 3. Localisation of the mutations introduced between exons 6A and 7 and their respective effect on the induction of splicing of exon 6B in myoblasts. Sequences implicated in base pairing according to the computer predicted secondary structure are indicated as Ia, Ib, IIa, IIB, IIIa and IIIb (Ia and Ib correspond to the structure shown in figure 2).

2) sequences acting negatively are located between the branch point and the 3′ end of the 6A-6B intron.

Since the initial computer analysis predicted other secondary structures, we applied the same methodology to test the possible base pairing between sequences of the 6A-6B intron on the one hand and sequences of either exon 6B or of the 6B-7 intron on the other hand. As shown on figure 3 none of the mutations located downstream of the middle of exon 6B induce splicing of exon 6B. On the other hand, all the mutations introduced in the 3′ end region of the 6A-6B intron were very efficient in inducing splicing of exon 6B in myoblasts. This region is very unusual since it is very polypyrimidine rich and also because the branch point is located at position -105 instead of the usual -20 to -30 (Goux-Pelletan, Libri et al. 1990). Interestingly, some recent publications have described the isolation and the characterization of a factor(s) which binds such a polypyrimidine rich region (Gil, Sharp, Jamison and Garcia-Blanco 1991; Patton, Mayer, Tempst and Nadal-Ginard 1991) and which could be involved in the regulation of alternative splicing.

To summarize, our mutational analysis does not support the hypothesis of a large secondary structure which would sequester exon 6B in a myoblast environment. Rather, we would like to propose that two mechanisms are superimposed: 1) a very limited secondary structure localized around the 5′ end of exon 6B somehow prevents the acceptor splice site of exon 6B from being recognized and 2) an intronic polypyrimidine rich region which acts negatively possibly through the binding of specific factor(s). It is interesting to note that a very similar region exists in the β tropomyosin gene of the rat (Helfman and Ricci 1989; Helfman, Roscigno, Mulligan, Finn and Weber 1990). As we shall see later, the same region may also have a positive effect as a determinant of branch point strength.

Splice sites associated with exons 6A and 6B are involved in a competition mechanism in myotubes: 1) exon 6A has a sub-optimal donor site.

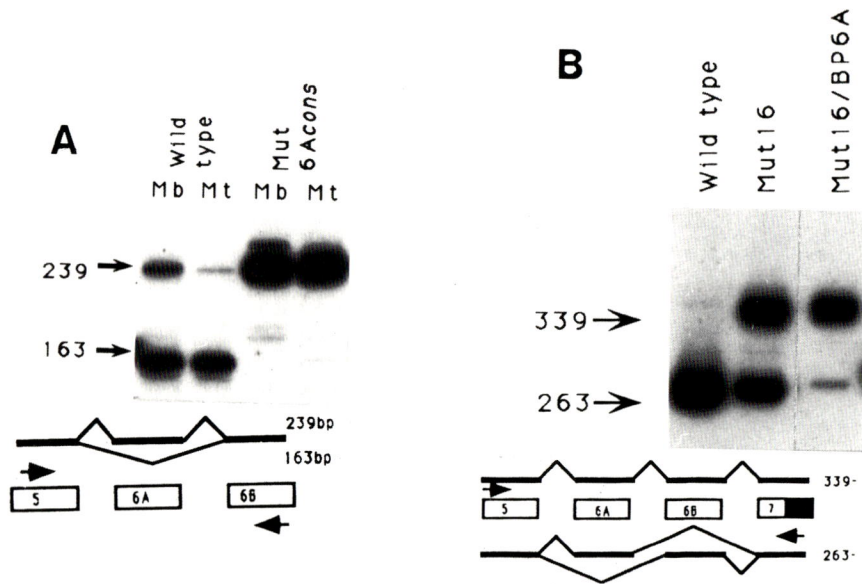

Fig. 4. A. Effect of mutations affecting the donor splice site of exon 6A. Amplifications were performed using oligonucleotides complementary to exons 6B and 5. B. Effect of mutations affecting the branch point associated with exon 6A. Amplifications were performed using oligonucleotides complementary to the SV 40 sequence and exon 5.

Alternative splicing is often related to the existence of non consensus donor sites. It is striking to note that the donor sites of exons 6A and 6B match poorly the consensus AG:GUAAGU since their sequence is respectively AG:GUACTG and AG:GUAUGA. To test this hypothesis, we have introduced three changes in the donor splice site of exon 6A (Mut6Acons). As a result of this mutation, all transcripts which contain exon 6B also contain exon 6A as shown by the presence of a 239 bp product and the absence of the 163 bp product when amplification is performed between exons 5 and 6B (figure 4A, compare wild type, lane Mt with Mut6Acons, lane Mt; the presence of the 239 and 163 bp products in wild type, lane Mb, is due to the fact that in our cell system the stringency of the control is not a 100%). Thus, by changing the donor site of exon 6A into a consensus donor site, we have abolished, in myotubes, the alternative nature of exon 6A.

2) exons 6B and 6A have competing branch sites.

The two major known determinants of branch site strength are the sequence and the pyrimidine content of the region immediatly downstream (Reed 1989) (Smith, Porro, Patton and Nadal-Ginard 1989). In the chicken β tropomyosin gene, the branch point has an abnormal position (Goux-Pelletan, Libri, D'Aubenton-Carafa, Fiszman and Brody 1990) and is followed by a highly pyrimidine rich region (Libri, Lemonnier et al. 1989). As discussed earlier, the pyrimidine stretch exerts a negative control on splicing of exon 6B in myoblasts, however, it could also have a positive effect in myotubes by specifying a stronger branch site than the one upstream of exon 6A. To test this hypothesis, we have duplicated, upstream of exon 6A, the branch site associated with exon 6B together with the following 30 nt (Mut16BP6A). This mutation was not introduced in a wild type minigene but in a minigene which bears a mutation in exon 6B (Mut16) so that splicing of exon 6B readily occurs in myoblasts. Fig. 4B shows the result of amplifications made between the SV 40 sequence and exon 5. When Mut16 is used to transfect myoblasts, two products are detected (263 and 339 bp) which correspond respectively to transcripts containing exons 5, 6B, 7 and exons 5, 6A, 6B, 7. When the double mutant is used, a different picture is obtained since we now predominently detect the 339 bp product. To make sure that this result was not an artefact due to the use of a double mutant, we have also engineered a duplication of the 6B branch site in an otherwise wild type background and we have obtained the same result when we analysed the transcripts made in differentiated myotubes (data not shown). So, by changing the branch site of exon 6A for the branch site of exon 6B, we have again abolished the dispensable nature of exon 6A and have made it become a constitutive exon.

Taken together, these results show that the branch site upstream of exon 6B is stronger than the one upstream of exon 6A, and that competition between the two sites is at least one of the elements responsible for the preferential use of exon 6B in myotubes. A very similar situation has been described in the alternative splicing of exons 2 and 3 of the rat α tropomyosin gene. In this case, exon 3 is the default choice while exon 2 is exclusively used in smooth muscle cells. The branch point associated with exon 3 is abnormally located (Smith and Nadal-Ginard 1989), is followed by a long polypyrimidine rich region and is also much stronger than the branch point associated with exon 2 (Mullen, Smith, Patton and Nadal-Ginard 1991).

Conclusions

We would like to propose that the regulation of alternative splicing of the β tropomyosin involves a double mechanism. In myoblasts, splicing of exon 6B is under a negative regulation which involves two cis-acting elements: a secondary structure around the 5' end of exon 6B and intron sequences localized between the branch point and the acceptor site associated with exon 6B. These intron sequences are possibly interacting with trans acting factor(s) one of which may be the polypyrimidine track binding protein (PTB) (Gil, Sharp et al. 1991; Patton, Mayer et al. 1991). In myotubes, this negative regulation is abolished and both exons become accessible to the splicing apparatus. Under these conditions, one can hypothesize that the stronger branch site associated with exon 6B will be selected together with the donor splice site of exon 5 which is stronger than the splice site of exon 6A (L. Balvay, unpublished observation) and this will lead to the splicing of exon 5 to exon 6B instead of exon 6A. Because of this process or for some independent reason, the donor splice site of exon 6B will be selected instead of the donor splice site of exon 6A for the splicing to exon 7.

D. Libri is supported by a fellowship from the Association des Myopathes de France. This work was supported by grants from the CNRS, the INSERM, the CEA, the Ministère de l'Industrie et de la Recherche, the Fondation pour la Recherche Médicale Française, the ARC, the Ligue Francaise contre le Cancer and the Association des Myopathes de France.

References

Gil, A., Sharp, P. A., Jamison, S. F. and Garcia-Blanco, M. A. (1991). 'Characterization of cDNAs encoding the polypyrimidine tract-binding protein.' *Genes & Dev.* **5**, 1224-1236.

Goux-Pelletan, M., Libri, D., d'Aubenton-Carafa, Y., Fiszman, M., Brody, E. and Marie, J. (1990). 'In vitro splicing of mutually exclusive exons from the chicken b-tropomyosin gene: role of the branch point location and very long pyrimidine stretch.' *EMBO J.* **9**, 241-249.

Helfman, D. M. and Ricci, W. M. (1989). 'Branch point selection in alternative splicing of tropomyosin pre-mRNAs.' *Nucleic Acids Res.* **17**, 5633-5650.

Helfman, D. M., Roscigno, R. F., Mulligan, G. J., Finn, L. A. and Weber, K. S. (1990). 'Identification of two distinct intron elements involved in alternative splicing of β-tropomyosin pre-mRNA.' *Genes & Dev.* **4**, 98-110.

Libri, D., Lemonnier, M., Meinnel, T. and Fiszman, M. Y. (1989). 'A single gene codes for the β subunit of smooth and skeletal muscle tropomyosin in the chicken.' *J. Biol. Chem.* **264**, 2935-2944.

Libri, D., Marie, J., Brody, E. and Fiszman, M. Y. (1989). 'A subfragment of the β tropomyosin gene is alternatively spliced when transfected into differentiating muscle cells.' *Nucleic Acids Res.* **17**, 6449-6462.

Libri, D., Mouly, V., Lemonnier, M. and Fiszman, M. Y. (1990). 'A nonmuscle tropomyosin is encoded by the smooth/skeletal β-tropomyosin gene and its RNA is transcribed from an internal promoter.' *J. Biol. Chem.* **265**, 3471-3473.

Libri, D., Piseri, A. and Fiszman, M. Y. (1991). 'Tissue specific splicing in vivo of the b-tropomyosin gene: dependence on an RNA secondary structure.' *Science.* **252**, 1842-1845.

Mullen, M. P., Smith, C. W. J., Patton, J. G. and Nadal-Ginard, B. (1991). 'α-tropomyosin mutually exclusive exon selection: competition between branch point/polypyrimidine tracts determines exon choice.' *Genes & Dev.* **5**, 642-655.

Patton, J. G., Mayer, S. A., Tempst, P. and Nadal-Ginard, B. (1991). 'Characterization and molecular cloning of polypyrimidine tract-binding protein: a component of a complex necessary for pre-mRNA splicing.' *Genes & Dev.* **5**, 1237-1251.

Reed, R. (1989). 'The organization of 3′ splice-site sequences in mammalian introns.' *Genes and Dev.* **3**, 2113-2123.

Smith, C. W. J. and Nadal-Ginard, B. (1989). 'Mutually exclusive splicing of a tropomyosin exons enforced by an unusual lariat branch point location: implication for constitutive splicing.' *Cell.* **56**, 749-758.

Smith, C. W. J., Porro, E. B., Patton, J. G. and Nadal-Ginard, B. (1989). 'Scanning from an independently specified branch point defines the 3′ splice site of mammalian introns.' *Nature.* **342**, 243-247.

Printed in Great Britain © *Society for Experimental Biology 1992*   363

THE CYTOPLASMIC TRANSLATIONAL INHIBITORY RNA OF CHICK EMBRYONIC MUSCLE: CLONING AND SEQUENCE OF A SUBSPECIES WHICH IS ANTISENSE RNA

Z-C. ZHENG, G-J. CAO, PAUL McCARTIN, Y. DU, ARPAD MOLNAR and SATYAPRIYA SARKAR

Department of Anatomy and Cellular Biology, Tufts University School of Veterinary Medicine, Boston, Massachusetts 02111, USA

Summary

A cytoplasmic translation inhibitory ribonucleoprotein (iRNP), about 10-12S in size and containing a heterogenous class of RNA (iRNA) in the 60-140 nucleotide size range, has been previously isolated from 13-14 day old chick embryonic leg and breast muscle. Both iRNA and iRNP are potent inhibitors of mRNA translation *in vitro*. A cDNA library for iRNA was prepared in pUC18 vector. The 114 nucleotide insert of one clone was subcloned in pBluescript vector. The *in vitro* transcript of this insert (only the 5'-3' orientation and not the reverse orientation) showed inhibition of muscle poly(A)$^+$ mRNA translation *in vitro*. The derived sequence of the iRNA indicates that it is a new sequence in which 27 nucleotides at the 3' and 5' ends can form base paired stem. Hybridization experiments and sequence analysis indicate that this iRNA has complementary (antisense) sequence to a highly conserved domain of the 28S eukaryotic rRNA and to a 5' leader segment of a muscle mRNA. These results suggest that iRNA mediated translational inhibition may act as a novel mechanism of controlling cellular mRNA levels in embryonic muscle and thus, acts in conjunction with the transcriptional control of myogenesis.

Introduction

It is currently believed that the tissue-specific and developmentally regulated expression of members of multigene families coding for various isoforms of contractile proteins that occurs during myogenesis, is primarily regulated at the transcriptional level and involves both *cis* and *trans* acting elements (for a review see Saidapet, Munro, Valgeirsdottir and Sarkar, 1982; Kedes and Stockdale, 1989). However, evidence for the role of posttranscriptional events which may act as an additional fine 'tune-up' control of the transcriptional regulation of myogenesis has recently appeared in literature (for review see Sarkar, Eller, Raychowdhury, Stedman and Wu, 1989). Thus, the cell cycle-mediated increase in myosin heavy chain mRNA stability during terminal differentiation (Medford,

Key words: antisense RNA, translation inhibitory RNA, translational control.

Nguyen and Nadal-Ginard, 1983), the appearance of transcribed and nontrans-lated forms of muscle-specific mRNAs in fusion-blocked cultured skeletal muscle cells (Endo and Nadal-Ginard, 1987) and the presence of a novel class of translation inhibitory cytoplasmic 10S RNP particle (iRNP) containing a heterogeneous low molecular weight RNA (iRNA) in the 65-150 nucleotide size range in chick embryonic muscle (Sarkar, Mukherjee and Guha, 1981) indicate that a diverse number of cytoplasmic events may also be involved in the regulation of myogenesis. Both iRNA and iRNP act as potent inhibitors of mRNA translation *in vitro* (Sarkar, Mukherjee and Guha, 1981; Mukherjee and Sarkar, 1981) suggesting that they may be involved in translational regulation of myogenesis *in vivo*. However, their physiological role remains to be understood.

In order to gain further insight into the structure and mechanism of action of individual iRNA species, we have attempted to clone the iRNA using recombinant DNA methodology. In this report we show the nucleotide sequence of an iRNA-cDNA clone and the biological activity of the RNA transcribed *in vitro* from this clone. The properties of this single iRNA subspecies suggest that it may act as a natural antisense RNA to a component of the translation system. The relevance of these results to a possible *in vivo* regulatory role of iRNA that augments the transcriptional control of myogenesis, is also discussed.

Results

Cloning of iRNA

Using total iRNA isolated from the iRNP particles of chick embryonic muscle (Sarkar, Mukherjee and Guha, 1981) a cDNA library for iRNA was prepared in pUC18 vector using random priming of total iRNA (Sambrook, Fritsch and Maniatis, 1989). The positive clones were identified by screening with 3'-end labeled radioactive total iRNA. The size of the inserts obtained by restriction enzyme digestion of the purified plasmid DNA isolated from the positive clones was examined. The majority of the clones contained inserts whose size was less than 50 nucleotides. This is presumably due to the fact that iRNA is highly sensitive to RNAse (Sarkar, Mukherjee and Guha, 1981) and full length cDNA may not be easily obtained by random priming.

The 114 nucleotide insert of a strongly positive clone was selected for further characterization. The insert was subcloned in the pBluescript vector for DNA sequencing and preparation of *in vitro* transcripts. The derived ribonucleotide sequence of the insert from this clone is shown in Fig. 1 (top panel). Sequence analysis with the data base (Genbank - release 66.0) indicates that it is a new sequence in which 27 nucleotides underlined at the 3' and 5' ends are capable of forming base paired stem, the remaining of the molecule having a possible loop structure (Fig. 1, top and middle panels). However, it is possible to generate computer predicted secondary structures within this loop domain (results not shown).

The derived sequence of the 114 nucleotide clone was also compared to the

5′ AUCCAUUUUCAGGGCUAGUUGAUUCGGCAGGUCAGUUGUUACACACUCCUUAGCGGGUUCCGACUUCCAUGGCCACCGUCCUGCUG-
UCUAGAUCAACCAGCCCUGAAAAUGGAU 3′

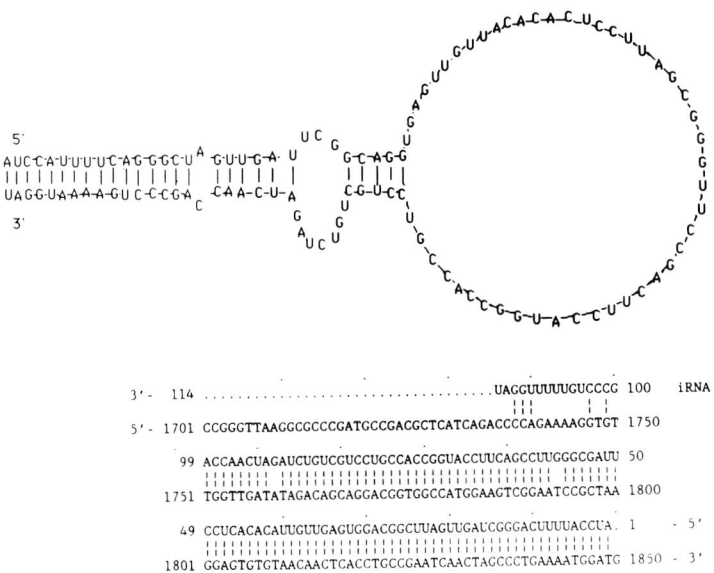

Fig. 1. Sequence and structural features of the 114 nucleotide iRNA. For details see also text. Top panel: The derived ribonucleotide sequence. The nucleotides underlined at the 5′ and 3′ ends can form base paired stem. Middle panel: The proposed secondary structure of iRNA. Bottom panel: The antisense complementarity of the iRNA to a domain of mouse 28S rRNA.

database for possible complementarity (antisense) using the FASTA program (Devereux, Haeberli and Smithies, 1984) of the GCG software package. The iRNA shows a high degree of complementarity to a conserved region of 28S eukaryotic ribosomal RNAs from a diverse number of species including primates, rodents, invertebrates and plants. A representative analysis shown with this domain (nucleotides 1850-1701) of mouse 28S ribosomal RNA gene (4712 bp nucleotide sequence available in the database) is shown (Fig. 1, bottom panel). The antisense homology was restricted to the first 103 nucleotides from the 5′ end of the iRNA and the degree of the complementarity usually ranged 80-96 percent in most cases. Interestingly, we have also observed that a 13 nucleotide overlapping part of iRNA (nucleotides 64-76 in the loop region; Fig. 1, middle panel) shows antisense homology to a 5′ leader segment of myosin light chain 2 mRNA (result not shown). This observation raises the possibility that this iRNA species may interact with RNA components of the translation system including both 28S rRNA and mRNA.

Biological activity of in vitro *transcripts of the 114 nucleotide clone*

In order to test the authenticity of the 114 nucleotide iRNA clone we prepared

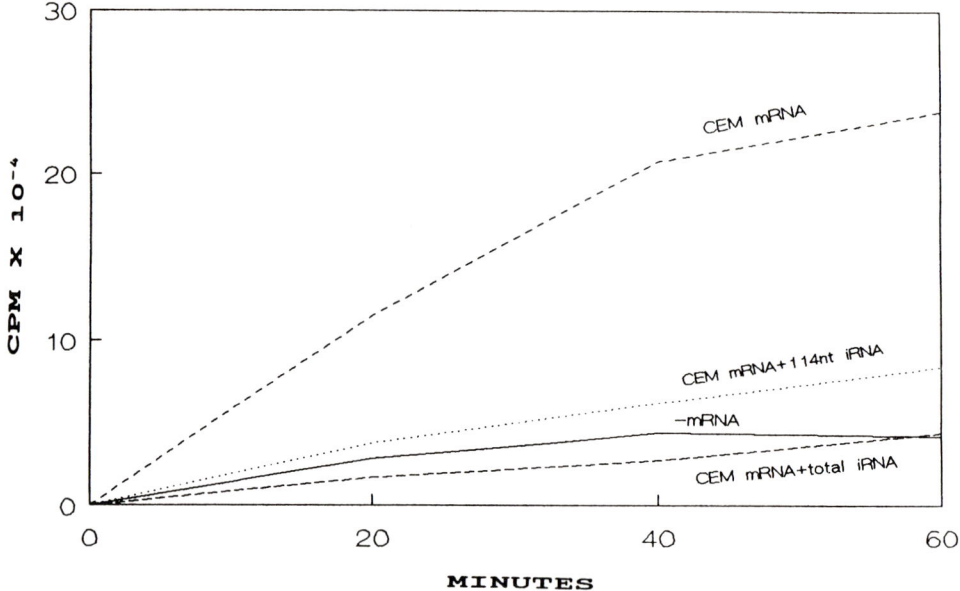

Fig. 2. The biological activity of the T7 RNA polymerase directed *in vitro* transcript
(5'-3' orientation of the 114 nucleotide iRNA). For details see also text. Translation
assays were carried out as previously described (Sarkar, Mukherjee and Guha, 1981)
using 0.2 μg of chick embryonic muscle (CEM) poly(A)$^+$ mRNA. The amount of
iRNA used: 100 ng for total iRNA and 200 ng for the 144 nucleotide *in vitro* transcript.
The ordinate represents CCl_3COOH insoluble (95°C) counts of ^{35}S-methionine
incorporated per ul of the reaction mixture.

in vitro transcripts using T7 (5'-3' orientation) and T3 (3'-5' orientation)
polymerase directed transcription of the pBluescript subclone. The orientation of
the transcripts was also confirmed by hybridization with the radiolabeled cDNA
probe, which was singly labeled at the 5' end and autoradiographic analysis of the
hybrids after electrophoresis. The biological activity of the *in vitro* transcripts was
tested by measuring amino acid incorporation in translation assays using the
nuclease treated rabbit reticulocyte lysate system (Sarkar, Mukherjee and Guha,
1981) programmed with chick embryonic muscle poly(A)$^+$ mRNA (Fig. 2). It was
observed that both total iRNA and the 5'-3' transcript inhibited the translation of
muscle poly(A)$^+$ mRNA (Fig. 2). In contrast, the 3'-5' transcript was totally
inactive (results not shown). The lower activity obtained with 200 ng of the 5'-3'
transcript, as compared to that obtained with 100 ng of total iRNA, is not
surprising, since the transcript represents a single molecular species of iRNA. The
possibility that the 114 nucleotide iRNA may selectively inhibit the translation of
mRNA coding for a specific myofibrillar protein, was also tested. The translation
products of incubations programmed with chick embryonic muscle poly(A)$^+$
mRNA in the presence and absence of different concentration of total iRNA and
the 114 nucleotide transcript were examined by autoradiography. The results
shown in Fig. 3 indicate that the translation of mRNAs coding for the majority of

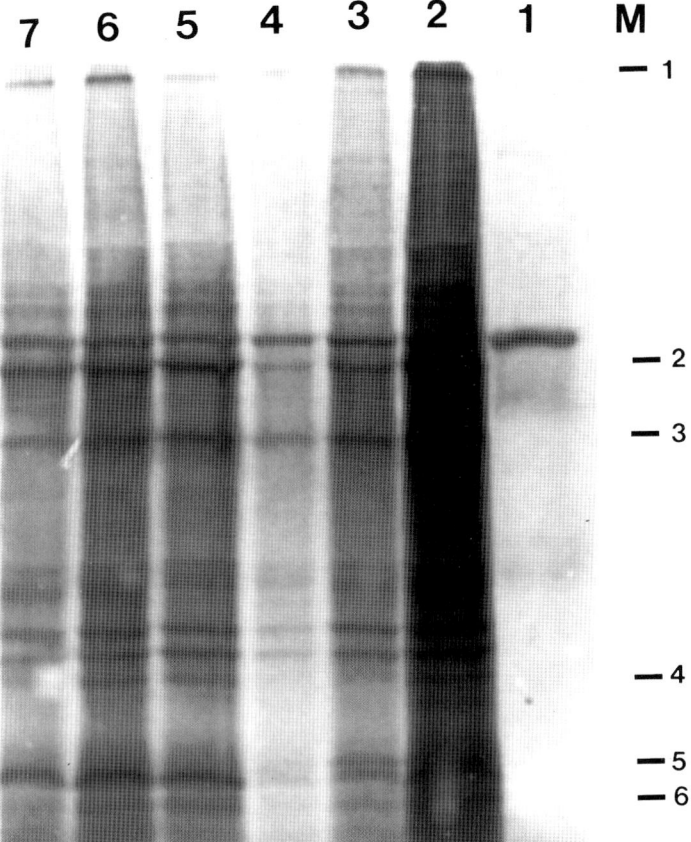

Fig. 3. Autoradiography of *in vitro* translation products obtained with muscle poly(A)$^+$ mRNA in the presence and absence of iRNA. For details see also text and legends to Fig. 2. The numbers at the right indicate the positions of the following myofibrillar protein bands run as markers (M). 1, myosin heavy chain; 2, actin; 3, a mixture of alpha and beta tropomyosin; 4, troponin I; 5, troponin C; 6, myosin light chains. Lane 1, nuclease treated lysate. Lane 2, lysate + 200 ng muscle poly(A)$^+$ mRNA. Lanes 3-4, incubations same as lane 2 and containing 10 and 100 ng of total iRNA, respectively. Lanes 5-7, incubations same as lane 2 and containing 20, 100 and 200 ng of the T7 RNA polymerase directed iRNA transcript, respectively. Electrophoresis was carried out using 10% acrylamide.

the protein bands is inhibited by the 114 nucleotide iRNA transcript. However, quantitation by densitometric scanning of the autoradiograms of the gel bands shown in Fig. 3 indicates that the translation of several mRNA species appears to be inhibited more than others (results not shown). Therefore, the possibility that individual iRNA species may modulate the translation pattern of specific mRNAs should not be excluded.

Hybridization of iRNA with cellular RNAs

The possibility that the 114 nucleotide iRNA interacts with cellular RNA

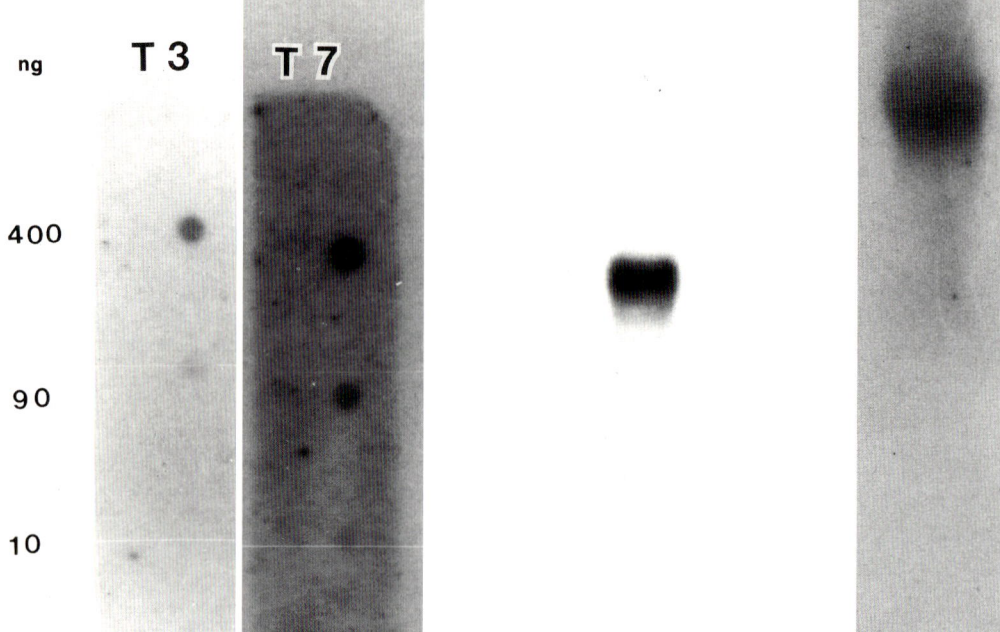

Fig. 4. Hybridization of 114 nucleotide iRNA with chick embryonic muscle RNAs. For details see also text. Hybridization of radiolabeled probes was carried out in 1M NaCl, 1% SDS, 5× Denhardt's solution, 10% dextran sulfate and 10 μg/ml denatured salmon sperm DNA at 60°C overnight. The blots were washed with 2× SSC at room temperature, 2× SSC and 1% SDS at 60°C and 0.1× SSC at room temperature. The filters were then dried and autoradiographed. Right panel: hybridization with radiolabeled 114 nucleotide cDNA insert as probe in Northern blot of chick embryonic muscle ribosomal RNAs. The single hybridization band corresponds to 28S chick muscle rRNA. Middle panel: hybridization of total iRNA isolated from iRNP particles in Northern blot using labeled 114 nucleotide cDNA insert as probe. Portion of the gels containing hybridized bands is shown in right and middle panels. Left panel: hybridization of indicated amounts of muscle poly(A)$^+$ mRNA in dot blots with radiolabeld *in vitro* iRNA transcripts (T7 RNA polymerase, 5'-3' orientation; T3 RNA polymerase, 3'-5' orientation).

components was next examined by hybridization experiments. Radiolabeled cDNA insert or the appropriate *in vitro* transcripts were used as the hybridization probes. The unbound and bound fraction of total cellular RNA of chick embryonic muscle after two cycles of oligo-dT-cellulose chromatography were used as the source of ribosomal and poly(A)$^+$ mRNA, respectively. Representative results are shown in Fig. 4. The 114 nucleotide iRNA hybridized with the 28S rRNA in Northern blots (right panel). Other cellular RNAs such as 18S rRNA and 4S RNA did not show any significant level of hybridization above the background level. Stringent hybridization conditions were used in these studies in order to eliminate and/or minimize non-specific hybridization. When total iRNA isolated from the 10-12S iRNP particles was analyzed in Northern blots, a strong hybridization band of about 120 nucleotide size range was obtained (middle panel) indicating that the 114 nucleotide iRNA clone is derived from a cellular RNA species present in the iRNP particles of chick embryonic muscle. Northern blot analysis of oligo-dT-cellulose purified chick embryonic muscle poly(A)$^+$ mRNA showed weak hybridization signals with the 114 nucleotide iRNA probe. The ability of iRNA to hybridize with poly(A)$^+$ mRNA was therefore, tested in dot blots using radiolabeled T7 (5'-3' orientation) and T3 (3'-5' orientation) polymerase directed *in vitro* transcripts as probes. Only the T7 polymerase transcript which showed biological activity hybridized with the poly(A)$^+$ mRNA (left panel). The intensity of the hybrid spots was dependent on the amount of mRNA used. The T3 polymerase transcript (3'-5' orientation) which is devoid of biological activity, gave background level of hybridization. These results, considered together with the biological activity of the 5'-3' transcript of the iRNA clone, indicate that the 114 nucleotide iRNA represents a cellular RNA entity. Furthermore, this iRNA is capable of interacting strongly with 28S rRNA and weakly with poly(A)$^+$ mRNA.

Discussion

We have previously shown that iRNA and iRNP represent a family of novel class of cytoplasmic macromolecules present in chick embryonic muscle and they are potent inhibitors of mRNA translation *in vitro* (Sarkar, Mukherjee and Guha, 1981; Mukherjee and Sarkar, 1981). The inhibition by iRNA is due to a specific effect on the initiation phase of protein synthesis. It involves the blocking of mRNA binding to ribosomes (Winkler, Lashbrook, Hershey, Mukherjee and Sarkar, 1983; Sarkar, 1984), which is a rate limiting step in the cellular translation process (Lodish, 1976). Interestingly, the inhibitory effect of iRNA is not mediated through the phosphorylation of the alpha subunit of the eukaryotic initiation factor eIF2 (Sarkar, 1984), which causes an initiation block. This phosphorylation is catalyzed by the well-known heme controlled protein kinase and the double stranded RNA stimulated protein kinase, which are known to operate in many eukaryotic cells (Hershey, 1989). The iRNA effect, therefore, reflects a novel type of translational regulation which is distinct from that mediated through the eIF2 phosphorylation pathway.

Regarding the specificity of inhibition of mRNA translation by iRNA, one iRNA subspecies (about 150 nucleotide size) purified by HPLC inhibits mRNA translation in a discriminatory manner, muscle poly(A)$^+$ mRNA being inhibited more strongly than non-muscle mRNA (Dasgupta, Pluskal and Sarkar, 1984; Sarkar, Eller, Raychowdhury, Stedman and Wu, 1989). This suggests that individual iRNA components may be involved in modulation of cellular RNA pattern in a selective manner. In view of microheterogenity at the 3' and 5' ends of several iRNA species, as determined by preliminary RNA sequencing (Dasgupta, Eller, Zolnay, Jayabaskaran, Stedman and Sarkar, 1986), precise information on the structure, function and mechanism of action of individual iRNA species can only be obtained by the use of well-characterized iRNA clones. The authenticity of the 114 nucleotide iRNA species, reported here, is shown by the biological activity of the *in vitro* transcript with the 5'-3' orientation and not the reverse orientation. The antisense complementarity of this iRNA to a highly conserved domain of 28S rRNA is intriguing. These results suggest that this domain of 28S rRNA may play a role in the initiation phase of the translation process, and the 114 nucleotide iRNA inhibits mRNA translation by specific interaction with this domain of 28S rRNA. This iRNA may also interact with mRNA during the translation process, as suggested by the presence of a more limited antisense complementarity detected in the 5' leader segment of a specific mRNA e.g., the myosin light chain 2 mRNA and the hybridization of the *in vitro* 5'-3' transcript with muscle poly(A)$^{+}$ mRNA. Whether or not multiple RNA-RNA interactions involving different segments of the 114 nucleotide iRNA play a role in its biological activity remains to be established.

However, the possible interactions of the 114 nucleotide iRNA with 28S rRNA and mRNA are likely to produce a quantitative modulation rather than an absolute mRNA-discriminatory effect in its biological activity. Thus, the *in vivo* function of iRNA may involve efficient regulation of the intracellular levels of functional mRNA which is actively translated. Since binding of mRNA to the ribosomes is a key rate-limiting step in protein synthesis (Lodish, 1976), the modulation of this step by iRNA is likely to yield striking alterations in cellular translation patterns. The iRNA activity which may serve as a cellular 'regulator of individual mRNA' for the translation process may be controlled by parameters such as intracellular concentration of iRNA and mRNA, affinity of iRNA and mRNA for the initiation site or event involving the antisense domain of 28S rRNA and/or mRNA, and the role of the protein moieties of iRNP in the function of iRNA. The iRNA activity may also lead to destabilization of the untranslated mRNAs. This may be physiologically relevant to control of myogenesis in embryonic muscle. Various isoforms of myofibrillar proteins coded by members of multigene family, are coexpressed in embryonic striated muscle (Kedes and Stockdale, 1989). The iRNA activity may serve as translational regulator of mRNA transcripts coding for such isoforms. The posttranscriptional control involving iRNA acting at a specific target step in the initiation phase of peptide synthesis, according to this view, actually augments the transcriptional control which is known to operate during myogenesis.

This work was supported by United States Public Health Service Grant 5PO1 HD23681 and funds from the Tufts University School of Veterinary Medicine. This article is dedicated to Paul M. Doty.

References

Dasgupta, S., Eller, M. S., Zolnay, S., Jayabaskaran, C., Stedman, H. and Sarkar, S. (1986). Cytoplasmic translation inhibitory RNA of chick embryonic muscle: possible role in myogenesis as antimessenger RNA. In Molecular Biology of Muscle Development (eds. C. Emerson, D.A. Fischman, B. Nadal-Ginard and M.A.Q. Siddiqui) pp. 591-603, New York, Alan R. Liss.

Dasgupta, S., Pluskal, M. G. and Sarkar, S. (1984). The cytoplasmic 4S translation inhibitory RNA species of chick embryonic muscle: fractionation of biologically active subspecies by high performance liquid chromatography. *Prep. Biochem.* **14**, 331-347.

Devereux, J., Haeberli, P. and Smithies, O. (1984). A comprehensive set of sequence analysis programs for VAX. *Nucleic Acids Research* **12**, 387-395.

Endo, T. and Nadal-Ginard, B. (1987). Three types of muscle-specific gene expression in fusion-blocked rat skeletal muscle cells: Translational control in EGTA-treated cells. *Cell* **49**, 515-526.

Hershey, J. W. B. (1989). Protein phosphorylation controls translation rates. *J. Biol. Chem.* **264**, 20823-20826.

Kedes, L. H. and Stockdale, F. E., eds (1989). 'Cellular and molecular biology of muscle development.' New York, Alan R. Liss.

Lodish, H. F. (1976). Translational control of protein synthesis. *Ann. Rev. Biochem.* **45**, 39-72.

Medford, R. M., Nguyen, H. T. and Nadal-Ginard, B. (1983). Transcriptional and cell cycle-mediated regulation of myosin heavy chain gene expression during muscle cell differentiation. *J. Biol. Chem.* **258**, 11063-11073.

Mukherjee, A. K. and Sarkar, S. (1981). The translational inhibitory 10S cytoplasmic ribonucleoprotein of chick embryonic muscle: dissociation and reassociation. *J. Biol. Chem.* **256**, 11301-11306.

Saidapet, C. R., Munro, H. N., Valgeirsdottir, K. and Sarkar, S. (1982). Quantitation of muscle-specific mRNA by using cDNA probes during chicken embryonic muscle development *in ovo. Proc. Natl. Acad. Sci. USA.* **79**, 3087-3091.

Sambrook, J., Fritsch, E. F. and Maniatis, T. (1989). Molecular Cloning: A Laboratory Manual. Second edition, Colds Spring Harbor Laboratory.

Sarkar, S. (1984). Translational control involving a novel cytoplasmic RNA and ribonucleoprotein. *Progress in Nucleic Acid Res. and Mol. Biol.* **31**, 267-293.

Sarkar, S., Eller, M. S., Raychowdhury, M. K., Stedman, H. and Wu, Q. L. (1989). The cytoplasmic translation inhibitory RNA of chick embryonic muscle: resolution of multiple biologically active subspecies and mechanism of action. In Cellular and Molecular Biology of Muscle Development (eds. L. H. Kedes and F. E. Stockdale) pp.555-570, New York, Alan R. Liss.

Sarkar, S., Mukherjee, A. K. and Guha, C. (1981). A ribonuclease resistant cytoplasmic 10S ribonucleoprotein of chick embryonic muscle: a potent inhibitor of cell-free protein synthesis. *J. Biol. Chem.* **256**, 5077-5086.

Winkler, M. M., Lashbrook, C., Hershey, J. W. B., Mukherjee, A. K. and Sarkar, S. (1983). The cytoplasmic translation inhibitory RNA species of chick embryonic muscle: effect on mRNA binding to 43S initiation complex. *J. Biol. Chem.* **258**, 15141-15145.

INDEX OF SUBJECTS